Springer Series in Molecular Biology

Springer Series in Molecular Biology

J. F. T. SPENCER, DOROTHY M. SPENCER, A. R. W. SMITH (Eds.)
Yeast Genetics · Fundamental and Applied Aspects (1983)

EUGENE ROSENBERG (Ed.)
Myxobacteria · Development and Cell Interactions (1984)

AHARON RAZIN, HOWARD CEDAR, ARTHUR D. RIGGS (Eds.)
DNA Methylation · Biochemistry and Biological Significance (1984)

TERRELL L. HILL
Cooperativity Theory in Biochemistry · Steady-State and Equilibrium Systems (1985)

ROGER L. P. ADAMS, ROY H. BURDON
Molecular Biology of DNA Methylation (1985)

BOYD HARDESTY, GISELA KRAMER (Eds.)
Structure, Function and Genetics of Ribosomes (1986)

ARNOLD W. STRAUSS, IRVING BOIME, GÜNTHER KREIL (Eds.)
Protein Compartmentalization (1986)

THEODOR WIELAND
Peptides of Poisonous Amanita Mushrooms (1986)

AKIRA WAKE, HERBERT R. MORGAN
Host-Parasite Relationships and the Yersinia Model (1986)

TERRELL L. HILL
Linear Aggregation Theory in Cell Biology (1987)

GRAHAM W. PETTIGREW, GEOFFREY R. MOORE
Cytochromes c · Biological Aspects (1987)

GEOFFREY R. MOORE, GRAHAM W. PETTIGREW
Cytochromes c · Evolutionary, Structural and Physicochemical Aspects (1990)

Geoffrey R. Moore · Graham W. Pettigrew

Cytochromes c

Evolutionary, Structural and Physicochemical Aspects

With 149 Figures

Springer-Verlag
Berlin Heidelberg New York London
Paris Tokyo Hong Kong Barcelona

Dr. GEOFFREY R. MOORE
University of East Anglia
School of Chemical Sciences
Norwich NR4 7TJ
U.K.

Dr. GRAHAM W. PETTIGREW
Department of Biochemistry
Royal (Dick) School
of Veterinary Studies
Summerhall
Edinburgh EH9 1QH
U.K.

ISBN-13: 978-3-642-74538-6 e-ISBN-13: 978-3-642-74536-2
DOI: 10.1007/978-3-642-74536-2

Softcover reprint of the hardcover 1st edition 1990

2131/3020-543210

Dedicated to
A. E. Moore
and to the memory of
R. J. Moore

Foreword

Of all proteins open to study by scientists cytochrome *c* has proved one of the most irresistible. It is in a popularity class with lysozyme, haemoglobin and ribonuclease. A major reason for the intensity of study of cytochrome *c* is its easy availability from many sources, another is its stability and solubility in water, and a third is its easily defined functional significance. Since it is really a prokaryote or organelle (mitochondrion or chloroplast) protein it has been an obvious target for genetic and evolutionary analyses. In fact it was one of the first, if not the first, protein to be looked at by site specific mutagenesis. Of course its crystal structure was determined very early and it has proved to be one of the most instructive proteins for NMR studies. Its function, electron transfer, is one which theoreticians and experimentalists can tackle equally with clear cut approaches to the nature of intermediates. Again it is a haem-protein so that optical and magnetic analyses of spin-states and excited states alone have occupied an army of scientists for more than fifty years.

The huge range of the studies on this, the simplest protein of the respiratory chain, has presented the authors of this book with a mammoth task. Their book does give a great number of the essential references, the collection of which is no small task, but they are not presented as an ill-digested compilation. Rather the book gives under the different chapter headings an authoritative appreciation of how far our understanding of many problems has reached. Cytochrome *c* is shown to be a model protein not just for electron-transfer kinetic studies, but also for redox potentials, for protein/protein binding and for internal mobility as well as for the genetic and evolutionary analyses already mentioned. Despite the depth of knowledge we have in these areas, there is still more to cytochrome *c*, since it has several conformational states and it can be used as a model protein for groove-opening reactions. In this sense the book should not be read as giving a completed picture of the nature of this protein, but as a source for the initiation of many new experiments and theories.

R. J. P. WILLIAMS

Preface

Cytochromes *c* play a central role in biological electron transport systems and they have been a popular subject for study in the general areas of protein chemistry and redox reactions. Our knowledge of cytochromes *c* has advanced through the contributions of disciplines as different as, for example, physical chemistry and microbial physiology. A comprehensive review must therefore deal with this diversity of interest and be intelligible to the chemist and biologist alike. To achieve such a synthesis is a daunting task but we feel it is an important one in providing a basis for the development of future research.

Our approach to reviewing the cytochrome *c* field has resulted in the production of two books - the present volume and its companion, *Cytochrome c: Biological Aspects* published in 1987. These two books are complementary. Their separate publication reflects the practical difficulties in dealing with a large quantity of material, and a correspondingly bulky manuscript, and the two volumes should be seen as a single review. We are conscious of the criticism levelled at the first book that it is far too expensive, and we have sought, therefore, to come to grips with recent developments in word-processing to produce a camera-ready manuscript. This will, we hope, result in a less costly book than *Cytochrome c: Biological Aspects*, and one with a shorter delay between the submission and publication of the book. The three year gap between the publication of the two books does not mean the present book misses out the last three years of published work. We have reviewed as much of the relevant literature as possible up to February 1990.

Our original view was that the two volumes would correspond roughly to a biologist's view of cytochrome *c* and a chemist's view. However, as several readers of the first book have pointed out, that volume contains a significant chemical element, and, as is readily apparent form the contents page of the present book, this one contains a substantial biological contribution. This is inevitable and it emphasises that contributions to the elucidation of protein function come from many different disciplines, amongst which a clear demarcation does not exist.

As with *Cytochrome c: Biological Aspects* the present book is directed at both the graduate student and the research worker. To this end, it is detailed in its treatment of the recent research literature and incorporates much methodological material. However, it is our hope that within the complexity of cytochrome *c* we have found unifying themes which will be of interest to the more general reader.

We would like to thank all our colleagues who critically read sections of the manuscript, particularly Prof. A. J. Thomson and R. J. P. Williams who provided invaluable help in many areas, Maxine Pettigrew for typing part of the manuscript and Kate Hesketh Moore, who not only turned a draft into a manuscript but also the manuscript into a book.

GEOFFREY R. MOORE
GRAHAM W. PETTIGREW

Contents

Chapter 3 Amino Acid Sequences of Cytochromes *c*

Chapter 4 The Structures of Class I Cytochromes *c*

Chapter 5 The Structures of Class II, Class III and Class IV Cytochromes *c*

Chapter 6 Evolution

Chapter 7 Redox Potentials

Chapter 8 Electron-Transfer Mechanisms

Chapter 1 Stereochemical and Physicochemical Properties of Hemes

1.1 Introduction

An extreme view of hemeproteins is that they are only coordination complexes of iron, and while this takes little account of their biochemical versatility it does focus on the obvious fact that not only is the heme central to the function of a hemeprotein, it also determines many of the protein's physicochemical properties. Accordingly the heme and axial ligands of cytochromes *c* have been widely studied. The results of these studies are brought together in this chapter and in Chapters 2 and 7.

The present chapter deals with the stereochemical and physicochemical properties of hemes other than their redox potentials, which are dealt with in Chapter 7, and their spectroscopic properties, which are dealt with in Chapter 2. In order to provide a base upon which to build a description of the interplay between heme and a polypeptide we include a brief account of the electronic structure of iron and the effect of heme and axial ligands upon it.

1.2 Electronic Structure of Iron

Iron (0) has the electronic structure $1s^2 2s^2 2p^6 3s^2 3p^6 3d^6 4s^2$ (where the symbols 1s, 2p etc. are the orbitals, and the superscripts are the electron occupancies). This is sometimes written as $[Ar]3d^6 4s^2$ to emphasise that it is the occupancy of the higher energy 4s and 3d orbitals, superimposed on an argon core, that determines the electronic properties, and ultimately the chemical behaviour, of iron.

Oxidation States. In hemeproteins iron exists in two common oxidation states: the ferrous state (Fe(II): $[Ar]3d^6 4s^0$); and the ferric state (Fe(III): $[Ar]3d^5 4s^0$). In addition there is the less common ferryl state (Fe(IV):

[Ar]$3d^4 4s^0$), which is invariably found as Fe(IV)=O and occurs during the catalytic cycle of some enzymes, such as catalases and peroxidases. For the simpler electron transfer proteins only the ferric and ferrous states are important. This selection is a result of the relative stabilities of the different oxidation states. The standard Fe(IV)/Fe(III) reduction potential for the metal ions in aqueous solution is estimated to be ~2 volts while the corresponding Fe(III)/Fe(II) potential is 0.77 volts (Phillips & Williams, 1966: vol. 2 p171). Although these potentials can be greatly varied by a heme-protein (Chapter 7) the Fe(IV)/Fe(III) potential remains strongly positive.

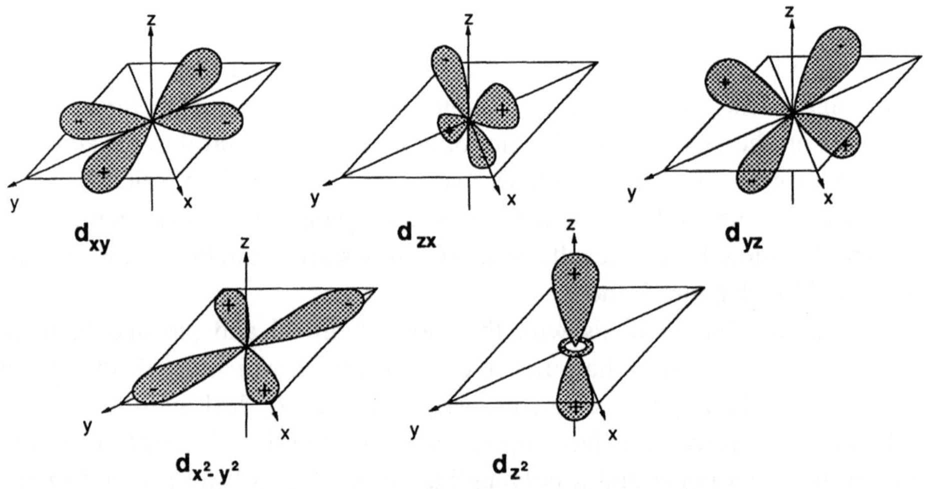

Fig. 1.1 Spatial arrangement of the d-orbitals. In a metalloporphyrin the porphyrin ring is in the xy plane. The axial ligands are in the z-direction. + & - represent the relative signs of the wavefunction ψ (see Chapter 8.2.2)

Crystal Field Theory and Spin States. There are five 3d orbitals each capable of being occupied by two electrons, and it is the distribution of electrons within these orbitals that determines the electronic properties of iron. The electron distribution is governed by the ligands coordinated to the iron and so any account of the electronic structure should also include a description of the bonding interactions. There are a number of descriptive approaches to bonding in transition metal complexes and for our present purposes the relatively simple Crystal Field Theory (CFT) satisfactorily accounts for much of the behaviour we wish to describe. Note that although CFT is not a realistic model, it is equivalent to molecular orbital theory in terms of splitting patterns of electronic energy levels. Hence, to a first approximation, we can understand magneto-optical spectra using CFT, but not parameters such as hyperfine interactions which depend upon electron delocalisation.

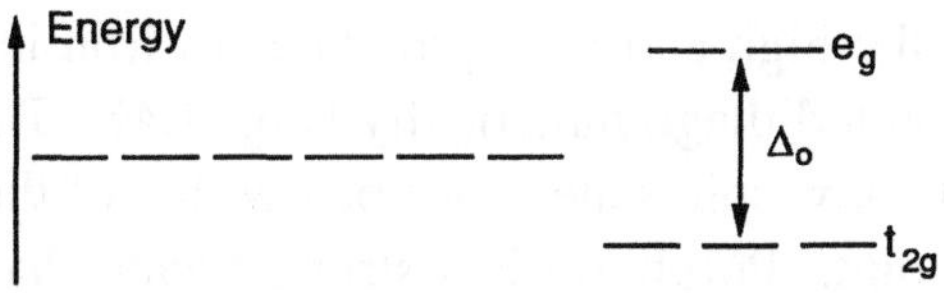

Fig. 1.2 Effect of an octahedral crystal field on the d-orbital splitting. Δ_o is the crystal field stabilisation energy. e_g and t_{2g} are symmetry labels. The $d_{x^2-y^2}$ and d_{z^2} orbitals comprise the e_g set, and the d_{xz}, d_{yz} and d_{xy} orbitals comprise the t_{2g} set. The absolute energies of both the e_g and t_{2g} sets are raised from the free ion values by the crystal field. Following usual practice (Cotton & Wilkinson, 1980; Phillips & Williams, 1966) this energy change is not represented in this figure

CFT treats the ligands as point charges and assumes that the bonding between the iron and its ligands is entirely electrostatic in origin. Thus it makes no provision for covalency. CFT then considers the effect a ligand has on the energy of an orbital. Since the d-orbitals are not spatially equivalent (Fig. 1.1) they are differentially affected by ligands. In a six-coordinate octahedral complex the orbitals lying along the Cartesian axes ($d_{x^2-y^2}$ and d_{z^2}) are destabilized by the incoming ligands more than the orbitals lying between the axes (d_{xy}, d_{xz} and d_{yz}) (Fig. 1.2). This is because of the greater repulsion between the d-electrons and the ligand electrons in the former set of orbitals. These orbitals are often called the e_g set with the remaining three comprising the t_{2g} set (e_g and t_{2g} are symmetry labels). Electrons in the t_{2g} orbitals are at a lower energy than electrons in the e_g orbitals.

There are two major energy terms that govern the distribution of electrons in d-orbitals: the strength of the d-orbital splitting (given the symbol Δ_o), which depends upon the electrostatic field created by the ligands, and the energy required to pair electrons in the same orbital (P). The pairing energy is always unfavourable but what is important is whether it requires less energy to pair electrons in the t_{2g} orbitals than it does to keep them unpaired by occupying the e_g orbitals. This is illustrated by the distributions for Fe(III) shown in Fig. 1.3. Where there is a weak crystal field the d-orbital splitting is smaller than the pairing energy and the electrons therefore remain unpaired and enter separate orbitals. This distribution is the high-spin configuration.

However, where there is a strong crystal field, the electrons enter the t_{2g} orbitals and pair to produce the low-spin configuration. In this state the unpaired electron is shared by the three t_{2g} orbitals. Fe(II) can also exist in the high-spin ($t_{2g}^4\ e_g^2$) and low-spin ($t_{2g}^6\ e_g^0$) states.

The description of the high and low spin-states of iron in terms of the field strength can be represented diagrammatically (Fig. 1.4). The field strength at which the energy of the low spin state drops below that of the high spin state is called the crossover point. Porphyrin is a strong ligand that places iron close to its spin-state crossover point so that relatively small energy differences between the axial ligand components of the field can cause the spin-state to change.

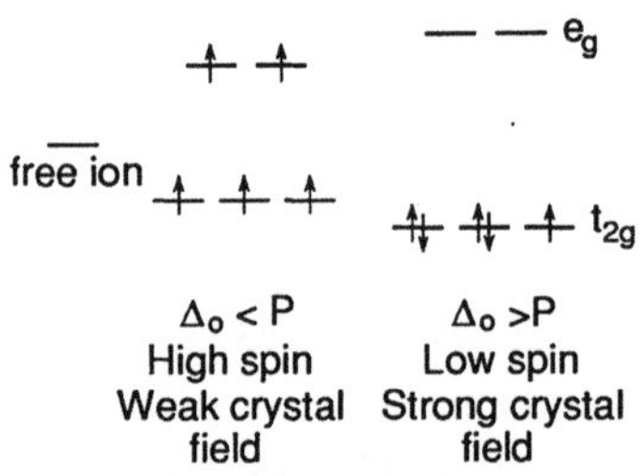

Fig. 1.3 Spin-states of octahedral Fe(III). The ↑'s represent electrons occupying the d-orbitals. In the low-spin state the unpaired electron is shared by the three t_{2g} orbitals

An intermediate spin-state for Fe(III), in which two electrons are paired and three unpaired ($t_{2g}^4\ e_g^1$), can also be imagined. However, this configuration does not arise for Fe(III) in a regular octahedral field (Fig. 1.3) because, as a result of the orbital degeneracy, single occupation of the e_g set cannot occur (Griffith, 1961: p237). As Maltempo & Moss (1976) point out, '...if it is helpful energetically to bring one electron down from the degenerate $d_{x^2-y^2}$ and d_{z^2} orbitals then it will be even more helpful to bring down both.' Nevertheless, the intermediate spin-state does occur, but only when there is a distortion of the crystal field.

Tetragonal Crystal Fields. The degeneracy of the e_g and t_{2g} orbital sets can be lifted by distortion of the octahedral field. The major distortion in hemeproteins involves the axial ligand components of the crystal field which are generally different to the in-plane contributions of the porphyrin. As a result of this difference the symmetry of the field is lowered from cubic to tetragonal.

The first case we consider is when there is a weak crystal field. Since porphyrin is a strong-field ligand we attribute this weakness to a reduction in the axial ligand field component. This can be viewed as resulting from a lengthening of the Fe-axial ligand bond length. A consequence of this is that although the energy of the $d_{x^2-y^2}$ orbital remains high (because it is largely affected by the porphyrin) the energy of the d_{z^2} orbital is decreased. Thus the degeneracy of the e_g orbital set has been lifted and this allows the intermediate spin-state to occur (Fig. 1.5). The orbital degeneracy of the t_{2g} set is also lifted with d_{xz} and d_{yz} being at a different energy to d_{xy}. In general, both the case when d_{xy} is higher in energy and that when it is lower in energy can

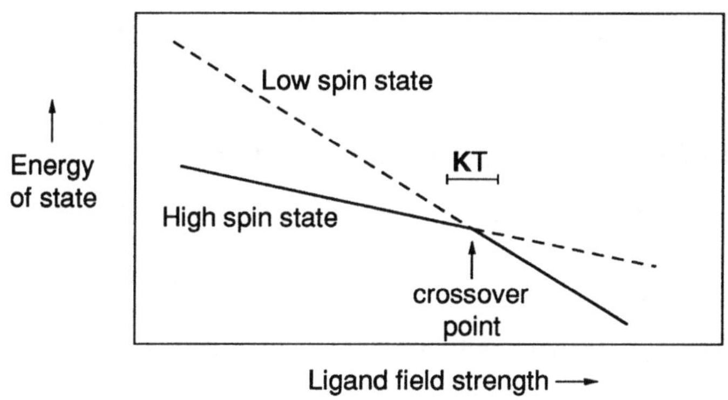

Fig. 1.4 The spin-state crossover point for Fe(II) and Fe(III) in an octahedral field. The electronic ground-state (solid line) is determined by the ligand field strength. At certain field strengths the energy separation between the two states is of the order of the thermal energy **KT** so that an equilibrium between the two may be observed

occur and it is not possible to predict, from the simple CFT used here, which occurs in practice. Griffith (1961: p371) suggests that the pattern of Fig. 1.5 occurs in metalloporphyrins.

The increased energy of the d_{xz}/d_{yz} pair compared to d_{xy} is not expected from a simple electrostatic treatment which considers the z-component of the ligand field to be weaker than the xy-component, and this splitting of the t_{2g} set can be seen as a reflection of the inadequacy of the CFT in dealing with real molecules in which covalent bonding occurs. The d_{xy} orbital lying in, or almost in, the porphyrin plane cannot combine with the porphyrin Np_z orbitals to form a π-bond, and the Np_x and Np_y orbitals are unavailable for π-bonding because they are involved in σ-bonds with carbon atoms of the porphyrin. However, the d_{xz} and d_{yz} orbitals extend out of the porphyrin plane and they can interact with the porphyrin Np_z orbitals (Fig. 1.6). They may also interact with orbitals of the axial ligands. It is not a simple matter to predict whether these interactions will stabilise or destabilise the d_{xz}/d_{yz} pair relative to the d_{xy} orbital but they are additional effects which could oppose the stabilisation due to a decrease in the axial electrostatic component of the ligand field. Griffith (1957) suggests the interaction is unfavourable because of electron-electron repulsion.

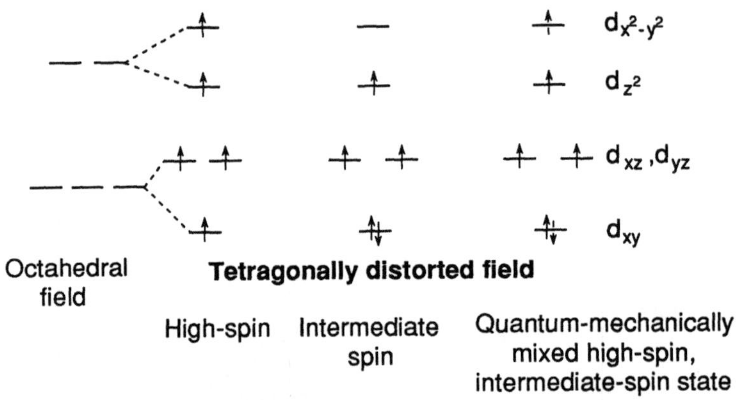

Fig. 1.5 Tetragonal distortion and the intermediate spin-state. Only an axial distortion is shown; for a rhombic distortion see Fig. 1.7. The electron occupancies for a quantum-mechanically admixed state cannot properly be represented by such orbital diagrams. Such states arise when the pure high-spin and intermediate-spin states are close in energy and can be coupled together (Maltempo & Moss, 1976). The diagram represents this by showing partial occupancy of the $d_{x^2-y^2}$ orbital and partial shared occupancy of the d_{xy} orbital

The second case we consider is when there is a strong crystal field. In this case all we are concerned with is the relative energies of the t_{2g} orbitals because the e_g orbitals are not occupied (Fig. 1.3). As with the weak-field case considered previously, with only an axial distortion the d_{xz}/d_{yz} pair remain degenerate and are of higher energy than the d_{xy} orbital. An in-plane, or rhombic, distortion lifts the degeneracy of the d_{xz}/d_{yz} pair as shown in Fig. 1.7. The unpaired electron is located mainly in d_{yz} but because the energy separation of the three orbitals is not great there is some orbital mixing so that the unpaired electron is not exclusively located in d_{yz}. As in the weak-field case the exact cause of the splitting of the t_{2g} orbitals is not known but similar arguments about the spatial disposition of the d_{xz}/d_{yz} orbitals and ligand orbitals apply. Either distortion of the porphyrin stereochemistry (section 1.3) or axial ligand effects (Fig. 1.6) could remove the degeneracy. This orbital splitting pattern (Fig. 1.7) has been shown to be valid for hemoglobin-azide (Gibson & Ingram, 1957; Griffith, 1957) and mitochondrial cytochrome *c* (Mailer & Taylor, 1972; Taylor, 1977; Palmer, 1979).

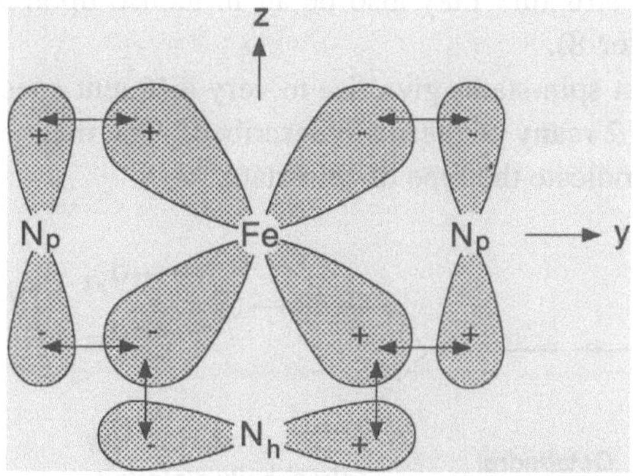

Fig. 1.6 Possible interactions between the Fe d_{yz} and ligand Np_z orbitals. Possible d-p interactions are indicated by ↔. The porphyrin (N_p)-Fe interactions are fixed by the stereochemistry but the axial ligand (N_h)-Fe interaction is modulated by rotation around the Fe-His bond (Fig. 1.12). The Fe d_{xz} orbital is similarly affected by the ligand Np_z orbitals but the d_{xy} orbital lies in the porphyrin plane and is unaffected by them

Mixed Spin-States. In the previous sections we have described the magnetically pure high, low and intermediate spin-states. Two additional cases are important. The first is a thermal mixture of spin-states. In this case the energy difference between the the high-spin and low-spin states is small compared to the thermal energy (Fig. 1.4) so that in an ensemble of molecules some will be in a high-spin state and others will be in a low-spin state. This situation was first described for methemoglobin (Williams, 1959; 1961a; George *et al.*, 1961; Antonini & Brunori, 1971).

The second case is the quantum-mechanically admixed spin-state. In this case each molecule in an ensemble has the same spin-state configuration but that state is not a pure high or intermediate spin-state. This cannot adequately be described by the simple CFT treatment but it can be viewed as arising when the energy separation of the $S = {}^3/_2$ and $S = {}^5/_2$ states of Fe(III) is small (Fig. 1.5). Now, instead of thermal energy leading to a given molecule adopting one of the two states, the two states in all of the molecules couple together (via spin-orbit coupling) to generate a mixed state. The realisation that this occurs and the application of the concept to hemeproteins, is due to the elegant work of Maltempo (1974).

Determination of Spin-States. It is important for the spin-state of a hemeprotein to be precisely defined because, as we shall see, it is a useful guide to the type of axial ligands and stereochemistry of the porphyrin. The iron electronic structure may also be an influence upon the rate of electron-transfer (Chapter 8).

The different spin-states give rise to very different spectroscopic properties and in Chapter 2 many of these are described. The magnetic properties of the iron may also indicate the type of spin-state.

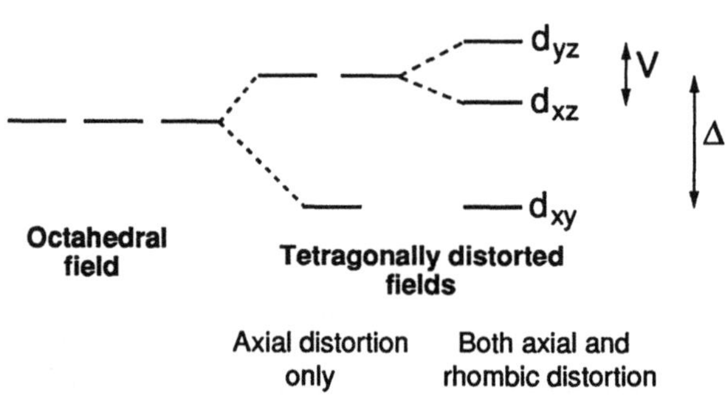

Fig. 1.7 Tetragonal distortion and the e_g orbitals of low-spin Fe(III). V and Δ are energy terms that determine the orbital splitting. They can be measured and used to assist identification of axial ligands (Chapter 2.2.3.2 & 2.3.2)

Magnetic Moments. The orbital splitting patterns and their electron occupancies (Figs. 1.3 & 1.5) satisfactorily account for most of the magnetic properties of cytochromes *c*. Each electron has an associated spin which gives rise to a magnetic moment, but when electrons are paired their spins are opposed, and hence their magnetic moments cancel each other. Therefore to a first approximation it is only necessary to count the number of unpaired electrons to determine an approximate value of the magnetic moment. Conversely measurement of the magnetic moment indicates the number of unpaired electrons.

Magnetic moments are not measured directly. The parameter that is measured is the magnetic susceptibility, which relates the magnetism of a sample to the strength of an applied magnetic field (B_o). Provided that $KT >> \mu_{eff}.\beta.B_o$ (where **K** is Boltzmann's constant and β is the Bohr magneton), the effective magnetic moment (μ_{eff}) of a sample and its magnetic susceptibility (χ) are related by:

$$\mu_{eff} = 2.84\sqrt{\chi.T} \qquad (1.1)$$

where T is the absolute temperature. It is straightforward for the magnetic properties of a sample in the solution state to be determined with a high-field NMR spectrometer but this is generally limited to a narrow temperature range. If low temperature data are needed, more specialised equipment is required (Moss, 1978).

Magnetic moments of various hemeproteins are given in Table 1.1 along with the expected magnetic moments for Fe(II) and Fe(III) compounds. Differences between the predicted spin-only values and the observed values for some spin-states may be due to contributions to the magnetic moment from the orbital motion of the electrons that is not taken into account in calculating μ_s. There is a negligible orbital contribution to high-spin Fe(III) compounds.

It can be seen from Table 1.1 that low-spin and high-spin compounds are readily distinguished from each other. The only problems that occur are distinguishing between high-spin and intermediate-spin systems and systems where spin-state mixtures occur. In general, these may be resolved by consideration of spectroscopic data (Table 2.3) or from the temperature dependence of the magnetic susceptibility.

1.3 Stereochemistry of Metalloporphyrins

Metalloporphyrin stereochemistry in small compounds has been extensively studied by X-Ray Diffraction (XRD) (Hoard, 1971; Scheidt, 1977; Scheidt & Reed, 1981; Scheidt & Gouterman, 1983) and an important conclusion of this work is that the porphyrin ligand possesses a relatively high degree of plasticity. Although the extensive electronic delocalization of porphyrin imposes a substantial planarity on the ligand, significant deformations normal to the heme plane occur (Fig. 1.8). In addition, the size of the porphinato core, as given by the radius of the central hole - the Ct-N distance (Fig. 1.9) - is not fixed. The effective size of the metal ion and the M-N bond lengths are other important stereochemical parameters.

In many porphyrins the M-N and Ct-N distances are the same with the metal sitting in the centre of the hole, but in some porphyrins M-N>Ct-N and the core is domed (Fig. 1.8) with the metal out of plane. Doming is generally observed for five-coordinate metalloporphyrins and also occurs in some six-coordinate systems, usually high-spin ones.

Doming is associated with the effective size of the metal and it has often been stated that high-spin Fe(II) and Fe(III) are too large to be accommodated

Table 1.1 Magnetic moments of iron compounds:

1.1a Spin-only magnetic moments and range of μ_{eff}

		number of unpaired electrons	μ_s [a]	μ_{eff}
low - spin	Fe(III)	1	1.73	1.7 - 2.3
high - spin	Fe(III)	5	5.92	~5.9
intermediate - spin	Fe(III)	3	3.87	b
low - spin	Fe(II)	0	0	0
high - spin	Fe(II)	4	4.90	5.0 - 5.5
intermediate - spin	Fe(II)	2	2.83	b

a μ_s is the theoretical spin-only magnetic moment, given by: $2\sqrt{n/2(n/2+1)}$, where n is the number of unpaired electrons. μ_{eff} is the effective magnetic moment observed at room temperature (17 - 23°C). The μ_{eff} ranges in Table 1.1a are from Cotton & Wilkinson (1980) and Phillips & Williams (1966). The units of μ_s and μ_{eff} are Bohr magnetons (BM).

b It is expected that there will be significant orbital contributions to the μ_{eff} of intermediate-spin compounds. However, few have been characterised. Only Fe(II)TPP (where TPP = tetraphenylporphyrin), with μ_{eff} = 4.4 BM (Collman *et al.*, 1975), and [Fe(III)TPP(C[CN]$_3$]$_n$, with μ_{eff} = 5.4.BM (Summerville *et al.*, 1978), are considered to exist in the intermediate-spin state. Other characterised porphyrin complexes are in the quantum mechanically admixed high-spin, intermediate-spin state (Reed *et al.*, 1979, Scheidt & Gouterman, 1983).

1.1b Magnetic moments of selected hemeproteins

	Protein	pH	μ_{eff}	Spin State		Refs.
horse	ferricytochrome *c*	7	2.29 [c]	>97%	low	e
D. vulgaris	ferricytochrome c_3	6	2.01	>97%	low	f
R. rubrum	ferricytochrome *c'*	6.5	5.4 [c d]	62%/38%	high/int.	g h
		10.1	5.9	>97%	high	g
	ferrocytochrome *c'*	6.5	5.2	>97%	high	g
C. vinosum	ferricytochrome *c'*	7	5.15 [c d]	32%/68%	high/int	h i
horse	methemoglobin	12	4.45 [d]	50%/50%	high/low	j
	metmyoglobin	11	5.11 [d]	70%/30%	high/low	j
	deoxymyoglobin	7	5.46	>97%	high	k

c Tasaki *et al.*, (1967) have measured the magnetic susceptibility of horse ferricytochrome *c* and *R. rubrum* ferricytochrome *c'* over the temperature range 4.2 - 150 K. Maltempo *et al.*, (1974) have determined the magnetic susceptibility of *Chromatium* ferricytochrome *c'* at 1.4 - 4.2K (and pH 7) to be 3.4 BM. Values in the table were determined at room temperature.

d The high-spin, intermediate-spin mixtures are quantum mechanical admixed states and the high-spin, low-spin mixtures are thermally admixed states.

[e]Ångstrom *et al.*(1982); [f]McDonald *et al.*(1974); [g]Emptage *et al.*(1981); [h]Maltempo (1974); [i]Ehrenberg & Kamen (1965): [j]George *et al.*(1961); [k]Taylor (1939).

in the central porphinato hole while their low-spin counterparts can fit. (Their respective ionic radii are: Fe(II) ~0.78Å & 0.61Å; and Fe(III) ~0.65Å & ~0.53Å). Indeed this has been one of the central features of a proposed triggering mechanism for the cooperativity of hemoglobin (Williams, 1961a; Hoard, 1971; Perutz, 1970). However, this now appears to be too simple a view, at least as far as hemeproteins are concerned. A combination of the flexibility of the porphyrin, and stereochemical constraints imposed by the side chains packed around the heme and the attached polypeptide, make Fe(II) and Fe(III) porphyrins in many hemeproteins adopt similar stereochemistries. Table 1.2 summarises some of the data in support of this; the maximum iron displacement found in the Table is for high-spin Fe(II) of deoxyerythrocruorin and the Fe(III) of cytochrome *c′*. These are much smaller than the displacements normally observed in simple porphyrin complexes, which are in the range 0.4 - 0.5Å.

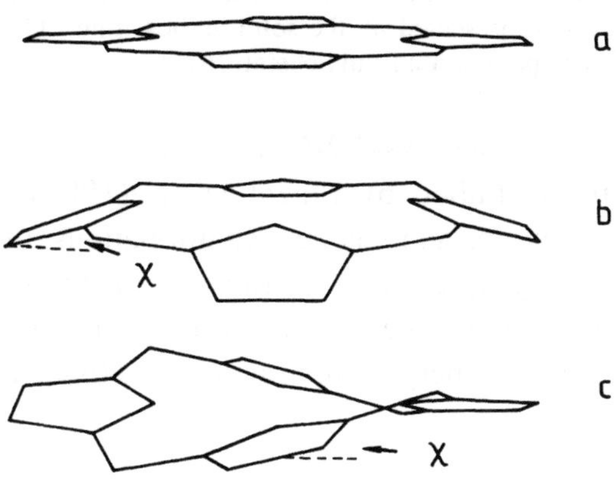

Fig. 1.8 Porphyrin stereochemistry. Metalloporphyrins may be planar (**a**), domed (**b**), or ruffled (**c**). Resonance Raman spectroscopy can distinguish between the deformed conformations by measuring the pyrrole tilt angle, χ (Chapter 2.2.4). (Adapted from Spiro *et al.*, 1978)

Heme Stereochemistries in Cytochromes c. In all highly refined XRD structures of cytochrome *c* the heme is found to be non-planar and can be described as being ruffled (Fig. 1.8) or saddle-shaped (Fig. 4.12). The cause of this distortion does not appear to lie with the thioether linkages because, although in mitochondrial cytochrome *c* it is the pyrrole rings bearing the thioethers that are tilted more than the others (Takano & Dickerson, 1981a;b), in *D. vulgaris* (Miyazaki) cytochrome c_3 only one of the four hemes is similarly distorted, the others being distorted by tilting of a thioether-bearing pyrrole and its adjacent propionate-bearing pyrrole (Higuchi *et al.*, 1984).

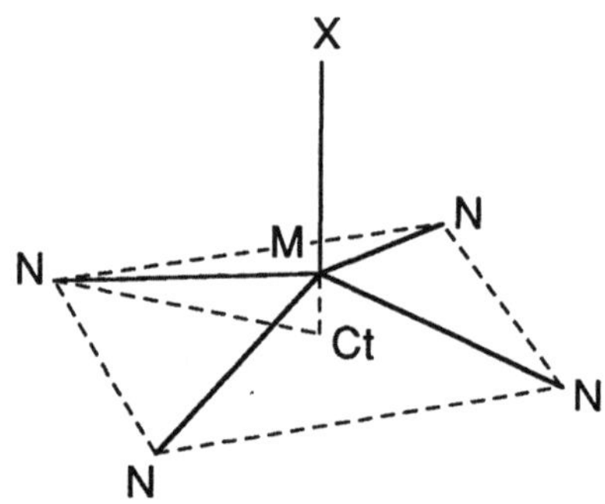

Fig. 1.9 The square-pyramidal coordination geometry for some metalloporphyrins. The metal is located at M and is bonded to the four porphyrin nitrogens (N). An axial ligand (X) bond is represented by the solid vertical line. Ct is the centre of the porphyrin. (Reproduced with permission from Scheidt, 1977)

The stereochemistry of the Cys-?-?-Cys linkages of classes I, II and III cytochromes *c* has been studied by Finzel *et al.* (1985), who show that although the main-chain atoms adopt helical conformations, the proteins divide into two groups depending upon the conformations of the Cys closest to the C-terminus and the His ligand. The two arrangements maintain preferred side chain torsional angles and an energetically favourable His-heme interaction.

1.4 Properties of Ligands

Ionization States of Porphyrins. There are a variety of ionisation states accessible to the porphyrin nucleus (Fig. 1.10). The pyrrole-type NH groups are weak acids with pK_a's >14 and the imino-type nitrogen atoms are weak bases whose pK_a's vary over the range 0 - 7 (Phillips, 1963). The interpretation of these data in terms of intrinsic properties of the porphyrin nucleus is complicated by the effect of ionized substituent groups in altering the observed pK_a's from their intrinsic values by an electrostatic field effect. A similar electrostatic effect is observed for the redox-state dependence of the substituent pK_a's in some proteins (Chapter 7.4). Phillips (1963) estimates that the intrinsic values of the imino pK_a's are ~5 and ~7.5.

Table 1.2 Porphyrin stereochemistry and ligand binding geometry of hemeproteins

Protein		Spin-state	Fe to heme plane	Fe to N-plane	His N^{ε} to N-plane	Fe to His N^{ε}	Met S^{δ} to N-plane [b]	Fe to Met S^{δ}	Refs.
			DISTANCES (Å) [a]						
Aquomet-erythrocruorin	Fe^{3+}	high	0.08	0.16	2.4	2.25	—	—	c
Cyanomet-erythrocruorin	Fe^{3+}	low	0.06	0.13	2.2	2 1	—	—	c
Deoxy-erythrocruorin	Fe^{2+}	high	0.17	0.23	2.4	2.20	—	—	c
Carbonmonoxy-erythrocruorin	Fe^{2+}	low	0.01	0.11	2.2	2.10	—	—	c
Ferricytochrome *c'*	Fe^{3+}	high/int.	0.2	0.2	2.2	2.00	—	—	
Ferrocytochrome *c*	Fe^{2+}	low	0.06	0.07	1.90	1.97	2.32	2.39	d
Ferricytochrome *c*	Fe^{3+}	low	0.03	0.02	2.00	2.00	2.27	2.22	d
Ferrocytochrome *c*-551	Fe^{2+}	low	0.03	0.04	2.02	1.97	2.35	2.31	e
Ferricytochrome *c*-551	Fe^{3+}	low	0.03	0.03	2.02	1.99	2.36	2.32	e
Ferricytochrome c_3	Fe^{3+}	low							
Heme 1			0.02	0.00	*	1.94,1.95	—	—	f
Heme 2			0.00	0.01	*	2.12,1.95	—	—	f
Heme 3			0.11	0.05	*	2.02,1.88	—	—	f
Heme 4			0.07	0.06	*	2.06,2.06	—	—	f

[a] The heme plane is defined by the four pyrrole ring atoms and the first atoms attached to them, or by the twenty four atoms forming the porphyrin core.
The N-plane denotes the plane through the four pyrrole nitrogens.
The difference between the Fe to heme plane distance and the Fe to N-plane distance is a measure of the extent of doming, and is given the symbol M-Ct in Fig. 1.9.

* Not reported; — shows that Met S^{δ} is not a ligand.

[b] The geometries of the bound Met residues of cytochromes *c* and blue copper proteins are similar (Chakrabarti, 1989).

[c] Steigemann & Weber (1979);

[d] Takano & Dickerson (1981a;b);

[e] Matsuura *et al.* (1982);

[f] Higuchi *et al.* (1984).

The species P^{2-} (Fig. 1.10) is that occuring in hemeproteins. Thus the net charge of the porphyrin nucleus of a ferroheme is 0. Protonation of the porphyrin nucleus in a metalloporphyrin causes displacement of the iron and this has been exploited to produce metal-free porphyrin cytochrome *c* (section 1.5). Such a protonation step may also be involved in the acid denaturation of some hemeproteins (Chapter 4.4.1).

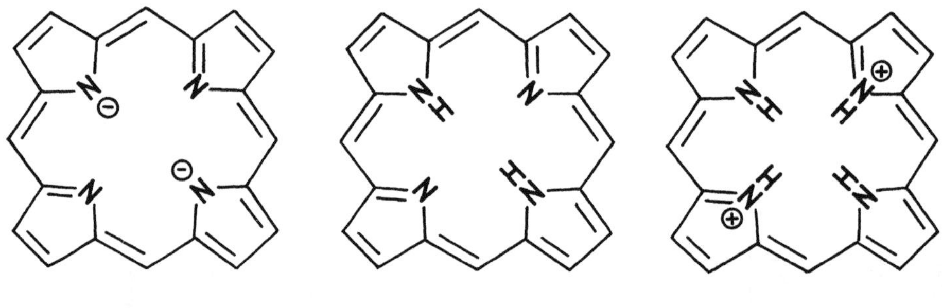

Fig. 1.10 Ionisation states of porphyrin. There are seven possible species ranging from P^{2-} to PH_6^{4+}, three of which are shown. These are (left to right): P^{2-}, PH_2 and PH_4^{2+} (Phillips, 1963)

The pK_a's of the propionic acid substituents of protoporphyrin IX have not been determined but Theorell (1938) reported both propionates of porphyrin *c* in H_2O to have pK_a's of 5.6. This requires confirmation in view of the analysis which follows. Since the carboxyl group is separated from the macrocycle by two CH_2 groups, inductive effects from the porphyrin will be slight and the intrinsic pK_a should be close to that found for monopyrrole carboxylic acids, namely ~4.8. It is possible that the ionisation of the second carboxylic acid will be affected by the presence of the first carboxylate but it is difficult to estimate how important this will be because the carboxylates are free to move and an electrostatic field effect is strongly dependent upon the dielectric properties of the solvent (Chapter 7.4). In aqueous succinic acid the pK_a's are 1.4 pH units apart and even in heptanedioic acid the pK_a's are 1.1 pH units apart. Consideration of these effects has led Phillips (1963) to suggest a value of 4.8 for the first, and 5.7 for the second, carboxylic acid pK_a of a porphyrin in H_2O.

The effect of metal ligation to a porphyrin on the carboxylic acid pK_a's is not known but it may be sizable. The magnitude of such an effect will again depend upon the solvent properties.

Ligand Field Strengths. Porphyrins are strong-field ligands but by themselves they are not strong enough to generate the low-spin states of Fe(II) and Fe(III). This requires the addition of particular axial ligands to increase the ligand field strength. As a general rule, the ligand fields of 5-coordinated hemes are rather weak, and this led to the widely held view that the low-spin state required 6-coordination. However, this is not the case since a low-spin 5-coordinated heme complex containing a sulfhydryl axial ligand has been described (English *et al.*, 1984). Nevertheless, the vast majority of low-spin systems are 6-coordinate.

Table 1.3 Axial ligation of characterised hemeproteins

Protein	Fe	Ligands	Spin - state	Refs.
class I cytochromes *c*	II/III	His, Met	low	d
E. coli cytochrome b_{562}	II/III	His, Met	low	e f
	III	His^-, Met	low	g
cytochrome b_5	II/III	His, His	low	e
cytochrome c_3	II/III	His, His	low	h
cytochrome *c'*	III	His	int./high	h
	III	His^-	high	h
	II	His	high	h
cytochrome *c* peroxidase [a]	III	His, H_2O	low/high	i j
hemoglobin	III	His, H_2O	high	k l
	III	His, OH^-	low/high	k l
deoxyhemoglobin	II	His	high	k m
hemoglobin M (Milwaukee)	III	His, Glu^-	high [b]	k n
hemoglobin M (Iwate)	III	His, Tyr^-	high [b]	k n o
hemoglobin M (Boston)	III	Tyr^-	high	k m
catalase	III	Tyr^-	high	p q
cytochrome P-450 (camphor)	III	Cys^-	low/high [c]	r s

[a] The spin-state and axial ligation of yeast cytochrome *c* peroxidase is complicated because both the acid form (presumably with a H_2O ligand) and the alkaline form (presumably with a OH^- ligand) have thermal spin-state equilibria. In the former case this is between a high-spin ground-state and low-spin upper state and in the latter case it is the reverse. The pK for the ionisation is ~6.3 and the crystals for XRD study were grown from a solution at pH 6.0.

[b] These spin-state designations are based on visible spectroscopic data only. It is possible that there is a thermal high-spin, low-spin admixture although one which is predominantly high-spin.

[c] The precise proportions of the high-spin and low-spin character depends upon the bound substrate (Fisher & Sligar, 1985). With camphor at 25°C ferricytochrome P-450 is 94% high-spin.

[d]see Chapters 2 & 4; [e]Xavier *et al.*(1978); [f]Mathews (1985); [g]Moore *et al.*(1985b); [h]see Chapters 2 & 5; [i]Yonetani (1976); [j]Finzel *et al.*(1984); [k]Fermi & Perutz (1981); [l]George *et al.*(1964); [m]Taylor (1939); [n]Perutz *et al.*(1972); [o]Motokawa *et al.*(1964; [p]Fita & Rossman (1985); [q]Torii *et al.*(1970); [r]Poulos *et al.*(1986); [s]Sharrock *et al.*(1976).

The number of different axial ligands found in hemeproteins is relatively small. Based on a survey of proteins characterised by XRD (Table 1.3) it appears that, in general, electron transfer proteins and normal globins have neutral ligands at about pH 7 but at higher pH, and in mutant hemoglobins, anionic ligands are also found. Some heme enzymes also have anionic ligands. Only histidine and methionine have been unambiguously identified as ligands in cytochromes *c* though lysine ligation has been suggested to occur at alkaline pH in some cases (Chapter 2.3.2.4) and at pH 7 in cytochrome *f* (Chapter 2.3.2.3).

The different ligation schemes alter the ligand field to different degrees so that both high and low-spin states are commonly found. However, a quantitative comparison of the field strength of different ligands is impossible, even for simple transition metal complexes. Part of the problem is that the ligand field strengths do not reflect just electrostatic effects; and an additional complication in hemeproteins is that ligand field strengths can be modulated by environmental effects of the protein. Nevertheless, a qualitative guide to field strengths can be constructed from spectroscopic data. An estimate of the energy separation Δ_0 can be determined by measurement of the energy at which an electron is excited from the t_{2g} orbital set to the e_g set. Such a d-d transition gives weak absorption bands in the visible region of the spectrum. The spectrochemical series of ligands constructed from such measurements gives the relative order of Δ_0 for a given metal ion, and it includes: $CO > CN^- > NH_3 > H_2O > OH^-$. This order is obviously not one of decreasing electrostatic field, another indication that CFT is not wholly adequate. The explanation for the series is best given in terms of molecular orbital theory (for details see Phillips & Williams, 1966 2: pp 397 - 401). Strong ligands are good σ-donors and π-acceptors while weak donor anions (such as halides and OH^-) and neutral saturated molecules (such as NH_3 and H_2O) are weak ligands.

Unfortunately the measurements of d-d absorption bands for metalloporphyrins needed to construct such a series for amino acid side chains have not been reported. This is because the visible region of the spectrum, where many of the d-d bands are expected to occur, is dominated by the $\pi \rightarrow \pi^*$ transitions of the porphyrin (Chapter 2.3.1.1). This prevents adequate observation of the much weaker d-d absorption bands. However, the energies of the charge transfer (CT) bands in the near infra-red region of the spectrum (Chapter 2.3.1.1) depend upon Δ_0, as well as V and Δ (Fig. 1.7), and give some insight into the order for heme axial ligands. Empirically the energy of the CT band decreases in the order:

CN^->imidazole>RNH_2>imidazolate>N_3^->SH^->phenolate>OH^- (Gadsby & Thomson, 1989). Note that imidazolate lies unexpectedly low, and its order with respect to imidazole is reversed from that expected. This is because this series is influenced by more than just Δ_o.

The spectrochemical series of ligands is not a rigorous guide to relative field strength and various additional factors have been described that may change the order of particular ligands in some cases (see Phillips & Williams, 1966 for further details). Thus the prediction of whether a given ligand will be strong or weak is not always possible to make. However, based on the studies summarised in Table 1.3 and the theoretical interpretation of the spectrochemical series we state as a general rule that His, Met, Cys and Lys will be strong ligands, and H_2O, Glu, Asp and Tyr will be weak ligands. OH^- is more difficult to place; in hemeproteins it is sometimes a strong ligand.

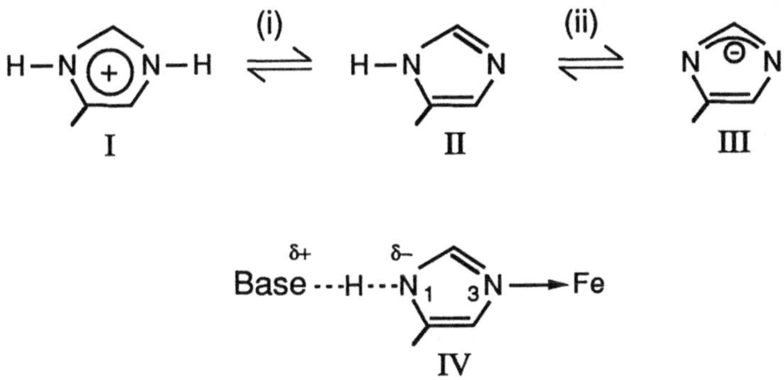

Fig. 1.11 Histidine ionisation and ligation. Histidine exhibits two side chain ionisations. Only forms II and III can bind to heme, though the neutral form is often hydrogen-bonded to give a ligand, IV, intermediate in character to II and III

Histidine Ligation. Histidine is the most common hemeprotein axial ligand. Ligation always occurs through N(3), probably because of unfavourable steric interactions for N(1) coordination (Figs. 1.11 & 1.12).

The imidazole ring of histidine exhibits two ionisations (Fig. 1.11). These have intrinsic pK_a's of 6.5 and >14, and complexation with metal ions lowers them (Theorell & Åkesson, 1941; Ehrenberg & Theorell, 1955; George *et al.*, 1964; Yagil, 1967; Martin, 1974). pK_a^i is lowered to <3.5 but the cationic form of histidine (I) does not bind to iron and has little relevance to cytochromes except when acid denaturation is considered (Chapter 4.4.1). Both the neutral form (II) and the histidinate anion (III) can ligate to iron.

The intrinsic value of pK_a^{ii} is of considerable interest but it has not been directly measured for a simple heme compound. However, studies have been made with small, non-porphyrin complexes of Co(II), Cu(II) and Co(III) (Table 1.4). These indicate that the intrinsic pK_a^{ii} for histidine bound to a dipositive metal ion is ~12 and for a tripositive metal ion is ~10. Although useful as a guide, these studies are not completely satisfactory: firstly, because they are not for porphyrin complexes of Fe(II)/Fe(III); secondly, because they take no account of possible trans effects from biologically important ligands; and thirdly, because few of the complexes have the same overall formal charge as does ferroheme or ferriheme.

Fig. 1.12 Histidine-heme orientation. (**a**) The dihedral angle, θ. The solid lines represent the Fe-N bonds of the heme. The broken line represents the histidine ring viewed edge-on. (**b**) Steric hindrance to an orientation of θ = 0°. The arrows indicate the points of non-bonded Van der Waals contact

Studies of cytochromes and imidazole complexes of globins have shown convincingly that the pK_a^{ii} of ligated histidine can vary over a wide range. In most proteins this pK_a is >8.8 but in a few cases it is as low as 7 (Table 1.4). The explanation usually given for this reduction is that the protein provides neighbouring positively charged groups to stabilise the histidinate form.

In many characterised hemeprotein structures the neutral histidine ligand is hydrogen-bonded to a base, and this imparts some histidinate character to the ligand (Fig. 1.11: IV), which is expected to increase the ligand field-strength (Salemme *et al.*, 1973b; Valentine *et al.*, 1979; Landrum *et al.*, 1980). This is a good example of the way in which a protein structure can affect the strength of a ligand. Hydrogen-bonding involving the N(1) hydrogen stabilises the neutral form of histidine, thus raising pK_a^{ii}, and this is consistent with the observation that in cytochromes *c′* and b_{562}, which have low pK_a^{ii} values (Table 1.4), the axial histidine is not hydrogen-bonded (Finzel *et al.*, 1985; Mathews, 1985).

Table 1.4 Histidinate and imidazolate pK_a's of small complexes and hemeproteins

Compound [a]	pK_a	Refs.
Histidine	14.4	Yagil (1967)
Imidazole	14.2	"
$[Co(His)_2]^0$	12.5	Martin (1974)
$[Cu(His)_2]^0$	11.7	"
$[Co(Im)_4]^{2+}$	12.5	"
$[Co(NH_3)_5Im]^{3+}$	10.0	Rowan *et al.* (1981)
$\alpha[Co(trien)(Im)_2]^{3+}$	9.6	"
$\beta[Co(trien)(Im)_2]^{3+}$	10.1	"
Ferricytochromes c_3	>11.0	see Chapters 2 & 5
Mitochondrial cytochromes *c*	>10.5	Gadsby *et al.* (1987)
Imidazole complex of metmyoglobin	10.45	George *et al.* (1964)
Ferricytochromes *c'*	7-9	see Chapters 2 & 5
E. coli ferricytochrome b_{562}	9.0	Moore *et al.* (1985b)
Imidazole complex of soybean leghemoglobin *a*	~7.0	Sievers *et al.* (1983)

[a] Im and His = neutral forms of imidazole and histidine; trien = triethylenediamine; α & β refer to different isomers.

Orientation of Bound Histidine. We have already noted the importance of histidine orientation relative to the heme plane as a stereochemical parameter (Fig. 1.6). In most hemeproteins and simple porphyrin complexes the histidine or imidazole has a similar orientation, bisecting the iron-pyrrole bonds (Fig. 1.12). There are a number of factors responsible for this.

The π-interactions described in Fig. 1.6 will influence the histidine orientation. Griffith (1957) suggests that these will be unfavourable but Timkovich (1979) points out that irrespective of whether they are favourable or unfavourable, the stereochemical consequence is the same when there is negligible rhombic distortion, namely the optimum dihedral angle θ (Fig. 1.12) will be 45°.

Steric factors also are important (Fig. 1.12). In a model where the histidine lies along the iron-pyrrole bond there will be substantial steric interaction between the histidine C-2 and C-4 hydrogens and the pyrrole nitrogens (Williams, 1961a; Collins *et al.*, 1972). This factor also favours a bisected histidine heme orientation.

In some porphyrin complexes θ~0° (Mashiko *et al.*, 1981) suggesting that there may be other electronic factors involved. Perhaps in these cases substantial rhombic distortions occur; this is certainly the case for the complex described by Mashiko *et al.*, (1981). In *R. molischianum* ferricytochrome *c'*, where the electronic and steric factors described above appear to be the only ones at work, θ is close to 45° (Table 1.5).

Table 1.5 Histidine - heme orientations of cytochromes *c*

Protein		θ	His N(1) - H hydrogen-bonded to carbonyls of:	Refs.
Tuna	ferrocytochrome *c*	47°	Pro 30	Takano & Dickerson 1981a,b
	ferricytochrome *c*	52°	"	
Ps. aeruginosa	ferrocytochrome *c*-551	44°	Pro 25	Matsuura *et al.*, 1982
	ferricytochrome *c*-551	45°	"	
R. molischianum	ferricytochrome *c'*	42.9°	not H-bonded	Finzel *et al.*, 1984
D. vulgaris (Miyazaki)	ferricytochrome c_3			
	Heme 1 (H70)	38.4°	Tyr 66	Higuchi *et al.*, 1984
	(H106)	32.4°	H_2O	"
	Heme 2 (H25)	-71.4°	Asn 21	"
	(H83)	-63.4°	Leu 97	"
	Heme 3 (H22)	40.2°	H_2O	"
	(H34)	35.2°	H_2O	"
	Heme 4 (H35)	-26.9°	Pro 36	"
	(H52)	37.4°	Ala 62	"

The final factor to be considered is the orienting effect of hydrogen-bonding involving the histidine N(1) hydrogen. This may overcome the other influences to impose an otherwise unfavourable θ on the structure. In most class I cytochromes *c*, where the histidine is hydrogen-bonded to the carbonyl of a proline, this does not appear to be the case however, and θ for these proteins is close to 45° (Table 1.5). The x-ray structure of a cytochrome lacking the proline is necessary to establish whether the hydrogen-bond may be responsible for orienting the peptide chain about the proline rather than the other way round. In *D. vulgaris* (Miyazaki) ferricytochrome c_3 some of the values of θ are far from 45° and this is presumably due, at least in part, to the hydrogen-bonding interactions.

Methionine Ligation. Methionine ligation through sulfur is common in cytochromes. The sulfur is sp^3 hybridised and forms two σ-bonds with side chain carbon atoms and a σ-bond with the iron. The remaining sp^3 electrons form a lone-pair that is hydrogen-bonded in structures of class I cytochromes *c*: to Tyr 67 O^{η} in mitochondrial cytochrome *c* (Takano & Dickerson, 1981a;b); and Asn 64 N^{δ} in *Ps. aeruginosa* cytochrome *c*-551 (Matsuura *et al.*, 1982). Whereas there are many small porphyrin complexes that appear to model histidine ligation in a hemeprotein, there are few complexes with thioether ligation. The reasons for this are discussed by Mashiko *et al.*, (1981) who describe the structure and properties of a number of metalloporphyrin complexes containing thioether ligands. Their Fe-S bond lengths are similar to those of cytochromes *c* (Table 1.2).

The possibility exists for back-bonding to occur between the d-orbitals of the iron and coordinated sulfur. This is illustrated in Fig. 1.13; electrons from Fe d-orbitals enter S d-orbitals to form a π-bond.

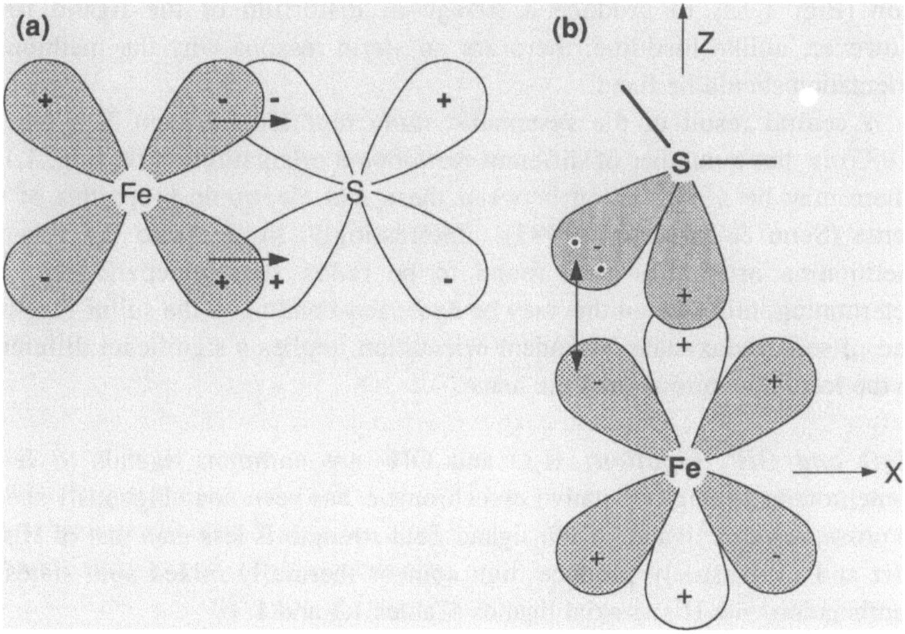

Fig. 1.13 Methionine ligation to heme. **(a)** Possible d-d back-bonding interaction. **(b)** Interaction of the sulfur lone-pair with the Fe d_{xz} orbital. The unshaded lobes in **(b)** represents the Fe d_{z^2} orbital (for clarity its torus has been omitted). A similar interaction may occur for d_{yz}.

The involvement of the 3d orbitals of sulfur in such bonding requires investigation as it may provide an explanation of the apparently higher stability of the Fe-S bond of ferrocytochrome *c* than that of ferricytochrome *c* (Chapter 4.5.4). This is because Fe(III) back-bonds to a much lesser extent than does Fe(II). It should be noted, however, that a variety of theoretical arguments can be levelled against this suggestion. These are concerned with the diffuse nature of the sulfur 3d orbitals and their large promotion energies (Mitchell, 1969).

Orientation of Bound Methionine. The orientation of the methionine ligand relative to the heme plane has been extensively studied in cytochromes *c*, mainly by Wüthrich and co-workers (Senn *et al.*, 1980; Senn & Wüthrich, 1985). The spectroscopic approach used by these workers is described in Chapter 2.3.2.1.

The interest in the orientation stems from the fact that, like histidine (Fig. 1.6), a methionine ligand can interact with the d_{xz} and d_{yz} orbitals of the iron (Fig. 1.13) to produce a tetragonal distortion of the ligand field. However, unlike histidine, there are no steric reasons why the methionine orientation should be fixed.

A central result of the systematic study reported by Senn & Wüthrich (1985) is that a number of different methionine orientations exist (Fig. 1.14). There may be a correlation between these and electronic properties of the heme (Senn & Wüthrich, 1985). Interestingly, in all cases but two the methionine orientation was found to be redox state independent. The determining influence on this may be hydrogen-bonding to the sulfur lone-pair and, if so, a redox-state dependent orientation implies a significant difference in the local structure around the heme.

H_2O and OH^- Ligation. H_2O and OH^- are common ligands to ferric hemeproteins, though no native cytochrome *c* has been unambiguously shown to possess such a ligand. Their ligand field strength is less than that of His or Met and they usually produce high spin or thermally mixed spin states in combination with His as axial ligands (Tables 1.1 and 1.3).

As with pK_a^{ii} of histidine, the pK_a for the ionisation of Fe(III)-H_2O can vary over a wide range; values of 6.5 - 9.0 have been reported (Antonini & Brunori, 1971; Yonetani, 1976). This range reflects variations in the Lewis acidity of the Fe(III) and the nature of the protein environment surrounding the H_2O/OH^-.

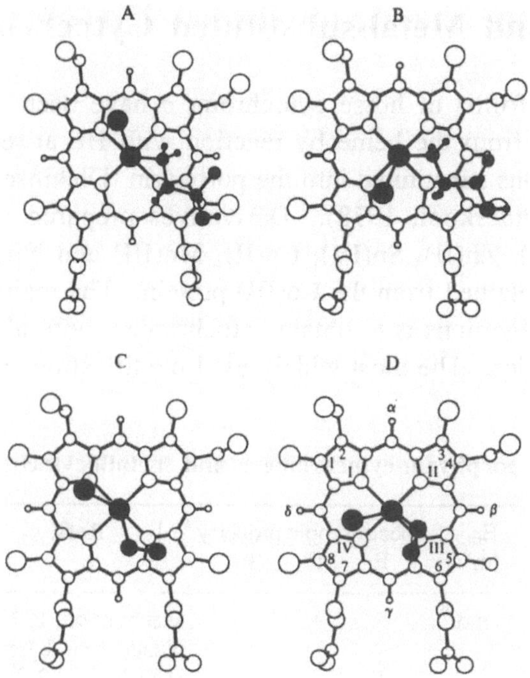

Fig. 1.14 Conformation of the methionine ligand of various cytochromes *c*. The view is perpendicular to the heme plane. The methyl and methylene groups are represented by equivalent spheres with the exception of βCH_2 and γCH_2 of methionine in the structures A and B where these methylene protons were stereospecifically assigned. **A** Mitochondrial ferrocytochrome *c* (*S. cerevisiae* iso-1 and iso-2, *C. krusei*, *Cr. oncopelti*, horse), *R. rubrum* cytochrome c_2, *S. platensis* ferrocytochrome *c*-553 and *E. gracilis* ferrocytochrome *c*-552. **B** Ferrocytochrome *c*-551 of *Ps. aeruginosa*, *Ps. mendocina*, *Ps. stutzeri* and *Rc. gelatinosa*. **C** *Ps. mendocina* ferrocytochrome c_5. **D** *D. vulgaris* and *D. desulfuricans* ferrocytochrome *c*-553. In structure **D** the meso-positions and the β positions of heme *c* are identified by α - δ, and by 1 - 8, respectively. The pyrrole rings are numbered I - IV. (Reproduced with permission from Senn & Wüthrich, 1985). [The proposed conformation of the methionine ligand of one of the hemes of *A. vinelandii* cytochrome c_4 is shown in Fig. 2.27].

Amide Ligation. There are no known examples of amide ligation to the iron of a hemeprotein, though amide coordination is known in simple inorganic compounds, such as in the formamide and dimethylacetamide complexes of Co(II) (Wayland *et al.*, 1966). Coordination through both the amide N and O atoms has been reported but the binding was weak, partly as a consequence of steric factors (Homer & Johnson, 1970). Therefore, it appears to be unlikely that amides will act as axial ligands in native cytochromes.

1.5 Metal-free and Metal-substituted Cytochrome *c*

A variety of modified forms of horse cytochrome *c* have been prepared in which iron is removed from the heme by reaction with HF at temperatures <0°C and other metal ions substituted into the porphyrin (Dickinson & Chien, 1974; Erecińska & Vanderkooi, 1978). Derivatives prepared in this way include those of Co(III), Zn(II), Sn(IV), Cu(II), Mn(III) and Ni(II). Co(II) cytochrome *c* can be obtained from the Co(III) protein. The main reason for carrying out these modifications is to obtain cytochrome *c* with altered redox or spectroscopic properties. The most widely used modifications are given in Table 1.6.

Table 1.6 Properties of porphyrin-cytochrome *c* and metallocytochromes *c*

Protein	Spin	$E_{m,7}$ (mV)	Spectroscopic property[a] E	F	P	Refs.
porphyrin	0	n.d.[b]	-	+	-	Fisher *et al.*, 1973 Vanderkooi & Erecińska, 1975
Fe(III) cytochrome *c*	½	260	+	-	-	see Chapters 2 & 6
Fe(II) cytochrome *c*	0		-	-	-	"
Co(III) cytochrome *c*	0	-140±20	-	n.d.	n.d.	Dickinson & Chien, 1975 a,b
Co(II) cytochrome *c*	½		+	n.d.	n.d.	"
Zn(II) cytochrome *c*	0	-	-	+	+	Vanderkooi *et al.*, 1976
Sn(IV) cytochrome *c*	0	n.d.[c]	-	+	+	"

[a] E = EPR; F = fluorescence; P = phosphorescence; n.d.= not determined; + indicates that the protein has a strong characteristic spectrum; - indicates the protein does not exhibit the spectrum or property designated.
[b] Porphyrin-cytochrome *c* anion radical can be generated by reduction with hydrated electrons. The radical reduces cytochrome oxidase (Veerman *et al.*, 1982).
[c] Sn(II) cytochrome *c* has not been reported.

An important requirement for the use of these derivatives as probes for native cytochrome *c* is that the structure should not be greatly changed by the modification. Such seems to be the case for the monomeric Co(III) and Zn(II) derivatives of horse cytochrome *c* as judged by NMR spectroscopy (Moore *et al.*, 1980a): there are only small conformational differences which appear to be on a similar scale to the native redox-state conformation change (Chapter 4.5.2). The Zn(II) derivative of *Hansenula* cytochrome *c* however, is dimeric (Thomas *et al.*, 1983).

The other proteins of Table 1.6 have not been studied by high-resolution structural methods but porphyrin-cytochrome *c* resembles the native protein in the fluorescence properties of Trp 59 (Vanderkooi & Erecińska, 1975) and in its hydrodynamic behaviour (Brems *et al.*, 1982).

Iron removal and metal insertion have not been reported for cytochrome *c* in which iron is five-coordinate, but significant conformational changes are expected to occur because of the different stereochemical requirements of the metal ions. For example, M-Ct distances (Å) for porphyrins containing five coordinate Zn(II), Mn(III), Fe(III), Fe(II), Co(III) and Co(II) are 0.33, 0.27, 0.56, 0.45, 0.42, 0.11, 0.13 respectively (Scheidt, 1977).

Cobalt cytochrome *c* has been used in cases where a paramagnetic reduced state and a diamagnetic oxidised state have been required (Dickinson & Chien, 1975a;b; Moore *et al.*, 1980a) and to study the effect of redox potential variation on the reaction of cytochrome *c* with its oxidase (Antalis & Palmer, 1982). The proteins with luminescent spectral properties have been used to study cytochrome *c* binding to proteins and membranes (Vanderkooi & Erecińska, 1975; Dixit *et al.*, 1982; McLendon *et al.*, 1985) and as redox inactive analogues of native cytochrome *c* (Veerman *et al.*, 1980). These topics have been discussed by Pettigrew & Moore (1987: Chapter 2).

Chapter 2 Spectroscopic Studies of Cytochromes

2.1 Introduction

Spectroscopy is one of the main tools used to classify cytochromes (Pettigrew & Moore, 1987: Chapter 1). However, the usefulness of spectroscopy goes far beyond the initial classification, and subsequent kinetic and thermodynamic characterization, made possible using, for example, visible spectroscopy. Many spectroscopic techniques are capable of providing detailed electronic and structural information. The purpose of the present chapter is to review some of the most common spectroscopic techniques in current use and to illustrate the kind of information they provide. However, we do not attempt a comprehensive review of the cytochrome *c* literature. Many spectroscopic studies of proteins have clearly defined biochemical or physicochemical objectives but there are others whose objectives point in other directions - for example, to assist development of a new method - or are not apparent at all. Studies of this latter type are not considered here.

The plan we have adopted is to classify particular spectroscopic techniques into groups depending upon the principal kind of information they provide; the groups are heme centred and protein centred techniques. By using this format we hope to emphasize that it is the electronic and magnetic properties of the heme that dominate many spectroscopic measurements. We then go on to consider various problems concerning cytochromes *c* that have been tackled by spectroscopic methods. Two of these, electronic and magnetic resonance spectroscopies, are given particular attention because of their application to a wide range of problems.

Table 2.1 lists the electronic and structural information that is required to characterise fully a cytochrome, and also which methods are the most revealing. It is clear that no technique on its own is sufficient. Currently only X-Ray Diffraction (XRD) is wholly satisfactory at determining protein structures, though even with this technique it is sometimes dependent on others. For example, determination of the x-ray structures of cytochromes *c* and b_{562} (Xavier *et al.*, 1978) was aided by the prior identification of the axial ligands by NMR (section 2.3.2.1). Also, as with any technique, it is necessary to exercise caution in interpreting x-ray data. The class I cytochromes *c* group contains a number of proteins whose x-ray structures were incorrect when originally reported (Chapter 4.3.1 & 4.5.2)). Nevertheless, XRD is the central structural technique. A detailed consideration of XRD is beyond the scope of this review but it should be noted that recent advances have considerably speeded up the time taken to collect and interpret XRD data, so that generally a structure can be obtained if suitable crystals of the protein are forthcoming (Eisenberg & Hill, 1989).

As structural tools, most spectroscopic techniques are less definitive than XRD: this is indicated by the commonly used description of XRD as a "non-sporting" method compared to spectroscopic methods, which are "sporting' because they often allow the quarry to escape! Only NMR approaches XRD in its power to determine protein structures (see e.g. Wright, 1989). However, the need for spectroscopy in cytochrome research is well-established. Many proteins are difficult to crystallize and thus the only structural information available for them comes from spectroscopy and model building on the basis of amino acid sequence analyses. Also, spectroscopy allows certain mechanistic properties to be investigated that are generally not accessible from XRD data. Examples are: the electronic structure of the heme; oxidation state of the iron; and ionisation state of particular functional groups.

2.2 Spectroscopic Methods

We give here an unsophisticated introduction to each of the spectroscopic techniques whose applications to cytochromes are described in the present chapter. We do this to help readers who are not familiar with spectroscopy and we do not concern ourselves overmuch with detailed theoretical descriptions of the techniques discussed, nor with the experimental methodologies. Good accounts of these can be found: in specific references given later; in the series of books edited by Dolphin ('The Porphyrins, 1973 - 1979); and in particular volumes of the 'Metal Ions in Biology' and 'Methods in Enzymology' series. Good general accounts of spectroscopy are provided by Cantor & Schimmel (1980) and Campbell & Dwek (1984).

Table 2.1 **Comparison of different techniques [a] in terms of the information [b] they yield about cytochromes**

Information	XRD	EXAFS	MS	MR methods EPR	NMR	Electronic spectroscopy UV	Vis.	NIR	Lm.	MOS	Vibrational spectroscopy IR	Ram.	Mössbauer spectroscopy
Fe oxidation state			***	**	**		**	**		**		**	***
Fe spin-state			***	**	*		**	**		**		**	**
Type of heme	***				*		***	**		**		*	
Axial ligand identification	***	**		**	**			***		***		*	
Shape of molecule	***												
Secondary structure	***				**	**							
Tertiary structure	***				(*)				*				
Dynamic features of structure	*				**				**				
Individual group ionisation states	(*)				***	**	(*)	(*)		(*)	**		

[a] The techniques are:
XRD, X-Ray Diffraction;
EXAFS, Extended X-ray Absorption Fine Structure;
MS, Magnetic Susceptibilities;
MR, Magnetic Resonance;
EPR, Electron Paramagnetic Resonance;
NMR, Nuclear Magnetic Resonance;
UV, Ultra-Violet;
Vis., Visible;
NIR, Near-Infra-Red;
Lm., Luminescence;
MOS, Magneto Optical Spectroscopy;
IR, Infra-Red;
Ram., Raman.

[b] The starred methods are generally applicable for the information required. The more *'s, the more reliable the method. (*) indicates cases where the information is given by a particular method in a limited number of cases.

2.2.1 General Aspects of Spectroscopy

Spectroscopy is the study of the interaction of electromagnetic radiation with matter. Three general phenomena may occur: radiation is absorbed, emitted or scattered. These phenomena, which form the basis of the techniques described in the present chapter, are briefly described below. A fourth class of effects, photochemical processes, may also occur. Relevant examples are the flash-photolysis of CO and O_2 bound to ferrohemeproteins (Gibson & Ainsworth, 1957) and the photodissociation of a methionine ligand of *Ps. aeruginosa* cytochrome *c* peroxidase (Greenwood *et al.*, 1984). Though useful tools for kinetic studies, and interesting in their own right, photochemical processes are not considered further in this chapter.

The actual measurements in most routine spectroscopic experiments are of the frequency and amount of radiation interacting with the sample. These, and other parameters, are described below.

Absorbance, Emission and Scattering. A convenient way to illustrate the interaction of radiation with matter is by an energy level diagram (Fig. 2.1). The sample can exist in at least two states which have different energies. When it is irradiated the sample may *absorb* energy and move from the lower to the higher energy level. Once in the higher energy level the sample can *emit* radiation. With both *absorbance* and *emission* the frequency of the radiation (ν) corresponds precisely to the energy gap (ΔU) between the states.

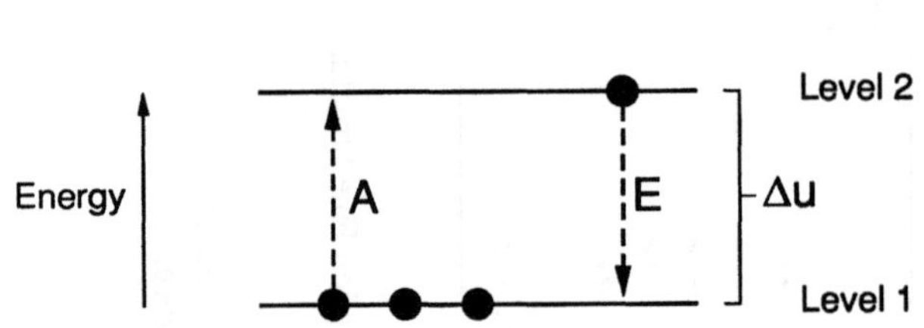

Fig. 2.1 Molecular energy levels and spectroscopy. Different molecules of a sample may possess slightly different energies. These differences arise because the molecules have different electronic, vibrational or magnetic configurations of their constituent atomic nuclei and electrons. The distribution of molecules with different energies can be represented by an energy level diagram: only certain energy values are allowed and hence there are discrete levels. Under normal conditions, most molecules have relatively low energy and so the lowest energy level is the most populated. Transitions between energy levels occur when a molecule in a lower energy level absorbs ΔU and moves to a higher energy level (**A**). A second kind of transition (**E**) is when a molecule in a higher energy level emits ΔU and moves to a lower level

The relationship is:

$$\Delta U = h\nu \tag{2.1}$$

where h is Planck's constant (6.63×10^{-34} J s).

Scattering is the name given to the deflection of radiation as a result of its interaction with the sample. The change in direction occurs because the radiation collides with electrons in the sample. If the scattered radiation gains or loses energy (i.e. its frequency is changed) then the scattering is termed 'inelastic' but if the frequency remains constant it is called 'elastic'. Elastic scattering is the phenomenon that ultimately gives rise to diffraction patterns, and hence XRD structures; this is beyond the scope of the present review. Inelastic scattering is the effect that gives rise to Raman spectroscopy, which is considered further in section 2.2.4.

The data for a spectroscopic experiment are usually presented in the form of a graph of the amount of radiation absorbed, emitted or scattered as a function of the frequency (Fig. 2.2a) or as the first derivative of the amount of

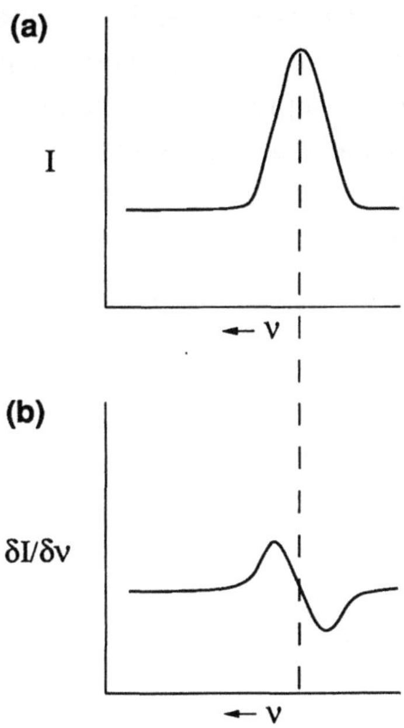

Fig. 2.2 Absorption and derivative spectra. The output of a spectrometer is usually presented in the form of a graph of signal intensity (**I**) against frequency (ν). The common convention has frequency increasing to the left. This form of presentation is the absorption mode spectrum, **(a)**. The first derivative of this, **(b)**, is also used in some spectroscopies, e.g. EPR

radiation against the frequency (Fig. 2.2b). With some spectroscopies, difference spectra are common. A difference spectrum is obtained when the spectrum of a sample under one set of conditions is subtracted from the spectrum of the same sample under a different set of conditions (e.g. a change in pH; see Fig. 2.38). This procedure improves the resolution of the spectrum and emphasises the signals that are affected by the changed conditions.

2.2.2 Spectroscopic Parameters

All spectroscopic methods are characterised by a number of common parameters. These include the intensity of the signal, its frequency and the lifetime of the sample in its excited state. Occasionally small splittings of the main spectroscopic signals are observed. Common features of these parameters are described in the present section.

Signal Frequencies. There is a plethora of frequency or energy scales used in spectroscopy. The relationships of some of these are given in Scheme 2.1.

In addition to the scales defined in Scheme 2.1, magnetic resonance spectroscopies use dimensionless numbers to represent the frequency and Mössbauer spectroscopy uses a velocity scale that represents the physical movement of the sample relative to part of the spectrometer. These are described further in sections 2.2.3 and 2.2.6.

Scheme 2.1 The frequency, energy, wavelength and wavenumber of radiation.

The frequency (ν) and wavelength (λ) of radiation are related by:

$$\nu = \frac{c}{\lambda} \tag{2.2}$$

where c is the speed of light (3×10^8 m sec^{-1}).

The relationship between frequency and energy is given by Eqn. 2.1 in the text. The wavenumber (ν') is the reciprocal of the wavelength in centimetres.

Thus for light with a wavelength of 550 nm:

λ = 550 nm

ν = 5.455×10^{14} Hz

ν' = 18,182 cm^{-1}

ΔU = 3.617×10^{-19} J [which is equivalent to 2.18×10^2 KJ mol^{-1} after multiplication by Avogadro's number (6.03×10^{23})].

Signal Intensities. The intensity of a spectroscopic signal relative to the noise level of the spectrum baseline depends upon the amount of sample, the relative populations of the energy levels, and the intrinsic probability of the spectral transition occurring.

The population difference between two energy levels is given by the Boltzmann distribution law (Scheme 2.2). This shows that as the energy separation between the levels increases, so does the population difference. Thus spectroscopies in which ΔU (and hence P_{lower}/P_{upper}) is very large tend to be more sensitive than spectroscopies in which ΔU is relatively small. An interesting comparison is visible electronic spectroscopy, which has a ΔU of 3.617×10^{-19} J (at 550 nm), and ^{1}H NMR,which has a ΔU of 3.32×10^{-25} J (at 500 MHz - section 2.2.3.1). A direct practical consequence of this is that far more material is required for a routine ^{1}H NMR spectrum than for a visible absorption spectrum.

Scheme 2.2 Boltzmann population distribution.

The intensity of a spectroscopic signal is related to the difference in the populations of the relevant energy levels (transition probabilities are also important parameters influencing signal intensities). This difference is given by the Boltzmann equation. For the system shown in Fig. 2.1, this equation is:

$$\frac{P_{lower}}{P_{upper}} = \exp\left(\frac{\Delta U}{\mathbf{KT}}\right) \tag{2.3}$$

where P_{lower} and P_{upper} are the populations of energy levels (1) and (2) respectively: ΔU is the energy difference between the levels; T is the absolute temperature; and **K** is Boltzmann's constant (1.38×10^{-23} J K^{-1}).

This population effect accounts for much of the sensitivity differences between spectroscopies. Thus, at 300 K:

(i) For ^{1}H NMR at 500 MHz:

$$\Delta U = 3.32 \times 10^{-25} \text{ J}; \quad \frac{P_{lower}}{P_{upper}} \approx 1.000080$$

(ii) For EPR at 9 GHz (x-band) and with g = 2 (section 2.2.3.2):

$$\Delta U = 5.97 \times 10^{-24} \text{ J}; \quad \frac{P_{lower}}{P_{upper}} \approx 1.00144$$

(iii) For IR of a C=O stretch at 1600 cm^{-1} (section 2.2.4):

$$\Delta U = 3.18 \times 10^{-20} \text{ J}; \quad \frac{P_{lower}}{P_{upper}} \approx 2167$$

(iv) For optical absorbance at 550 nm:

$$\Delta U = 3.617 \times 10^{-19} \text{ J}; \quad \frac{P_{lower}}{P_{upper}} \approx 9 \times 10^{37}$$

The transition probabilities for spectroscopic transitions are given by transition selection rules. Molecules possess a large number of excited states but transitions between them and the ground state are not always possible, even when the radiation is of the correct frequency. The reason for this is that not only must the frequency of the radiation be correct, but also there has to be a mechanism for the molecule to interact with the radiation. For example, in magnetic resonance spectroscopies the mechanism of interaction involves the magnetic component of the electromagnetic radiation, in electronic absorption spectroscopy it involves the electric component, and in Circular Dichroism it involves both the electric and magnetic components. The selection rules describe cases where interaction mechanisms exist, and hence the transitions have a relatively high probability of occurring. Such transitions are called 'allowed transitions'.

A detailed description of selection rules is beyond the scope of the present account and we only briefly consider relevant rules in sections 2.2.3 - 2.2.5.

Signal Lifetimes and Linewidths. Spectroscopic signals vary in linewidth from being very sharp (e.g. for proton NMR, linewidths of <10 Hz are common) to very broad (e.g. for the Mössbauer spectrum of ^{57}Fe, linewidths of 10^6 Hz are common). This is partly a consequence of the difference in the lifetimes of the samples in their respective excited states: a system with a long-lived excited state has a sharp signal.

The processes that give rise to the decay of a molecule in its excited state are relaxation processes. These follow first order kinetics. Both relaxation processes and excited-state lifetimes provide much information about the system under study, and some spectroscopy experiments are designed solely to measure these and not the frequency of the interacting radiation (e.g. Fig. 2.30).

Splitting of Spectroscopic Signals. Spectroscopic signals are often split into a number of components. The physical mechanism of the splitting varies with different spectroscopies but the underlying principle is the same: the excited state lies close in energy to at least one other excited state so that transitions between the ground state and two or more excited states occur. Fig. 2.3 illustrates this situation for the two cases where the transition probabilities between the ground state and the two excited states are either the same or different.

Fig. 2.3a shows the case when the two excited states have the same energy (they are 'degenerate') and Fig. 2.3b when the excited states have different energies (they are 'split' or 'non-degenerate').

Splitting of the ground state may also occur so that the transitions could be from a split ground-state to a degenerate excited-state (the spectrum would resemble Fig. 2.3b) or to a non-degenerate excited state. In the latter case the signal may be split into four, or more, lines.

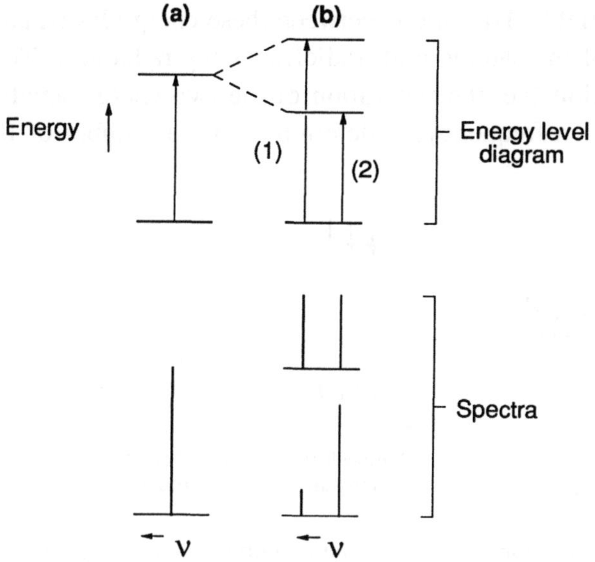

Fig. 2.3 Spectroscopic fine-structure. **(a)** illustrates a spectroscopic transition and the resulting absorption mode spectrum for a system involving two energy levels. **(b)** illustrates comparable cases when the higher energy level of **(a)** is split into two (note that there is a centre of gravity relationship for the excited states). The spectra for **(b)** are when transitions (1) and (2) have equal probabilities (upper spectrum) and when (2) has a greater probability than (1) (lower spectrum)

2.2.3 Magnetic Resonance Spectroscopies

In magnetic resonance spectroscopies the degeneracy of particular energy levels of a sample is lifted by placing the sample in a strong magnetic field. The two most common forms of this type of spectroscopy are Nuclear Magnetic Resonance (NMR) and Electron Paramagnetic Resonance (EPR) spectroscopies. EPR is also known as Electron Spin Resonance (ESR). As their names imply, these two techniques exploit the magnetic properties of nuclei and electrons. We consider them in turn.

2.2.3.1 NMR Spectroscopy

Some atomic nuclei, such as those of ^{1}H and ^{13}C, can be considered to resemble small bar magnets. Thus they interact magnetically with an external magnetic field (Fig. 2.4). The external field, B_0, forces the nuclear bar magnets to adopt one of two orientations: parallel and antiparallel to the direction of the applied field. Transitions between these energy levels are then induced by irradiation of the sample with radiofrequency radiation. The size of the magnetic interaction (i.e. the separation of the two energy levels) and hence the precise radiation frequency, is determined by a number of factors (Scheme 2.3).

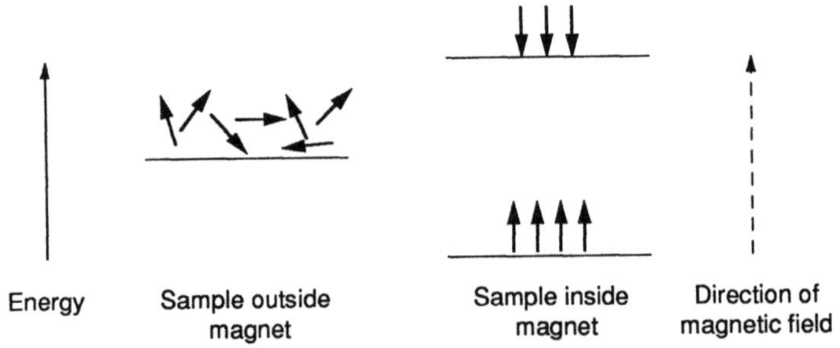

Fig. 2.4 Energy level diagram for ^{1}H NMR. The two orientations of hydrogen atoms in a strong magnetic field are often called the parallel (↑) and antiparallel (↓) orientations. In the antiparallel orientation there is a force acting on the hydrogens which tends to tip them into the parallel orientation. There is no corresponding force on the hydrogens in the parallel orientation and, therefore, the antiparallel hydrogens are slightly less stable than the parallel hydrogens

Eqn. 2.4 (Scheme 2.3) indicates that there are three contributions to the splitting of nuclear magnetic energy levels. The first term, which is called the Zeeman splitting, is the same for all nuclei of a given type. The second term (the chemical shift) is due to the magnetic properties of surrounding electrons which act to shield the atomic nucleus from the applied field, B_0. The shielding constant, σ, is a measure of the extent to which this occurs.

Since electrons are involved in chemical bonding their shielding influence varies considerably with the nature of the bond, and thus NMR is sensitive to small changes in the distribution of electrons within samples. An indication of this sensitivity is the ^{1}H NMR spectrum of tryptophan in Fig. 2.5: all the protons have individually resolved resonances. Because the resonance frequency of an NMR signal is directly proportional to B_0, a dimensionless

Scheme 2.3 NMR Spectroscopy.

^{1}H and ^{13}C nuclei interact with an external magnetic field, B_o, and adopt one of two orientations. These have different energies. The difference in energy between the orientations is given by:

$$\Delta U = \frac{h\gamma B_o}{2\pi} + \text{chemical shift term} \pm \text{spin-spin coupling term} \quad (2.4)$$

where γ is the magnetogyric ratio (this describes the magnetic properties of nuclei and is a constant for nuclei of a given type). The chemical shift and spin-spin coupling terms result from magnetic interactions within the sample. These provide information concerning the structure and properties of the sample.

Eqn. 2.4 can be recast as:

$$\nu = \frac{\gamma B_o}{2\pi}(1 - \sigma) \pm \text{spin-spin coupling term} \quad (2.5)$$

where σ is the shielding constant. σ depends upon the magnetic environment of nuclei; it is variations in σ that give rise to chemical shifts in NMR.

Since σ is a molecular constant and depends upon the magnetic field strength (Eqn. 2.4), NMR spectra are calibrated in a dimensionless number called the chemical shift, δ.

$$\delta = \frac{\nu_s - \nu_r}{\nu_r} \approx \sigma_r - \sigma_s \quad (2.6)$$

where the subscripts s and r refer to sample and reference respectively. For ^{1}H and ^{13}C NMR. $Si(CH_3)_4$ is the reference compound. Thus for a sample resonance 100Hz from the reference resonance in a 100MHz ^{1}H NMR spectrometer:

$$\delta = \frac{100}{100 \times 10^6} = 1 \times 10^{-6} = 1 \text{ ppm}$$

The spin-spin coupling term is far smaller than the chemical shift term and is generally not measured in conventional one-dimensional protein NMR unless the resolution of the spectrum is enhanced by mathematical manipulation.

number, called the chemical shift (δ), is used to represent the frequency scale (Scheme 2.3). This allows spectra obtained with different sizes of magnet to be compared easily.

Fig. 2.5 reveals that most of the tryptophan resonances consist of more than one line. This fine structure results from a splitting of the energy levels by a through-bond interaction between the atomic nuclei. This is called spin-spin coupling, and is the third contributor to Eqn. 2.4. For example, the C-4 proton resonance is split into two because the C-4 proton is affected by its C-5 neighbour three bonds away, and the C-5 proton resonance is split into three because it is affected by both the C-4 and C-6 protons. The C-2 proton is not within three bonds of another CH and so it appears as a singlet resonance.

Both the multiplicity of a signal and the separation between the components of the multiplet, the spin-spin çouping constant, provide valuable structural information.

The relative intensities of NMR signals reflect directly the number of nuclei contributing to the signals. Thus in Fig. 2.5 all the resonances have the same intensity (based on integrated areas of the peaks). This is equivalent to stating that the extinction coefficient for NMR lines is a constant for a particular type of nucleus, and this is in marked contrast to other spectroscopies, such as electronic and vibrational, where extinction coefficients vary considerably.

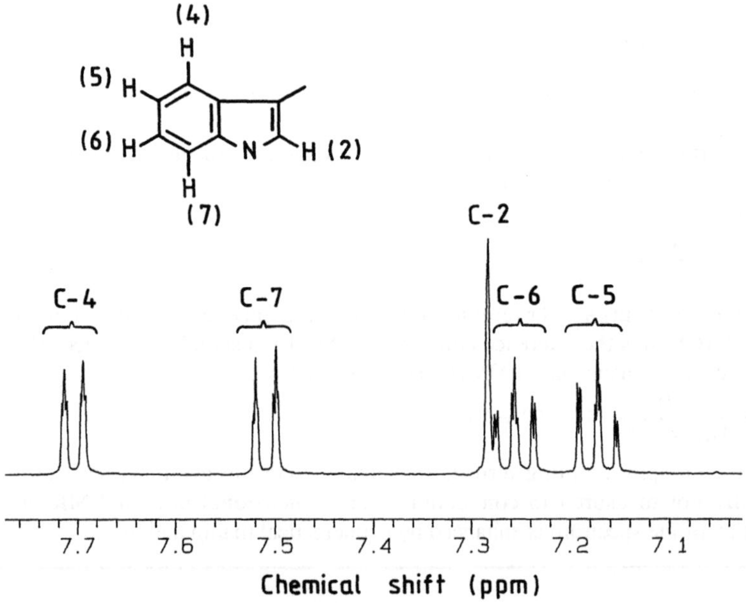

Fig. 2.5 Aromatic region of the ^1H NMR spectrum of tryptophan in D_2O. The large splitting of the C-4, C-5, C-6 and C-7 proton resonances is from hydrogens situated up to three bonds away from the indicated hydrogen. The additional small splittings are from hydrogens further than three bonds away

The linewidths of NMR signals cover a wide range because they are sensitive to number of different processes. Three are of particular importance for cytochromes:

1. As the molecular weight of a molecule increases so does the linewidth of most of its NMR signals. This is because, as the size increases, the molecule tumbles more slowly in solution. This leads to enhanced

relaxation and hence a broader signal. Resonances of groups which have a high degree of independent mobility within the protein (e.g. those of lysines on the protein surface) are exceptions to this. Membrane-bound cytochromes and their detergent-solubilized counterparts usually have large molecular masses and hence broad lines.

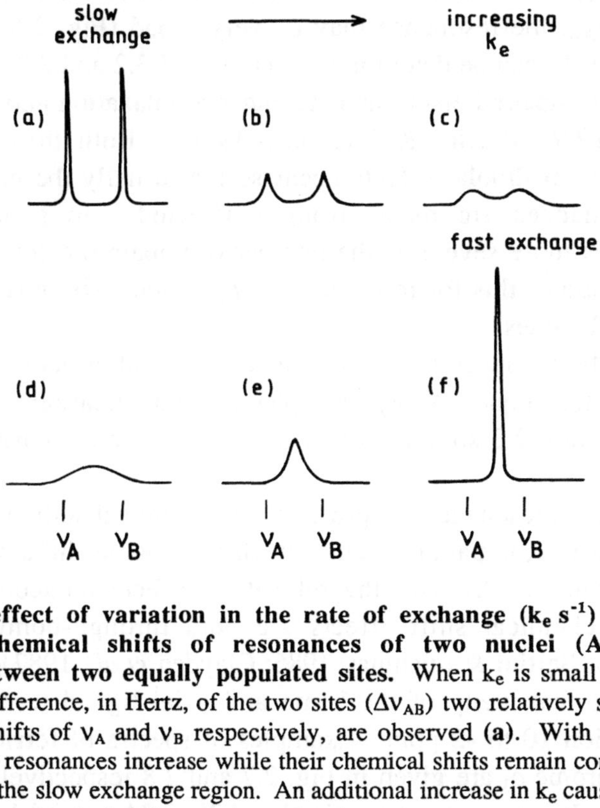

Fig. 2.6 The effect of variation in the rate of exchange (k_e s^{-1}) upon the linewidths and chemical shifts of resonances of two nuclei (A and B) that interchange between two equally populated sites. When k_e is small compared to the chemical shift difference, in Hertz, of the two sites ($\Delta\nu_{AB}$) two relatively sharp resonances, with chemical shifts of ν_A and ν_B respectively, are observed **(a)**. With increasing k_e the linewidths of the resonances increase while their chemical shifts remain constant **(b)**. These spectra illustrate the slow exchange region. An additional increase in k_e causes the linewidths to increase and the chemical shifts to vary so that the two resonances coalesce **(c)**, **(d)**. These spectra illustrate the intermediate exchange region. The chemical shift of the coalesced resonance is $^1/_2(\nu_A + \nu_B)$ and a further increase in k_e causes a decrease in the linewidth **(e)**, **(f)**. These spectra illustrate the fast exchange region. The resonance line-shapes were calculated for the following values of $k_e/(\nu_A - \nu_B)\nu_o$, where ν_o is the spectrometer frequency in MHz: **(a)** 0.13, **(b)** 0.52, **(c)** 1.25, **(d)** 2.2, **(e)** 6.3, and **(f)** 31. (Adapted from Becker, 1980)

2. Chemical exchange processes can lead to exceedingly broad lines (Fig. 2.6). The frequency difference ($\Delta\nu_{AB}$) between the resonances of a group in different magnetic environments measured in Hertz (note, not in δ ppm) and the rate of exchange between these environments (k_e) are the key parameters. Taking the simplest case of a two site exchange system with an

equal number of nuclei in each site (e.g. an ionisable group at its pK_a [Fig. 2.36 and 2.37]) then the NMR spectrum could appear as shown in Fig. 2.6. When the rate of exchange is faster than the reciprocal of the frequency difference only a single, relatively sharp signal is observed, but when $k_e<\Delta\nu_{AB}$, two sharp signals are observed. These two cases are the fast and slow exchange cases respectively. In the intermediate exchange region, when $k_e \sim \Delta\nu_{AB}$, the resonance may be very broad (Fig. 2.6(d)). In some circumstances k_e can be determined (sections 2.3.3.2 and 2.3.3.4).

3. The effect of unpaired electrons is to enhance relaxation and broaden lines (Wüthrich 1976; Bertini & Luchinat, 1986). Both through-bond and through-space or dipolar effects occur so that usually the heme and axial ligand resonances are most strongly affected. In general the line broadening is more severe as the number of unpaired electrons increases. Fig. 2.7 illustrates this for resonances of *Rps. palustris* ferricytochrome *c'* at various pH values.

Because NMR monitors individual nuclei it is both a heme-centred and a protein-centred technique. Many examples of its application to cytochromes are given in section 2.3 so we confine ourselves to some general comments here.

Most CH proton resonances of proteins have chemical shifts in the range 10 to -2 ppm. However, in paramagnetic cytochromes large shifts result from the unpaired electron(s). As with the relaxation effects induced by unpaired electrons, the chemical shift effects are both through-bond and dipolar (Wüthrich 1976; Bertini & Luchinat, 1986; Clayden *et al.*, 1987). These shifts can be so large that many of the heme and axial ligand resonances occur outside the region 10 to -2 ppm. Examples of spectra of ferricytochrome *c'* and ferricytochrome c_3 are given in Fig. 2.7 and 2.8 respectively. Spectra of class I ferricytochromes *c* are given in Fig. 2.29, 2.34 and 2.37.

All of these spectra show that resonances of the heme methyls are generally well resolved. Also, they show that as the number of unpaired electrons increases, the heme methyl chemical shifts generally increase (compare Fig. 2.7 with Fig. 2.32). The cause of the separation between the four heme methyl resonances for any given protein has not been established in detail but it is related to the orientation of the axial ligands with respect to the heme plane (Wüthrich, 1976; Senn *et al.*, 1980; Byrn *et al.*, 1983; Senn & Wüthrich, 1985).

Even without the precise assignment of the strongly shifted peaks to specific heme or axial ligand protons, the peaks can be used to monitor physicochemical properties of the protein such as the occurrence of heme-linked ionisations (Fig. 2.7 and 2.37), and the rate of electron self-exchange (Fig. 2.8).

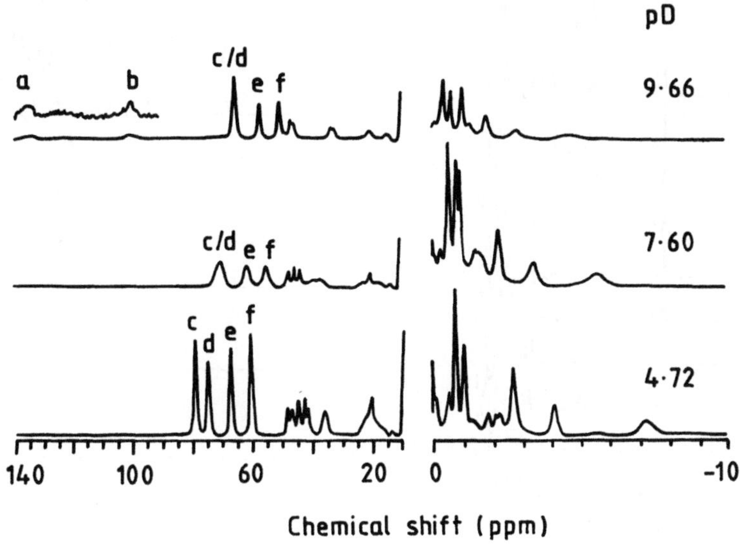

Fig. 2.7 The 360 MHz ^{1}H NMR spectrum of *Rps. palustris* ferricytochrome *c'* at 45°C and various values of pD. The 140 to 10 ppm region has an absorbance scale 10^{-1} times that of the 0 to -10 ppm region. Peaks **c**, **d**, **e** and **f** come from heme methyls; **a** and **b** probably come from the His ligand. The changes in linewidth for peaks **c**, **d**, **e** and **f** between pD 4.72 and 9.66, are a result of the spin-state change (section 2.3.2.5). (Reproduced from Jackson *et al.*, 1983)

Analysis of amino acid resonances in the region 10 to -2 ppm is considered in section 2.3.3 but we note here that it is usually not possible to carry out chemically useful experiments without first assigning these peaks to specific nuclei. This assignment problem is usually the rate limiting factor in a high-resolution NMR study of a cytochrome.

2.2.3.2 EPR Spectroscopy

EPR spectroscopy detects unpaired electrons in an analogous way to the NMR detection of magnetic atomic nuclei (Fig. 2.4). The central difference between the two techniques stems from the lower mass of the electron compared to the

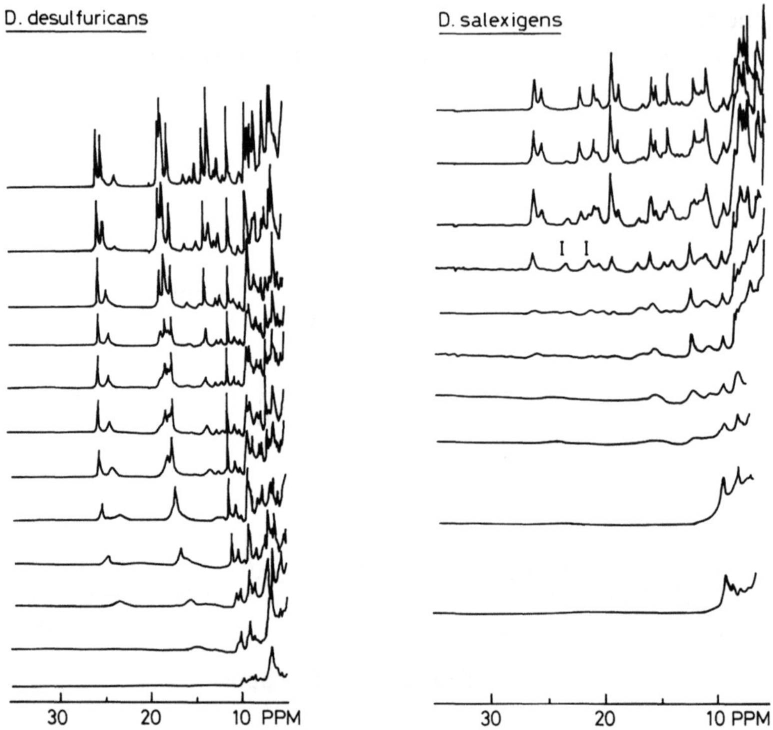

Fig. 2.8 High-frequency regions of the ^{1}H NMR spectra of tetraheme cytochromes c_3 at 25°C and pD 7. The bottom traces are the spectra of the fully reduced, diamagnetic, ferrocytochromes. The top traces are the spectra of the fully oxidised ferricytochromes. Many of the heme methyl resonances are clearly resolved in the ferricytochrome spectra. The addition of increasing amounts of reductant to the ferricytochromes (descending spectra) causes resonances to broaden and shift. This is a result of fast and intermediate chemical-exchange affecting the appearance of the spectra. In spectra of *D. salexigens* cytochrome c_3 (fourth from top) additional peaks marked I appear. These come from molecules containing both ferric and ferrous hemes with a slow rate of electron exchange. A full analysis of such spectra yields the intra- and inter-molecular exchange rates, and the relative redox potentials (Moura *et al.*, 1978; 1982; Santos *et al.*, 1984). (Spectra are unpublished data of G.R.Moore, I.Moura, J.J.G.Moura, H.Santos & A.V.Xavier)

mass of a nucleus. A consequence of this is that unpaired electrons have much higher magnetic moments than do atomic nuclei, and hence a stronger interaction with an external magnetic field (i.e. ΔU in Fig. 2.4 is larger). A practical result is that EPR requires magnets of lower field strength than does NMR, and an excitation radiation of higher frequency (in the microwave region of the electromagnetic spectrum).

Since EPR only detects unpaired electrons, with cytochromes only the iron or associated organic free radicals can be studied. Intrinsic free radicals have been detected only in those cytochromes that react with H_2O_2 , for example yeast cytochrome *c* peroxidase (YCCP) (Yonetani, 1976), and therefore we shall confine ourselves to the EPR characteristics of heme iron. However, it should be noted that cytochromes can be labelled with organic free radicals known as spin-labels. Such labelled cytochromes have been used to monitor the interaction of cytochromes with membranes and enzymes (e.g. Vanderkooi & Erecińska, 1974; Brown & Wüthrich, 1977). Also, the addition of NO to ferrohemeproteins leads to an EPR active iron-NO complex with distinctive spectroscopic properties (section 2.3.2.2). These make such modified proteins useful for studying intermolecular interactions.

The frequency scale in EPR is usually calibrated in the dimensionless number known as the g-factor. This is given by:

$$\nu = \frac{g\beta B}{h} \tag{2.7}$$

where ν is the (constant) operating frequency of the spectrometer, β is the electron Bohr magneton (9.273×10^{-26} J Tesla^{-1}), h is Planck's constant, and B is the value of the magnetic field (in Tesla) at the relevant point in the EPR spectrum. This method of defining the absorption frequency arises because it is generally easier to change the value of a magnetic field (at least, of the type used for EPR spectrometers) than it is to vary the frequency of a microwave source.

As with NMR, EPR spectra can be readily integrated to give the number of spins contributing to the signal and, in principle, EPR signals are affected by the spin-spin interactions between unpaired electron(s) and nuclear spins which lead to multiple signals. However, for cytochromes the linewidths are so broad that electron-nucleus spin-spin coupling is not observed. The broad lines partly reflect the rapid electron relaxation. This is so rapid that most cytochromes do not give observable EPR signals at room temperature and samples need to be cooled to liquid N_2 or liquid He temperatures to reduce the relaxation rate sufficiently to get an observable, although still broad, signal.

Because the energy gap, ΔU (Fig. 2.4), is much greater in EPR than it is in NMR, the population difference between the energy levels is greater and thus EPR is far more sensitive than NMR (Scheme 2.2). Hence EPR can be used to study cytochromes *in vivo* (e.g. Prince *et al.*, 1974; Corker & Sharpe, 1975).

Two general types of EPR spectra are observed for frozen solutions of monoheme cytochromes *c* in which the molecules are oriented randomly with respect to the magnetic field (Fig. 2.9). The single absorption band is very broad and so the spectra are presented in the derivative mode (compare Fig. 2.9 with Fig. 2.2b). The interpretation of these spectra starts with the identification of the g-factor with the magnetic moment of the iron, and with the observation that the g-factor has a shape (Palmer, 1979; 1983; 1985). This shape reflects the distribution of the unpaired electron(s) in space and hence the g-factor is determined by the orbitals in which the electron(s) are located.

The g-factor can have up to three g-values, as shown in Fig. 2.9. These values reflect the alignment of the spectrometer magnetic field with particular magnetic axes (designated x, y and z) of the sample. These axes specify the shape of the g-factor with respect to the molecular structure.

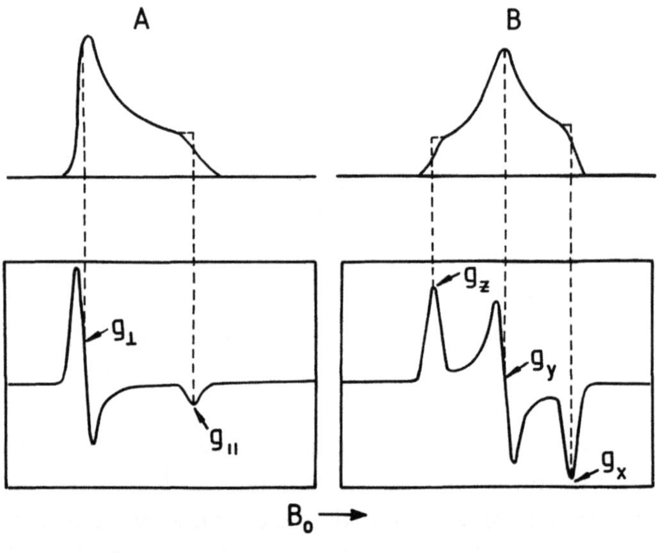

Fig. 2.9 EPR spectra of axial (A) and rhombic (B) ferric-hemes. The top traces are the absorption mode spectra and the lower traces are the corresponding derivative mode spectra. The positions of principal values of the g-factors are shown. EPR spectra are commonly reported in the derivative mode. (Adapted from Palmer, 1985)

When the value of g along the z-axis (i.e. g_z) is different from those along the x and y axes (g_x and g_y), and $g_y = g_x$, the system is *axial*. In such cases only two g-values are measured, with g_x and g_y being given by the g-value for the sample perpendicular to the applied magnetic field ($g_\perp$) and g_z corresponding to the g-value when the sample is parallel to the applied magnetic field ($g_\parallel$). When $g_z \neq g_y \neq g_x$, three g-values can be measured

(Fig. 2.9b) and the system is described as *rhombic*. The corresponding orbital energy level diagrams for axially and rhombically distorted low-spin Fe(III) hemes are given in Fig. 1.7.

Leaving aside for the moment intermediate spin-systems, ferricytochromes can be one of four types: axial and high-spin, rhombic and low-spin, rhombic and high-spin, and axial and low-spin. The former two are by far the most common. Axial high-spin systems have $g_\perp$ ~6 and $g_\parallel$ ~2, and rhombic low-spin systems have g_x, g_y and g_z in the range 1.0 to 3.8.

Cytochromes with a quantum-mechanically mixed S = $^3/_2$, $^5/_2$ spin-state (Chapter 1.2) have spectra similar to those of axial high-spin systems but with the $g_\perp$ value shifted to higher field (it is in the range 4 - 6) (see Fig. 2.22).

Ferrous Hemeproteins. Low-spin ferrous cytochromes are diamagnetic, and hence do not have an EPR spectrum, and high-spin ferrous systems are also expected to be EPR-silent as a result of the EPR selection rules. However, EPR spectra of ferrous myoglobin have recently been reported (Hendrich & Debrunner, 1988) though they are of exceedingly low intensity and very broad.

Low-spin Ferricytochromes. A detailed analysis of low-spin Fe(III) EPR spectra allows the energy terms associated with axial and rhombic distortions of the ligand field, the parameters Δ and V (Fig. 1.7), to be determined (Taylor, 1977; Palmer, 1983; 1985). The full details of these calculations are beyond the scope of the present review but we shall use their results in section 2.3.2.2 to assist identification of cytochrome axial ligands.

For an isolated heme Δ, V, g_x, g_y and g_z are related by the following equations (where λ is the spin-orbit coupling constant of Fe(III)):

$$\frac{V}{\lambda} = \frac{g_x}{g_z + g_y} + \frac{g_y}{g_z - g_x} \tag{2.8}$$

$$\frac{\Delta}{\lambda} = \frac{g_x}{2(g_z + g_y)} + \frac{g_z}{g_y - g_x} - \frac{g_y}{2(g_z - g_x)} \tag{2.9}$$

Since g_x is often not easily measured (because it is broad and of low intensity) the following relationship is frequently used to obtain it:

$$g_z^2 + g_y^2 + g_x^2 = 16 \tag{2.10}$$

The orientations of g_x, g_y and g_z with respect to the heme plane can be determined by single-crystal EPR measurements, as has been done for various mitochondrial cytochromes *c* (Hori & Morimoto, 1970; Mailer & Taylor, 1972; Taylor, 1977). g_z is almost coincident with the heme normal and g_x and g_y are in the heme plane.

High-spin Ferricytochromes. Detailed analysis of high-spin Fe(III) EPR spectra yielding various ligand-field parameters has been reported (Palmer, 1979, 1983, 1985). Further consideration of these is beyond the scope of the present review.

Multiheme Cytochromes. When a cytochrome possesses more than one heme there is the possibility of heme-heme interactions leading to g-values considerably different from those given above. Examples are the tetraheme *Nitrosomonas europaea* cytochrome *c*-554, which contains one pair of interacting hemes with g-values of 10.13 and 3.31 (Andersson *et al.*, 1986), and the hexaheme nitrite reductase from *Wolinella succinogenes,* which contains two pairs of interacting hemes with g-values of 10.3 and 3.7, and 4.8 and 3.2 (Blackmore *et al.*, 1987).

2.2.4 Vibrational Spectroscopies

Infrared (IR), Raman and Resonance Raman (RR) spectroscopies are all examples of vibrational spectroscopic methods. As this implies, they monitor the vibrational states of molecules.

Molecules are not static; their atoms are continually changing position so that their bond lengths change. Normally these changes are relatively small but if sufficient energy can be imparted to the molecule, the vibrations increase markedly. The energy level diagram in Fig. 2.10 illustrates this. The optimum bond length for a molecular fragment is the one of lowest potential energy. Compressing or stretching the bond leads to an increase in energy. Since only certain values of energy are allowed, the energy dependence on bond length can be represented as a series of energy levels. Transitions between these levels produce IR spectra. This is an absorption spectroscopy that obeys the Beer-Lambert law. Its frequency scale is normally given in units of cm^{-1} (Scheme 2.1; see Fig. 2.38).

Protein IR spectroscopy is not straightforward (Wharton, 1986). This is partly because all the bond vibrations contribute to the spectrum and partly because H_2O and D_2O have intense and complex IR spectra. Thus IR difference spectra are normally employed to achieve some simplification. We give two examples of this approach in section 2.3.3.

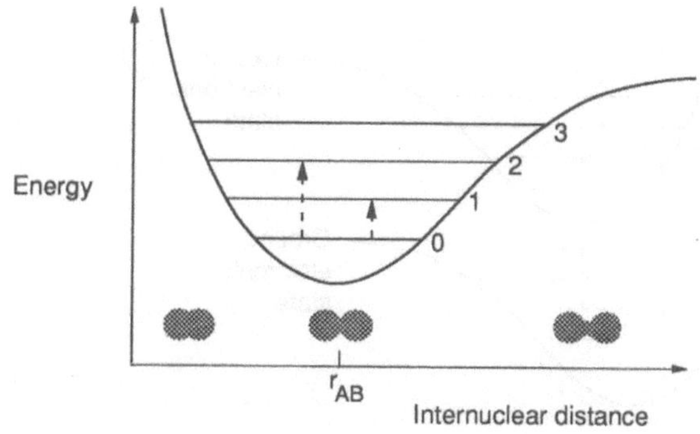

Fig. 2.10 Energy-level diagram for vibrational spectroscopy. The graph shows the energy of a diatomic molecular fragment as a function of the bond length. The optimum bond length is r_{AB}. The energy levels are shown as horizontal lines and the transitions that give rise to IR spectra as vertical lines. The energy levels are only allowed certain values of energy. These are indicated by the symbols 0, 1, 2 etc. Level 0 is the ground vibrational level. Bond compression and stretching are indicated by the relative positions of the shaded spheres

Raman spectroscopy is a form of vibrational spectroscopy that is not commonly used for cytochromes. However, the related RR is extensively used. When monochromatic light is shone on a sample some of it is scattered elastically; this is known as Rayleigh scattering. However, some of it is scattered inelastically (i.e. its frequency changes): this is known as Raman scattering and it is the basis of Raman spectroscopy. The RR extension of the basic Raman experiment is to use monochromatic light of a wavelength that corresponds to an electronic transition of the sample (section 2.2.5). This can lead to both a considerable enhancement and selection of the signal observed. Only vibrations associated with the selected electronic transition are enhanced and, since the heme α, β and Soret bands are normally used, it is the heme ring vibrations that are enhanced and monitored. Thus, in this application, RR is a heme centred technique.

The physical basis of RR is shown in simplified fashion in Fig. 2.11. On absorbing energy a molecule moves from the vibrational level indicated in the ground electronic state to a higher vibrational level of an excited electronic state (transition A). On emitting energy the molecule moves back to the ground electronic state but to the higher vibrational level (transition E). The frequency of the resulting signal is the sum of the frequency of the irradiating light plus the difference (ν_A - ν_E). As with IR the spectra are presented with frequency units of cm^{-1}.

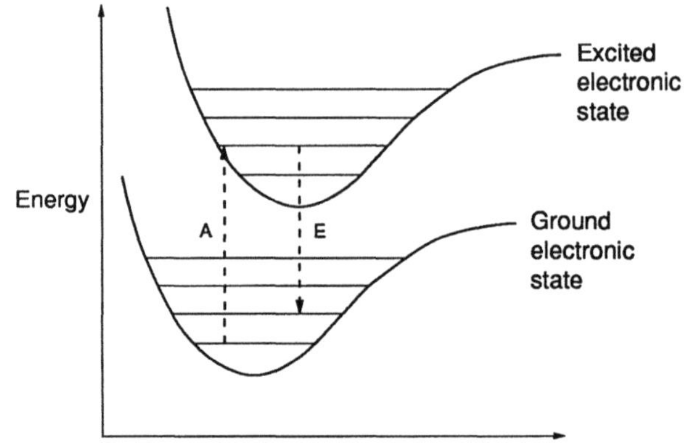

Fig. 2.11 Energy-level diagram for RR spectroscopy. **A** represents the transition from a vibrational level of the ground electronic state to a higher vibrational level of an excited electronic state. **E** represents the transition back to a vibrational energy level of the ground electronic state. The **E** transition is to a different vibrational level because during the experiment the molecule has changed its internuclear distance. The frequency of the RR signal is $\nu_A + (\nu_A - \nu_E)$

An example of a RR spectra is given in Fig. 2.12 for horse ferrocytochrome *c*. Some of the vibrations observed are sensitive to the amino acid sequence; for example, the 1313 cm^{-1} line is shifted by up to 1.6 cm^{-1} in a range of mitochondrial cytochromes *c*. Shelnutt *et al.* (1981) concluded that these shifts reflected structural differences between the proteins but they did not provide a detailed structural interpretation of the data.

More significant differences occur when the oxidation state, spin-state or axial ligation of the iron changes. However, analyses of RR data usually rely extensively on a comparison of spectra for different hemeproteins and 'model' heme complexes because a rigorous fundamental analysis of the signal frequencies is not always possible. Examples of this kind of comparative analysis are: the pH dependencies of class I cytochromes *c* and the axial ligation at extremes of pH (Kitigawa *et al.*, 1977b; Lanir *et al.*, 1979; Uno *et al.*, 1984; Davis *et al.*, 1988), and the nature of the heme ligation in cytochromes *c'* (Kitigawa *et al.*, 1977a; Teraoka & Kitigawa, 1980). Spiro *et al.* (1979) have attempted a fundamental analysis of RR spectra of high-spin ferric heme systems and they propose a method for characterising the extent of heme doming (Chapter 1.3) which would lead to a measurement of the angle χ in Fig. 1.8. Spiro (1983) has also reviewed RR spectra of hemeproteins, principally globins, and listed many assignments for the vibrational bands, including metal-axial ligand stretching frequencies. However, at present the EPR-MCD approach to axial ligand assignment described in section 2.3.2.2 provides a more reliable way of identifying ligands in cytochromes *c*.

As with other spectroscopies, RR spectra of multiheme proteins may show features indicative of heme-heme interactions. An example is provided by cytochrome c_3 (4-heme) (Verma *et al.*, 1988) whose RR spectra contain multiple heme bands proposed to arise from vibrational splitting of the heme signals resulting from heme-heme interaction.

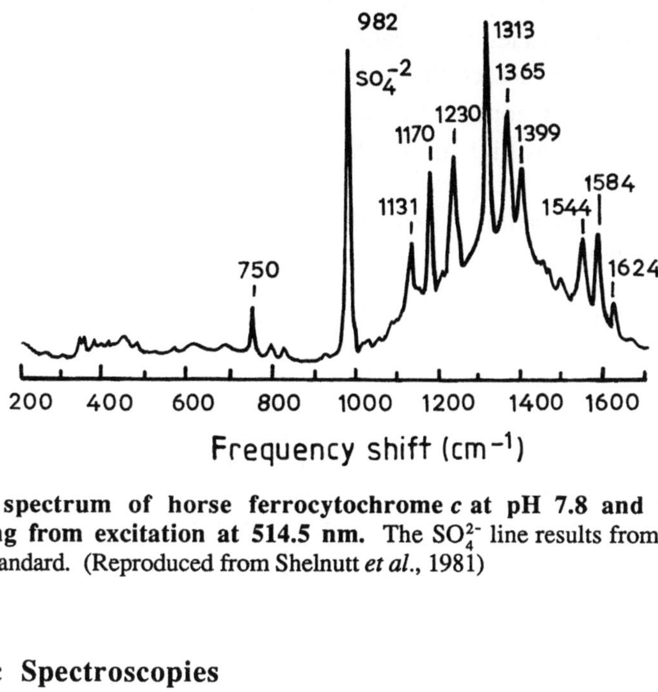

Fig. 2.12 The RR spectrum of horse ferrocytochrome *c* at pH 7.8 and room temperature resulting from excitation at 514.5 nm. The SO_4^{2-} line results from SO_4^{2-} added as an intensity standard. (Reproduced from Shelnutt *et al.*, 1981)

2.2.5 Electronic Spectroscopies

Molecules can exist in a variety of electronic states and electronic spectroscopy monitors the transitions between these. Examples of relevant electronic energy levels are those associated with the splitting of the iron d-orbitals in a ligand field (Chapter 1.2) and those involving molecular orbitals associated with particular functional groups. For cytochrome spectra, some of the most important molecular orbitals are the π-bonding and π^*-antibonding orbitals of the porphyrin (section 2.3.1). Transitions within peptide electronic states and within the π-levels of aromatic amino acids also provide useful information.

Conventional electronic spectroscopy is an absorption technique that obeys the Beer-Lambert law:

$$\text{Absorbance} = \log\left(\frac{I_o}{I_t}\right) = \varepsilon.c.l. \tag{2.11}$$

where I_o and I_t are, respectively, the radiation incident on, and transmitted by, the sample; ε is the extinction coefficient (measured in units of $M^{-1}cm^{-1}$); l is the length (in cm) of the sample through which the radiation travels; and c is the molar concentration of the sample. Since ε varies over a wide range for different transitions, c also needs to be varied to obtain spectra with a sufficiently high signal:noise ratio for analysis. Table 2.2 gives some typical extinction coefficients for bands appearing in cytochrome spectra. From these values it is easy to see that solutions of 10^{-6} M concentration give good heme spectra; but for comparable spectra in the weak charge-transfer band region, concentrations as high as 10^{-3} M may be needed.

Table 2.2 Approximate extinction coefficients for electronic absorbance bands of cytochromes

Band origin [a]		Wavelength, λ (nm) [b]	ε (mm^{-1} cm^{-1}) [b]
Tyr	π → π*	275	1.3
Trp	π → π*	280	5.0
Heme	π → π* (Soret)	390 - 430	100 - 200
Heme	π → π* (α/β)	520 - 560	10 - 16
Ligand	→ Fe CT	600 - 800	0.5 - 6.0

a See section 2.3.1.1.for band assignments.
b The values for λ and ε are taken from Smith & Williams (1970), Lemberg & Barret (1973), Bartsch (1978) and Wood (1984). These articles should be consulted for specific proteins.

The extinction coefficient varies because the selection rules governing electronic transitions show that many are forbidden. Thus π → π* transitions are generally allowed but n → π* transitions (which are the promotion of non-bonding or lone-pair electrons to π* orbitals) and d-d transitions (Chapter 1.4) are forbidden.

Typical electronic absorption spectra of a cytochrome are shown in Fig. 2.13. Because the bulk of the radiation absorbed is in the visible region, such spectra are often called optical spectra. The frequency scale is calibrated in units of wavelength (Scheme 2.1).

The absorption of the heme dominates the spectra; this produces the bands at ~410 nm (the Soret band) and between 500 and 600 nm (the β and α bands). The band at ~280 nm originates from the aromatic amino acids (four Tyr, four Phe and one Trp for horse cytochrome *c*) and the weak band at 695 nm results from an interaction between the heme iron and its ligands. This band is a charge-transfer (CT) absorbance (section 2.3.1). At longer wavelengths (i.e. further into the infra-red region) other CT bands occur. These can only be observed with the straightforward absorption experimental arrangement if the sample is dissolved in D_2O since H_2O absorbs in the Near-IR (NIR) and obscures the region. The CT bands can be readily detected by the MCD experiment described in section 2.2.5.2. Further into the blue region at ~190 - 230 nm are the absorbance bands that arise from the peptide bonds.

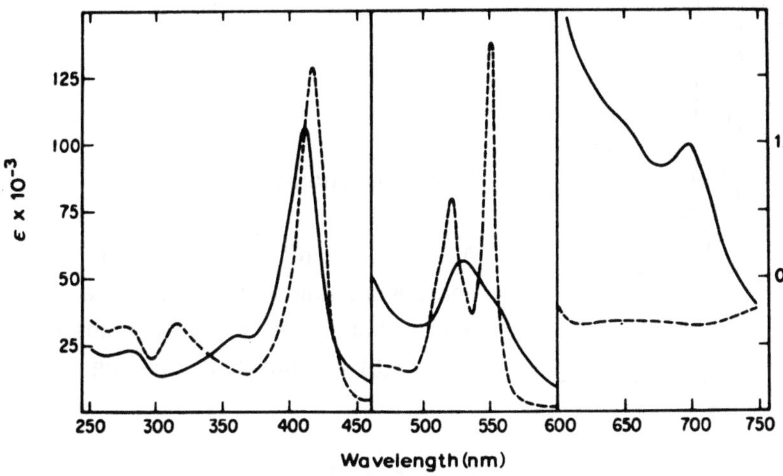

Fig. 2.13 Electronic absorption spectra of horse cytochrome *c* at pH 7.0. Solid line, ferricytochrome *c*; dashed line, ferrocytochrome *c*. Center panel, use the ordinate on right x20. (From Harbury and Marks, 1973)

The spectra in Fig. 2.13 are rather broad, as are virtually all such spectra at room temperature, because vibrational motions lead to a range of closely-spaced electronic transitions (Fig. 2.14). These electronic transitions occur more rapidly than nuclear motions (this is the basis of the Franck-Condon principle which is also relevant to the mechanism of electron transfer; see Chapter 8.2). They can therefore be represented by vertical lines, as in

Fig. 2.14. In this example we have assumed that the equilibrium bond length of the excited electronic state is slightly greater than that of the ground state. The most probable transition is then from the v = 0 vibrational level of the ground electronic state to the v' = 2 vibrational level of the excited electronic state; this is the so-called (2,0) transition. However, since there is some probability of finding the system at the extremities of the v = 0 level, other transitions can occur that lead to the (0,0), (1,0) etc. absorbancies. Thus the resulting absorbance band is a broad continuum rather than a sharp line

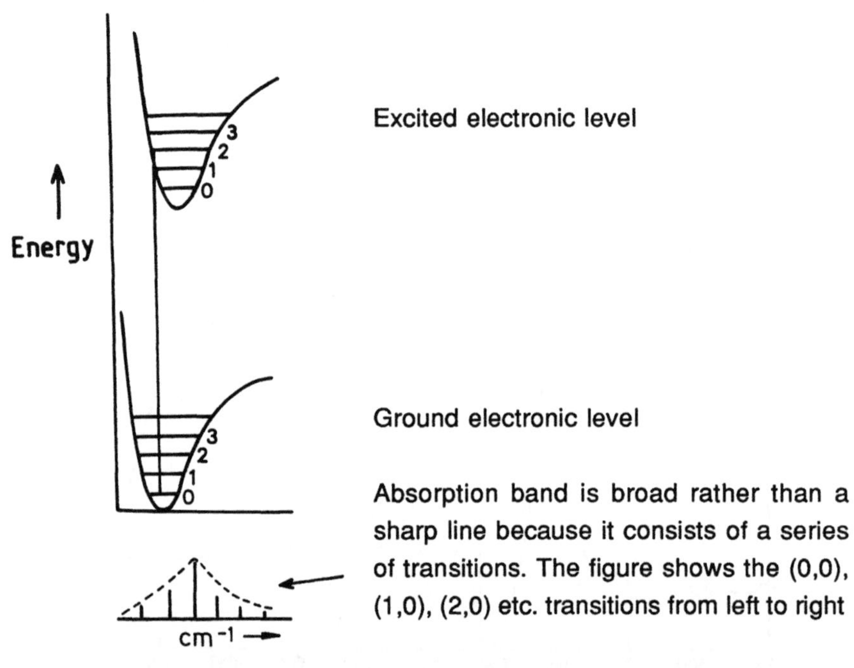

Fig. 2.14 Coupling between vibrational and electronic transitions. The most probable transition, the (2,0) transition, is shown. Other transitions also occur - (0,0), (1,0) etc - and because these are close together the band is broad rather than sharp. Reducing the vibrational motion sharpens the bands (see text)

(Fig. 2.14). Reducing the molecular vibrations therefore leads to a sharpening of the spectrum. This can be achieved by reducing the temperature and an example of this is given in Fig. 2.15 for mitochondrial cytochrome *c*. The improvement in resolution is dramatic.

There are many modifications of the basic absorption experiment but we only consider two of these: the CD and MCD experiments. We also consider the electronic emission spectroscopies: fluorescence and phosphorescence.

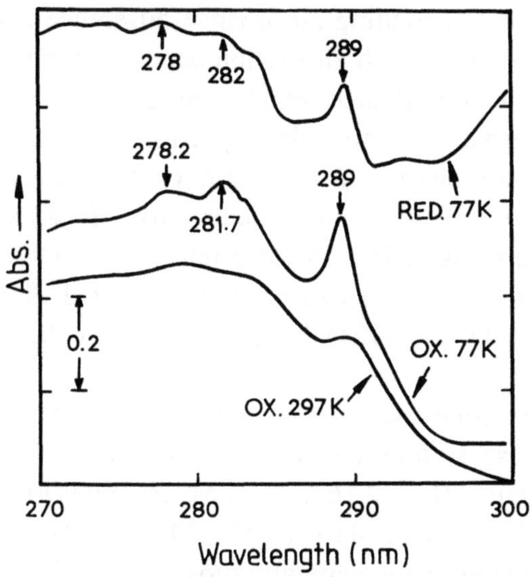

Fig. 2.15 Effect of temperature on the electronic absorption spectrum of Trp 59 in cytochrome *c*. Absorption spectra of 6.3 mM bovine heart ferri- and ferro-cytochrome *c* at 0.07 mm path length. Cytochrome *c* was dissolved in water-glycerol (1:1, v/v) containing sodium phosphate (pH 7). The signal at 289 nm is from a (0,0) transition of Trp 59. (Adapted from Strickland *et al.*, 1971)

2.2.5.1 Circular Dichroism (CD) Spectroscopy

Electromagnetic radiation is conveniently treated as though it behaves as two oscillating field components perpendicular to each other. One of these is a magnetic wave and the other is an electric wave. These can oscillate in any direction perpendicular to the direction of propagation. Such radiation is called unpolarised. However, the radiation can also be polarised so that the electric wave oscillates in only one plane. This is linearly polarised radiation. If two such waves, which oscillate in different planes but have identical frequencies and directions of propagation, are combined with a 90° phase difference, the resultant is circularly polarised radiation. The vector indicating the direction of polarisation of the electric component describes a helical path which can be left circularly polarised (lcp) or right circularly polarised (rcp).

If a transition within a molecule leads to spiral electron motion this transition will differentially absorb rcp and lcp radiation, depending upon the handedness of the spiral. The corresponding absorbance difference is circular dichroism, and it can be observed only with absorption bands associated with a chiral centre or heme. Heme absorbance bands are CD-active because buckling of the heme-plane gives rise to spiral electron motion. A CD spectrum is recorded as a plot against wavelength or frequency of the difference ($\Delta\varepsilon$) between the molar extinction coefficients for left (ε_l) and right (ε_r) circularly polarised light (see Fig. 2.16). A close relative of CD is Optical Rotatory Dispersion (ORD); their spectra can be interconverted directly by mathematical manipulation. CD and ORD bands are often called Cotton effects.

ORD and CD spectra have been used extensively to study a variety of cytochromes *c*. Many of these studies concentrated on the visible region dominated by the heme absorbance (Fig. 2.16A) (e.g. Myer, 1968; Flatmark & Robinson, 1968: Urry, 1968; Harbury & Marks, 1973, Myer & Pande, 1978). However, it is the far-UV region which is generally most informative.

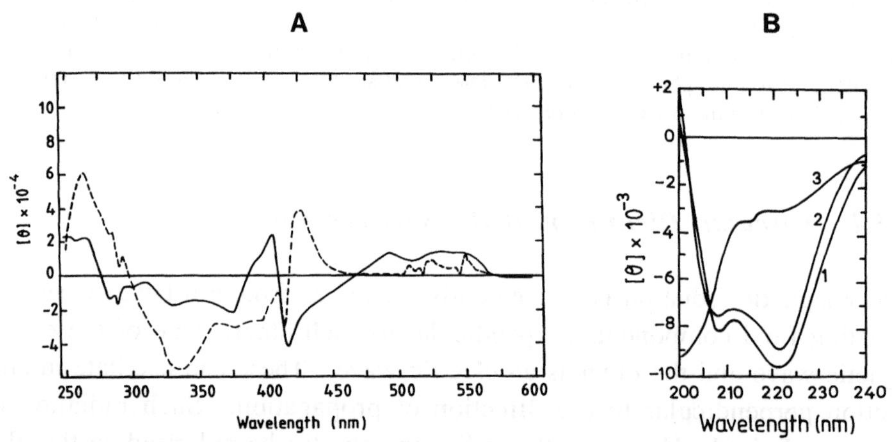

Fig. 2.16 Circular dichroism spectra of horse ferricytochrome *c*. (**A**) Spectra of ferricytochrome *c* (solid line) and ferrocytochrome *c* (broken line) at pH 7.0. (**B**) Far UV spectra of ferricytochrome *c* at pH 5 and pH 7 (curve 1), porphyrin-cytochrome *c* at pH 5 (curve 2), and apo-cytochrome *c* at pH 5 (curve 3). (Reproduced from Fisher *et al.*, 1973)

It is possible to calculate, with a reasonable degree of accuracy, the amount of secondary structure present in a protein from its far-UV CD spectrum. Hennessey & Johnson (1981) and Provencher & Glöckner (1981) calculate ~40% α-helix, 17 - 26% turns, 7 - 15% β-sheet and 25 - 40% random structure for horse cytochrome *c* from its CD spectrum compared to the values of

~38%, ~24%, ~0% and ~37% respectively they calculate from its x-ray structure.

Perhaps a more important use of the far-UV CD is in comparing related proteins. The spectra of ferricytochrome *c*, porphyrin cytochrome *c* and apocytochrome *c* in Fig. 2.16(B) show that removal of the iron does not greatly affect the amount of secondary structure present; but the removal of the heme abolishes most, if not all, of the secondary structure (Fisher *et al.*, 1973).

2.2.5.2 Magnetic Circular Dichroism (MCD) Spectroscopy

The origins of this effect are quite different from those of natural CD. Placing a sample in a magnetic field induces splitting of previously degenerate energy levels and many of the resulting levels differentially absorb circularly polarised light. Thus MCD involves a similar experimental arrangement to natural CD with the addition of a high-field magnet. An important distinction between CD and MCD spectroscopies is that whereas the former requires an optically active sample, the latter does not.

A magnetic field can have three effects on electronic transitions. These are designated A, B and C terms. The A and C terms are illustrated in Fig. 2.17. The A term involves the splitting of an excited state (j) by the magnetic field whereas the C term occurs when the ground state (a) is split by the magnetic field. In both cases the single transition in the absence of the field is replaced by two transitions in the presence of the field. A and C terms can be distinguished by the shape of their absorbance bands, and because the intensity of the C term band is temperature dependent. This is because the population difference between levels a' and a" influences the final intensity of the band, and this population difference is temperature dependent (Eqn. 2.3). The starting populations of levels j' and j" are both zero so the intensity of their absorption is the same.

B terms arise by a more complex process that leads to energy levels becoming mixed together. The signals have the same general appearance as those of the C term but their intensities are independent of temperature. Thus the two types of signal can be distinguished.

As indicated above, the temperature dependence of MCD spectra can provide useful information. In particular the magnetic properties of the ground state may be determined from the magnetisation curves derived from

MCD spectra over the temperature range 1.5 - 273 K (Thomson & Johnson, 1980; Foote *et al.*, 1989; Thomson, 1989).

MCD spectroscopy is an important technique for characterising cytochromes and we consider its application in detail in section 2.3.2.2. However, we note here that although tryptophan gives a strong MCD signal at ~290 nm, MCD of cytochromes has almost exclusively concentrated on the heme and CT bands.

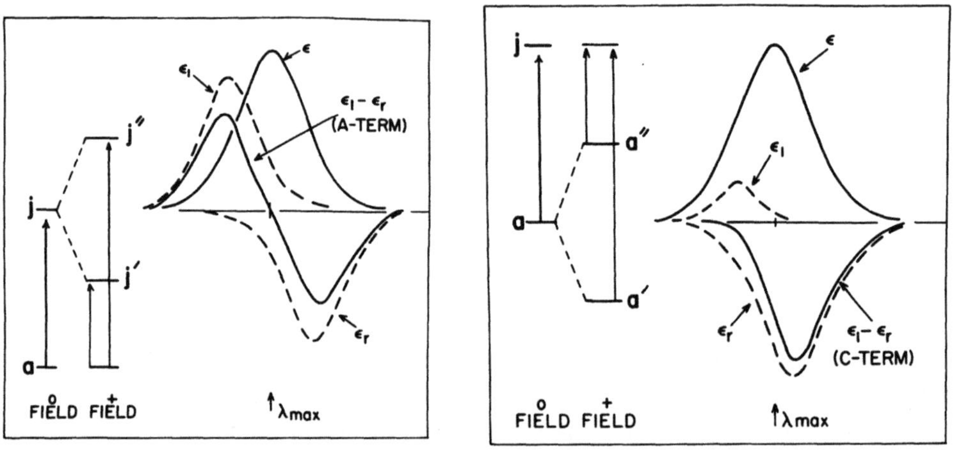

Fig. 2.17 Origins of the MCD A-term (left) and C-term (right) signals. In the absence of a magnetic field there is a single transition from the ground-state to the excited state, a → j, represented by ε. This transition absorbs left and right circularly polarized light equally and thus has no net CD effect. In the presence of a magnetic field the excited state may split (left) or the ground state may split (right). These cases give rise to the A-term and C-term MCD spectra repectively. The A-term signal appears because the transition a → j', which occurs with left circularly polarized (lcp) light, has an opposite sign, though equal intensity, to the transition a → j", which occurs with rcp. These two transitions, therefore, produce the signals labelled ε_l and ε_r. The difference, $\varepsilon_l - \varepsilon_r$, is the net MCD signal. The C-term signal is produced by the difference between the signals corresponding to the transitions a' → j and a" → j (i.e. ε_l and ε_r respectively). (Reproduced from Vallee & Holmquist, 1980)

2.2.5.3 *Fluorescence and Phosphorescence Spectroscopies*

Fluorescence and phosphorescence (also known as luminescence) spectroscopies are forms of emission spectroscopy. They require the excitation of a molecule by a monochromatic light beam in resonance with an absorbance band. They are usually exceedingly sensitive methods of detection and may require solutions of only ~10^{-6} M concentration.

Fluorescence and phosphorescence occur when a molecule undergoes a transition from an excited electronic state to its ground electronic state, releasing the excess energy as emitted radiation. If both the excited and ground electronic states have the same number of unpaired electrons (usually none) the process is termed fluorescence, whereas if the two states differ in their electron spin the emission is called phosphorescence. Emitted fluorescence radiation is usually at a lower frequency (longer wavelength) than the exciting radiation. This is because excitation generally puts the molecule into a high vibrational level of the excited state. Vibrational relaxation then occurs which puts the molecule in the ground vibrational level of the excited state. Fluorescence occurs from this ground vibrational level. Phosphorescence also occurs at longer wavelength than the exciting radiation. This is because the excited electronic state with unpaired electrons is of lower energy than the excited state to which the molecule is promoted by the incident radiation

In general the intrinsic luminescent properties of native cytochromes are too weak to be of much use in characterising the proteins. This is because heme is exceedingly efficient at depopulating excited electronic states, thus reducing the intensity of any luminescent signals. This process, known as quenching, can occur over distances up to 50Å (Cantor & Schimmel, 1980 pp 451-455). Therefore it is not surprising that the fluorescence of Trp 33 of tuna cytochrome *c* (G.R.Moore, unpublished data), and Trp 59 of mitochondrial cytochrome *c* (Fisher *et al.*, 1973; Tsong, 1974; Myer *et al.*, 1980a) is almost completely quenched. For Trp 59 of ferricytochrome *c* there is a 98% reduction in fluorescence intensity compared to free tryptophan. The porphyrin fluorescence of native cytochrome *c* is also almost completely quenched (Vanderkooi *et al.*, 1976).

In non-native cytochromes luminescence has been observed. Fisher *et al.* (1973) confirmed that the fluorescence of Trp 59 in native cytochrome *c* and porphyrin-cytochrome *c* was quenched but in apocytochrome *c* and unfolded ferricytochrome *c* it was not (Fig. 2.18). Vanderkooi & Erecińska (1975) elegantly demonstrated in a direct experiment that the porphyrin quenched the fluorescence of Trp 59 by showing that the fluorescence emission spectrum obtained upon excitation of porphyrin-cytochrome *c* at 280 nm (tryptophan) was the same as that obtained upon excitation at 500 nm (porphyrin). Thus tryptophan fluorescence is a good probe for investigating gross deformations of structure but is not useful for studying the conformations of small, structured cytochromes. It has been extensively used for investigating the folding and unfolding pathways of cytochrome *c* (Chapter 4.5.4) and the structural integrity of non-covalent complexes of cytochrome *c* (Chapter 4.6.3).

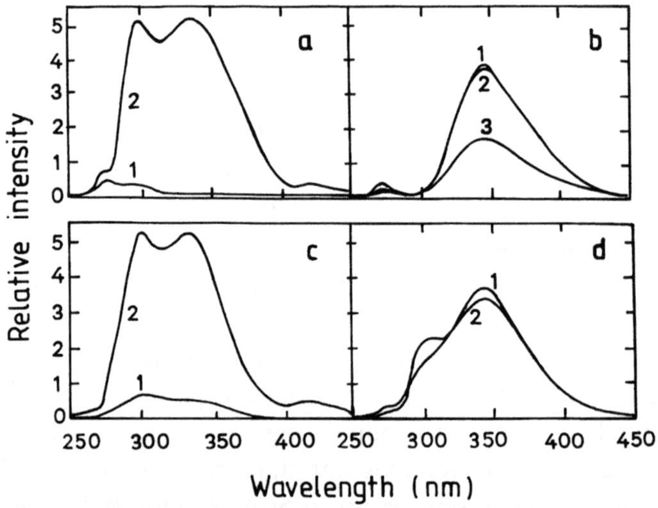

Fig. 2.18 Fluorescence emission spectra of horse cytochrome *c* in H_2O at pH 7 and 25°C obtained with excitation at 280nm. a ferricytochrome *c* in the absence (curve 1) and presence (curve 2) of 6 M guanidine HCl. **b** tryptophan in the absence (curve 1) and presence (curve 2) of 6 M guanidine HCl, and the presence of 6 M guanidine HCl and 0.04 M heme (curve 3). **c** porphyrin-cytochrome *c* in the absence (curve 1) and presence (curve 2) of 6 M guanidine HCl. **d** apocytochrome *c* in the absence (curve 1) and presence (curve 2) of 6 M guanidine HCl. (Adapted from Fisher *et al.*, 1973)

Metals with completely filled d-orbitals are poor quenchers and this has led some workers to characterise Zn(II) and Sn(IV) substituted horse cytochrome *c* (Table 1.6). Both of these derivatives have relatively intense fluorescent and phosphorescent spectra and these have been used to probe the interactions between the derivatives and other proteins and membranes (Vanderkooi *et al.*, 1977; Dixit *et al.*, 1982; Koloczek *et al.*, 1987; Alleyne & Wilson, 1987).

Derivatives of cytochrome *c* labelled with fluorescent dyes (such as tetramethyl-rhodamine) have been described and used to study the interaction of cytochrome *c* with membranes. In the experiment of Hochman *et al.* (1982) the fluorescence of labelled cytochrome *c* in mitochondria was measured after bleaching the fluorescence spectrum originating from one area of the mitochondrial membrane. From the rate of recovery of the fluorescence signal, the rate of diffusion of labelled cytochrome *c* within the system could be obtained. This technique, called Fluorescence Redistribution After Photo-bleaching (FRAP), is a powerful method for determining dynamic aspects of the interaction between cytochromes and membranes. Such interactions have been considered by Pettigrew & Moore (1987: Chapter 2).

An alternative approach for monitoring cytochrome-membrane interactions is to study the effect added protein has on the fluorescence of a dye within the membrane. Using this approach the interaction of mitochondrial cytochrome *c* with a variety of membranes has been studied (Kaminsky *et al.*, 1973a; Letellier & Schechter,1973; Teissie, 1981).

2.2.6 Mössbauer Spectroscopy

Mössbauer spectroscopy is a form of γ-ray spectroscopy (Bearden & Dunham, 1970; Münck, 1979). The ^{57}Fe atomic nucleus absorbs γ-rays and is excited from its ground state to a higher, degenerate energy level. For hemeproteins the degeneracy of the excited state is partially lifted (by electric quadrupole interactions) so that it is split into two. The observed spectrum is then a doublet with lines of equal intensities (compare Fig. 2.19b with Fig. 2.3).

The γ-rays are generated by the decay of radioactive ^{57}Co nuclei which produce excited state $^{57}Fe^*$ nuclei. The subsequent decay of the $^{57}Fe^*$ nuclei to ^{57}Fe releases γ-rays which bombard the sample. A geiger counter placed behind the sample is used to count the radiation and when the sample absorbs γ-rays the count falls (Fig. 2.19).

The frequency axis of the spectrum is calibrated in velocity at mm s^{-1}. This arises because in the experimental arrangement, the distance between the γ-ray source and the sample is continually changed. Thus the γ-rays produced by the source with a constant frequency (because they correspond to a nuclear emission process) are modulated by a Doppler shift so that by the time they have arrived at the sample their frequency has been altered. Since the Doppler shift, and hence frequency shift, are directly proportional to the relative velocity of the source and the sample, this latter parameter is used to provide a frequency scale. The velocity required to obtain a signal is called the chemical shift.

The major problem with Mössbauer spectroscopy of hemeproteins is that only ^{57}Fe, which is 2.2% naturally abundant, is detected and thus the method is relatively insensitive. One way to overcome this is to enrich the sample in ^{57}Fe, although this is not always straightforward for cytochromes *c*. A typical example is the study of *R. rubrum* cytochrome *c′* by Emptage *et al.* (1977). These workers used 93% enriched $^{57}FeCl_2$ in the growth medium and required ~12 - 15 mg of sample for their Mössbauer experiments. *Ps. aeruginosa* cytochrome *c*-551 enriched with ^{57}Fe has also been studied by Mössbauer spectroscopy (Dwivedi *et al.*, 1979).

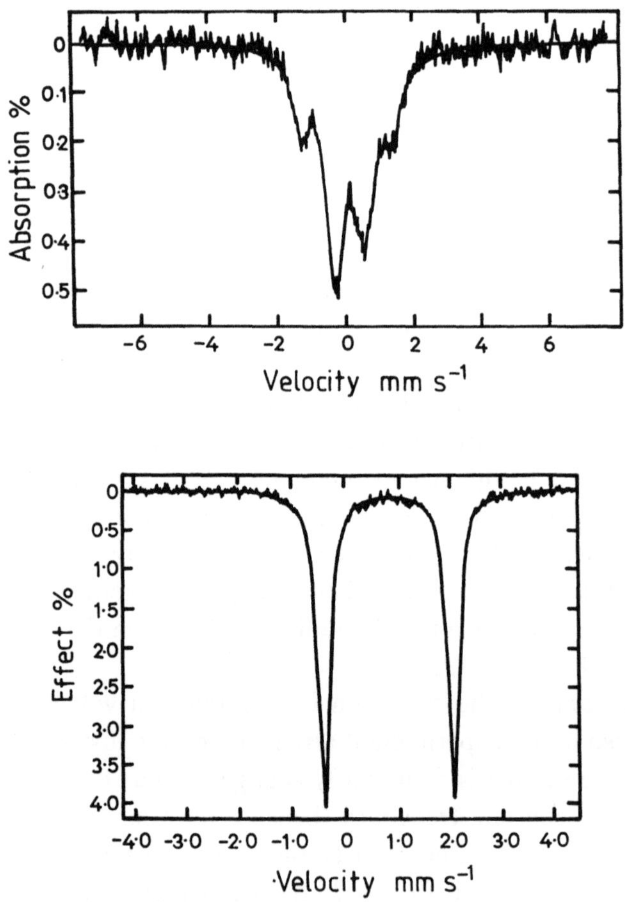

Fig. 2.19 Mössbauer spectra of *Ps. putida* cytochrome P-450. *Top*: The spectrum of the ferricytochrome in the presence of camphor at 200 K appears as two overlapping doublets. These arise from low-spin molecules (camphor bound) and high-spin molecules (no camphor bound). (Reproduced from Sharrock *et al.*, 1973). *Bottom*: The spectrum of the diamagnetic ferrocytochrome at 4.2 K. (Reproduced from Champion *et al.*, 1975). The different isomer chemical shifts and splitting constants for the three species are diagnostic for their iron oxidation states and spin-states

Another important experimental factor is that Mössbauer spectra are obtained on solid, often frozen samples, usually at low temperatures. This eliminates line broadening by nuclear recoil processes when the γ-rays are absorbed by the sample.

Mössbauer spectroscopy is an excellent method for determining the oxidation state of iron (from the chemical shift: section 2.3.1.3), but for many proteins there are other, more accessible methods (Table 2.1). This, combined with the problem of sensitivity, probably accounts for the limited number of reported Mössbauer studies of cytochromes *c*. Nevertheless, it is a useful method for probing cytochromes since not only the chemical shift but also the quadrupole splitting reflects the electronic structure of the iron, as the Mössbauer spectra of cytochrome P-450 in Fig. 2.19 reveal.

However, the Mössbauer chemical shift is not particularly sensitive to the iron environment. It can distinguish different oxidation states and spin-states (Fig. 2.19 & section 2.3.1.3) but not hemes in the same oxidation and spin-states but with different chemical environments. For example, Mössbauer spectra of the tetraheme ferricytochrome c_3 (Ono *et al.*, 1975) and ferrocytochrome c_3 (Utuno *et al.*, 1980) consist of a single doublet resonance, even though other spectroscopic methods - e.g. NMR (Fig. 2.8), and XRD (Chapter 5.3) - show the hemes have very different environments. Similarly, straightforward Mössbauer spectra of the tetraheme ferricytochrome *c*-554 from *Nitrosomonas europaea* fail to resolve signals from the three low-spin hemes. However, Mössbauer spectra of ferricytochrome *c*-554 placed in magnetic fields do allow some resolution of the spectrum. Andersson *et al.* (1986) analyse the magnetic Mössbauer data in terms of two low-spin heme signals which arise from magnetically interacting heme groups. This is consistent with the corresponding EPR data (section 2.2.3.2).

2.2.7 X-Ray Absorption Spectroscopy

Metal ions absorb x-rays and generate complex spectra in so doing (Fig. 2.20). Such spectra can be divided into two general regions. In the first, the absorption edge region, core electrons of the metal are excited to higher energy, unoccupied orbitals. These transitions are relatively sharp. The second region, the Extended X-ray Absorption Fine Structure (EXAFS) region, arises because some of the core electrons are given so much energy that they are ionised. These ionised electrons can be considered as spherical waves radiating from the metal ion, which are then scattered by atoms near to the metal. This scattering produces an interference pattern which is observed as an oscillation in the spectrum above the absorption edge (Fig. 2.20). Analysis of the oscillatory pattern can give information about the number and type of the atoms that are close neighbours of the metal, and the distance between these atoms and the metal. Thus, it may be used to identify ligand atoms. In practise the analysis is complex and requires the spectrum to be simulated from various possible coordination schemes (Cramer & Hodgson, 1979). Both the type of ligand atom and its bond length with the metal are important parameters for the simulation. It should also be noted that whilst the distinction between S and N or O is straightforward, N and O are impossible to distinguish. This is because N and O have similar atomic numbers and, hence, scattering properties.

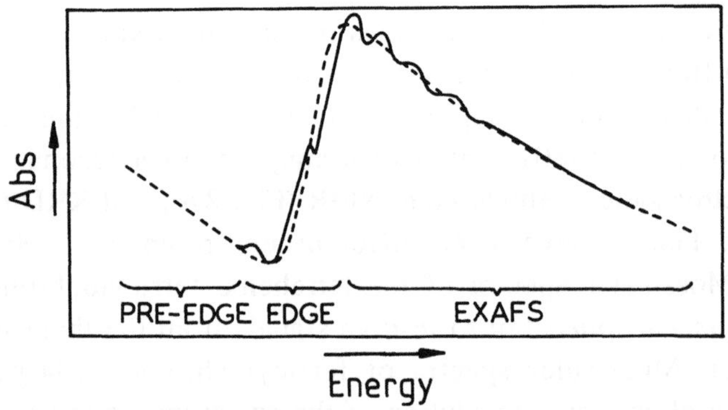

Fig. 2.20 Schematic representation of a typical x-ray absorption spectrum. The edge region extends over 25 to 50 eV and EXAFS is typically observable over several hundred to 1000 eV. The transition region between the edge structure and the EXAFS region may also provide useful structural information but generally, with proteins, this region is not analysed. (Adapted from Cramer & Hodgson, 1979)

For hemeproteins, where four of the ligand atoms are already known, EXAFS is a powerful method for distinguishing between S and N or O containing axial ligands. However, the studies with cytochrome *c* to date have been rather limited. EXAFS studies of mitochondrial cytochrome *c* and bacterial cytochromes c_2 (Labhardt & Yuen, 1979; Korszun *et al.*, 1982) provide compelling evidence that in solution these proteins do not undergo a significant change in their coordination geometries with a change in oxidation state of the iron; nor is there a significant difference in the metal-ligand bond lengths of a wide range of class I cytochromes *c*. EXAFS studies of other types of cytochrome *c* have not been reported.

It is not clear if the edge region of the spectrum is as useful as the EXAFS region. Labhardt and Yuen (1979) report that there is a redox-state shift of 2 eV in the transitions giving rise to the edge absorption of horse cytochrome *c*, and from this they concluded that the electron added on reduction was not extensively delocalised away from the iron. Colosimo *et al.* (1981) have also analysed the edge region of native cytochrome *c* and compared it with the corresponding spectra of carboxymethylated ferricytochrome *c*, which is high-spin and lacks Met 80 as a ligand (Schejter & Aviram, 1970; Wilson *et al.*, 1973), and native ferricytochrome *c* at pH 10, which is largely in the state IV form (Chapter 4.4.4). They found that the spectra were remarkably similar, considerably more so than the spectra of native oxidised and reduced cytochrome *c*. Thus the edge region does not appear to be particularly sensitive to the iron spin state or type of sixth ligand.

2.2.8 The Problems of Assignment and Interpretation

Two of the major difficulties with spectroscopy are the assignment of spectral features to specific molecular fragments and the interpretation of the relevant spectral parameters on a structural basis.

The assignment problem has varying degrees of complexity. For EPR of monoheme proteins it is non-existent, but for NMR it is often a major limiting factor. Even with electronic spectra, the assignment of signals to particular transitions is not always certain. Mis-assignment of signals is one of the major sources of error in spectroscopic studies.

The interpretation of spectroscopic data can also be problematic, As we have seen (Scheme 2.1), shifts in spectral features represent changes in molecular energies. Some of these can be very small; for example, a 10 nm shift in an absorbance band from 550 nm to 560 nm is equivalent to a shift in energy of 0.039 kJ mol^{-1}. Thus to provide a detailed and quantitative interpretation of the spectral shift requires a structural model to account for the change in energy. Such an analysis bears some resemblance to the analyses of pK_a values and redox potentials (Chapter 7) A shift of 1.1 pH units for a pK_a or 65 mV for a redox potential, is equivalent to 6.3 kJ mol^{-1} and, as we show later, even such a relatively large change in energy is not yet fully explicable with a structural model. Thus it is not surprising that mis-interpretation of spectroscopic data occurs.

One common analytical device is to compare spectral features of a given cytochrome with those for model compounds, including structurally well-characterised proteins and small porphyrin complexes. This method, of course, depends on the correct assignments and interpretations for the model compounds.

A further attempt to overcome some of the assignment and interpretation difficulties is to use a combined spectroscopic approach. For example, in combination, EPR, MCD and NMR can be used to identify axial ligands in circumstances where one of these techniques, used on its own, would not give a definite answer (e.g. cytochrome *f*: section 2.3.2.3).

Scheme 2.4 Saturation effects in spectroscopy.

Several important spectroscopic procedures are based on the alteration of energy level populations by irradiation of the sample before a spectrum is measured. Such techniques are commonly applied to cases where the population difference between levels is relatively small (Scheme 2.2). When the populations of two levels are equalised, the system is said to be 'saturated'. A saturated system does not produce an absorbance spectrum.

This approach is particularly powerful for protein NMR. In cases where there is a slow-exchange process (e.g. between oxidised and reduced cytochromes [Chapter 8.4] or slow flipping of aromatic amino acids [section 2.3.3.1]) saturation of a resonance arising from one of the exchanging forms may lead to a reduction in intensity of the corresponding resonance of the second form. This is saturation-transfer. From quantitative measurements of this effect the rate of exchange can be obtained.

A second NMR effect is the nuclear Overhauser enhancement (NOE). For this, saturation of a resonance leads to a change in the intensity of other resonances through a process of cross-relaxation that leads to the re-establishment of the Boltzmann distribution. The change in intensity of the affected resonances is related to the strength of the cross-relaxation between the nuclei of the resonance that is saturated and those of the resonances affected by the NOE.

Under certain experimental conditions the magnitude of the NOE (I) is related to the separation between the atomic nuclei (r) by $I \propto r^{-6}$. Therefore this effect provides direct structural information.

A related approach combines two spectroscopies in the same experiment. Examples of this are Electron Nuclear DOuble Resonance (ENDOR) and the Optical Detection of Magnetic Resonance (ODMR), which is also known as Paramagnetic Resonance Optically Detected (PROD). ENDOR combines NMR and EPR spectroscopies (Lubitz & Babcock, 1987) and PROD combines MCD and EPR spectroscopies (Barrett *et al.*, 1986). In both cases a spectrum is monitored whilst energy levels relevant to the second spectroscopy are saturated (Scheme 2.4). This directly relates various features in the two spectra. This approach may also provide information not directly attainable by the comparable conventional spectroscopies.

2.2.9 Spectroscopic Studies and the Variation of Temperature

Now we have surveyed the spectroscopic techniques listed in Table 2.1 we are in a position to compare a key feature of their experimental requirements. Both EPR and Mössbauer spectroscopies are carried out with frozen solutions, often at liquid He temperature. This may result in incorrect deductions being drawn about the protein at room temperature because, even if the protein is stable to freezing, its spin-state may be temperature-dependent (Chapter 1.2). Spectroscopies that can be carried out at both low temperatures and room temperature are essential for checking the validity of conclusions based on Mössbauer and EPR spectroscopies. Conventional electronic, MCD and RR spectroscopies can be used for this.

A possible additional complication is that glassing agents, such as ethylene glycol and glycerol, are frequently added to aqueous solutions in order to obtain a suitable frozen sample. Sometimes the solvent is 50% organic. These reagents do cause changes in some protein structures, and do perturb optical spectra of some cytochromes (Chapter 7.4.3). Thus functional tests (such as measurements of redox potentials) and/or overall protein structure investigations (such as far-UV CD measurements) should be carried out with such samples.

There is yet another potential complication to low-temperature studies. Thermodynamic parameters, such as pK_a values and redox potentials, may be temperature dependent; so even if there is not a change in spin-state, or complicating solvent effect, the properties of a protein at low temperature may be different from those at room temperature. This is of particular significance for redox potentials since EPR is a commonly used method for measuring these: e.g. for cytochrome c_3 (DerVartanian *et al.*, 1978; Xavier *et al.*, 1979; Gayda *et al.*, 1985).

2.3 Selected Applications of Spectroscopy to Cytochromes *c*

2.3.1 The Type of Heme, its Oxidation State and its Spin-State

The type of heme, its oxidation state, and its spin-state, can usually be determined from conventional electronic absorption spectra alone, although other spectroscopic techniques are frequently used to confirm the latter two properties. Also, conventional electronic absorption spectra are used to classify cytochromes (Pettigrew & Moore, 1987: Chapter 1). Thus a detailed description of the optical spectra of cytochromes is required to place this analysis on a sound footing. We attempt to provide this below.

2.3.1.1 Optical Spectra of Cytochromes

The absorbance bands occurring in optical spectra of cytochromes result from three different types of transition. These are:

1. *$\pi \rightarrow \pi^*$ transitions.* These involve the excitation of electrons from bonding (π) to antibonding (π^*) molecular orbitals that are located primarily on the porphyrin. They generally produce intense bands.
2. *d-d transitions.* These are transitions of electrons within the iron d-orbital set (Chapter 1.2) They are formally forbidden by spectroscopic selection rules and therefore, for cytochromes, are usually too weak to be observed.
3. *Charge-transfer (CT) transitions.* These result from the transfer of electrons between iron orbitals and orbitals of the axial ligands, and between iron orbitals and orbitals of the porphyrin. Intensities of CT bands vary over a wide range but they are normally less intense than allowed $\pi \rightarrow \pi^*$ transitions.

These three types of transition are illustrated by the energy level diagram of Fig. 2.21. This diagram represents the orbitals of the porphyrin, iron, and an axial ligand as being localised on each of the three groups. Only the relevant highest occupied and lowest unoccupied molecular orbitals of the ligands are shown. The pattern of the d-orbital splitting has been considered elsewhere (Chapter 1.2). The transitions can be considered in three groups.

***$\pi \rightarrow \pi^*$* Transitions 1 and 2.** Analysis of the absorbance bands resulting from the $\pi \rightarrow \pi^*$ transitions has attracted workers from many disciplines with the result that a sophisticated quantitative treatment of such spectra is

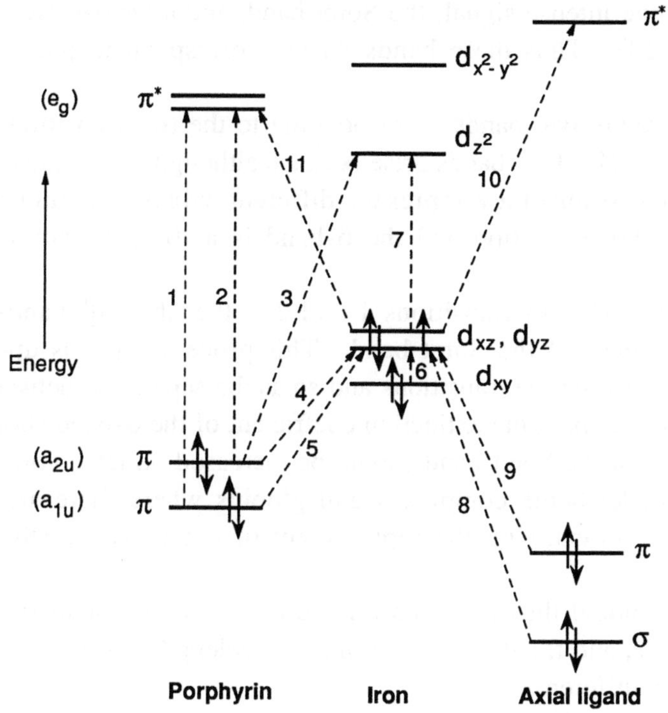

Fig. 2.21 Schematic diagram of the electron promotion transitions reponsible for the optical absorbance of ferricytochromes. σ, π and π* refer to bond types; a_{1u}, a_{2u} and e_g are symmetry labels; d_{xy},d_{xz}, d_{yz}, $d_{x^2-y^2}$ and d_{z^2} are the iron d-orbitals (Chapter 1.2). The pair of porphyrin π* orbitals are degenerate, as are the d_{xz}/d_{yz} pair of iron orbitals. The small arrows (↑ ↓) represent electrons located in the indicated orbitals. The broken lines indicate possible electron promotions. For example, transition 1 represents the promotion of a single electron from the a_{1u} orbital to a π* orbital. Other transitions are discussed in the text. The diagram assumes that the ferric iron is in a tetragonally distorted, strong ligand field (Chapter 1.2). The axial ligand orbitals are for illustrative purposes only; they do not represent an actual ligand. The relative order of the orbitals has been deduced from theoretical calculations allied to spectroscopic data (Adar, 1978; Gouterman, 1978; Spiro, 1983)

often possible. Most current analyses are based on the model of Gouterman, which contains a detailed molecular orbital description of porphyrins that quantitatively accounts for the appearance of many hemeprotein spectra (Adar, 1978; Gouterman, 1978; Makinen & Churg, 1983; Spiro, 1983). The details of this analysis are beyond the scope of the present review and we shall only describe some of its more important conclusions.

Both transitions 1 and 2 are allowed and should therefore be intense. However, they are mixed together under the influence of interelectronic repulsion to give one intense signal, the Soret band, and a weaker signal, the α/β band (Fig. 2.13). Thus these bands do not correspond to pure orbital transitions.

The reason there are two bands corresponding to the α/β transition is not apparent from Fig. 2.21. Two bands arise because although they represent the same electronic transition, they represent different vibrational transitions. The α-band is a (0,0) transition and the β-band is a (0,1) transition (Fig. 2.14).

A result of the mixing of transitions 1 and 2 is that the α/β bands gain intensity at the expense of the Soret band. This process depends upon the energy separation between the transitions and so as the separation between the α and Soret bands increases, the extinction coefficient of the α-band should be reduced, whilst that of the Soret band should be increased. Such a correlation is observed for model heme complexes and globins where there are large spectral changes accompanying the replacement of axial ligands (Smith & Williams, 1970).

The theoretical model thus accounts for the difference in intensity of the Soret and α/β bands, and the difference in their wavelengths. It also explains other spectroscopic differences.

1. The transitions result in the movement of electrons towards the periphery of the porphyrin plane. Thus the less electronegative the metal, the longer the wavelength of the transitions. Also, substituents on the porphyrin ring that are electron-withdrawing cause shifts to longer wavelengths. Examples of these effects are: the red-shift of the Soret band on reduction of ferricytochrome *c* (Fig. 2.13), and the differences in the α-band maxima of ferrohemes of type *c, b* and *a* (they occur at ~550, ~560 and ~600 nm respectively - Pettigrew & Moore, 1987: Chapter 1.3). As well as accounting for the frequency shifts, Gouterman's model can also account for some of the intensity changes.
2. Split absorbance bands can occur. The α/β bands are one example of an electronic transition split as a result of vibrational processes. Further splitting of the α-band is common (Pettigrew & Moore, 1987: Fig.1.8) whereas split Soret bands at room temperature are not common. One example is ferrocytochrome *c'* (Pettigrew & Moore, 1987: Fig. 1.6). It has not been established whether the split band in this case arises from vibrational effects or from a removal of the degeneracy of the porphyrin π^* orbitals. An indication of how subtle the factors may be that produce split bands, is given by the comparison of native horse ferrocyto-

chrome *c* with an analogue in which Tyr 67 has been replaced by Leu (Chapter 4.6.4). In the spectrum of the latter the α-band is split, whilst in the former it is not (Boon, 1981).

3. Short wavelength absorbance bands due to the heme occur at ~215 nm, ~325 nm, and occasionally, partway between 215 and 325 nm. All of these are $\pi \rightarrow \pi^*$ transitions but since they are weak and overlap with the peptide band and aromatic amino acid absorbance we do not consider them further.

d-d Transitions 6 and 7. As indicated previously, the absorbance bands resulting from these transitions are weak and usually not observed. However, bands of horse ferrocytochrome *c* in the region 800 - 860 nm ($\varepsilon = 200\ m^{-1}\ cm^{-1}$) have been proposed to arise from transition 7 (Eaton & Charney, 1969; Sutherland & Klein, 1972). Transition 6 does not occur in ferrocytochrome *c* because the d_{xz}/d_{yz} orbital pair contains four electrons.

CT Transitions 3, 4, 5, 8, 9, 10 and 11. These transitions fall into three groups: 11 is strictly disallowed; 3, 4 and 5 are allowed, and their absorbance bands have been identified in spectra of some proteins; and the status of 8, 9 and 10 is uncertain. The theoretical consensus is that 8, 9 and 10 are allowed for some but not for all ligand types. Also, their frequencies will be strongly dependent upon the ligand and difficult to predict (Spiro, 1983).

The bands corresponding to transitions 3, 4 and 5 have been identified in spectra of high-spin ferric systems near 600 nm, and in spectra of low-spin ferric systems between 1200 and 1500 nm (Smith & Williams, 1970; Gouterman, 1978; Makinen & Churg, 1983; Spiro, 1983). The situation with ferrous systems is less clear. In low-spin systems transitions 4 and 5 (and 8 and 9) cannot occur, because the d_{xz}/d_{yz} orbital pair contains four electrons, but in high-spin systems they can occur, along with transition 3. However, unequivocal band assignments have not been reported for ferrous proteins.

CT bands occur in the NIR region of the spectra of some cytochromes *c*, to as high a wavelength as 1950 nm. Their use in identifying axial ligands is described in section 2.3.2. In the remainder of the present section we consider two largely CT absorbance bands that are regarded as diagnostic of particular types of iron coordination, and consider the application of optical spectra to the classification of cytochromes.

The '695 Band'. Theorell & Åkesson (1941) first observed the weak 695 nm absorbance band of ferricytochrome *c* (Fig. 2.13) and noted that it was sensitive to pH. Subsequent experiments have confirmed that the 695 nm band, or a band close to 695 nm, is characteristic of class I cytochromes *c* and is sensitive to the conformational state of the protein (Schejter & George, 1964; Schechter & Saludjian, 1967; Chance *et al.*, 1968; Vinogradov, 1970; Sreenathan & Taylor, 1971; Pettigrew *et al.*, 1978; Pettigrew & Moore, 1987: Chapter 1.3; present work: Chapter 4.4.4 - 4.6). Coincident with the loss of the 695 band is the displacement of the methionine ligand. This, together with the observation that all cytochromes possessing a 695 band contain a methionine ligand - and this includes class II cytochromes *c* and *E. coli* cytochrome b_{562} (van den Branden *et al.*, 1975; Myer & Bullock, 1978; Moore *et al.*, 1982b; 1985b) - has led to the view that the 695 band is diagnostic of methionine coordination. This suggests that it arises from one of the following: a ligand-based transition (8, 9 or 10); a porphyrin-based transition (3, 4 or 5); or a d-d transition (6 or 7) (Smith & Williams, 1970). The current view is that it actually arises from a porphyrin → Fe(III) transition (Smith & Williams, 1970; Makinen & Churg, 1983). As these authors point out, a band is observed at 690 - 1000 nm for the low-spin proteins metmyoglobin-OH and metmyoglobin-N_3. Furthermore, these bands, like the 695 band of ferricytochrome *c*, behave similarly with polarised light; they are all z-polarised. Thus they probably arise from a common porphyrin → Fe(III) transition that does not depend on the axial ligand energy levels. Which porphyrin → Fe(III) transition is involved is not certain. Eaton & Hochstrasser (1967) propose transition 3 whilst Makinen & Churg (1983) suggest it is transition 4 or 5. Whichever assignment is correct, the correlation with methionine ligation is due to the influence of the axial ligands on the iron d-orbitals (Chapter 1.2).

Although the presence of the 695 band in spectra of cytochromes may be diagnostic for Met ligation, its absence does not mean Met is not a ligand. At a temperature of ~50°C, when the 695 band is not observed, ferricytochrome *c* retains its Met ligand according to NMR spectroscopy (Ångstrom *et al.*, 1982). Also ionisation of the axial His to histidinate causes a shift of the 695 nm band to shorter wavelengths where it may not be resolved from the α/β band in the optical spectrum (Moore *et al.*, 1985b).

The '640 Band'. There is a band at 600 - 640 nm in the spectra of high-spin ferrihemeproteins (e.g. ferricytochrome *c'*: Fig. 2.29) that was once regarded as diagnostic for the presence of H_2O as an axial ligand (Horio & Kamen, 1961; Maltempo & Moss, 1976; Maltempo, 1976). However, it is clearly established that the iron of cytochrome *c'* is five-coordinate (section 2.3.2.5; Chapter 5.2) and thus the '640 band' is not diagnostic of H_2O coordination.

The '640 band' may arise from a porphyrin → Fe(III) CT transition (Makinen & Churg, 1983). Alternatively, the band may not be a pure CT absorbance but a result of the α/β transitions mixing with a CT transition (Braterman *et al.*, 1964). This proposal explains the appearance of a band at ~500 nm for high spin ferric systems. Maltempo has proposed that the band should be present in spectra of proteins with a quantum-mechanically admixed S = $^3/_2$, $^5/_2$ state, as is observed (e.g. see Fig. 2.29), but that its intensity should be lower than that of a pure high-spin ferricytochrome *c*.

Classification of Cytochromes. The use of optical spectroscopy to classify cytochromes has been described elsewhere (Meyer & Kamen, 1982; Pettigrew & Moore, 1987: Chapter 1). Small shifts in absorbance bands and minor variations in extinction coefficients serve to distinguish many different cytochromes *c*. In some hands this is a remarkably useful approach to relating cytochromes (Meyer & Kamen, 1982), though it does appear to have a fragile theoretical base. For example, the algal ferrocytochromes *c*-553 have an α-band absorbance between 552 and 554 nm that is asymmetrically shaped. By contrast, ferrocytochromes c_5 have a symmetric α-band at 554 nm (Pettigrew & Moore, 1987: Table 1.5.). Though the absorbance ratios (α/β) are slightly different, the main classification criteria are the shape and wavelength of the α-band. These reflect the coupling between electronic and vibrational states for the mixed 1 and 2 (0,0) transitions. The small differences in the two kinds of α-band could result from a minor perturbation of the heme geometry, or from a change in the polarity of the heme environment. Thus the optical classification rests on subtle structural influences that are currently poorly understood. Therefore, if classification is based on optical spectra alone, errors are to be expected. Fortunately, it is possible to classify cytochromes with a range of criteria (Pettigrew & Moore, 1987: Chapter 1).

2.3.1.2 *The Spin-State of Iron*

The spin-state of a cytochrome can be determined by a variety of methods (Table 2.1). Usually the distinction between high-spin and low-spin systems is relatively straightforward, and often electronic spectra are all that is required to achieve this. The distinction between the quantum-mechanically admixed $S = {}^3/_2, {}^5/_2$ state (Chapter 1.2) and the pure high-spin state is not so straightforward, and for this a wide variety of methods needs to be employed together. Mössbauer and electronic spectroscopies on their own are not sufficient to discriminate between them (Maltempo & Moss, 1976; Emptage *et al.*, 1977), though EPR (Fig. 2.9 & 2.22; Maltempo & Moss, 1976) and NMR (La Mar *et al.*, 1981) are sometimes able to do so. A potential complication with EPR spectroscopy is that extraneous iron complexes in the form of impurities in the spectrometer or in the sample often give a signal at g = 4.3, which is close to the region where the mixed spin-state system may absorb.

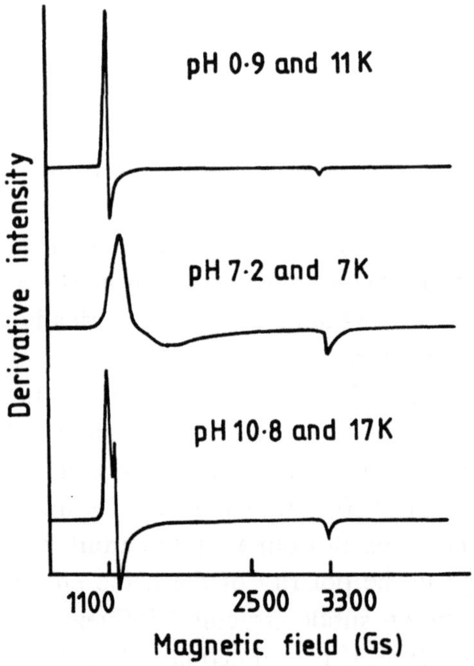

Fig. 2.22 EPR spectra of *Chromatium* ferricytochrome *c'*. The low pH and high pH spectra indicate high-spin states with g~6 and ~2 (they are axially and rhombically split respectively). The pH 7.2 spectrum is composed of two signals, both with quantum-mechanically mixed $S = {}^3/_2, {}^5/_2$ spin-states. These have g-values of 4.75 and 1.99, and 5.27 and 1.99. (Reproduced from Maltempo & Moss, 1976)

Magnetic susceptibility measurements are capable of distinguishing between different spin-states (Chapter 1.2) but when a magnetic moment in the range 3 to 5.7 BM is obtained for a ferric protein, spectroscopic data are required to establish whether the system is a quantum-mechanically admixed spin-state or a thermal mixture of states.

Table 2.3 summarises distinguishing features of the different ferric and ferrous spin-states.

Table 2.3 Magnetic and spectroscopic indicators of oxidation and spin-states

Ferric:		
$S = {}^5/_2$	μ	= 5.9
	EPR spectrum:	Type A (Fig. 2.9)
	$g_{\|\|}$	~ 6
	CT Absorbance band at	~ 600 - 640 nm
$S = {}^1/_2$	μ	= 1.7 - 2.3
	EPR spectrum:	Type B (Fig. 2.9)
	g_z	=2.8 - 3.5
	No CT Absorbance band at	~600 - 640 nm
$S = {}^3/_2, {}^5/_2$	μ	= ? (Table 1.1)
	EPR spectrum:	resembles Type A
	$g_{\|\|}$	= a6.0 + b4.0 (where a & b are coefficients of the ${}^5/_2$ and ${}^3/_2$ states, respectively and (a + b = 1).
Ferrous:		
$S = 2$	μ	= 5.0 - 5.5
	α/β absorbance bands:	not resolved from each other (Fig. 2.23)
$S = 0$	μ	= 0
	α/β absorbance bands:	clearly resolved (Fig. 2.13)
$S = 1$; $S = 1,2$ mixtures	These have not been observed for hemeproteins, therefore spectroscopic indicators have not been defined.	
Ferryl:		
$S = 1$	(section 2.3.1.3)	

Redox-Linked Spin-State Changes in Cytochromes c. The spin-state of most cytochromes *c* does not change with a change in the oxidation state. Notable exceptions are the cytochromes *c'*, which have a high-spin state in the reduced form and a $S = {}^3/_2, {}^5/_2$ state in the oxidised form at pH values below the pK of the axial histidine (Table 5.1). Three other cytochromes *c* have recently been suggested to undergo redox-linked spin-state changes.

A monoheme cytochrome *c* of M_r~8.2 kD from *Wolinella succinogenes* is completely low-spin in the reduced form with histidine and methionine axial ligation, but in the oxidised form it is a mixture of the high and low spin-states

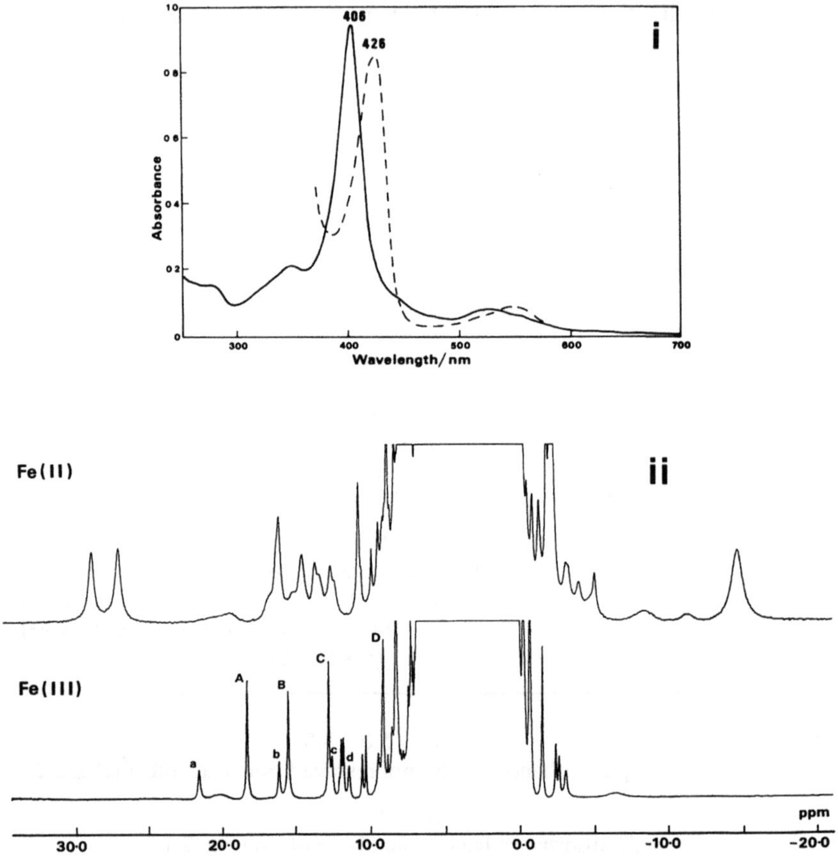

Fig. 2.23 Spectra of a cytochrome *c*-555 from *Methylophilus methylotrophus*. The optical absorption spectra at pH 6 (i) indicate the oxidised protein (solid line) is low-spin but the reduced protein (broken line) is high-spin. The ^{1}H NMR spectra (ii) confirms the spin-state analysis. (Reproduced from Santos & Turner, 1988)

(Moura *et al.*, 1988). The key evidence for this is NMR and magnetic susceptibility data: for example, at pH 7.5 the magnetic moment is 3.38 BM. The electronic spectrum of the oxidised protein was shown to contain bands at ~619 nm and ~695 nm, consistent with some high-spin character and suggesting that methionine ligation is retained by at least some of the cytochrome molecules. The spin-state equilibrium was pH dependent with the low-spin character increasing with increasing pH. The pK for the transition was ~7.3.

A 15 kD monoheme cytochrome *c* from *Methylophilus methylotrophus* is even more intriguing. This protein is high-spin in the reduced form and low-spin in the oxidised form (Santos & Turner, 1988). The electronic absorption and NMR spectra are characteristic of these states (Fig. 2.23), as are the magnetic moments: 2.2 BM and 5.2 BM for the ferric and ferrous forms respectively. Santos & Turner suggest that the change in spin-state results from a conformation change that includes a change in axial ligation. The ligation of the oxidised protein is suggested to be bis-histidine by EPR-MCD studies (M.Berry & A.J.Thomson, personal communication), but that of the ferrous protein has not yet been established.

Further studies of both these unusual proteins may assist analyses of the conformation change contribution to protein redox potentials (Chapter 7.7) and self exchange electron transfer rates (Chapter 8.4).

The iron spin-states of the tetraheme cytochrome *c*-554 from *Nitrosomonas europaea* have proved difficult to establish. The EPR spectrum of the fully oxidised form is unusual due to heme-heme interactions (Andersson *et al.*, 1986) and thus this gives no information concerning spin-states. However, Mössbauer spectra over the temperature range 4.2 to 250 K reveal that the ferricytochrome at pH 7 contains 75% low-spin iron and 25% high-spin iron. A similar ratio was found for the fully reduced cytochrome (Andersson *et al.*, 1986). Since RR studies at 277 K (Andersson *et al.*, 1984) suggest that in the fully reduced protein the hemes are six coordinate and low-spin while in the fully oxidised protein they are five-coordinate with an indeterminate spin-state, further work is needed to establish firmly the spin-states of this protein.

2.3.1.3 The Oxidation State of Iron

There are many ways to determine the oxidation state of iron. Some of these are summarised in Table 2.1 and exemplified by Tables 1.1 and 1.3, and section 2.3.1.1.

The most challenging experimental situation occurs when the oxidation state of iron is to be determined for an enzyme during its reaction cycle. Cytochrome *c* peroxidases are good examples of this case. These heme containing enzymes catalyze the oxidation of ferrocytochrome *c* to ferricytochrome *c* by H_2O_2. Yeast cytochrome *c* peroxidase (YCCP) contains a single *b*-heme and *Pseudomonas* cytochrome *c* peroxidase (PCCP) contains two *c*-hemes. H_2O_2 removes two reducing equivalents from both proteins. In both cases one reducing equivalent is removed from a heme to generate a ferryl oxene species [Fe(IV)=O]; the second reducing equivalent is removed from the second heme of PCCP to produce a ferriheme, or from a group, or set of groups of YCCP to generate a protein based free radical (Pettigrew & Moore, 1987). The evidence indicating the formation of Fe(IV)=O is largely spectroscopic.

The ferryl ion contains four d electrons which can, in theory, be arranged to give both a high-spin and a low-spin state. In a tetragonally distorted octahedral field, d_{xy} is significantly lower in energy than the degenerate, or almost degenerate, d_{xz} and d_{yz} orbitals (Fig. 1.5), and thus the ferryl d^4 state is low-spin with two unpaired electrons (the electron distribution is two in d_{xy}, and one each in d_{xz} and d_{yz}). This state therefore has a net spin of $S = 1$. As a consequence it should be EPR silent, as is found experimentally (Greenwood *et al.*, 1988).

Mössbauer spectroscopy provides compelling evidence for the d^4 configuration (Moss *et al.*, 1969; Lang *et al.*, 1976; Harami *et al.*, 1977; Schulz *et al.*, 1984). The chemical shift of the Fe(IV) state is lower than that of the Fe(III) and Fe(II) states. Also the net spin can be shown to be integral by the effect on the Mössbauer spectrum of an external magnetic field applied to the sample. Integral spins only occur for d^6 and d^4 configurations which means that the chemical shift is needed only to distinguish between Fe(II) and Fe(IV) states. This is readily achieved.

The major problem with Mössbauer spectroscopy is its lack of sensitivity (section 2.2.6). Thus, whilst it has been possible to carry out Mössbauer studies of YCCP (Lang *et al.*, 1976) there has been insufficient material for a Mössbauer study of PCCP. Greenwood *et al.* (1988) have overcome this problem by using compound I of YCCP, myoglobin-peroxide, and compound II of horse-radish peroxidase - all of which contain the Fe(IV)=O species - to provide an MCD signature of the ferryl ion. The key feature of this diagnostic method is the temperature dependence of the spectrum (Foote *et al.*, 1989).

The presence of the Fe(IV)=O double bond is clearly indicated by vibrational spectroscopy. For example, in compound II of horse-radish peroxidase the Fe(IV)=O stretching vibration appears in RR spectra at 779 cm^{-1}, when $H_2{}^{16}O_2$ is used to generate the intermediate, but at 743 cm^{-1} when $H_2{}^{18}O_2$ is used. This shift is consistent with the change in mass of the Fe(IV)=O unit (Terner *et al.*, 1985).

The formal oxidation state of oxygen in the Fe(IV)=O species is -2 and though O^{2-} does bind to Fe(II) and Fe(III) it has not been found to do so for heme iron. Thus the RR feature at 779 and 743 cm^{-1} may be indicative of a ferryl oxene species. Note that the stretching frequency is very different to that of O_2 bound to ferrous hemes (Barlow *et al.*, 1973) so there should be no confusion with this configuration. However, it is similar to the O-O stretching frequency of O_2^{2-} bound in some binuclear metalloproteins (Loehr *et al.*, 1974).

2.3.2 The Identity and Stereochemistry of Axial Ligands

2.3.2.1 Histidine-Methionine Axial Ligation

The key work identifying the thioether linkages and histidine ligand to the heme of class I cytochromes *c* involved a combination of electronic spectroscopy and chemical studies (Theorell, 1938; Theorell & Åkesson, 1941; Margoliash *et al.*, 1961; Harbury & Marks, 1973). The question of the identity of the sixth ligand was resolved by Harbury *et al.* (1965) and McDonald *et al.* (1969).

Harbury *et al.* (1965) showed that, in the presence of N-acetylmethionine, electronic spectra of short, ferriheme containing peptides made from cytochrome *c* (Chapter 4.6.3) possessed a weak 695 nm absorption band comparable to that of the native protein. Since alternative ligands did not generate this absorption, this result strongly suggested that methionine was the sixth ligand of native ferricytochrome *c*. The NMR studies of McDonald *et al.* (1969) showed that methionine was also the sixth ligand in the ferrocytochrome. These workers argued that resonances of axial ligands of diamagnetic hemes would be shifted to low-frequency in NMR spectra because of the heme ring-current and they identified the pattern of resonances shown in Fig. 2.24(b). The relative intensities of the peaks indicate the resonance at -3.25 ppm comes from a methyl group and, since the only common amino acid side chain containing a methyl whose resonance is a singlet is methionine, this experiment demonstrated convincingly that methionine was the axial ligand. Subsequent NMR experiments confirmed methionine as the axial ligand of ferricytochrome *c* (Wüthrich, 1969; Redfield & Gupta, 1971).

The general pattern of methionine resonances in the NMR spectrum of Fig. 2.24(b) is now used routinely to identify methionine ligands in heme proteins. Examples outside the monoheme class I cytochrome *c* group include: cytochrome b_{562} (Xavier *et al.*, 1978); some members of class II cytochromes *c* (Moore *et al.*, 1982b); one heme *c* of *Pseudomonas* cytochrome *c* peroxidase

(Villalain *et al.*, 1984); and cytochrome cd_1 (Timkovich *et al.*, 1985). The range observed for the methyl chemical shift is -2.7 to -3.7 ppm (Moore & Williams, 1977; Moura *et al.*, 1988), reflecting the different types of sidechain in the immediate environment of the methionine and the variations in its stereochemistry.

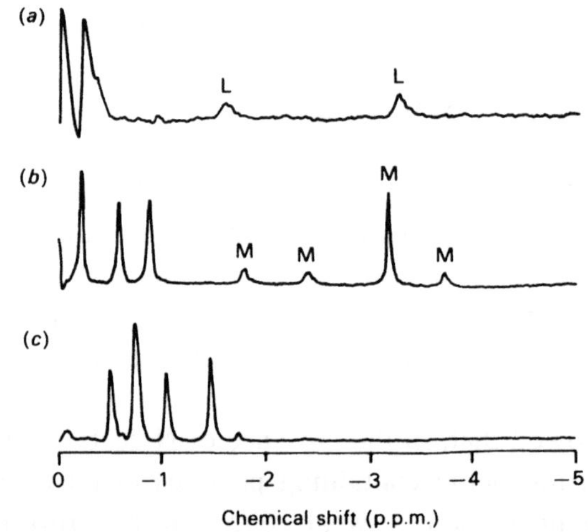

Fig. 2.24 Low frequency regions of the ^{1}H NMR spectra of (a) ferrocytochrome *f*; (b) ferrocytochrome *c*; (c) ferrocytochrome b_5. Resonances in the region -1.8 to -4 ppm come from axial ligands. Histidine ligands do not have resonances in this region [note **(c)** - cytochrome b_5 has bis-histidine ligation]. Methionine ligand resonances (indicated by **M**) do appear in this region. The ligand resonances of cytochrome *f* (indicated by L) may be diagnostic of lysine ligation. (Reproduced from Rigby *et al.*, 1988)

The corresponding resonances of the paramagnetic ferricytochromes are not so useful as a routine diagnostic for methionine ligation. In these cases the chief influence on the chemical shifts is the paramagnetism of the iron. Since the distribution of the unpaired electron is different in different ferricytochromes, the ligand methionine resonances have considerably different chemical shifts: for example, the methionine methyl resonance of class I ferricytochromes *c* have chemical shifts covering the range -1.6 to 24.0 ppm (Moore, 1985).

Resonances of histidine ligands are also strongly affected by the heme and in NMR spectra of ferrocytochrome *c* the ring proton resonances can often be resolved in the low-frequency region (McDonald & Phillips, 1973; Moore &

Williams, 1977). However, in spectra of ferricytochromes the histidine resonances generally cover a wide chemical shift range with many of the peaks appearing at high-frequency (La Mar *et al.*, 1976; Moore & Williams, 1984; Santos & Turner, 1987).

Stereochemistry of Histidine and Methionine Ligands. The stereochemistry of heme ligands is usually determined by XRD studies. However, Wüthrich and coworkers have described a spectroscopic approach for determining the stereochemistry of the histidine and methionine ligands of class I cytochromes *c*. This method makes use of NMR of the diamagnetic reduced form and CD of the 695 nm absorption band of the ferric form (Senn *et al.*, 1980).

The NMR experiments are based on the nuclear Overhauser effect (Scheme 2.4). Saturation of the methionine ligand methyl resonances cause heme meso resonances to appear in the difference spectrum (Fig. 2.25). For mitochondrial ferrocytochrome *c*, meso resonances α and δ are affected first, but for *Ps. aeruginosa* ferrocytochrome *c*-551 it is meso resonances γ and δ that are perturbed. Not only do these data show the two methionine ligands have different chirality, they also define their stereochemistries. So far, four different heme-methionine orientations have been found (Fig. 1.14).

Similar NMR experiments identify the relative orientation of the histidine ligand, which is the same for both mitochondrial cytochrome *c* and *Pseudomonas* cytochrome *c*-551 (Senn & Wüthrich, 1985). The ring proton resonances of the histidine ligand appear when heme meso-protons α and γ are saturated but not when protons β and δ are saturated. This indicates that the imidazole ring plane is roughly perpendicular to the heme plane and oriented along a line through the meso protons α and γ (Fig. 1.14).

Analogous NOE experiments for the ferric protein are not so revealing because the unpaired electron provides a potent nuclear relaxation mechanism. However, CD of the 695 nm band indicates the methionine chirality. For mitochondrial cytochromes *c* there is a strong negative Cotton effect at the 695 nm band but for *Ps. aeruginosa* cytochrome *c*-551 there is a weak positive Cotton effect (Fig. 2.25). The XRD structures show that these proteins have different methionine chirality and they have been use to calibrate the spectral responses. Thus CD measurements can now be used to determine methionine chirality (Senn & Wüthrich, 1985).

EPR studies may yield information about the relative orientation of axial ligands. This is obtained when the EPR spectrum is 'ramp-shaped' and has a

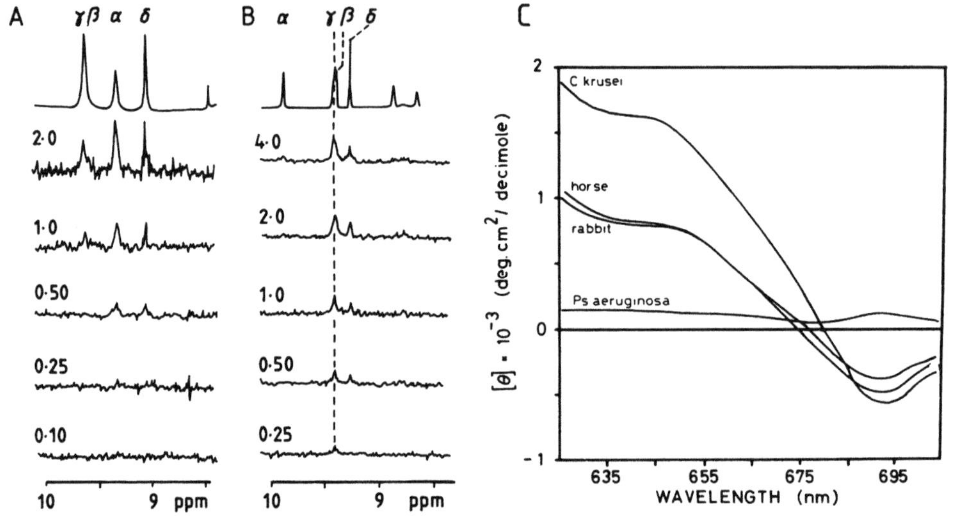

Fig. 2.25 Spectroscopic determination of the chirality of methionine ligands. **A & B** show the region from 8.5 to 10.0 ppm of the ^{1}H NMR spectra at 360 MHz of reduced horse cytochrome *c* and *Ps. aeruginosa* cytochrome *c*-551, respectively. The top traces show the normal spectra. The assignments of the meso-proton resonances are indicated. For each species a series of difference spectra, obtained with preirradiation of the εCH_3 resonance of the axial methionine, are shown. The length of the preirradiation time, in seconds, is indicated with each trace. **C** shows the CD spectra, plotted as molar ellipticity vs. wavelength, of *Candida krusei* ferricytochrome *c*, rabbit ferricytochrome *c* and *Ps. aeruginosa* ferricytochrome *c*-551. The protein concentrations were 0.7, 0.8, 0.6 and 1.0 mM, pH 6.7, T = 22°C. These NMR and CD spectra indicate that the methionine chirality of mitochondrial cytochrome *c* is different from that of *Ps. aeruginosa* cytochrome *c*-551 (see text). (Reproduced from Senn *et al.*, 1980)

high g_z value (e.g. see Fig. 2.26), both of which are diagnostic of a HALS system (section 2.3.2.2). The relationship between EPR spectra and axial ligand stereochemistry has been established with small porphyrin complexes containing bis-imidazole coordination (Carter *et al.*, 1981; Gadsby & Thomson, 1986; Walker *et al.*, 1986; Scheidt *et al.*, 1987) and it appears that complexes in which the imidazole planes are almost parallel exhibit HALS-type EPR. His-Met coordination may also give rise to HALS-type spectra and Gadsby *et al.* (1989) propose that this arises from an orthogonal relative orientation of the two ligands (Fig. 2.26).

Where the heme-methionine orientation is redox-state independent, which is so in most cases studied by Senn & Wüthrich (1985), there appears to be no correlation between the orientation and functional properties such as redox

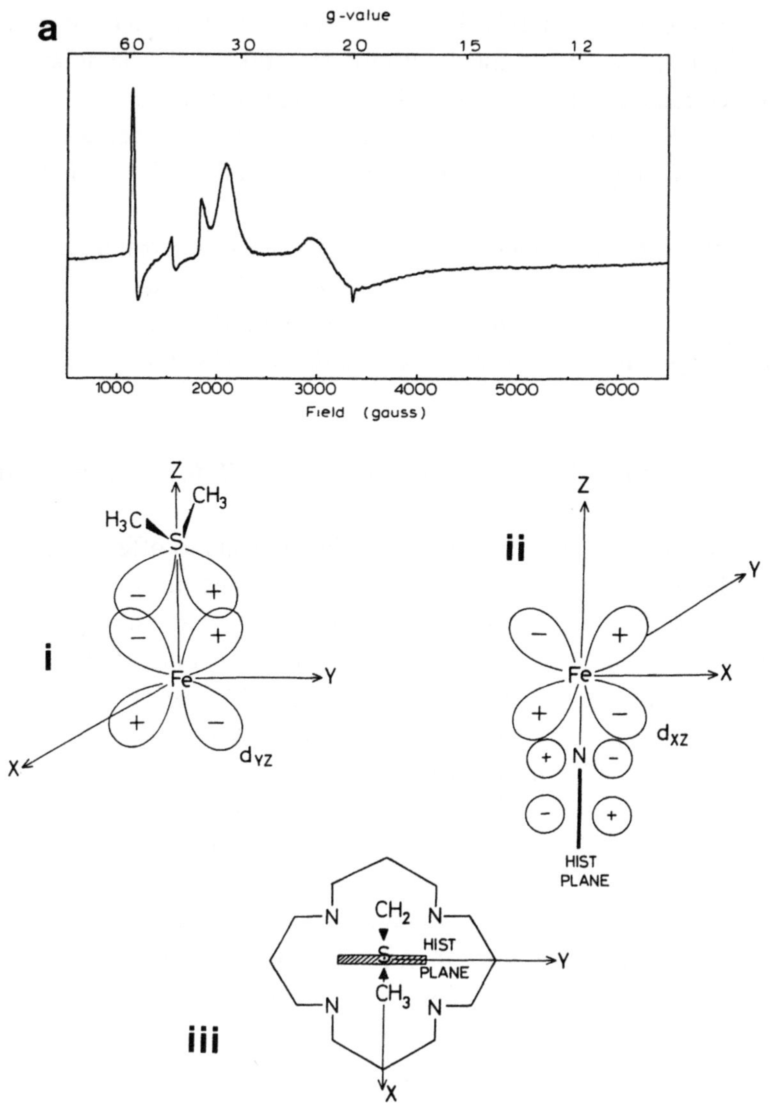

Fig. 2.26 His-Met ligation of a HALS heme of cytochrome c_4**. a** The EPR spectrum of oxidised *A. vinelandii* cytochrome c_4, pD 7.2, 50 mM Hepes buffer and temperature 10 K. **b** Representation of the interactions between the iron d_{xy}, d_{yz} orbitals with the lone-pair orbitals of the sulfur atom of the coordinated methionine residue and the π-orbitals of the coordinated histidine of the HALS heme of cytochrome c_4. For clarity the axes in **i** and **ii** are rotated 90° about the z-axis. **iii** is a plan view of the porphyrin ring showing the proposed 90° orientation of the histidine plane and the methionine axis, CH_3-S-CH_2-. The orientation of the ligand planes relative to the pyrrole nitrogen atoms is arbitrary. The position shown minimises the steric interaction between the pyrrole N-atoms and the axial ligands. (Reproduced from Gadsby *et al.*, 1989)

potentials and electron-transfer kinetics. However, there may be a correlation between the orientation and the electronic structure of the heme as manifested by the magnitude of the heme methyl chemical shifts in the oxidized form: for mitochondrial cytochrome *c*, methyls 8 and 3 are most affected by the paramagnetism; but in cytochrome *c*-551 it is methyls 1 and 5 that are most affected.

In two cases, the monoheme cytochrome *c*-553 from *Desulfovibrio vulgaris* (Hildenborough) and *D. desulfuricans* (Norway 4), the chirality of the methionine ligand is different in the oxidised and reduced forms (Senn *et al.*, 1983). The implications of this for the redox potential are discussed in Chapter 7.

2.3.2.2. *The EPR-MCD Approach to Identifying Axial Ligands*

The approach that is described in this section appears to offer a reliable way to identify the commonly found axial ligand types in low-spin ferricytochromes *c*. MCD spectra in the Near-IR (NIR) wavelength region are central to this analysis (Gadsby & Thomson, 1989; Simpkin *et al.*, 1989).

Fig. 2.27 shows the NIR-MCD spectrum of horse ferricytochrome *c* at different pH values over the wavelength range 800 - 2000 nm. The intense absorbancies at 1725 nm and 1465 nm represent porphyrin → Fe(III) CT transitions (probably 4 or 5 in Fig. 2.21 since 3 should be at lower wavelength) for His-Met and His-Lys axial coordination respectively. The correlation of the band wavelength with the ligand type comes from studies of a range of hemeproteins and model complexes with defined axial ligation. An example is the comparison of ferricytochrome *c* at high pH and the butylamine adduct of ferric-leghemoglobin (Gadsby *et al.*, 1987). Table 2.4 summarises the NIR-MCD band assignments for low-spin ferricytochromes *c*.

The relationship between EPR and NIR-MCD data comes about because the EPR spectra are influenced by values of V and Δ, which reflect the relative energies of the d_{xy}, d_{yz} and d_{xz} iron orbitals (Fig. 1.7), and NIR-MCD monitors transitions into the d_{xz}/d_{yz} orbital pair (Fig. 2.21: transitions 8 and 9). Not only are the EPR g-values and NIR band wavelengths related, as indicated by Table 2.4, but so also are the intensities of the NIR-MCD bands. The details of the theoretical models describing these relationships are beyond the scope of this review and Thomson and Gadsby (1989) should be consulted for further details.

Table 2.4 reveals that the NIR band can be used to distinguish between the common forms of low-spin ligation except bis-histidine and His-Lys: the similarity of their spectra reflecting the fact that both His and Lys are N-donors. This is one reason that EPR spectroscopy is used to assist the ligand identification. Another reason is that EPR is routinely available in many laboratories while NIR-MCD is currently available in only a comparatively small number of laboratories. Thus although NIR-MCD is a powerful method, it is not applied to all uncharacterised cytochromes under investigation.

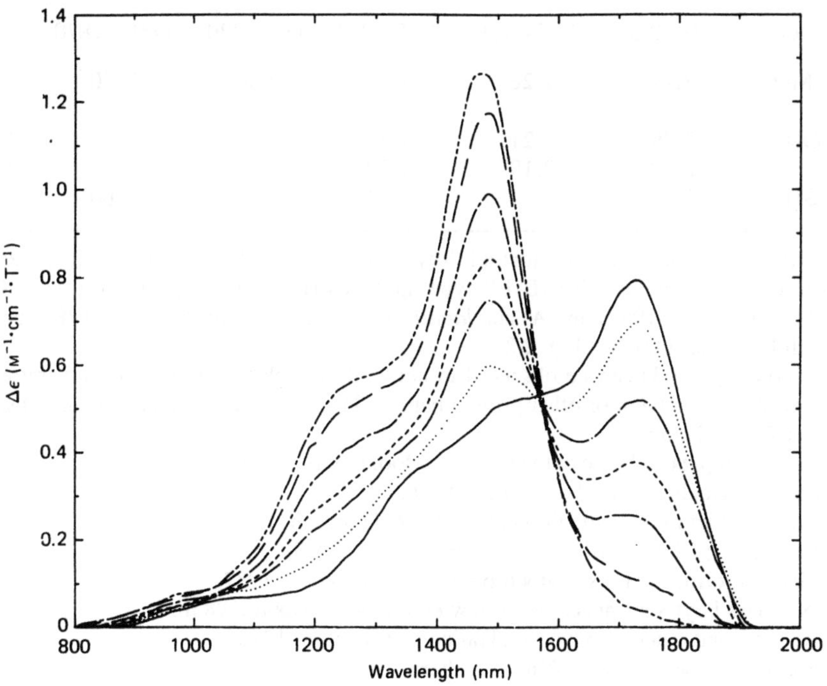

Fig. 2.27 pD dependence of the axial ligation of horse ferricytochrome *c*. NIR-MCD spectra all at 25°C and at the following pD values: —— 7.8, ········ 8.6, — · — · 9.1, ---------9.3, — - — - 9.7, ——— 10.2 and — -- — -- 11.0. The peaks at 1725 nm and 1465 nm titrate with a pK of 9.3. (Reproduced from Gadsby *et al.*, 1987)

The main EPR measurement is the determination of g-values. However, from these, values of the energy terms Δ and V can be obtained (section 2.2.3.2: Taylor, 1977; Palmer, 1983). These may aid the identification of axial ligands since they are strongly influenced by the strength and shape of the ligand field produced by these groups.

Table 2.4 EPR-MCD characteristics of selected axial ligation schemes for low-spin ferricytochromes

Ligation Scheme	EPR g-values [a]			MCD CT bands [b] (nm)		Refs. and proteins
	g_z	g_y	g_x			
His, His	2.96	2.27	1.5	—	1480	c
	2.97	2.29	1.56	—	1510	d
His, His^-	2.82	2.29	1.69	—	1350	e
His, Lys	3.55	1.7	n.d	—	1520	f
	3.3 - 3.5	2.05 - 1.8	n.d	—	1480	g
His, Met	3.0 - 3.3	2.05 - 2.3	1.15 - 1.4	690 - 720	1750 - 1870	c f g h
His^-, Met	2.79	2.26	1.67	680	1550	c
His, OH^-	2.79	2.21	1.76			i
	2.62	2.19	1.82			j
His, SH^-	2.56	2.24	1.84		1200	k

[a] The EPR g-values are for normal low-spin spectra (Fig. 2.9) in all cases but cytochrome *f* (f). Proteins with HALS type spectra may have different g-values. Cytochrome *f* (Rigby *et al.*, 1988) and *A. vinelandii* cytochrome c_4 (Gadsby *et al.*, 1989) contain hemes with HALS spectra (see Fig. 2.27)

[b] Corresponding bands can be observed in conventional electronic spectra but frequently these are masked by solvent or other protein bands. The — indicates a band is not observed in the range 680 - 720 nm.

[c] Imidazole complex of horse cytochrome *c*; Moore *et al.*, 1985b.

[d] *D. vulgaris* cytochrome c_3; Foote *et al.*, 1984.

[e] Imidazole adduct of leghemoglobin; Gadsby & Thomson, 1982.

[f] Cytochromes *f*; Rigby *et al.*, 1988, Simpkin *et al.*, 1989.

[g] Alkaline conformer of horse cytochrome *c*; Gadsby *et al.*, 1987.

[h] Various class I cytochromes *c*; Pettigrew *et al.*, 1978; Moore,1985b.

[i] Lysine-modified cytochromes *c* at alkaline pH; Morton, 1973.

[j] Alkaline conformer of myoglobin; Ehrenberg, 1962.

[k] SH^- adduct of myoglobin; Gadsby & Thomson, 1989.

Blumberg and Peisach (1971) proposed that a plot of V/Δ against Δ/λ (where λ is the spin orbit coupling constant; taken to be 420 cm^{-1} for free Fe(III) - Palmer, 1983) would be a useful method of correlating data and they coined the name 'Truth Diagram' for such a plot. They found that a wide range of hemeproteins fall into five groups on this diagram, corresponding to their type of axial ligation. However, it is now clear that this approach is not an absolute indicator of ligand type and needs to be used in conjunction with other methods. Palmer (1983) discusses the 'Truth Diagram' in some detail and gives cases where the assumptions underlying its use may be invalid.

One of the cases where the 'Truth Diagram' doesn't work is with Highly Anisotropic Low-spin hemeproteinS (HALS). (Although this name is commonly used in the literature, it is not always appropriate. Many of the systems designated as HALS type are really highly axial and non-rhombic.) These had not been identified at the time the 'Truth Diagram' was proposed but they appear to form a large group of proteins (Salerno, 1984; Palmer, 1983; 1985) including cytochrome *f* (Siedow *et al.*, 1980; Rigby *et al.*, 1988) and *A. vinelandii* cytochrome c_4 (Fig. 2.26) (Gadsby *et al.*, 1989). The distinguishing feature of highly axial and non-rhombic systems is a high g_z value and ramp-shaped g_z signal (Fig. 2.26). Such a combination arises when V is close to zero (Salerno, 1984). On this basis Gadsby *et al.* (1989) propose that the His-Met orientation of one of the hemes of *A. vinelandii* cytochrome c_4 is as shown in Fig. 2.26b. This is very different to the His-Met orientation of other class I cytochromes *c* (Fig. 1.14) that have more normal EPR spectra.

In summary then, EPR spectra cannot be used to identify axial ligand types with confidence solely on the basis of the observed g-values or derived V and Δ terms. The range of g-values commonly found for particular ligation schemes is shown in Table 2.4. Fortunately the major NIR-MCD problem, distinction between His-Lys and His-His ligation, is reasonably reliably made by EPR. Therefore, the combined EPR-MCD approach is the best general spectroscopic approach currently available to determine the ligation of low-spin ferric hemeproteins.

The use of EPR and MCD to investigate hemeproteins is an active area with studies underway of the visible-MCD region (Vickery *et al.*, 1976; Siedow *et al.*, 1980; Sutherland, 1978; Holmquist, 1978), MCD bands, shapes and signs (Gadsby & Thomson, 1982, 1986; 1989; Gadsby *et al.*, 1987; 1989) and the development of model complexes to assist the correlation of EPR and MCD data (Yoshimura & Ozaki, 1984; Gadsby & Thomson, 1986). These studies may lead to an unequivocal spectroscopic approach, not only to the identification of axial ligands, but also to the determination of their relative geometries.

The ligation states of proteins with high-spin or quantum-mechanically admixed spin-states are not easily determined by the EPR-MCD approach described above. Studies of various cytochromes *c'* by NIR-MCD (Rawlings *et al.*, 1977) and EPR (Maltempo & Moss, 1976) reveal some of the problems with interpretation of the spectra of high-spin and S = $^3/_2$, $^5/_2$ spin-states.

Recently Gadsby and Thomson (1989) have proposed that the EPR-MCD approach can be applied to the cyanide adducts of five-coordinate proteins allowing identification of the fifth ligand. Consistent data were obtained for three His coordinated proteins.

2.3.2.3 *Axial Ligation of Cytochrome f and Cytochrome c_1*

The presence of the standard Cys-?-?-Cys-His heme binding site in the sequences of cytochrome *f* (Willey *et al.*, 1984b) defines histidine as the fifth coordination ligand of the iron. Since the iron is low spin, the sixth ligand must be strong field and therefore is either lysine, histidine or methionine.

Several lines of evidence appear to exclude methionine. Ferricytochrome *f* does not have a 695 nm band (Gray 1978), and the NMR spectrum of the ferrocytochrome lacks the diagnostic peaks for methionine ligation (Fig. 2.24). The MCD data (Rigby *et al.*, 1988; Simpkin *et al.*, 1989) are consistent with a nitrogenous base as sixth ligand (Table 2.4) and the EPR spectra (Table 2.4) and RR spectra (Davis *et al.*, 1988) favour lysine. There are several unvaried lysines on the sequence but only one unvaried histidine. Lysine therefore seems to be the most likely sixth ligand for cytochrome *f*.

In contrast to cytochrome *f*, cytochrome c_1 contains a methionine ligand. However, until recently, the spectroscopic data supporting methionine ligation has been weak. The position has changed since Simpkin *et al.* (1989) have used the EPR-MCD approach to show clearly that cytochrome c_1 contains a His-Met ligated heme. Prior to this study the main piece of evidence in favour of methionine ligation was that the optical spectrum of ferricytochrome c_1 possesses a weak 695 nm band (Kaminsky *et al.*, 1975). However, at 77 K this band is absent from the spectrum of bovine ferricytochrome c_1 at pH values ≤ 5.8 (Kaminsky *et al.*, 1975). Furthermore, on reduction of bovine and *Rhodobacter sphaeroides* cytochromes c_1, the 695 nm band is replaced by a weak and broad absorption band at ~670 nm (Yu & King, 1972; Yu *et al.*, 1986). Thus, although the 695 nm band may be the counterpart of the band indicative of methionine ligation in other cytochromes (section 2.3.1), there are sufficient differences between the spectral behaviour of cytochrome c_1 and methionine ligated class I cytochromes *c* to make this assignment unclear.

The EPR study of Tervoort and van Gelder (1983) complicated the issue further. These workers found g_z and g_y values for ferricytochrome c_1 of 3.36 and 2.04 respectively; values which are consistent with both His-Met and

His-Lys coordination (Table 2.4). However, Tervoort and van Gelder reject histidine as a ligand because the central signal in the EPR spectrum of nitrosyl ferrocytochrome c_1 at pH 10.5 is split into three sharp lines only, rather than the expected nine observed with cytochrome *c* (Kon, 1969) and other hemeproteins (Palmer, 1983), and they proposed the axial ligation to be Met-Cys.

The EPR experiment with nitric oxide has been investigated extensively. The sharp lines come from the unpaired electron on $NO^{\bullet}$, which is coupled to the nitrosyl N atom to give three lines and should also be coupled to the histidine ligand N(3) atom to give 3 x 3 = 9 lines. However, Kon (1975) showed that a three line spectrum occurs when there is only a small coupling constant between the histidine ligand and $NO^{\bullet}$. Such a situation frequently occurs and may result from a particular orientation of the bound histidine (F.Kadir, A.J.Thomson & M.T.Wilson, personal communication). Thus the data of Tervoort and van Gelder (1983) for cytochrome c_1 do not rule out a histidine ligand.

In summary then we have the remarkable situation in that two related classes of proteins have different heme ligation schemes. This is all the more striking because the methionine proposed to be a ligand in cytochrome c_1 is matched by an unvaried methionine in cytochrome *f* according to the sequence alignments of Chapter 3.3.

2.3.2.4 The Alkaline Isomer of Mitochondrial Ferricytochrome c

Mitochondrial ferricytochrome *c* undergoes a transition at alkaline pH from the native (Type III) form to the Type IV form, with a pK of 8 - 10 depending on the source of the protein (Table 4.4). This transition was first detected spectroscopically by the loss of the 695 nm absorption band with increasing pH (Theorell & Åkesson, 1941). Now virtually all the spectroscopic methods discussed in section 2.2 have been applied to help clarify the nature of this transition. In this section we address the question of the identity of the axial ligands at high pH. In Chapter 4.4.4, the nature of the triggering ionisation, and kinetic and chemical modification studies of the transitions are described.

The EPR and NMR spectra of horse ferricytochrome *c* indicate that at high pH ferricytochrome *c* does not comprise a single species (Fig. 2.28); multiple resonances are observed indicating at least two species. However, one of these is in large excess over the other(s) and is taken to be Type IV

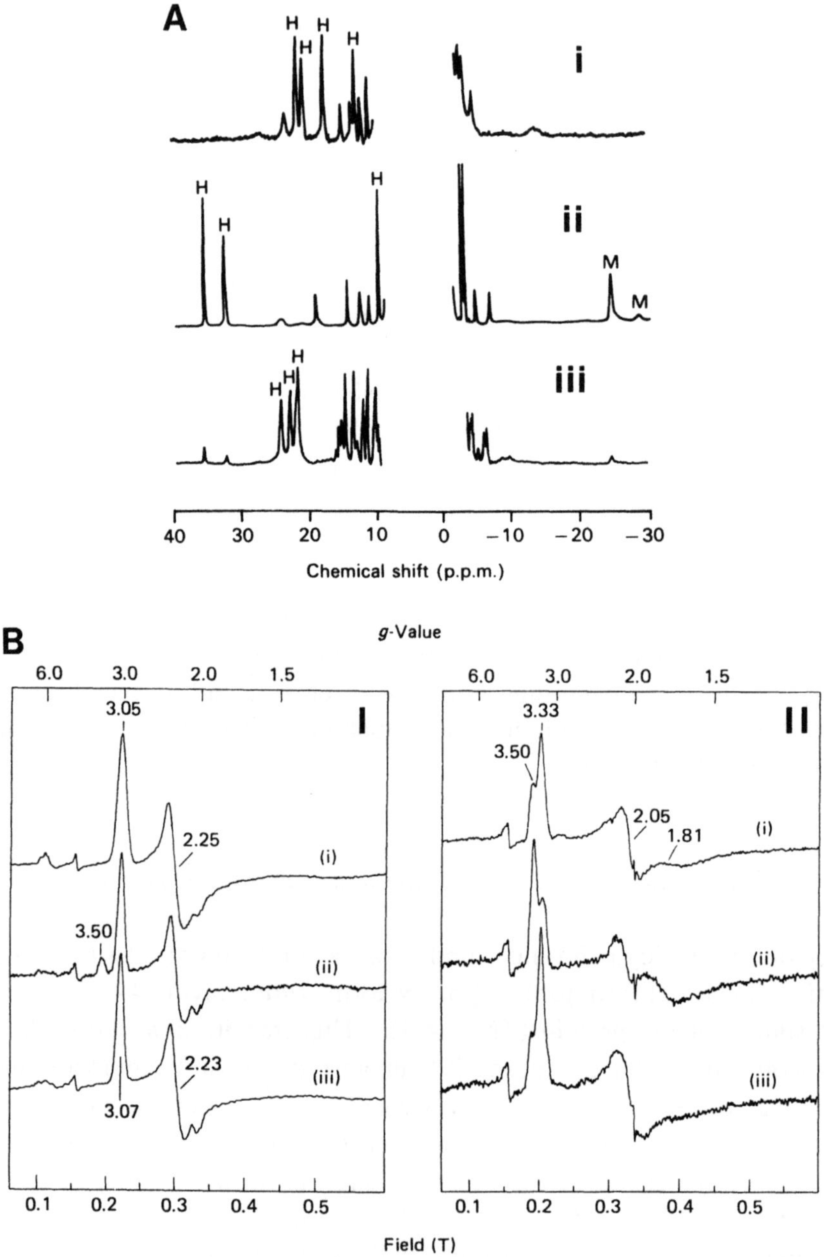

Fig. 2.28 Spectroscopic studies of the alkaline form of ferricytochrome *c*. A ^{1}H NMR spectra at 25°C of **i** charlock ferricytochrome *f* at pD 7.5; **ii** horse ferricytochrome *c* at pD 7.0; **iii** horse ferricytochrome *c* at pD 10.8. Resonances indicated by **H** come from heme methyls and those indicated by **M** come from the axial methionine ligand.

ferricytochrome *c*. This complication does not markedly affect the EPR-MCD analysis.

Like the native form of ferricytochrome *c* the alkaline isomer is low-spin. Thus both axial ligands contribute a strong field. The available spectroscopic and chemical evidence strongly suggests histidine remains a ligand and thus the question becomes: what is the sixth ligand?

The loss of the 695 nm absorption band is consistent with the replacement of the coordinated methionine. However, partly as a consequence of the chemical modification studies and partly because the 695 nm band can apparently be lost even though methionine remains a ligand (section 2.3.1.1), Wooten *et al.* (1981) decided to obtain independent data concerning the methionine ligand. These workers used ^{2}D NMR to study horse ferricytochrome *c* in which the methionine ligand methyl had been specifically enriched with ^{2}D. The pH dependence of the NMR spectrum showed unequivocally that methionine was displaced as a ligand with a pK~9. Previous ^{1}H NMR studies of native ferricytochrome *c* had been similarly interpreted (Gupta & Koenig, 1971; Morishima *et al.*, 1977) but these suffered in comparison to the work of Wooten *et al.* in monitoring only the methyl resonance of the methionine ligand in the Type III ferricytochrome *c*. Wooten *et al.* monitored the methyl resonance of both Type III and Type IV proteins and showed that only the Type III resonance was significantly affected by the paramagnetism of the iron.

EPR and NIR-MCD studies indicate that the sixth ligand of Type IV ferricytochrome *c* is a nitrogenous base (Brautigan *et al.*, 1977; Kobayashi *et al.*, 1981; Gadsby *et al.*, 1987). As usual, the question of bis-histidine or lysine coordination requires detailed spectral analysis, and the comparison of the spectral data with those for better characterised systems. Gadsby *et al.* found that the visible and NIR-MCD spectrum of Type IV ferricytochrome *c* was remarkably similar to the spectrum of the soyabean ferric-leghemoglobin-*n*-butylamine adduct, after adjusting the latter spectrum to take account of the different type of heme. Gadsby *et al.* took this result to indicate that both proteins had the same axial ligation, namely histidine-amine. Further

(Reproduced from Rigby *et al.*, 1988). **B** EPR spectra of horse ferricytochrome *c* at pD 6.6 in phosphate buffer (**I**) and pD 11.0 in cyclohexylaminopropane sulfonic acid buffer (**II**). Samples were (**i**) without organic cosolvents, (**ii**) with 50% ethanediol, and (**iii**) with 50% glycerol. (Reproduced from Gadsby *et al.*, 1987). The presence of multiple species for horse ferricytochrome *c* is indicated by the two g_z values (at 3.33 and 3.50) and the multiple sets of NMR resonances between 10 and 25 ppm support for this ligand assignment is provided by the close similarity of spectral data between ferricytochrome *f* and Type IV ferricytochrome *c* (Table 2.4). The remaining problem is to identify which of the many possible lysines is the ligand.

2.3.2.5 Axial Ligation of Cytochromes c'

Cytochromes *c'* have been studied with most of the techniques described in section 2.2. The main question that has been investigated is: what is the origin of the spectroscopic pK_a at alkaline pH (Fig. 2.29)? Since it had been suggested that it was due to the ionisation of coordinated water, a secondary question has been: is water an axial ligand?

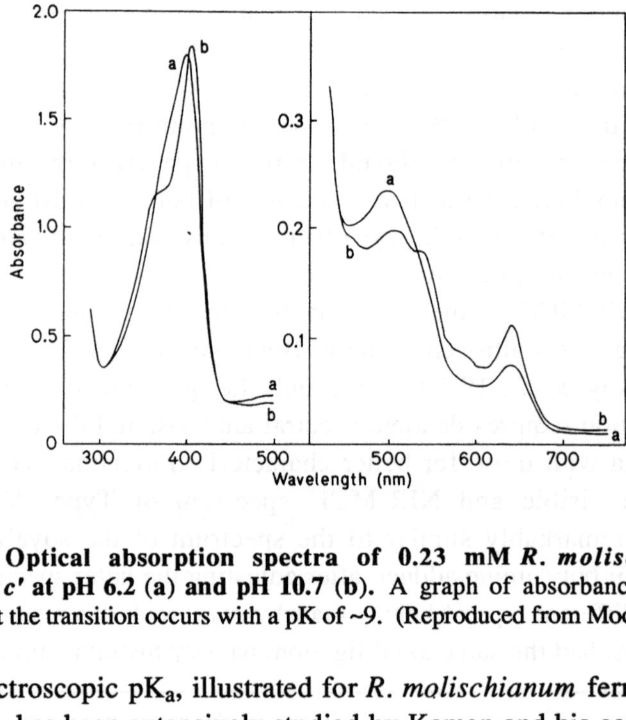

Fig. 2.29 Optical absorption spectra of 0.23 mM *R. molischianum* ferricytochrome *c'* at pH 6.2 (a) and pH 10.7 (b). A graph of absorbance change against pH shows that the transition occurs with a pK of ~9. (Reproduced from Moore *et al.*, 1982b)

The spectroscopic pK_a, illustrated for *R. molischianum* ferricytochrome *c'* in Fig. 2.29, has been extensively studied by Kamen and his colleagues. They demonstrated a fundamental difference between ferricytochrome *c'* and the high-spin metmyoglobin in that, with increasing pH, the ~640 nm absorption of ferricytochrome *c'* increased whereas it decreased for metmyoglobin (Horio & Kamen, 1961). With metmyoglobin the ionisation of the coordinated H_2O leads to a situation in which the fully high-spin protein is converted into a system which is a thermal admixture of high and low spin states (Chapter 1.2). This means that the magnetic moment and spectroscopic properties arising from the high-spin state, such as the ~640 nm absorption and g~6 EPR signal, are decreased in intensity; and spectral properties arising from the low-spin state, such as the g~3 EPR signal, are increased.

Cytochrome *c'* behaves very differently. Its magnetic moment increases rather than decreases with increasing pH (Table 1.1), the ~640 nm absorption increases (Fig. 2.29) and a g~3 EPR signal does not appear (Fig. 2.22). All of these data indicate that the cytochrome *c'* transition is different to the metmyoglobin transition. Thus they argue against H_2O ionisation.

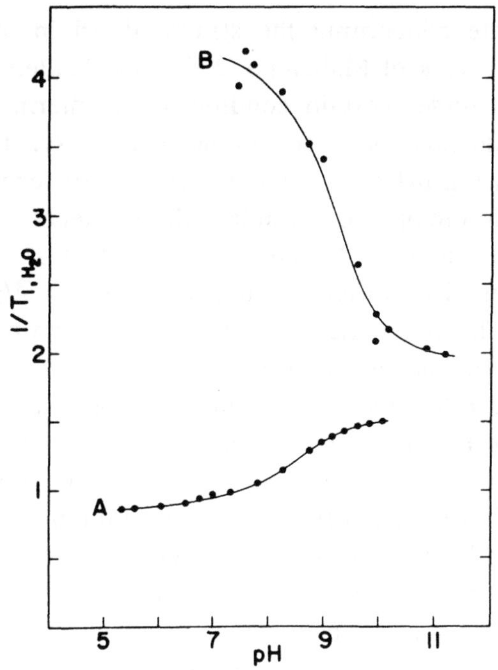

Fig. 2.30 H_2O ^1H NMR spin-lattice relaxation rates (sec^{-1}) of *R. rubrum* ferricytochrome *c'* (A) and metmyoglobin (B). These experiments measure the spin-lattice relaxation ($^1/T_1$) of the bulk water (see text). (Reproduced from Moore *et al.*, 1982b)

The NMR experiments shown in Fig. 2.30 are consistent with the absence of H_2O as a ligand at pH 7. In these experiments the NMR spin-lattice relaxation rate ($^1/T_1$) of solvent water was measured in the presence of various proteins. If H_2O binds to the iron and exchanges rapidly with free H_2O, then the NMR relaxation rate of the bulk H_2O will be considerably increased compared to that for free H_2O not exchanging with a paramagnetic centre (Bertini & Luchinat, 1986). As a general rule in such a system, the H_2O relaxation rate will decrease as the magnetic moment of the iron decreases. The data (Fig. 2.30) for myoglobin (Fabry & Eisenstadt, 1971) and cytochrome *c′* (Emptage *et al.*, 1981; Moore *et al.*, 1982b) are consistent with metmyoglobin having a bound H_2O that ionises to OH^- with increasing pH, and with cytochrome *c′* not having a coordinated H_2O. The trends in the graphs mirror the changes in magnetic moment (Table 1.1). Thus these spectroscopic measurements suggested that the iron of cytochrome *c′* was five-coordinate (Moore *et al.*, 1982b), a view subsequently confirmed by the x-ray studies (Chapter 5.2).

There has been much debate concerning the structural origin of the spectroscopic pK but the elegant work of Maltempo (1974) and Maltempo & Moss (1976), establishing that under certain conditions the intermediate $S = {}^3/_2, {}^5/_2$ mixed spin-state could exist, has resolved the controversy. It was the observation that with increasing pH the high-spin character of ferricytochrome *c′* increased, that led Maltempo to formulate the existence of the mixed intermediate spin-state (Chapter 1.2) and to propose that at pH 7, different ferricytochromes *c′* contained different amounts of the $S = {}^3/_2$ and $S = {}^5/_2$ states (Table 1.1). The magnetic and spectroscopic properties predicted by this model fit the experimental data well.

Landrum *et al.* (1980) proposed that the ionising group responsible for the spectroscopic transition was the histidine ligand and this was shown to be consistent with the x-ray structure of *R. molischianum* ferricytochrome *c′*, which reveals that the ligand, His 122, is not hydrogen-bonded through its N1 hydrogen, and with the expected ligand field strength (Weber, 1982: Chapter 1.4). However, it has not been conclusively demonstrated in cytochrome *c′* that the histidine ionises, because the accompanying spin-state change makes the comparison of cytochrome *c′* with other proteins and model porphyrin complexes difficult. However, the ligand histidine of the structurally homologous *E. coli* cytochrome b_{562} (Chapter 6.10.2) has been shown convincingly to ionise with a pK~9 (Moore *et al.*, 1985b) strongly supporting the proposal of Landrum *et al.*

2.3.3 The Protein Structure

2.3.3.1 Determination of Protein Tertiary Structure

Many spectroscopic studies of cytochromes investigate specific aspects of the protein structure but only one is capable, in principle, of providing a structure for the entire molecule. This is NMR spectroscopy, and we consider its structural application below.

An important area where spectroscopy has contributed to the development of protein structures is in the study of oriented membrane samples. Because of the motional properties of cytochromes in such samples, NMR is not a particularly useful technique for these (section 2.2.3.1) but other methods provide some structural information as we describe below. We have considered elsewhere in the present book the application of spectroscopy to the identification of axial ligands (section 2.3.2), CD to characterise the amount of secondary structure present (section 2.2.5.1), fluorescence to investigate tryptophan residues (section 2.2.5.3), and solvent perturbation spectroscopy to identify the solvent exposure of aromatic groups (Chapter 7.4.3).

NMR Spectroscopy. NMR spectra of proteins reflect directly the proteins' tertiary structures. This is illustrated in Fig. 2.31 for tuna cytochrome *c*. The spectrum calculated from the amino acid composition of the protein (**a**), assuming there is no organised secondary or tertiary structure, is considerably different from that of either ferrocytochrome *c* (**b**), or ferricytochrome *c* (**c**). These differences are because resonances have shifted from their chemical shift values in small peptides. Not only do resonances shift, however, but also, in some cases, their spin-spin coupling constants (Scheme 2.3) change. The mechanisms causing these changes are both through-bond and through-space effects. Provided resonances are unambiguously assigned to specific nuclei in the amino acid sequence, and their conformation-dependent chemical shifts and coupling constants measured, certain regions of the protein can be structurally characterised to high-resolution. An example of the structural use of conformation dependent shifts is the identification of methionine as a ligand of ferrocytochrome *c* (section 2.3.2.1). However, a complete tertiary structure cannot be obtained by this approach because the chemical shifts and coupling constants of many resonances are insufficiently sensitive to the protein conformation to be used as reliable probes.

The differences between the spectra of ferricytochrome *c* and ferrocytochrome *c* (Fig. 2.31) largely arise because the ferric protein is paramagnetic whilst the ferrous protein is diamagnetic (Moore *et al.*, 1980a). The paramagnetism of ferricytochrome *c* gives rise to a strong through-space effect from which internuclear bond angles and distances can be obtained (Clayden *et al.*, 1987). To date only tuna cytochrome *c* has been investigated by an analysis of such paramagnetic effects and the results indicate that the solution structure of the protein is not greatly altered from its x-ray structure (Williams *et al.*, 1985b; c). However, there are small structural differences (Chapter 4.2.2 & 4.5.2).

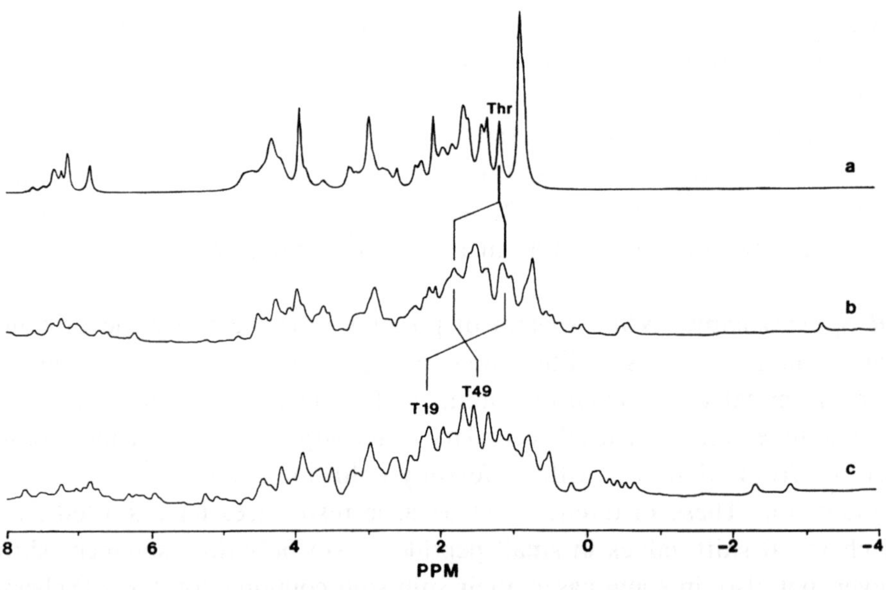

Fig. 2.31 Conformation-dependent shifts of NMR resonances of tuna cytochrome *c* at pH 7.0 and 25°C. a Simulated 300-MHz ^{1}H NMR spectrum in which all conformation-dependent shifts have been averaged to zero. The appearance of the spectrum depends only on the amino acid composition. **b** Experimental 300-MHz ^{1}H NMR spectrum of ferrocytochrome *c* showing the effects of diamagnetic conformation-dependent shifts. **c** Experimental 300-MHz ^{1}H NMR spectrum of ferricytochrome *c* showing the effects of both diamagnetic and paramagnetic conformation-dependent shifts. The positions of the methyl resonances of Thr 19 and Thr 49 are illustrated

A different approach to protein structure determination with NMR is the use of the Nuclear Overhauser Enhancement (NOE: Scheme 2.4). Wüthrich (1986) has reviewed its application as a structural method and shown that it can be used to obtain tertiary structures of proteins containing less than 100 amino acids (e.g. tendamisat; see Billeter *et al.*, 1989). A recent summary of 2D-NMR determination of protein structures is given by Wright (1989).

The NOE effect has been used to investigate the structures of class I cytochromes *c*. There have been three general approaches:

1. The NOE profiles of particular resonances are monitored to investigate small regions of the protein structure. An example is the region around the heme propionates of mitochondrial cytochrome *c* (Fig. 2.32). The NOE difference spectra show that heme methyl-8, heme propionate-7 and Trp 59 (together with Leu 35 and Leu 64) are close in space with a geometry consistent with the x-ray structure (Fig. 2.32b) (Moore & Williams, 1984). Analogous experiments on mutant *S. cerevisiae* iso-1 cytochrome *c* show that the replacement of Trp 59 by phenylalanine creates a similar structural unit to that shown in Fig. 2.32b. Also, the replacement of Arg 38 by a variety of residues, and Tyr 48 by phenylalanine (residues 38 and 48 also interact with HP-7), does not perturb the Trp 59/HP-7/heme methyl-8 unit (Cutler *et al.*, 1989; A.M.Davis, G.Guillemette, A.G.Mauk & G.R.Moore, unpublished data).
2. All the NOE profiles of a protein are systematically identified and analysed with a computer-graphics procedure to generate a complete tertiary structure. Such an analysis is not routine (Wüthrich, 1986) and no cytochrome structures have been determined by this approach. However, data has been collected for horse cytochrome *c* and an *ab initio* structure determination is now in progress (Feng *et al.*, 1989; Wand *et al.*, 1989).
3. This approach combines elements of the previous two. All the NOE profiles of a protein are determined and they are compared with those of a closely related protein that has been structurally characterised by XRD. This protein pair may be, for example, a wild-type protein and a single site mutant of it. The differences in the NOE profiles are used in conjunction with the x-ray structure and model-building to derive a structure for the original protein. Such a study has been reported for a mutant of yeast cytochrome *c* in which Phe 82 has been replaced by tyrosine (Pielak *et al.*, 1988b).

Before any useful structural analysis can be made of chemical shifts, coupling constants, or NOE effects, resonances must be specifically assigned. This is not a trivial matter. Horse cytochrome *c* contains 660 protons that give rise to >500 resonances, most of which overlap. Thus even for a

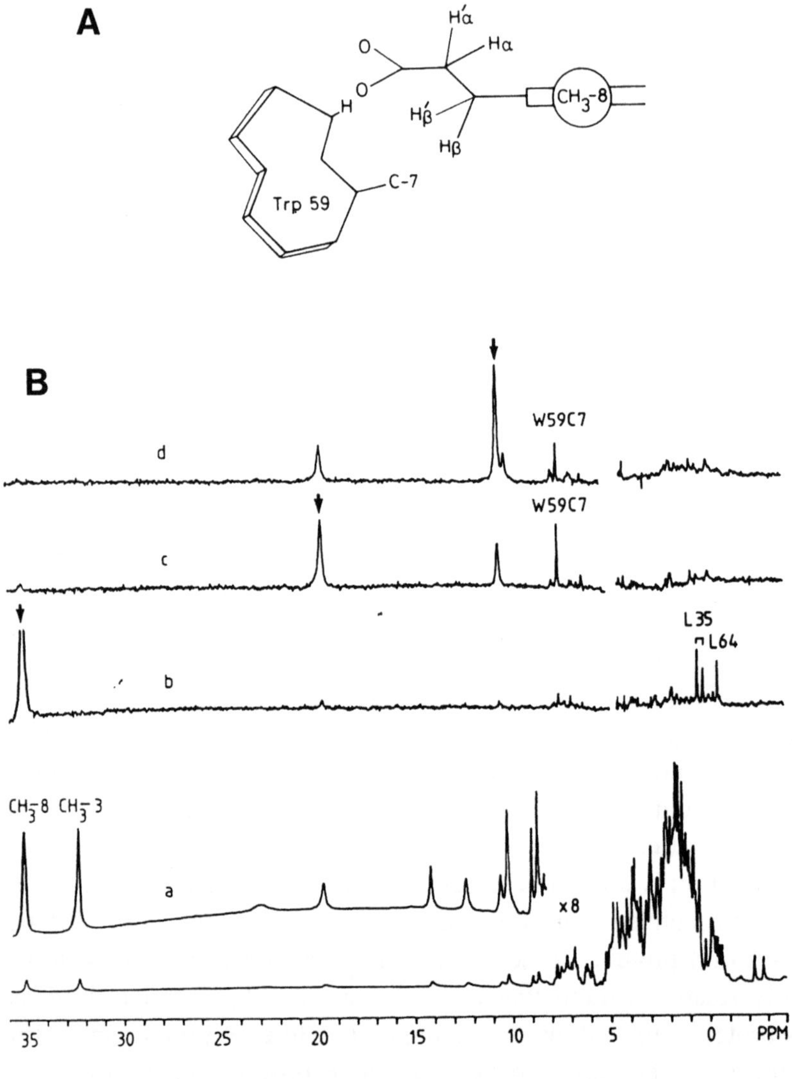

Fig. 2.32 The structural relationship between Trp 59 and the heme of cytochrome *c*. **A** Schematic representation of the relative orientations of Trp 59 and HP-7. The distances between Trp 59 C-7 and protons of the heme propionates are: Hα, 4.97Å; H'α 5.44Å; Hβ, 2.62Å; H'β, 3.69Å. **B** 300 MHz ^{1}H NMR spectra of 4 mM tuna ferricytochrome *c* in D_2O at 27°C. **a** The normal spectrum. **b** - **d** Nuclear Overhauser enhancement difference spectra resulting from the 0.5 s pre-irradiation of **b** CH_3-8, **c** and **d** HP-7 β CH resonances. The resonance labels are: L35, Leu 35; W59, Trp 59; L64, Leu 64. (Reproduced from Moore & Williams, 1984)

relatively small cytochrome, resolution of resonances is a problem (Fig. 2.31). An example of a well-resolved proton ^{1}H NMR spectrum is given in Fig. 2.36; the assignments of all the CH resonances in the aromatic region of the spectrum of *Ps. mendocina* ferrocytochrome *c*-551 are indicated. For horse cytochrome *c*, which is the cytochrome best characterised by NMR, the efforts of many laboratories have established the assignments of >90% of the CH proton resonances in both oxidation states (Kowalsky, 1965; McDonald & Phillips, 1973; Wüthrich, 1969; Redfield & Gupta, 1971; Cohen *et al.*, 1974; Robinson *et al.*, 1983a; Williams *et al.*, 1985a; Moore *et al.*, 1985a; Keller & Wüthrich, 1978;1981: Feng *et al.*, 1989; Wand *et al.*, 1989). The heme resonances were assigned by a procedure generally applicable to small diamagnetic ferrocytochromes *c* (Keller & Wüthrich, 1978; Senn *et al.*, 1980), but many of the amino acid resonances were not. Various sequence-dependent assignment procedures in protein NMR have been developed (Wüthrich, 1986; Englander & Wand, 1987) but these may not be generally applicable to small proteins. They certainly do not work for proteins in which the resonance linewidths are substantially greater than the coupling constants. This is generally the case when the M_r is >15 kD or where the heme iron is high-spin.

In summary, our view is that at its present state of development, although NMR can be used to give a complete structure it is most productively applied to characterise particular features of cytochromes. Examples of this approach are given throughout Chapter 2 and include: characterisation of pH dependent features (Fig. 2.7, 2.29, 2.30, 2.36 and 2.37); exchange processes (Fig. 2.8 and 2.37); local structural effects (Fig. 2.32); interprotein complexes (Fig. 2.34 and 2.35); and heme ligation and spin-state (Fig. 2.23, 2.24, 2.25, 2.28 and 2.30). We stress this point since accurate structures of proteins such as water-soluble cytochromes *c* are relatively easily obtained by x-ray techniques because of the ease with which they can be crystallised.

Oriented Membrane Samples and Crystalline Proteins. A considerable amount of structural information can be gained from samples containing symmetrically placed molecules, such as crystalline proteins or oriented membranes. The principle of this approach is that many spectroscopic transitions are anisotropic, for example, EPR g-values (section 2.2.3.2). This means that a molecule aligned along the main axis of the magnetic field (the z axis) may have a different EPR spectrum to one aligned at an angle to this axis. For frozen solutions there is a random distribution of molecules and only the principal values of the g-factor can be obtained. However, if an organised system is studied by EPR then it can be rotated in the magnetic field, and its EPR spectrum as a function of the rotation angle may give information about the alignment of the g-factor within the sample.

An example is frozen, oriented cytochrome oxidase (Blasie *et al.*, 1978). The sample is a multilayer membraneous material that, when oriented so that the plane of the multilayer is parallel to the z-axis of the magnet, gives a different EPR spectrum to that when it is oriented so that its plane is perpendicular (Fig. 2.33). The virtual disappearance of the signal at g = 3, which is the g_z signal of heme *a*, reveals the alignment of the g-factor within the sample. Since g_z is coincident, or almost so, with the Fe-axial ligand direction (and g_x, g_y are in the porphyrin plane: Palmer, 1985), the EPR data indicate that heme *a* is oriented, in the multilayer membrane, so that its porphyrin plane is perpendicular to that of the membrane. Analogous experiments have shown that both hemes *a* and a_3 of cytochrome oxidase, and hemes *b* and c_1 of ubiquinol cytochrome *c* reductase are similarly aligned in oriented mitochondrial multilayers (Blaisie *et al.*, 1978; Erecinska *et al.*, 1978), raising the possibility that such an arrangement is a common feature of membraneous cytochromes. The orientation of the primary donor in crystalline *Rps. palustris* photosynthetic reaction centres is a good example of this EPR approach applied to a single crystal (Gast *et al.*, 1983).

A related EPR approach exploits the relaxation properties of the heme signals. In the presence of additional paramagnets (either intrinsic, as in a multicentre protein, or extrinsic, such as added Dy(III) compounds) the linewidth of a heme EPR signal and its saturation properties (Scheme 2.4) may be modified. From the perturbation of these parameters information regarding the relative positions of the paramagnets can be obtained, as shown by a recent study of bovine ubiquinol cytochrome *c* oxidoreductase (Ohnishi *et al.*, 1989).

Optical spectroscopy with polarised light (section 2.2.5) is a complementary approach to the use of EPR for organised systems. The optical transitions described in Fig. 2.21 are polarised (Makinen & Churg, 1983): for example, the axial ligand → Fe(III) CT transitions are z-polarised, and the Soret band transition is x,y-polarised. Thus spectroscopy of organised systems with polarised light can reveal how the chromophores are arranged in the sample. For membraneous cytochrome oxidase, confirmation that the heme planes are perpendicular to the plane of the membrane was obtained by this method (Blaisie *et al.*, 1978). *Pseudomonas* cytochrome cd_1 is an example where the relative heme orientation was determined by polarised spectroscopy of single crystals (Makinen *et al.*, 1983); hemes *c* and d_1 were found to be oriented perpendicular to each other.

As yet there have been relatively few examples of this general approach applied to cytochromes, but as the efforts to characterise membrane proteins intensifies, we anticipate this approach will be used increasingly. Its application to the *Rps. viridis* reaction centre cytochrome is described in Chapter 5.4.

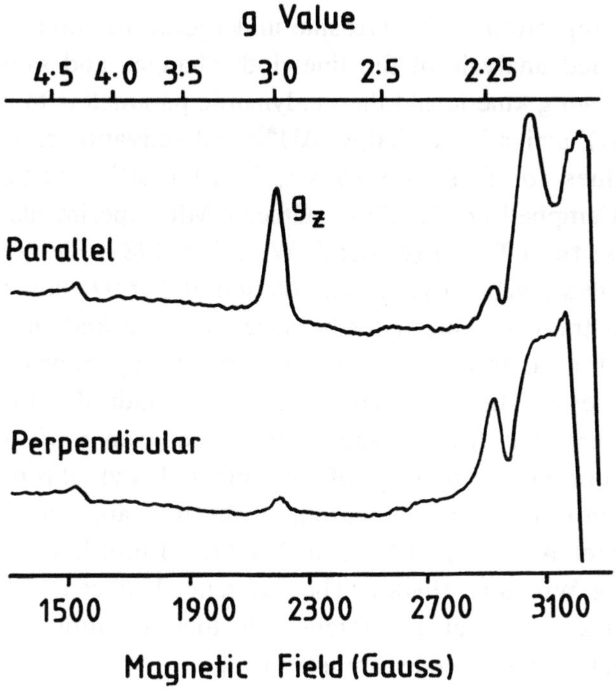

Fig. 2.33 EPR spectra of ferricytochrome *c* oxidase in frozen oriented membrane multilayers at 19.5 K. The spectra were obtained with samples where the plane of the multilayer is oriented parallel and perpendicular to the direction of the applied magnetic field. (Reproduced from Blasie *et al.*, 1978)

2.3.3.2 Dynamic Aspects of Protein Structure

As we describe more fully in Chapter 4.5, tertiary structures of proteins are not static but dynamic, with continual movements of main chain and side chain atoms. This 'breathing' of the structure can be investigated spectroscopically and we give two examples of particular dynamic processes occurring in mitochondrial cytochrome *c* that have been characterised by NMR methods. The significance of these motions for the physiological function and physicochemical properties of cytochrome *c* is not yet clear (Chapter 4.5.5).

Mobility of Aromatic Amino Acid Side Chains. With increasing temperature the aromatic region of the ^{1}H NMR spectrum of horse ferrocytochrome *c* undergoes a number of small changes (Campbell *et al.*, 1976; Moore & Williams, 1980a;c). The changes we are concerned with here arise from the four ring protons of a tyrosine undergoing rotation about its c_β - c_γ bond. A detailed analysis of the linewidth changes and saturation effects allows the following kinetic and thermodynamic parameters to be determined: rotation rate (k_r); activation enthalpy (ΔH^*); and activation entropy (ΔS^*). At 25°C, the values for these are ~8 s^{-1}, 97 kJ mol^{-1}, and 96 J K^{-1} mol^{-1} respectively (Campbell *et al.*, 1976). Other NMR experiments show that the resonances arise from Tyr 97 (Keller & Wüthrich, 1981; Eley *et al.*, 1982a).

The parameters given above for the rotation of Tyr 97 show that its motion is severely constrained. This arises because atoms packed close to it have to move in order for the group to rotate. The activation parameters relate to this 'breathing' process. The same tyrosine in mitochondrial cytochromes *c* and bacterial cytochrome c_2 has restricted motion (Cookson *et al.*, 1978) showing that it is a common property of the class I cytochrome *c* structures (Chapter 4). For tuna ferrocytochrome *c* the relevant kinetic and thermodynamic parameters are: k_r ~10 s^{-1}; ΔH^* is 112 kJ mol^{-1}; and ΔS^* is 95 J k^{-1} mol^{-1} (Moore & Williams, 1980c). The difference between ΔH^* for horse and tuna cytochromes *c* indicates differences in their breathing modes, but the structural aspects of these modes are unknown.

The rotations of Phe 10, Phe 46, Tyr 48 and Tyr 67 of horse cytochrome *c* are similarly restricted but Phe 36, Tyr 74 and Phe 82 have unrestricted rotation, with the rate for Phe 82 being $\geq 10^4$ s^{-1} (Moore *et al.*, 1982a; Williams *et al.*, 1985c).

Tryptophan side chains may also have some degree of motion, though because of their bulk this motion is more likely to be a slight wobble than a complete rotation about the c_β - c_γ bond. NMR studies show that Trp 33 of tuna cytochrome *c* is mobile but that Trp 59 of horse and tuna cytochromes *c* and tryptophans 56 and 77 of *Ps. aeruginosa* cytochrome *c*-551 are immobile (Moore *et al.*, 1977; Moore & Williams, 1980c).

Exchange of NH Protons for Deuterons. NH protons of a protein exchange with protons or deuterons in H_2O molecules. Thus if a fully protonated protein is dissolved in D_2O, then as time progresses, ^1H NMR signals arising from the NH protons diminish in intensity as the protons are replaced by deuterons. From this spectral change the rate of exchange, and hence the corresponding activation parameters, can be obtained. Those NH protons with the slowest rate of exchange are generally located in parts of the protein which are less dynamic than other parts. A more detailed account of the NH exchange process is given in Chapter 4.5.4.

Patel and Canuel (1976) first used NMR to monitor the NH exchange of horse ferrocytochrome *c* but the information they obtained was limited, partly because the NH resonances had not been assigned to specific groups in the sequence. The subsequent assignment of one of their resonances to Trp 59 (Moore *et al.*, 1980b) led to the conclusion that the exchange rate of the tryptophan indole NH was relatively slow, consistent with its internal, hydrogen-bonded location (Chapter 4). It is also markedly oxidation-state dependent: at pD 7 and 45°C the exchange half-life for the ferric protein is 20 min., while for the ferrous protein it is 480 min. This dependence is probably a consequence of the redox-state change in the conformation and dynamic properties of cytochrome *c* (Chapter 4.5).

Englander and his co-workers have recently assigned most of the peptide NH resonances of horse ferrocytochrome *c* (Feng *et al.*, 1989; Wand *et al.*, 1986; 1989) and described the NH exchange behaviour of the N-terminal helix (Wand *et al.*, 1986). The exchange rates ranged from 3.6×10^{-2} to 3.6×10^5 s^{-1} (at pD 7 and 20°C), with the only substantial oxidation-state changes being observed for Cys 14 and Ala 15. Their rates increased from 0.0038 and 0.038 hr^{-1} respectively, in the ferrous form, to 0.083 and 0.44 hr^{-1} in the ferric form. This trend is consistent with other studies of the dynamic properties of cytochrome *c* (Chapter 4.5). As with the aromatic group mobilities, a detailed structural interpretation of these NH exchange data is lacking.

Hydroxyl protons also undergo chemical exchange with the solvent and thus are rarely observed by NMR. However, Feng *et al.* (1989) report that the hydroxyl resonance of Thr 49 of horse ferricytochrome *c* is observed, and thus this has a slow hydrogen exchange rate. This is consistent with the x-ray structure which shows that this hydroxyl is hydrogen-bonded to heme propionate-6 (Fig. 4.4).

Roder *et al.* (1988) have built on the NH assignment and exchange rate work with an elegant procedure that uses proton exchange to investigate the folding pathways of cytochrome *c*. Their work indicates that α-helices form and interact together at an early stage of the folding process.

2.3.3.3 Interprotein Complexes

Most interprotein complexes involving cytochromes *c* are governed largely by electrostatic interactions which make crystallisation of the complex by the normal salting-out procedure difficult to attain. Thus spectroscopic methods have been employed, in conjunction with chemical modifications and computer-graphics (Pettigrew & Moore, 1987), to investigate such complexes. The information gained from the spectroscopic studies has led to the following: a definition of the interacting protein surfaces; identification of specific interacting groups; measurement of the distances between the complexed redox centres; and measurement of protein-protein association constants.

The change in absorbance of the Soret band as a result of complex formation has been widely used to determine associated thermodynamic parameters since Erman and Vitello (1980) introduced the method with a study of the interaction between cytochrome *c* and yeast cytochrome *c* peroxidase. The small absorbance differences are exceedingly difficult to measure accurately (for example, with the cytochrome c-b_5 complex at 10^{-5} M the change in absorbance at 416 nm is only ~0.04 [Mauk *et al.*, 1982]) and recently the CD and MCD of the Soret absorbance have been used instead (Michel *et al.*, 1989a,b). These produce larger spectral changes and are easier to measure accurately.

NMR has been used extensively to investigate the interaction surfaces of the complexed proteins. Gupta and Yonetani (1973) were the first to demonstrate that ^{1}H resonances of the heme of horse ferricytochrome *c* were shifted by complex formation (Fig. 2.34), in their case with yeast cytochrome *c* peroxidase (YCCP). Subsequently, it has been shown that complexation of cytochrome *c* with polyglutamate (Boswell *et al.*, 1980), cytochrome b_5 (Eley & Moore, 1983), and flavodoxin (Hazzard & Tollin, 1985) produces similar shifts, as does chemical modification of the lysines of cytochrome *c* (Boswell *et al.*, 1980; Falk *et al.*, 1981). The similarity of these spectral perturbations indicate that they result from changes in the charge distribution close to the heme (Boswell *et al.*, 1980) and thus they do not give any detailed information about the structures of the interprotein complexes. Therefore the structures of the YCCP-cytochrome *c* complex proposed by Gupta and Yonetani (1973) and Satterlee *et al.* (1987) on the basis of their NMR experiments are not, in fact, supported by the NMR data. Currently the major use of these NMR perturbations is to confirm complex formation has taken place and to determine apparent association constants (Eley & Moore, 1983; Thomas *et al.*, 1987).

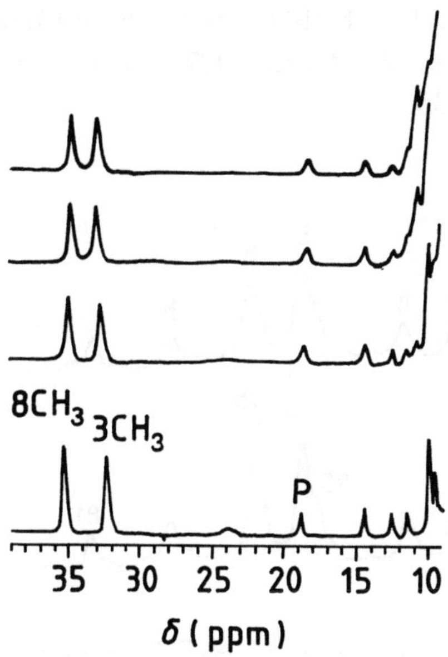

Fig. 2.34 Effect of yeast cytochrome *c* peroxidase on heme NMR signals of horse cytochrome *c* at 23°C and pD 6.6. The YCCP:cytochrome *c* molar ratios are (from top to bottom): 1.4, 1.0, 0.5, 0.0. The resonances of heme methyls-3 and -8, and heme propionate-7 (P) are both shifted and broadened by complex formation. (Reproduced from Satterlee *et al.*, 1987, after Gupta & Yonetani, 1973)

A variety of non-physiological complexes of cytochromes *c* have been studied by NMR including those between: mitochondrial cytochrome *c* and flavodoxins (Hazzard & Tollin, 1985; Tollin *et al.*, 1987), mitochondrial cytochrome *c* and plastocyanin (King *et al.*, 1985), and cytochromes c_3 and ferredoxins (Moura *et al.*, 1977; Guerlesquin *et al.*, 1985). In all cases the ferricytochrome heme resonances were affected by complex formation.

Studies of amino acid NMR peaks have been more productive in defining interprotein complexes. A major difficulty with this approach is that since the overall size and shape of the protein complexes are so different from those of the uncomplexed proteins, the peaks of an NMR spectrum of the complexes are considerably broader than those of the sum of the spectra of the individual proteins. A number of approaches have been taken to overcome this problem, and the most generally useful has been to raise the ionic strength of the solution so that the complex is formed only by a small fraction of the molecules present. The composite NMR spectrum is then formed by the weighted average of the spectra for the free and bound proteins in fast exchange (Fig. 2.6). An example of this approach applied to the cytochrome *c*-b_5 interaction (Pettigrew & Moore, 1987: Chapter 2) and using ^{13}C NMR spectra, is given in Fig. 2.35.

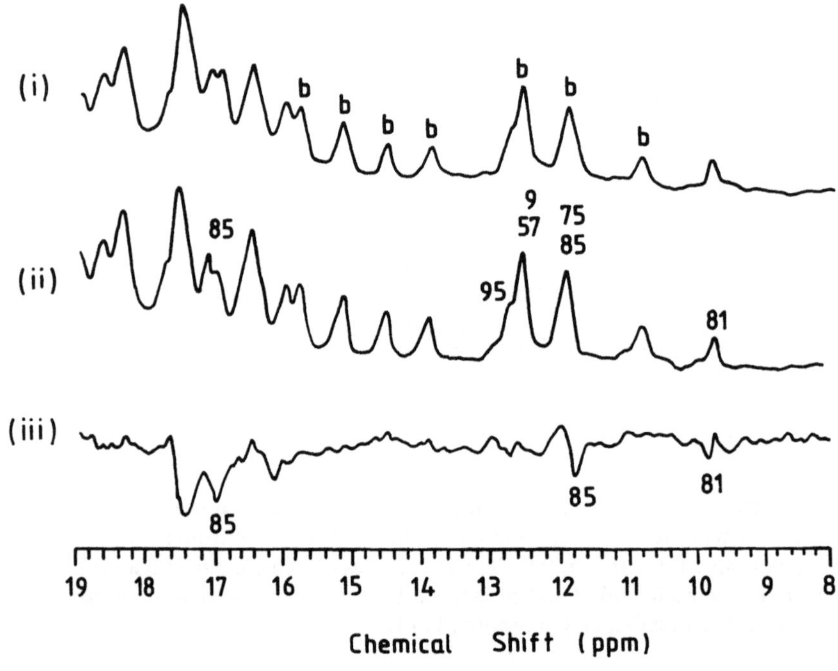

Fig. 2.35 Part of the methyl carbon region of the ^{13}C NMR spectra of ferri-cytochromes *c* and b_5 at pD 7.5 (50 mM phosphate) and 25°C. (i) Spectrum of a mixture of cytochromes *c* and b_5. **(ii)** Sum of the spectra of cytochromes *c* and b_5. **(iii)** The difference spectrum [(ii) - (i)]. The resonance labels are: **b**, cytochrome b_5; **9**, **57**, **75**, **81**, **85** and **95**, resonances of isoleucines 9, 57, 75, 81, 85 and 95, respectively, of cytochrome *c*. (Reproduced from Burch *et al.*, 1988)

Complex formation at a relatively high ionic strength is accompanied by small shifts of some resonances but most are unaffected; and the linewidths of the peaks are not appreciably changed (Fig. 2.35). Thus the difference spectrum between those of the mixture of proteins and the sum of the individual spectra reveal just those resonances specifically affected by the complex formation. This spectrum supports a model for the complex in which Ile 81 and Ile 85 of cytochrome *c* are close to, or in, the protein interface region and isoleucines 9, 57, 75 and 95 are not directly affected by the complexation (Burch *et al.*, 1988, 1990). This pattern is consistent with the Salemme (1976) computer graphics model.

The general procedure employed in Fig. 2.35 can be applied to both ^{1}H and ^{13}C NMR. ^{13}C NMR has been advocated for the study of protein complexes because of its increased sensitivity to minor structural perturbations, and because it has a greater chemical shift dispersion compared to ^{1}H NMR (200 ppm compared to 15 ppm) for amino acid resonances not bound to paramagnetic iron (Burch *et al.*, 1988). Its major drawback is its lack of sensitivity; a typical 1-D ^{1}H spectrum of horse ferricytochrome *c* can be obtained with 10 mg of sample in 10 mins of spectrometer time whilst the natural abundance ^{13}C spectrum of ferricytochrome *c* used to construct Fig. 2.35(ii) required 150 mg of sample and 14 hrs of spectrometer time.

An alternative approach to improving the resolution of ^{1}H NMR spectra of protein complexes has been described by Eley & Moore (1983). In their experiment the competition between one protein and small paramagnetic reagents for a second protein is monitored. The paramagnetic reagents considerably perturb NMR spectra and, provided the interactions between the small reagents and proteins are known, this experiment can identify the protein - protein interaction surfaces. To date, this approach has been applied only to the cytochrome c-b_5 complex, where it provides considerable support for Salemme's model, as far as the cytochrome *c* surface is concerned, but disagrees with some aspects of the proposed cytochrome b_5 surface (Eley & Moore, 1983; Hartshorn *et al.*, 1987).

NMR has given direct information concerning some of the lysines of mitochondrial cytochrome *c* involved in interprotein complexes. Trimethyl-lysine residues have strong ^{1}H signals from the nine equivalent methyl protons, especially when the side chains have motion independent from the protein tumbling (Eley *et al.*, 1982b). When complex formation involves a salt-bridge link to a trimethyl-lysine, its methyl ^{1}H resonance shifts and/or broadens (Eley & Moore, 1983). At least two of the three trimethyl-lysines of *Hansenula anomala* cytochrome *c* (residues 55, 72 and 73) were shown to be involved in the binding of flavocytochrome b_2 by this method (Thomas *et al.*, 1987).

Although methylated lysines can be readily monitored by ^1H NMR, non-methylated lysines usually cannot. However, chemically modified lysines can be monitored. Thus some of the ^1H NMR resonances of the additional methyl groups of N-acetimidylated cytochrome *c* are perturbed by complex formation with cytochrome b_2 (Burch *et al.*, 1990). Since the additional methyls are attached directly to the lysines this approach may lead to the identification of the key lysines involved in interprotein salt-bridge formation.

In principle, FT-IR should allow the formation of lysine---carboxylate ion pairs to be monitored via changes in the carbonyl stretching frequency. Again, the cytochrome *c*-b_5 interaction provides an example. Holloway and Mantsch (1988) observed that small spectral differences at ~1650 cm^{-1} and 1575 cm^{-1} accompanied complex formation. They ascribed these to changes in the secondary structure of cytochrome *c* and the formation of three lysine---glutamate salt-bridges respectively. Neither of these conclusions are firmly based. A reduction in the α-helix content of cytochrome *c* is inconsistent with the NMR studies, and the proposal of three intermolecular salt-bridges involving glutamates assumes that the extinction coefficients and frequencies for all the glutamates are identical. This may not be the case. The observed difference at ~1650 cm may just be a consequence of imperfect subtraction of the spectra of the individual proteins from that of the complex (see Fig. 2.38).

The interheme separation of protein-protein complexes can be obtained from luminescence spectroscopy by a method that exploits the quenching properties of iron-porphyrins (section 2.2.5.3). In these experiments one of the binding proteins is native and the other is a luminescent derivative, such as that obtained by replacement of the Fe with Zn (Table 1.6). The quenching of the luminescence of this derivative by complexation reflects the distance between the two porphyrins. Obtaining a value for the distance requires various assumptions to be made (Cantor & Schimmel, 1980) but it has been estimated that the uncertainty in the calculated distance is only ~20% (McLendon *et al.*, 1985). This approach has been applied to complexes of cytochrome *c* with YCCP (Leonard & Yonetani, 1974), cytochromes b_5 and b_2 (McLendon *et al.*, 1985; Vanderkooi *et al.*, 1980) and cytochrome *c* oxidase (Vanderkooi *et al.*, 1977). The interheme separations obtained for the first two kinds of complex agree remarkably well with those proposed by Salemme (1976) and Poulos & Kraut (1980) for their hypothetical structures of the complexes (Pettigrew & Moore, 1987: Fig. 2.20).

2.3.3.4 *Group Ionisation Constants*

Generally spectroscopic methods only give information about the ionisation states of particular groups if the ionisations perturb the spectrum and can be monitored directly. This is because difference spectroscopy, in which the spectrum of the ionised form is subtracted from that of the unionised form, is often needed to resolve the signals of interest, and because absolute absorption frequencies are not always indicative of charge states. Exceptional cases are metal ligands, such as histidine or histidinate (section 2.3.2).

The main ionisable groups in proteins are carboxylates, amines, histidines and tyrosines. Tyrosines generally have high pK_a values that can be monitored by optical spectroscopy (Stellwagen, 1964) or ^{13}C NMR (Boswell *et al.*, 1983). Lysine pK_a values can in principle be determined by NMR but this is not generally feasible because of the complex appearance of lysine signals, their occurrence in crowded spectral regions, and the relatively large number of lysines present in most cytochromes. N-terminal amines can be monitored if their ionisation affects proton NMR signals of the N-terminal amino acid. The pK_a of the N-terminus of *Chl. limicola* cytochrome *c*-555 was determined to be 7.7 by this method (G.R.Moore & G.W.Pettigrew, unpublished data).

NMR is particularly useful for determining the pK values of histidine residues and we illustrate this below for two bacterial cytochromes *c*. Carboxylic acid pK values are difficult to measure by 1D-NMR because, with one exception, the NMR signals of groups carrying carboxylic acids are not well-resolved from other resonances. The exception is the heme propionic acid substituents of paramagnetic hemes, as illustrated below for several cytochromes. IR provides an alternative approach to the study of carboxylic acid ionisations, but with this technique the major problem is one of assignment to specific groups. This approach is illustrated below for the ionisation of a heme propionic acid substituent of *Pseudomonas* ferricytochrome *c*-551 and for the alkaline isomerisation of horse ferricytochrome *c*.

NMR and Non-Ligand Histidines. The ring proton CH NMR signals of histidines are generally readily observed unless histidine is bound to a paramagnetic metal ion. The CH between the two ring nitrogens, the C-2 proton, usually resonates at a higher frequency than does the C-4 proton (Fig. 2.36A. As the pH changes so that the protonation state of the histidine changes, the resonances shift markedly, allowing the pK to be determined

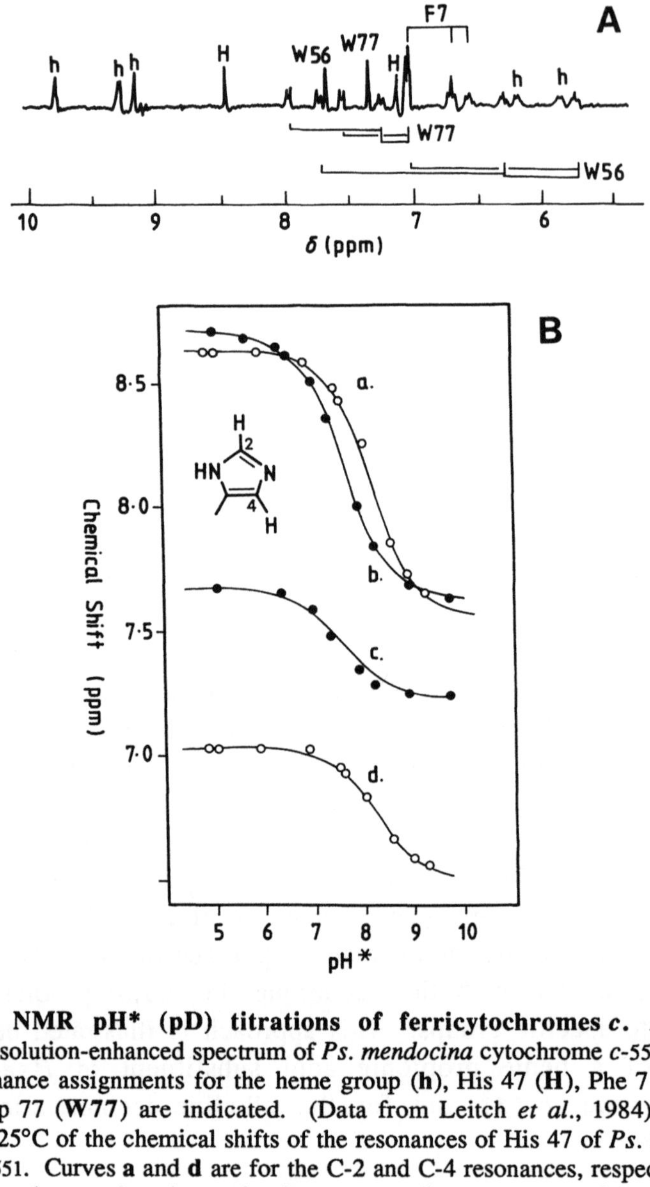

Fig. 2.36 ^{1}H NMR pH* (pD) titrations of ferricytochromes *c*. A Aromatic region of the resolution-enhanced spectrum of *Ps. mendocina* cytochrome *c*-551 at 25°C and pD 7.2. Resonance assignments for the heme group (**h**), His 47 (**H**), Phe 7 (**F7**), Trp 56 (**W56**) and Trp 77 (**W77**) are indicated. (Data from Leitch *et al.*, 1984). **B** The pD dependence at 25°C of the chemical shifts of the resonances of His 47 of *Ps. stutzeri* (221) cytochrome *c*-551. Curves **a** and **d** are for the C-2 and C-4 resonances, respectively, of the ferrous protein, and curves **b** and **c** are for the corresponding resonances of the ferric protein. The solid lines are theoretical curves for one proton ionisations with pK_a values of 8.2 (curves **a** and **d**) and 7.6 (curves **b** and **c**). (Reproduced from Leitch *et al.*, 1984)

(Fig. 2.36B). Many cytochrome histidine pK values have been determined in this way (Table 4.3).

An alternative NMR approach is to employ ^{14}N (or ^{13}C) NMR to monitor the histidine ionisation. This is usually unnecessary for most proteins but Yu & Smith (1988) found it necessary for His 42 of *R. rubrum* ferrocytochrome c_2. This was because Smith (1979) had found that the C-2 proton resonance of His 42 was not observable over all of the relevant pH range. Broadening of histidine resonances does occur sometimes at pH values close to the pK_a because of exchange processes (Fig. 2.6) (e.g. in yeast cytochrome *c*: Robinson *et al.*, 1983a), but it is not clear that this was the case in the *R. rubrum* study.

Although the quality of protein NMR spectra decreases markedly with increasing protein molecular weight, histidine resonances often remain sharp and, even with relatively large proteins such as hemoglobin (M_r 64,000), histidine pK_a values can be determined (Ho & Russu, 1987).

NMR and Heme Propionates. The ^{1}H NMR spectrum of *Chl. limicola* ferricytochrome *c*-555 contains the usual paramagnetically shifted resonances of the heme group and axial ligands (Fig. 2.37), and with a change in pH many of these resonances shift (e.g. P1, M1 and M4). Some also broaden at pH values close to the pK_a of 6.3 (e.g. M1). These pH induced shifts are relatively small (compare Fig. 2.37 with Fig. 2.7) and both their size and pattern - methionine ligand resonances largely unaffected, heme propionic acid and methyl resonances affected with both high-frequency and low-frequency shifts observed - have been found for other class I cytochromes *c* (Moore *et al.*, 1980c; 1984a; Leitch *et al.*, 1984). Similar behaviour has also been found for class II and class III cytochromes *c* (Chapter 5.2.3 & 5.3.2).

The chemical shift variation (Fig. 2.37) gives the pK_a, and the linewidth change provides a measure of the proton on-off rate (k_e). At the pK_a the excess linewidth, $\Delta\nu^{e}_{1/2}$, resulting from the exchange process (Wüthrich, 1976) is given by:

$$\Delta\nu^{e}_{1/2} = \frac{2\pi\,(\delta_A - \delta_B)^2}{k_e} \qquad (2.12)$$

where δ_A and δ_B are the frequencies in Hertz of the fully ionised and unionised species.

For *Chl. limicola* ferricytochrome *c*-555 k_e is ~10^{-5} sec^{-1}, indicating that the ionisation is not accompanied by a substantial conformational change, which would be expected to have a much slower rate. This conclusion is also supported by the main protein NMR spectrum which is little affected by the ionisation (Moore *et al.*, 1986b).

Studies of *Ps. aeruginosa* ferricytochrome *c*-551, which shows similar NMR behaviour to Fig. 2.37, have failed to detect the ionisation by visible spectroscopy (Moore *et al.*, 1980c), EPR (Chao *et al.*, 1979) and NIR-MCD (J.Peterson & A.J.Thomson, personal communication).

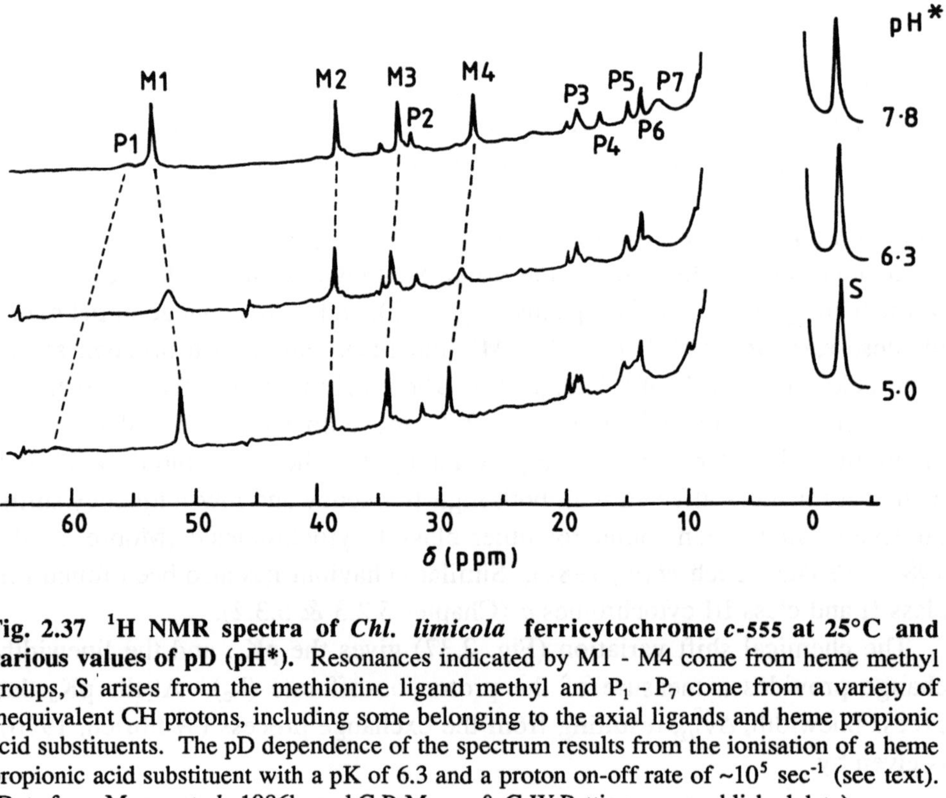

Fig. 2.37 ^{1}H NMR spectra of *Chl. limicola* ferricytochrome *c*-555 at 25°C and various values of pD (pH*). Resonances indicated by M1 - M4 come from heme methyl groups, S arises from the methionine ligand methyl and P_1 - P_7 come from a variety of inequivalent CH protons, including some belonging to the axial ligands and heme propionic acid substituents. The pD dependence of the spectrum results from the ionisation of a heme propionic acid substituent with a pK of 6.3 and a proton on-off rate of ~10^5 sec^{-1} (see text). (Data from Moore *et al.*, 1986b; and G.R.Moore & G.W.Pettigrew, unpublished data)

The situation is more difficult with class I ferrocytochromes *c* because these are diamagnetic. Nevertheless, it has been possible in some cases to follow heme propionic acid ionisations by NMR. Usually the resonances that enable the ionisation to be monitored are those from side chains that hydrogen-bond to the propionic acid: tryptophan in the case of *Rm. vannielii* ferrocytochrome c_2; and histidine in some Pseudomonad ferrocytochromes *c*-551 and *Chl. limicola* cytochrome *c*-555. The resolved heme resonances are usually insensitive to the ionisation. However, with *Chl. limicola* cytochrome *c*-555 the resonance of the heme meso-δ proton, which is the proton between the two propionic acid substituents (Fig. 1.14), shifts as the ionisation occurs.

Ionisation of the heme propionic acids of some ferrocytochromes may be detected by optical spectrophotometry because the α-bands of some proteins, such as that of *Ps. aeruginosa* cytochrome *c*-551 (Fig. 7.14), are shifted to the red by the ionisation However, such a shift has not been observed for all ferrocytochromes with an ionising heme propionic acid, although where it has been observed it has been shown to correlate with the propionic acid ionisation.

The significance of the charge states of the heme propionic acid substituents of class I cytochromes *c* with respect to redox state conformational changes and redox -potentials is described in Chapters 4.3.2 and 7.6 respectively.

FT-IR and Carboxylic Acids. Infra-red spectroscopy is widely used to study small carboxylic acids since the carbonyl stretching frequency is markedly dependent upon whether the acid is ionised or not (Timasheff *et al.*, 1973; Parker, 1971). There is a major problem with proteins however; namely, the peptide carbonyl absorbance is so intense that it masks the carboxylic acid carbonyl absorbance. Therefore pH difference spectra are required to monitor the ionisation (Timasheff *et al.*, 1973).

Fig. 2.38 shows the pH difference FT-IR spectra of *Ps. aeruginosa* ferricytochrome *c*-551 over the pH range 5.5 to 7.8. The feature at ~1570 cm^{-1} titrates with a pK of 6.15 and has been assigned to the ionised HP-7 (Table 4.2) (Tonge *et al.*, 1989). A reciprocal pH-dependent feature representing the loss of the absorbance of the unionised form is not obvious but a pH dependent feature at ~1660 cm^{-1} is seen. This latter signal is coincident with an amide absorbance band and it has been proposed to arise largely from this; its appearance in the difference spectrum results from imperfect cancellation. The lack of an identified signal for the unionised acid may be because it overlaps with the amide signal. The usual frequency for carboxylic acid absorbance is ~1710 cm^{-1} but when the acid is involved in hydrogen-bonding the signal is often shifted to a lower frequency (Koeppe & Stroud, 1976; Tonge *et al.*, 1989). The signal of the ionised carboxylate is less sensitive to hydrogen-bonding effects and thus appears in Fig. 2.38 at about the normal frequency.

This FT-IR approach is generally applicable to proteins, and it has been applied to mitochondrial ferricytochrome *c* at high pH. Tonge *et al.* (1989) observed an absorbance band at ~1570 cm^{-1}, that behaved in the same way as the ~1570 cm^{-1} band of ferricytochrome *c*-551 (Fig. 2.38), for native and fully trifluoroacetylated ferricytochrome *c*. The pK values governing the appearance of this band, ~9.4 and ~10.4 respectively, are coincident with the pK values of the alkaline transitions of these two proteins (Table 4.4). Thus Tonge *et al.* (1989) concluded that a carboxylic acid group was affected by the transition. However, as with the bacterial protein (Fig. 2.38), no feature identifiable as arising from the unionised carboxylic acid was observed. This weakens the proposal of Tonge *et al.* (1989) that they had monitored the ionisation of a carboxylic acid which triggered the alkaline isomerisation. This topic is discussed further in Chapter 4.4.

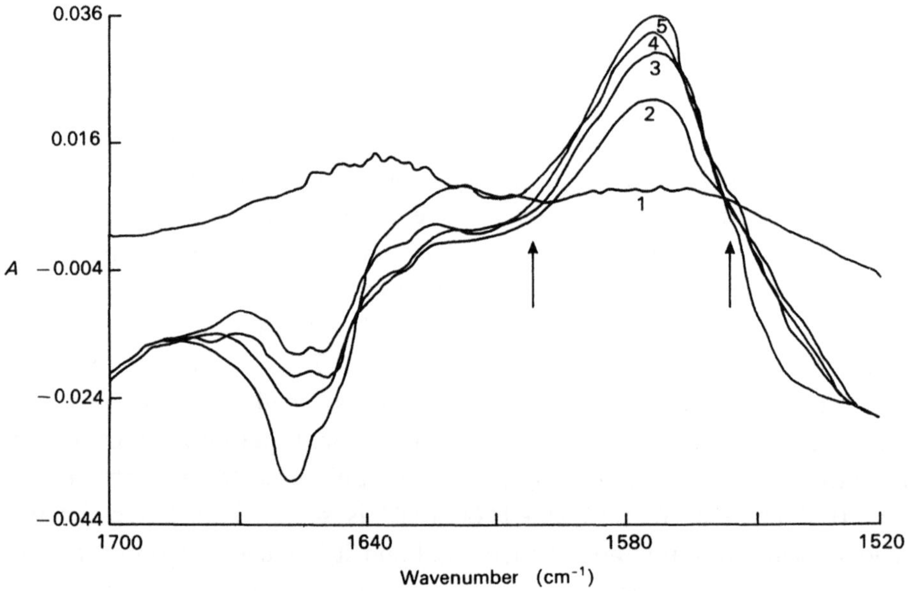

Fig. 2.38 IR difference spectra of *Ps. aeruginosa* ferricytochrome *c*-551. The pD value of the reference spectrum was 5. The reference spectrum was subtracted from sample spectra at pD values of 5.5 (**1**), 6.15 (**2**), 6.6 (**3**), 7.0 (**4**) and 7.8 (**5**) to generate the difference spectra. A graph of the integrated intensity between the arrows against pD yields a pK of 6.14 ± 0.11. (Reproduced from Tonge *et al.*, 1989)

Moss *et al.* (1990) have provided support for the interpretation of Tonge *et al.* (1989). Moss and his co-workers monitored the electrochemical reduction and oxidation of horse cytochrome *c* at pH 7 in a thin-layer cell by FT-IR, and found that the redox-state difference spectrum contained a number of bands resulting from changes in vibrational frequencies accompanying the redox-state conformational change. They assigned a band at 1693 cm^{-1} to an unionised and hydrogen-bonded carboxylic acid, which could be the counterpart of the 1570 cm^{-1} band observed by Tonge *et al.* (1989). It should be noted that unlike the pH difference spectra of Tonge *et al.*, the redox difference spectra of Moss *et al.* did not contain major bands due to amide carbonyls, and thus resolution of the carboxylic acid band at 1693 cm^{-1} in their spectra was possible.

Although the examples used in this section are heme propionic acid groups and not amino acid carboxylic acids, the FT-IR technique is applicable to both cases. An example of its use to monitor amino acid carboxylates of cytochromes is given in section 2.3.3.3.

Chapter 3 Amino Acid Sequences of Cytochromes *c*

3.1 Introduction

This chapter presents alignments for *c*-type cytochrome sequences, and matrices defining their degree of similarity. The sequences are recorded in the single letter code (Table 3.1) with a single space between each residue except where gaps are required for optimal alignment. Where possible gaps or insertions have been placed on the basis of crystallographic structures. Otherwise the use of gaps has been kept to a minimum since it is not statistically justified to insert gaps for a small increase in identities (Needleman and Wunsch, 1970). Regions of the sequence for which information is tentative are shown in parenthesis.

Table 3.1 The one letter code for the amino acids

A	Alanine	I	Isoleucine	S	Serine
B	Aspartic acid or asparagine	K	Lysine	T	Threonine
C	Cysteine	L	Leucine	V	Valine
D	Aspartic acid	M	Methionine	W	Tryptophan
E	Glutamic acid	N	Asparagine	X	Non-standard
F	Phenylalanine	P	Proline	Y	Tyrosine
G	Glycine	Q	Glutamine	Z	Glutamic acid or glutamine
H	Histidine	R	Arginine		

The heme attachment site (which almost always has the form Cys-?-?-Cys-His) is boxed, together with the fifth and sixth coordination ligands of the iron. In several cases these ligands are known from crystallography, while in other cases they are identified on the basis of sequence homology with other cytochrome groups. Pyrrolidone carboxylic acid is found at the N-terminus of some cytochromes *c* and is denoted by <Q. It is presumed to arise from the cyclisation of glutamine. An acetylated N-terminus is denoted by @.

In comparing sequences to obtain a matrix of % identities, the total number of comparisons was obtained by summing:

(a) the number of core residues common to the whole set of sequences,
(b) the number of additional residues common to the pair under consideration,
(c) the number of gaps shared by a pair of sequences,
(d) the number of gaps where one sequence differs from the other (gaps were scored as one point of comparison no matter the size).

The number of identities in a pairwise comparison was obtained by summing:

(a) the number of identical residues,
(b) the number of identical gap regions.

3.2 Amino Acid Sequences of Mitochondrial Cytochromes *c*

Cytochrome *c* amino acid sequences are shown in Fig. 3.1 and the variability of mitochondrial cytochrome *c* is shown in Fig 3.2. The organisms are: **(1)** human, **(2)** chimp, **(3)** rhesus monkey, **(4)** spider monkey, **(5)** horse, **(6)** donkey, **(7)** zebra, **(8)** cow, **(9)** pig, **(10)** sheep, **(11)** camel, **(12)** guanaco, **(13)** great whale, **(14)** elephant seal, **(15)** dog, **(16)** hippopotamus, **(17)** bat, **(18)** rabbit, **(19)** guinea pig, **(20)** rat, **(21)** mouse, **(22)** mouse testis, **(23)** kangaroo, **(24)** chicken, **(25)** turkey, **(26)** emu, **(27)** ostrich, **(28)** king penguin, **(29)** pekin duck, **(30)** pigeon, **(31)** snapping turtle, **(32)** rattlesnake, **(33)** bullfrog, **(34)** tuna, **(35)** bonito, **(36)** carp, **(37)** dolphin fish, **(38)** dogfish, **(39)** pacific lamprey, **(40)** starfish, **(41)** garden snail, **(42)** freshwater prawn, **(43)** common brandling worm, **(44)** fruit fly, **(45)** fruit fly, **(46)** fruit fly, **(47)** mediterranean fruit fly, **(48)** screw-worm fly, **(49)** blowfly, **(50)** adult housefly, **(51)** larval housefly, **(52)** silkworm moth, **(53)** tobacco horn worm moth, **(54)** locust, **(55)** honeybee, **(56)** *S. cerevisiae* iso-1, **(57)** *S.cerevisiae* iso-2, **(58)** *Debaromyces kloeckeri*, **(59)** *Candida krusei*, **(60)** *Hansenula anomala*, **(61)** *Schizosaccharomyces pombe*, **(62)** *Neurospora crassa*, **(63)** *Humicola lanuginosa*, **(64)** *Ustilago sphaerogena*, **(65)** *Crithidia oncopelti*, **(66)** *Crithidia fasciculata*, **(67)** *Euglena gracilis*, **(68)** *Enteromorpha intestinalis*, **(69)** *Tetrahymena pyriformis*, **(70)** love-in-a-mist, **(71)** mung bean, **(72)** cauliflower, **(73)** rape, **(74)** pumpkin, **(75)** hemp, **(76)** European elder, **(77)** Indian mallow seed, **(78)** cotton seed, **(79)** castor bean, **(80)** tomato,

(**81**) potato, (**82**) rice, (**83**) maize, (**84**) cuckoo-pint, (**85**) sesame seed, (**86**) leek, (**87**) box elder, (**88**) niger, (**89**) sunflower seed, (**90**) nasturtium, (**91**) parsnip, (**92**) wheat germ, (**93**) buckwheat seed, (**94**) spinach, (**95**) ginkgo, (**96**) *Chalmydomonas reinhardtii.*

Amino acid sequence (**1**) is from Matsubara and Smith (1963); (**2**) is from Needleman and Margoliash (unpublished results cited in Dickerson and Timkovich, 1975); (**3**) is from Rothfus and Smith (1965); (**4**) is from Borden, Tarr, Vensel, Dethmers, Kottke, Stanford and Margoliash (unpublished results cited in Osheroff *et al.*, 1980); (**5**) is from Margoliash *et al.* (1961);(**6**) is from Walasek and Margoliash (unpublished results cited in Dickerson and Timkovich, 1975); (**7**) is from Gürtler and Horstmann (1971); (**8**) is from Nakashima *et al.* (1966); (**9**) is from Stewart and Margoliash (1965); (**10**) is from Smith and Margoliash (1964); (**11**) is from Sokolovsky and Moldovan (1972); (**12**) is from Niece *et al.* (1977); (**13**) is from Goldstone and Smith (1966); (**14**) is from Augusteyn *et al.* (1972); (**15**) is from McDowall and Smith (1965); (**16**) is from Thompson *et al.* (1978); (**17**) is from Strydom *et al.* (1972); (**18**) is from Needleman and Margoliash (1966); (**19**), (**20**) and (**21**) are from Carlson *et al.* (1977); (**22**) is from Hennig (1975); (**23**) is from Nolan and Margoliash (1966); (**24**) is from Chan and Margoliash (1966b); (**25**), (**28**), (**29**), (**30**) and (**48**) are from Chan, Tulloss and Margoliash (unpublished results cited in Dickerson and Timkovich, 1975); (**26**) is from Augusteyn (1973); (**27**) is from Howard *et al.* (1974); (**31**) is from Chan *et al.* (1966); (**32**) is from Bahl and Smith (1965); (**33**) is from Chan, Walasek, Barlow and Margoliash (unpublished results cited in Dickerson and Timkovich, 1975); (**34**) is from Kreil (1963, 1965); (**35**) is from Nakayama *et al.* (1971); (**36**) is from Gürtler and Horstmann (1970); (**37**) is from R.P.Ambler and J.V.Bannister (personal communication); (**38**) is from Goldstone and Smith (1967); (**39**) is from Nolan *et al.* (1973); (**40**) is from Lyddiatt and Boulter (1976a); (**41**) is from Brown *et al.* (1972); (**42**) is from Lyddiatt and Boulter (1976b); (**43**) is from Lyddiatt and Boulter (1976c); (**44**) is from Nolan, Weiss, Adams and Margoliash (unpublished results cited in Dickerson and Timkovich, 1975); (**45**) is from Limbach and Wu (1985) and Swanson *et al.* (1985); (**46**) is from Limbach and Wu (1985); (**47**) is from Fernandez-Sousa *et al.* (1975); (**49**) is from Shaw *et al.* (1978); (**50**) and (**51**) are from Yamanaka *et al.* (1980); (**52**) is from Chan and Margoliash (1966a);(**53**) is from Chan (1970); (**54**) is from Lyddiatt and Boulter (1977); (**55**) is from Inoue *et al.* (1985); (**56**) is from Narita and Titani (1969); Yaoi (1967) and Lederer *et al.* (1972); (**57**) is from Montgomery *et al.* (1980);

```
                                    1                 10
(1,2)                              @G D V E K G K K I F I M K C S Q C H
(3)                                @G D V E K G K K I F I M K C S Q C H
(4)                                @G D V E K G K R I F I M K C S Q C H
(5)                                @G D V E K G K K I F V Q K C A Q C H
(6,7)                              @G D V E K G K K I F V Q K C A Q C H
(8,9,10)                           @G D V E K G K K I F V Q K C A Q C H
(11,12,13)                         @G D V E K G K K I F V Q K C A Q C H
(14)                               @G D V E K G K K I F V Q K C A Q C H
(15)                               @G D V E K G K K I F V Q K C A Q C H
(16)                               @G D V E K G K K I F V Q K C A Q C H
(17)                               @G D V E K G K K I F V Q K C A Q C H
(18)                               @G D V E K G K K I F V Q K C A Q C H
(19,20,21)                         @G D V E K G K K I F V Q K C A Q C H
(22)                               @G D A E A G K K I F V Q K C A Q C H
(23)                               @G D V E K G K K I F V Q K C A Q C H
(24,25)                            @G D I E K G K K I F V Q K C S Q C H
(26)                               @G D I E K G K K I F V Q K C S Q C H
(27)                               @G D I E K G K K I F V Q K C S Q C H
(28)                               @G D I E K G K K I F V Q K C S Q C H
(29)                               @G D V E K G K K I F V Q K C S Q C H
(30)                               @G D I E K G K K I F V Q K C S Q C H
(31)                               @G D V E K G K K I F V Q K C A Q C H
(32)                               @G D V E K G K K I F T M K C S Q C H
(33)                               @G D V E K G K K I F V Q K C A Q C H
(34)                               @G D V A K G K K T F V Q K C A Q C H
(35)                               @G D V A K G K K T F V Q K C A Q C H
(36)                               @G D V E K G K K V F V Q K C A Q C H
(37)                               @G D V A K G K K V F V Q K C A Q C H
(38)                               @G D V E K G K K V F V Q K C A Q C H
(39)                               @G D V E K G K K V F V Q K C S Q C H
(40)                                G Q V E K G K K I F V Q R C A Q C H
(41)                                G Z A Z K G K K I F T Q K C L Q C H
(42)                               *G D V E K G K K I F V Q R C A Q C H
(43)                      G G I P A G D V E K G K T I F K Q R C A Q C H
(44)                        G V P A G D V E K G K K L F V Q R C A Q C H
(45)                        G V P A G D V E K G K K L F V Q R C A Q C H
(46)                            G S G D A E N G K K I F V Q K C A Q C H
(47)                        G V P A G D V E K G K K L F V Q R C A Q C H
(48)                        G V P A G D V E K G K K I F V Q R C A Q C H
(49)                        G V P A G D V E K G K K I F V Q R C A Q C H
(50)                        G V P A G D V E K G K K I F V Q R C A Q C H
(51)                        G V P A G D V E K G K K L F V Q R C A Q C H
(52)                        G V P A G N A E N G K K I F V Q R C A Q C H
(53)                        G V P A G N A D N G K K I F V Q R C A Q C H
(54)                        G V P Q G D V E K G K K I F V Q R C A Q C H
(55)                        G I P A G D P E K G K K I F V Q K C A Q C H
(56)                      T E F K A G S A K K G A T L F K T R C L Q C H
(57)                      T G F K P G S A K K G A T L F K T R C Q Q C H
(58)                    P A P Y E K G S E K K G A N L F K T R C L Q C H
(59)                    P A P F E Q G S A K K G A T L F K T R C A Q C H
(60)                    P A P F K K G S E K K G A T L F K T R C L Q C H
(61)                        P Y A P G D E K K G A S L F K T R C A Q C H
(62)                        G F S A G D S K K G A N L F K T R C A Q C H
(63)                A K G G S F E P G D A S K G A N L F K T R C A Q C H
(64)                        G F E D G D A K K G A R I F K T R C A Q C H
(65)             @P X A R E P L P P G D A A K G E K I F K G R A A Q C H
(66)            P P X A R A P L P P G D A A R G E K L F K G R A A Q C H
(67)                               @G D A E R G K K L F E S R A A Q C H
(68)               @S T F A B A P P G B P A K G A K I F K A K C A Z C H
(69)          G P K E P E V T V P E G D A S A G R D I F D S Q C S A C H
(70)               @A S F D E A P A G N S A.S G E K I F K T K C A Q C H
```

```
                           1                 10
(71)      @A S F D E A P A G N S K S G E K I F K T K C A Q C H
(72,73)   @A S F D E A P P G N S K A G E K I F K T K C A Q C H
(74)      @A S F N E A P P G N S K A G E K I F K T K C A Q C H
(75)      @A S F D E A P P G N S K A G E K I F K T K C A Q C H
(76)      @A S F A E A P P G N P K A G E K I F K T K C N Q C H
(77)      @A S F Q E A P P G N A K A G E K I F K T K C A Q C H
(78)      @A S F Q E A P P G N A K A G E K I F K T K C A Q C H
(79)      @A S F N E A P P G N V K A G E K I F K T K C A Q C H
(80)      @A S F D E A P P G N P K A G E K I F K T K C A Q C H
(81)      @A S F G E A P P G N P K A G E K I F K T K C A Q C H
(82)      @A S F S E A P P G N P K A G E K I F K T K C A Q C H
(83)      @A S F S E A P P G N P K A G E K I F K T K C A Q C H
(84)      @A S F A E A P P G N P K A G E K I F K T K C A Q C H
(85)      @A S F D E A P P G N V K S G E K I F K T K C A Q C H
(86)      @A T F A E A P P G N Q K A G E K I F K L K C A Q C H
(87)      @A S F A E A P P G N P A A G E K I F K T K C A Q C H
(88)      @A S F A E A P A G D A K A G E K I F K T K C A Q C H
(89)      @A S F A E A P A G D P T T G A K I F K T K C A Q C H
(90)      @A S F A E A P A G D N K A G D K I F K N K C A Q C H
(91)      @A S F A E A P P G D K D V G G K I F K T K C A Q C H
(92)      @A S F S E A P P G N P D A G A K I F K T K C A Q C H
(93)      @A T F S E A P P G N I K S G E K I F K T K C A Q C H
(94)      @A T F S E A P P G N K D V G A K I F K T K C A Q C H
(95)      @A T F S E A P P G D P K A G E K I F K T K C A Q C H
(96)       S T F A E A P A G D L A R G E K I F K T K C A Q C H
```

Fig. 3.1a The amino acid sequences of mitochondrial cytochromes *c*: residues -11 to 18. @ indicates the N-terminal amino group is acetylated, and * indicates it is blocked with an unidentified group

	20 — 30 — 40 — 50
(1,2)	T V E K G G K H K T G P N L H G L F G R K T G Q A P G Y S Y T A
(3)	T V E K G G K H K T G P N L H G L F G R K T G Q A P G Y S Y T A
(4)	T V E K G G K H K T G P N L H G L F G R K T G Q A S G Y S Y T E
(5)	T V E K G G K H K T G P N L H G L F G R K T G Q A P G F T Y T D
(6,7)	T V E K G G K H K T G P N L H G L F G R K T G Q A P G F S Y T D
(8,9,10)	T V E K G G K H K T G P N L H G L F G R K T G Q A P G F S Y T D
(11,12,13)	T V E K G G K H K T G P N L H G L F G R K T G Q A V G F S Y T D
(14)	T V E K G G K H K T G P N L H G L F G R K T G Q A P G F S Y T D
(15)	T V E K G G K H K T G P N L H G L F G R K T G Q A P G F S Y T D
(16)	T V E K G G K H K T G P N L H G L F G R K T G Q S P G F S Y T D
(17)	T V E K G G K H K T G P N L H G L F G R K T G Q A P G F S Y T D
(18)	T V E K G G K H K T G P N L H G L F G R K T G Q A V G F S Y T D
(19,20,21)	T V E K G G K H K T G P N L H G L F G R K T G Q A A G F S Y T D
(22)	T V E K G G K H K T G P N L W G L F G R K T G Q A P G F S Y T D
(23)	T V E K G G K H K T G P N L N G I F G R K T G Q A P G F T Y T D
(24,25)	T V E K G G K H K T G P N L H G L F G R K T G Q A E G F S Y T D
(26)	T V E K G G K H K T G P N L N G L F G R K T G Q A E G F S Y T D
(27)	T V E K G G K H K T G P N L D G L F G R K T G Q A E G F S Y T D
(28)	T V E K G G K H K T G P N L H G I F G R K T G Q A E G F S Y T D
(29)	T V E K G G K H K T G P N L H G L F G R K T G Q A E G F S Y T D
(30)	T V E K G G K H K T G P N L H G L F G R K T G Q A E G F S Y T D
(31)	T V E K G G K H K T G P N L N G L I G R K T G Q A E G F S Y T E
(32)	T V E K G G K H K T G P N L H G L F G R K T G Q A V G Y S Y T A
(33)	T C E K G G K H K V G P N L Y G L I G R K T G Q A A G F S Y T D
(34)	T V E N G G K H K V G P N L W G L F G R K T G Q A E G Y S Y T D
(35)	T V E N G G K H K V G P N L W G L F G R K T G Q A E G Y S Y T D
(36)	T V Z B G G K H K V G P N L W G L F G R K T G Q A P G F S Y T B
(37)	T V E Q G G K H K V G P N L W G I F G R K T G Q A D G Y S Y T D
(38)	T V E N G G K H K T G P N L S G L F G R K T G Q A Q G F S Y T D
(39)	T V E K A G K H K T G P N L S G L F G R K T G Q A P G F S Y T D
(40)	T V E K A G K H K T G P N L N G I L G R K T G Q A A G F S Y T D
(41)	T V E A G G K H K T G P N L S G L F G R K Q G Q A P G F A Y T D
(42)	S A Q A N L K H K T G P N L N G L F G R Q T G Q A S G Y V Y T D
(43)	T V D K G G P H K T G P N L H G I F G R A T G Q A A G F A Y T D
(44)	T V E A G G K H K V G P N L H G L I G R K T G Q A A G F S Y T N
(45)	T V E A G G K H K V G P N L H G L I G R K T G Q A A G F A Y T D
(46)	T Y E V G G K H K V G P N L G G V V G R K C G T A A G Y K Y T D
(47)	T V E A G G K H K V G P N L H G L I G R K T G Q A A G F S Y T N
(48)	T V E A G G K H K V G P N L H G L F G R K T G Q A A G F A Y T N
(49)	T V E A G G K H K V G P N L H G L F G R K T G Q A P G F A Y T N
(50)	T V E A G G K H K V G P N L H G L F G R K T G Q A A G F A Y T D
(51)	T V E A G G K H K V G P N L H G L F G R K T G Q A A G F A Y T D
(52)	T V E A G G K H K V G P N L H G F Y G R K T G Q A P G F S Y S N
(53)	T V E A G G K H K V G P N L H G F F G R K T G Q A P G F S Y S N
(54)	T V E A G G K H K V G P N L H G L I G R K T G Q A A G F A Y T N
(55)	T I E S G G K H K V G P N L Y G V Y G R K T G Q A P G Y S Y T D
(56)	T V E K G G P H K V G P N L H G I F G R H S G Q A E G Y S Y T D
(57)	T I E E G G P N K V G P N L H G I F G R H S G Q V K G Y S Y T D
(58)	T V E E G G P H K V G P N L H G V V G R T S G Q A Q G F S Y T D
(59)	T I E A G G P H K V G P N L H G I F S R H S G Q A E G Y S Y T D
(60)	T V E K G G P H K V G P N L H G I F G R Q S G K A E G Y S Y T D
(61)	T V E K G G A N K V G P N L H G V F G R K T G Q A E G F S Y T E
(62)	T L E E G G G N K I G P A L H G L F G R K T G S V D G Y A Y T D
(63)	S V E Q G G A N K I G P N L H G L F G R K T G S V E G Y S Y T D
(64)	T L G A G E P N K V G P N L H G L F G R K S G T V E G F S Y T D
(65)	T G A K G G A N G V G P N L F G I V N R H S G T V E G F A Y S K
(66)	T A N Q G G A N G V G P N L Y G L V G R H S G T I E G Y A Y S K
(67)	S A Q K G V N S T G P S L W G V Y G R T S G S V P G Y A Y S N
(68)	T V B A G A G H K Q G P N L N G A F G R T S G T A A G F S Y S A
(69)	A I E G D S T A A P V L G G V I G R K A G Q E K F A Y S K
(70)	T V D Q G A G H K Q G P N L H G L F G R Q S G T V A G Y S Y S A

```
            20                  30                  40                  50
(71)      T V D K G A G H K Q G P N L N G L F G R Q S G T T A G Y S Y S T
(72,73)   T V D K G A G H K Q G P N L N G L F G R Q S G T T A G Y S Y S A
(74)      T V D K G A G H K Q G P N L N G L F G R Q S G T T P G Y S Y S A
(75)      T V D R G A G H K Q G P N L N G L F G R Q S G T T A G Y S Y S A
(76)      T V D K G A G H K Q G P N L N G L F G R Q S G T T A G Y S Y S A
(77)      T V E K G A G H K Q G P N L N G L F G R Q S G T T P G Y S Y S A
(78)      T V D K G A G H K Q G P N L N G L F G R Q S G T T A G Y S Y S A
(79)      T V E K G A G H K Q G P N L N G L F G R Q S G T T A G Y S Y S A
(80)      T V E K G A G H K Q G P N L N G L F G R Q S G T T A G Y S Y S A
(81)      T V D K G A G H K E G P N L N G L F G R Q S G T T A G Y S Y S N
(82)      T V D K G A G H K Q G P N L N G L F G R Q S G T T P G Y S Y S T
(83)      T V E K G A G H K Q G P N L N G L F G R Q S G T T A G Y S Y S A
(84)      T V E K G A G H K Q G P N L N G L F G R Q S G T T A G Y S Y S A
(85)      T V D K G A G H K Q G P N L N G L F G R Q S G T T P G Y S Y S A
(86)      T V E K G A G H K Q G P N L N G L F G R Q S G T A A G Y S Y S A
(87)      T V D K G A G H K Q G P N L N G L F G R Q S G T T A G Y S Y S A
(88)      T V E K G A G H K Q G P N L N G L F G R Q S G T T A G Y S Y S A
(89)      T V E K G A G H K Q G P N L N G L F G R Q S G T T A G Y S Y S A
(90)      T V D K G A G H K Q G P N L N G L F G R Q S G T T A G Y S Y S A
(91)      T V E L G A G H K Q G P N L N G L F G R Q S G T T A G Y S Y S A
(92)      T V D A G A G H K Q G P N L H G L F G R Q S G T T A G Y S Y S A
(93)      T V E K G A G H K Q G P N L N G L F G R Q S G T T A G Y S Y S A
(94)      T V D L G A G H K Q G P N L N G L F G R Q S G T A A S Y S Y S A
(95)      T V E K G A G H K Q G P N L H G L F G R Q S G T T A G Y S Y S T
(96)      V A E K G G G H K Q G P N L G G L F G R V S G T A A G F A Y S K
```

Fig. 3.1b The amino acid sequences of mitochondrial cytochromes *c*: residues 19 to 50

```
                                  60                  70                  80
(1,2)           A N K N K G I I W G E D T L M E Y L E N P K K Y I P G T K M I F
(3)             A N K N K G I T W G E D T L M E Y L E N P K K Y I P G T K M I F
(4)             A N K N K G I I W G E D T L M E Y L E N P K K Y I P G T K M I F
(5)             A N K N K G I T W K E E T L M E Y L E N P K K Y I P G T K M I F
(6,7)           A N K N K G I T W K E E T L M E Y L E N P K K Y I P G T K M I F
(8,9,10)        A N K N K G I T W G E E T L M E Y L E N P K K Y I P G T K M I F
(11,12,13)      A N K N K G I T W G E E T L M E Y L E N P K K Y I P G T K M I F
(14)            A N K N K G I T W G E E T L M E Y L E N P K K Y I P G T K M I F
(15)            A N K N K G I T W G E E T L M E Y L E N P K K Y I P G T K M I F
(16)            A N K N K G I T W G E E T L M E Y L E N P K K Y I P G T K M I F
(17)            A N K N K G I T W G E A T L M E Y L E N P K K Y I P G T K M I F
(18)            A N K N K G I T W G E D T L M E Y L E N P K K Y I P G T K M I F
(19,20,21)      A N K N K G I T W G E D T L M E Y L E N P K K Y I P G T K M I F
(22)            A N K N K G I V W S Z Z T L M Z Y L Z B P K K Y I P G T K M I F
(23)            A N K N K G I I W G E D T L M E Y L E N P K K Y I P G T K M I F
(24,25)         A N K N K G I T W G E D T L M E Y L E N P K K Y I P G T K M I F
(26)            A N K N K G I T W G E D T L M E Y L E N P K K Y I P G T K M I F
(27)            A N K N K G I T W G E D T L M E Y L E N P K K Y I P G T K M I F
(28)            A N K N K G I T W G E D T L M E Y L E N P K K Y I P G T K M I F
(29)            A N K N K G I T W G E D T L M E Y L E N P K K Y I P G T K M I F
(30)            A N K N K G I T W G E D T L M E Y L E N P K K Y I P G T K M I F
(31)            A N K N K G I T W G E E T L M E Y L E N P K K Y I P G T K M I F
(32)            A N K N K G I I W G D D T L M E Y L E N P K K Y I P G T K M V F
(33)            A N K N K G I T W G E D T L M E Y L E N P K K Y I P G T K M I F
(34)            A N K S K G I V W N N D T L M E Y L E N P K K Y I P G T K M I F
(35)            A N K S K G I V W N E N T L M E Y L E N P K K Y I P G T K M I F
(36)            A N K N K G I V W B Z Z T L M E Y L E N P K K Y I P G T K M I F
(37)            A N K S K G I V W G D D T L M E Y L E N P K K Y I P G T K M I F
(38)            A N K S K G I T W Q Q E T L R I Y L E N P K K Y I P G T K M I F
(39)            A N K S K G I V W N Q E T L F V Y L E N P K K Y I P G T K M I F
(40)            A N R N K G I T W K N E T L F E Y L E N P K K Y I P G T K M V F
(41)            A N K G K G I T W K N Q T L F Q Y L E N P K K Y I P G T K M V F
(42)            A N K A K G I T W Q A D T L D V Y L E N P K K Y I P G T K M V F
(43)            A N K S K G I T W T K D T L Y E Y L E N P K K Y I P G T K M V F
(44)            A N K A K G I T W Q D D T L F E Y L E N P K K Y I P G T K M I F
(45)            A N K A K G I T W N E D T L F E Y L E N P K K Y I P G T K M I F
(46)            A N I K K G V T W T E G N L D E Y L K D P K K Y I P G T K M V F
(47)            A N K D K G I T W Q D D T N E E Y L E N P K K Y I P G T K M I F
(48)            A N K A K G I T W Q D D T L F E Y L E N P K K Y I P G T K M I F
(49)            A N K A K G I T W G D D T L F E Y L E N P K K Y I P G T K M I F
(50)            A N K A K G I T W N E D T L F E Y L E N P K K Y I P G T K M I F
(51)            A N K S K G I T W G D D T L F E Y L E N P K K Y I P G T K M I F
(52)            A N K A K G I T W G D D T L F E Y L E N P K K Y I P G T K M V F
(53)            A N K A K G I T W Q D D T L F E Y L E N P K K Y I P G T K M V F
(54)            A N K D K G I T W D E N T L F I Y L E N P K K Y I P G T K M V F
(55)            A N K G K G I T W N K E T L F E Y L E N P K K Y I P G T K M V F
(56)            A N I K K N V L W D E N N M S E Y L T N P X K Y I P G T K M A F
(57)            A N I N K N V K W D E D S M S E Y L T N P K K Y I P G T K M A F
(58)            A N K K K G V E W T E Q D L S D Y L E N P X K Y I P G T K M A F
(59)            A N K R A G V E W A E P T M S D Y L E N P X K Y I P G T K M A F
(60)            A N I K K A V E W S E Q T M S D Y L E N P X X Y I P G T K M A F
(61)            A N K D R G I T W B Z Z T L F A Y L E N P X K Y I P G T K M A F
(62)            A N K Q K G I T W D E N T L F E Y L E N P X K Y I P G T K M A F
(63)            A N K Q A G I T W N E D T L F E Y L E N P X K F I P G T K M A F
(64)            A N K K A G Q V W E E E T F L E Y L E N P X K Y I P G T K M A F
(65)            A N A D S G V V W T P E V L D V Y L E N P X K F M P G T K M S F
(66)            A N A E S G V V W T P D V L D V Y L E N P X K F M P G T K M S F
(67)            A N K N A A I V W E E E T L H K F L E N P K K Y V P G T K M A F
(68)            A B K B K T A D W B Z B T L Y D Y L L N P X K Y I P G T K M V F
(69)            G M K G S G I T W N E K H L F V F L K N P S K H V P G T K M A F
(70)            A N K N K A V N W E E K T L Y D Y L L N P X K Y I P G T K M V F
```

									60										70										80			
(71)	A	N	K	N	M	A	V	I	W	E	E	K	T	L	Y	D	Y	L	L	N	P	X	K	Y	I	P	G	T	K	M	V	F
(72,73)	A	N	K	N	K	A	V	E	W	E	E	K	T	L	Y	D	Y	L	L	N	P	X	K	Y	I	P	G	T	K	M	V	F
(74)	A	N	K	N	R	A	V	I	W	E	E	K	T	L	Y	D	Y	L	L	N	P	X	K	Y	I	P	G	T	K	M	V	F
(75)	A	N	K	N	M	A	V	T	W	Q	E	K	T	L	Y	D	Y	L	L	N	P	X	K	Y	I	P	G	T	K	M	V	F
(76)	A	N	K	N	M	A	V	N	W	E	E	K	T	L	Y	D	Y	L	L	N	P	X	K	Y	I	P	G	T	K	M	V	F
(77)	A	N	K	N	M	A	V	N	W	G	E	N	T	L	Y	D	Y	L	L	N	P	X	K	Y	I	P	G	T	K	M	V	F
(78)	A	N	K	N	M	A	V	Q	W	G	E	N	T	L	Y	D	Y	L	L	N	P	X	K	Y	I	P	G	T	K	M	V	F
(79)	A	N	K	N	M	A	V	Q	W	G	E	N	T	L	Y	D	Y	L	L	N	P	X	K	Y	I	P	G	T	K	M	V	F
(80)	A	N	K	N	M	A	V	N	W	G	E	N	T	L	Y	D	Y	L	L	N	P	X	K	Y	I	P	G	T	K	M	V	F
(81)	A	N	K	N	M	A	V	T	W	G	E	N	T	L	Y	D	Y	L	L	N	P	X	K	Y	I	P	G	T	K	M	V	F
(82)	A	N	K	D	M	A	V	I	W	E	E	N	T	L	Y	D	Y	L	L	N	P	X	K	Y	I	P	G	T	K	M	V	F
(83)	A	N	K	N	K	A	V	V	W	E	E	N	T	L	Y	D	Y	L	L	N	P	X	K	Y	I	P	G	T	K	M	V	F
(84)	A	N	K	N	M	A	V	I	W	E	E	S	T	L	Y	D	Y	L	L	N	P	X	K	Y	I	P	G	T	K	M	V	F
(85)	A	N	K	N	M	A	V	I	W	G	E	N	T	L	Y	D	Y	L	L	N	P	X	K	Y	I	P	G	T	K	M	V	F
(86)	A	N	K	N	M	A	V	I	W	E	E	N	T	L	Y	D	Y	L	L	N	P	X	K	Y	I	P	G	T	K	M	V	F
(87)	A	N	K	N	M	A	V	N	W	G	Y	N	T	L	Y	D	Y	L	L	N	P	X	K	Y	I	P	G	T	K	M	V	F
(88)	A	N	K	N	K	A	V	A	W	E	E	N	S	L	Y	D	Y	L	L	N	P	X	K	Y	I	P	G	T	K	M	V	F
(89)	A	N	K	N	M	A	V	I	W	E	E	N	T	L	Y	D	Y	L	L	N	P	X	K	Y	I	P	G	T	K	M	V	F
(90)	A	N	K	N	K	A	V	L	W	E	E	A	T	L	Y	D	Y	L	L	N	P	X	K	Y	I	P	G	T	K	M	V	F
(91)	A	N	K	N	K	A	V	L	W	A	D	N	T	L	Y	D	Y	L	L	N	P	X	K	Y	I	P	G	T	K	M	V	F
(92)	A	N	K	N	K	A	V	E	W	E	E	N	T	L	Y	D	Y	L	L	N	P	X	K	Y	I	P	G	T	K	M	V	F
(93)	A	N	K	N	K	A	V	T	W	G	E	D	T	L	Y	E	Y	L	L	N	P	X	K	Y	I	P	G	T	K	M	V	F
(94)	A	N	K	N	K	A	V	I	W	S	E	D	T	L	Y	E	Y	L	L	N	P	X	K	Y	I	P	G	T	K	M	V	F
(95)	G	N	K	N	K	A	V	N	W	G	E	Q	T	L	Y	E	Y	L	L	N	P	X	K	Y	I	P	G	T	K	M	V	F
(96)	A	N	K	E	A	A	V	T	W	G	E	S	T	L	Y	E	Y	L	L	N	P	K	K	Y	M	P	G	N	K	M	V	F

Fig. 3.1c The amino acid sequences of mitochondrial cytochromes *c*: residues 51 to 82

	90 100
(1,2)	V G I K K K E E R A D L I A Y L K K A T N E
(3)	V G I K K K E E R A D L I A Y L K K A T N E
(4)	V G I K K K E E R A D L I A Y L K K A T N E
(5)	A G I K K K T E R E D L I A Y L K K A T N E
(6,7)	A G I K K K T E R E D L I A Y L K K A T N E
(8,9,10)	A G I K K K G E R E D L I A Y L K K A T N E
(11,12,13)	A G I K K K G E R A D L I A Y L K K A T N E
(14)	A G I K K T G E R A D L I A Y L K I A T K E
(15)	A G I K K T G E R A D L I A Y L K K A T K E
(16)	A G I K K K G E R A D L I A Y L K Q A T N E
(17)	A G I K K S A E R A D L I A Y L K K A T K E
(18)	A G I K K K D E R A D L I A Y L K K A T N E
(19,20,21)	A G I K K K G E R A D L I A Y L K K A T N E
(22)	A G I K K K S E R E D L I K Y L K Q A T S S
(23)	A G I K K K G E R A D L I A Y L K K A T N E
(24,25)	A G I K K K S E R V D L I A Y L K D A T S K
(26)	A G I K K K S E R A D L I A Y L K D A T S K
(27)	A G I K K K S E R A D L I A Y L K D A T S K
(28)	A G I K K K S E R A D L I A Y L K D A T S K
(29)	A G I K K K S E R A D L I A Y L K D A T A K
(30)	A G I K K K A E R A D L I A Y L K Q A T A K
(31)	A G I K K K A E R A D L I A Y L K D A T S K
(32)	T G L S K K K E R T N L I A Y L K E K T A A
(33)	A G I K K K G E R Q D L I A Y L K S A C S K
(34)	A G I K K K G E R Q D L V A Y L K S A T S
(35)	A G I K K K G E R Q D L V A Y L K S A T S
(36)	A G I K K K G E R A D L I A Y L K S A T S
(37)	A G I K K K G E R Q D L I A Y L K S A T S
(38)	A G L K K K S E R Q D L I A Y L K K T A A S
(39)	A G I K K E G E R K D L I A Y L K K S T S E
(40)	A G L K K Q K E R Q D L I A Y L E A A T K
(41)	A G L K B Z T E R V H L I A Y L Z Z A T K K
(42)	A G L K K A N E R A D L I A Y L K Q A T N L
(43)	A G L K N Z K Z R A N L I A Y L E Q E T K
(44)	A G L K K P N E R G D L I A Y L K S A T K
(45)	A G L K K P N E R G D L I A Y L K S A T K
(46)	A G L K K A E E R A D L I A F L K S N K
(47)	A G L K K P N E R G D L I A Y L K S A T K
(48)	A G L K K P N E R G D L I A Y L K S A T K
(49)	A G L K K P N E R G D L I A Y L K S A T K
(50)	A G L K K P N E R G D L I A Y L K S A T K
(51)	A G L K K P G D R A D L I A Y L K S A T A
(52)	A G L K K A N E R A D L I A Y L K E S T K
(53)	A G L K K A N E R A D L I A Y L K Q A T K
(54)	A G L K K P E E R A D L I A Y L K E S T K
(55)	A G L K K P Q E R A D L I A Y I E Q A S K
(56)	A G L K K E K D R N D L I T Y L K K A C E
(57)	A G L K K E K D R N D L I T Y M T K A A K
(58)	A G L K K A K D R N D L I T Y L V K A T K
(59)	A G L K K A K D R N D L V T Y M L E A S K
(60)	G G L K K E K D R N D L V T Y L A N A T K
(61)	A G F K K P A D R N N V I T Y L K K A T S E
(62)	A G L K K D K D R N D I I T F M K E A T A
(63)	A G L X K N K D R N D L I T Y L K E A T K
(64)	A G L K K E K D R N D L V T Y L R E E T K
(65)	A G I K K P Q E R A D L I A Y L E N L K
(66)	A G M K K P Q E R A D V I A Y L E T L K G
(67)	A G I X A K K D R Q D I I A Y M K T L K D
(68)	A G L K K P Z B R A D L I A F L K D A T A
(69)	A G L P A D K D R A D L I A Y L K S V
(70)	P G L X K P Q E R A D L L A Y L K E S T A

```
                         90                  100
(71)        P G L X K P Q D R A D L I A Y L K E S T A
(72,73)     P G L X K P Q D R A D L I A Y L K E A T A
(74)        P G L X K P Q D R A D L I A Y L K E A T A
(75)        P G L X K P Q D R A D L I A Y L K E S T A
(76)        P G L X K P Q D R A D L I A Y L K Q S T A
(77)        P G L X K P Q D R A D L I A Y L K E S T A
(78)        P G L X K P Q D R A D L I A Y L K E S T A
(79)        P G L X K P Q D R A D L I A Y L K Q A T A
(80)        P G L X K P Q E R A D L I A Y L K E A T A
(81)        P G L X K P Q E R A D L I A Y L K E A T A
(82)        P G L X K P Q E R A D L I S Y L K E A T S
(83)        P G L X K P Q E R A D L I A Y L K E A T A
(84)        P G L X K P Q E R A D L I A Y L K E S T A
(85)        P G L X K P Q E R A D L I A Y L K E A T A
(86)        P G L X K P Q D R A D L I A Y L K E S T A
(87)        P G L X K P Q D R A D L I A Y L K Q S T A S
(88)        P G L X K P Q E R A D L I A Y L K A S T A
(89)        P G L X K P Q E R A D L I A Y L K T S T A
(90)        P G L X K P Q D R A D L I A Y L K E S T A
(91)        P G L X K P Q D R A D L I A Y L K H A T A
(92)        P G L X K P Q D R A D L I A Y L K K A T S S
(93)        P G L X K P Q E R A D L I A Y L K D S T Z
(94)        P G L X K P Q D R A D L I A Y L K D S T Q
(95)        P G L X K P Q E R A D L I S Y L K Q A T S Q E
(96)        A G L K K P E E R A D L I A Y L K Q A T A
```

Fig. 3.1d The amino acid sequences of mitochondrial cytochromes *c*: residues 83 to 104

(58) is from Sugeno *et al.* (1971); **(59)** is from Narita and Titani (1968) and Lederer (1972); **(60)** is from Becam and Lederer (1981); **(61)** is from Simon-Becam *et al.* (1978); **(62)** is from Heller and Smith (1966) and Lederer and Simon (1974); **(63)** is from Morgan *et al.* (1972) and Lederer and Simon (1974); **(64)** is from Bitar *et al.* (1972); **(65)** is from Pettigrew (1972); **(66)** is from Hill and Pettigrew (1975); **(67)** is from Pettigrew (1973) and Lin *et al.* (1973); **(68)** is from Meatyard and Boulter (1974); **(69)** is from Tarr and Fitch (1976); **(70)** is from Brown and Boulter (1973b); **(71)** is from Thompson *et al.*(1970) and Thompson *et al.*(1971a); **(72)** and **(93)** are from Thompson *et al.* (1971d); **(73)** is from Richardson *et al.* (1971); **(74)** is from Thompson *et al.* (1971c); **(75)** is from Wallace *et al.* (1973); **(76)**, **(87)**, **(90)** and **(91)** are from Brown and Boulter (1974); **(77)** and **(78)** are from Thompson *et al.* (1971a); **(79)** and **(85)** are from Thompson *et al.* (1971b); **(80)** is from Scogin *et al.* (1972) and Brown and Boulter (1975); **(81)** is from Martinez *et al.* (1974) and Brown and Boulter (1975); **(82)** is from Mori and Morita (1980); **(83)**, **(84)** and **(88)** are from Boulter (unpublished results cited in Dickerson and Timkovich, 1975); **(86)** is from Brown and Boulter (1973a); **(89)** is from Ramshaw *et al.* (1970); **(92)** is from Stevens *et al.* (1967) and Delange *et al.* (1969); **(94)** is from Brown *et al.* (1973); **(95)** is from Ramshaw *et al.* (1971); and **(96)** is from Amati *et al.* (1988). Modifications to sequences **(36)**, **(39)**, **(70)**, **(71)**, **(72)**, **(74)**, **(75)**, **(77)**, **(78)**, **(79)**, **(80)**, **(85)**, **(86)**, and **(95)** were taken from Dickerson and Timkovich (1975). The amidation states of residues 52 and 54 of sequence **(82)** are the opposite of that given by Mori and Morita (1980). The present amide assignment fits the chemical data as well as the original assignment and is more consistent with the other sequences of Fig. 3.1.

Dickerson and Timkovich (1975) give a matrix of identities for 60 sequences, and phylogenetic trees based on the data of Fig. 3.1 are given in Chapter 6.

Some of the information contained in Fig. 3.1 can be condensed into a variability diagram as shown in Fig. 3.2. Variability is defined as the number of different amino acids at a given position divided by the frequency of the most common amino acid at that position (Wu and Kabat, 1970). The variability analysis was done excluding *Tetrahymena pyriformis* cytochrome *c* because this protein appears anomalous in several ways. A reflection of its unusual status is that, unlike all the other cytochromes in Fig. 3.1, the *Tetrahymena* protein has an acidic pI.

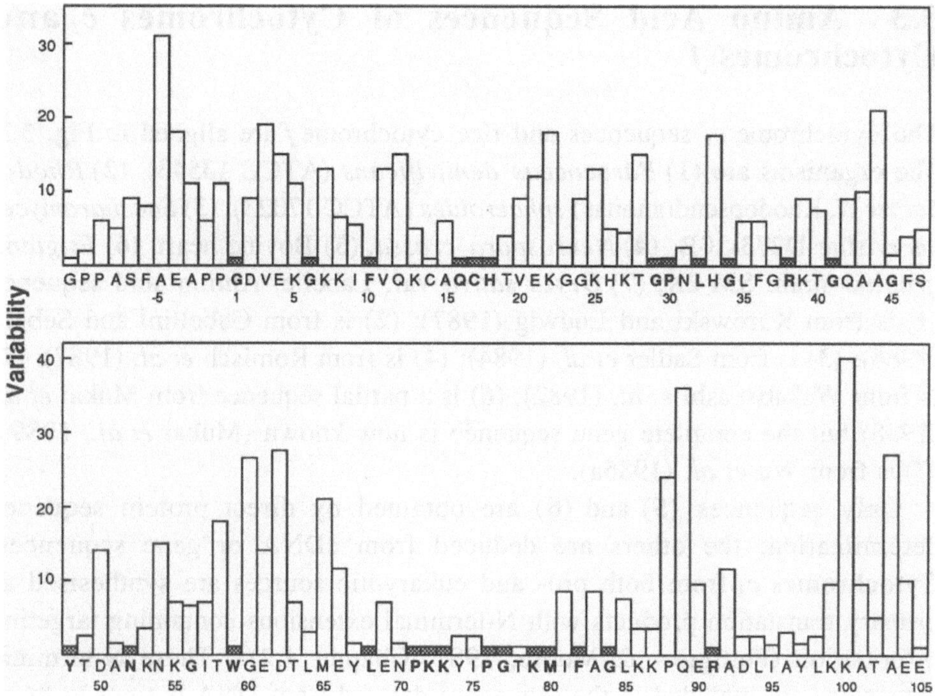

Fig. 3.2 The variability of amino acid residues in mitochondrial cytochromes *c*. The sequence compilation of Fig. 3.1 excluding *Tetrahymena* cytochrome *c* was used to calculate the variabilities. Unvaried residues are indicated by the stippled boxes and bold letters. The amino acid sequence is a hypothetical sequence of the most common amino acids occuring at each site

As Fig. 3.1 shows, some sites are more variable than others. It is generally presumed that the variability of a site is related to the role played by the amino acid occupying that site, although it is not always clear how variability and function are related. As we shall see (Chapter 4.6) many of the unvaried residues have been replaced by semisynthetic and genetic methods without significantly altering the structure and redox properties of cytochrome *c*. Nevertheless, it is notable that 26 amino acids are unvaried.

3.3 Amino Acid Sequences of Cytochromes c_1 and Cytochromes *f*

The cytochrome c_1 sequences and rice cytochrome *f* are aligned in Fig. 3.3. The organisms are (**1**) *Paracoccus denitrificans* (ATCC 13543), (**2**) *Rhodobacter* (f. Rhodopseudomonas) *sphaeroides* (ATCC 17023), (**3**) *Saccharomyces cerevisiae* D273-10B, (**4**) *Neurospora crassa*, (**5**) Bovine heart, (**6**) *Euglena gracilis* strain SM-2K, (**7**) *Oryza sativa* var. Labelle. Amino acid sequence (**1**) is from Kurowski and Ludwig (1987); (**2**) is from Gabellini and Sebald (1986); (**3**) is from Sadler *et al.* (1984); (**4**) is from Romisch *et al.* (1987); (**5**) is from Wakabayashi *et al.* (1982); (**6**) is a partial sequence from Mukai *et al.* (1988) but the complete gene sequence is now known (Mukai *et al.*, 1989); (**7**) is from Wu *et al.* (1986a).

Only sequences (**5**) and (**6**) are obtained by direct protein sequence determination; the others are deduced from cDNA or gene sequences. Cytochromes c_1 from both pro- and eukaryotic sources are synthesised as primary translation products with N-terminal extensions containing targeting information (Pettigrew & Moore, 1987: Chapter 4.2). These N-terminal extensions are included in the sequences derived from DNA but not in those derived from studies on the mature protein. In the former cases, the start of the mature protein can be identified either by N-terminal analysis or by prediction of signal cleavage sites. For example, the mature N-terminus is predicted to be glutamine for *Paracoccus* cytochrome c_1 but this cannot be verified by N-terminal analysis because the N-terminus is blocked, presumably by pyrrolidone carboxylic acid. In both yeast and *Neurospora* cytochromes c_1, however, the N-terminii of the mature proteins have been directly identified (Sadler *et al.*, 1984; Tsugita *et al.*, 1979). (C) in the *Euglena* sequence is deduced to be cysteine by homology but it was not positively identified. It is, however, confirmed by the gene sequence (Mukai *et al.*, 1989).

The present alignment differs from published alignments in some important ways. The most important proposed gap is at position 158 (bovine) because this allows alignment of Met 160 (bovine) and the strongly homologous regions just C-terminal to this residue. Met 160 is the only conserved methionine in the alignment and thus is a possible 6th iron ligand (Chapter 2.3.2.3). The gap at position 158 (bovine) is a feature of the alignments of Gabellini and Sebald (1986) and Romisch *et al.* (1987) but not of the alignment of Kurowski and Ludwig (1987). The former alignments include a further gap at residue 131 (bovine) whereas placing this gap at position 121, as in the present alignment, leads to greater matching of the pro- and eukaryotic sequences.

The *Paracoccus* cytochrome c_1 is unique in having an N-terminal extension which contains 40% Ala, 14% Pro and 38% acidic amino acids (Kurowski and Ludwig, 1987). The acidic nature of this region is reminiscent of the 'hinge protein' associated with mitochondrial cytochromes c_1 (Pettigrew & Moore, 1987: Chapter 2.2) but there is no sequence homology.

* in Fig. 3.3 indicates acidic residues that are protected in the complex of bovine cytochrome c_1 with horse cytochrome *c* (Stonehuerner *et al.*, 1985) or are in the region to which horse cytochrome can be cross-linked (Broger *et al.*, 1983).

An uncharged region near the C-terminus (boxed) may be a membrane anchor. It is of interest to note that the bacterial cytochromes c_1 and the chloroplast cytochromes *f* which are synthesised by insertion from one side of the membrane to the other (Pettigrew & Moore, 1987: Chapter 4.2) have larger anchors (20 - 21 residues) than the mitochondrial cytochromes c_1 which are attached to the inner membrane from the cytoplasmic side. It may be that the stretch of 15 uncharged amino acids in the latter case is not long enough to be transmembrane.

The matrix of % identities is shown in Table. 3.2.

Cytochrome *f* sequences are highly conserved (Fig. 3.4) and show enough homology to the cytochromes c_1 to allow alignment (Fig. 3.3, sequence 7). They are longer by virtue of a 71 residue insertion at position 193 (bovine). The sequence of cytochrome *f* has been determined from rice (Wu *et al.*, 1986a); wheat (Willey *et al.*, 1984a); spinach (Alt & Hermann, 1984); pea (Willey *et al.*, 1984b); and *Vicia faba* (Ko & Strauss, 1987). The entire nucleotide sequences of the chloroplast genomes of liverwort (Ohyama *et al.*, 1986) and tobacco (Shinozaki *et al.*, 1986) have been determined.

The matrix of identities is shown in Table 3.3.

(1) *Paracoccus denitrificans* Q D A S T A P G T T A P A G S S Y H T N
E A A P A A A D T A P A A E A A D E P A A E E A E A G E A E V T E E P
A A T E T P A E E P A A D E P A A T E E P D A E A E P A A E E A Q A T
T E E A P A E E P A A E E P A A E E P A E E P A A D A P A E E A A A E
E A P A E P E A A A E E P A A E E P E A T E E E A P A E E A A A E E A
P A E E V V E D E A A A D H G D A A A Q E A G D

1 10
(1) *Paracoccus denitrificans* S H A A A H I E D I S F S F E G
(2) *Rhodobacter sphaeroides* N S N V Q D H A F S F E G
(3) *Saccharomyces cereviseae* M T A A E H G L H A P A Y A W S H N G
(4) *Neurospora crassa* M T P A E E G L H A T K Y P W V H E Q
(5) Bovine S D L E L H P P S Y P W S H R G

(7) Rice Y

20 30 HEME 45
(1) P F G K F D Q H Q L Q R G L Q V Y T E V C S A C H G L R Y V P L
(2) I F G K F D Q A Q L R R G F Q V Y S E V C S T C H G M K F V P I
(3) P F E T F D H A S I R R G Y Q V Y R E V C A A C H S L D R V A W
(4) W L K T F D H Q A L R R G F Q V Y R E V C A S C H S L S R V P Y
(5) L S S L D H T S I R R G F Q V Y K Q V C S S C H S M D Y V A Y

(6) *Euglena gracilis* ....E V Y E Q V F A P(C)H S L S F I K..
(7) P I F A Q Q G Y E N P R E A T G R I V C A N C H L A N K P V D

50 60 70 80
(1) R T L A D E G G P Q L P E D Q V R A Y A A N F D I T D P E T E
(2) R T L S D D G G P Q L D P T F V R E Y A A G L D T I I D K D S G
(3) R T L V G V S H T N E E V R N M A E E F E Y D D E P D E Q G N P
(4) R A L V G T I L T V D E A K A L A E E N E Y D T E P N D Q G E I
(5) R H L V G V C Y T E D E A K A L A E E V E V Q D G P N E D G E M
* * *

(7) I E V P Q A V L P D T V F E A V L R I P Y D M Q L K Q V L A N

90 100 110
(1) E D R P R V P T D H F P T V S G E G M G P D
(2) E E R D R K E T D M F P T R V G D G M G P D
(3) K K R P G K L S D Y I P G P Y P N E Q A A R A A N Q G A L P P D
(4) E K R P G K L S D Y L P D P Y K N D E A A R F A N N G A L P P D
(5) F M R P G K L S D Y F P K P Y P N P E A A R A A N N G A L P P D

(7) G K K G G L N V G A V L I L P E G F E L A P P

120
(1) L S L M A K A R A G F H G P Y G T G L S Q L F N G I G G P E Y I
(2) L S V M A K A R A G F S G P A G S G M N Q L F K G I G G P E Y I
(3) L S L I V K A R H G G C D Y I
(4) L S L I V K A R H G G C D Y I
(5) L S Y I V R A R H G G E D Y V

(7) R I S P E L K E K I G N L S

130 140 150
(1) H A V L T G Y D G E E K E E A G A V L Y H N A A F
(2) Y R Y V T G F P E E N P A C A P E G I D G Y Y Y N E V F Q V G G
(3) F S L L T G Y P D E P P A G V A L P P G S N Y N P Y F P G G S
(4) F S L L T G Y P D E P P A G A S V G A G L N F N P Y F P G T G
(5) F S L L T G Y C E P P T G V S L R E G L Y F N P Y F P G Q A

(7) F Q S Y R P N K K N I L V I G P V P G K K Y S E I V F P I L S

```
                                             160                 170
(1)                                AGNWIQ[M]AAPLSDD QVT
```

```
                                          160                 170
(1)                              A G N W I Q|M|A A P L S D D   Q V T
(2)     V P D T C K D A A G I K T T H G S W A Q|M|P P A L F D D   L V T
(3)     I                                     A|M|A R V L F D D   M V E
(4)     I                                     A|M|A R V L Y D G   L V D
(5)     I                                     G|M|A P P I Y N E   V L E
   <--------------------------------------------'
(7)     P D P                                 A M K K D V H F L K Y P I


                          180                 190                 200
(1)     Y E D G T P A T V D Q M A T D V A A F L M W T A E P K M M D R K
(2)     Y E D G T P A T V D Q M G Q D V A S F L M W A A E P K L V A R K
(3)     Y E D G T P A T T S Q M A K D V T T F L N W C A E P E H D E R K
(4)     Y E D G T P A S T S Q M A K D V V E F L N W A A E P E M D D R K
(5)     F D D G T P A T M S Q V A K D V C T F L R W A A E P E H D H R K
            * *
(7)     Y V G G N R G R G   Q I Y P D G S K S N N T V A E I V L Q D P L
                                                    | |
                  Y N A T S T G V V R K I L R K E K G G Y E I S
                  I V D A S D G R Q V I D L I P P G P E L L V S
                  E G E S I K L D Q P L T S N P N V G G F G Q G D


                      210                   220                 230
(1)     Q V G F V S V I F L I V L A A L L Y L T N K K   L W Q P I K H P
(2)     Q M G L V A V V M L G L L S V M L Y L T N K R   L W A P Y K R Q
(3)     R L G L K T V I I L S S L Y L L   S I W V K K F K W A G I K T R
(4)     R M G M K V L V V T S V L F A L   S V Y V K R Y K W A W L K S R
(5)     R M G L K M L L M M G L L L P L   V Y A M K R H K W S V L K S R

(7)     R V Q G L L F F F A S V I L A Q V F L V L K K K Q F E K V Q L Y


                      240
(1)     R K P E
(2)     K A
(3)     K F V F N P P K P R K
(4)     K I V Y D P P K R P P P A T N L A L P Q Q R A K S
(5)     K L A Y R P P K

(7)     E M N F
```

Fig. 3.3 The amino acid sequences of the cytochromes c_1

1. *Paracoccus denitrificans*					
59	2. *Rhodobacter sphaeroides*				
38	38	3. *Saccharomyces cereviseae*			
37	38	66	4. *Neurospora crassa*		
32	33	58	60	5. Bovine	
12	9	17	17	11	7. Rice (*f*)

Table 3.2 Matrix of % identities for the cytochromes c_1

```
                1                 10                  20┌─HEME──┐
(1) Rice        Y P I F A Q Q G Y E N P R E A T G R I V[C]A N[C H]L A
(2) Wheat       Y     F
(3) Spinach     Y     F
(4) Pea         Y     F
(5) Vicia faba  Y     F
(6) Tobacco     Y     F
(7) Liverwort   F     Y

         30                  40                  50
(1)  N K P V D I E V P Q A V L P D T V F E A V L R I P Y D M Q L K Q
(2)  S       D           A V     D             L R             L
(3)  N       D           A V     D             V R             L
(4)  N       D           A V     D             V R             V
(5)  N       D           A I     D             V R             V
(6)  N       E           A V     D             V R             L
(7)  K       D           S V     N             V K             I

     60                  70                  80                  90
(1)  V L A N G K K G G L N V G A V L I L P E G F E L A P P D R I S P
(2)              K   G                                   P D   I S
(3)              K   G                                   P D   I S
(4)              K   A                                   P H   L S
(5)              K   A                                   P D   L S
(6)              R   G                                   P D   I S
(7)              K   S                                   S D   I P

                     100                 110                 120
(1)  E L K E K I G N L S F Q S Y R P N K K N I L V I G P V P G K K Y
(2)  E L       I       A     S   R P D   K     L               K
(3)  E M       M       S     S   R P N   Q     L               Q
(4)  Q I       I       S     S   R P T   K     L               K
(5)  E I       I       S     S   R P T   K     I               K
(6)  E M       I       S     S   R P N   K     L               Q
(7)  E M       I       F     P   S N D   K     L               K

                 130                 140                 150
(1)  S E I V F P I L S P D P A M K K D V H F L K Y P I Y V G G N R G
(2)      I V         S         T K K D A H           I         N
(3)      I T         A         T K K D V H           I         N
(4)      I T         S         T K R D V Y           L         N
(5)      I T         S         T K R D V Y           I         T
(6)      I T         S         T K K D V H           I         N
(7)      M V         S         T N K E A H           I         N

             160                 170                 180
(1)  R G Q I Y P D G S K S N N T V Y N A T S T G V V R K I L R K E K
(2)                            T   Y   A T S T   I   R K I L
(3)                            T   Y   S T A T   I   K K I V
(4)                            N   S   A T A T   V   K Q I I
(5)                            N   Y   A T A T   V   N K K I
(6)                            T   Y   A T A A   I   S K I I
(7)                            T   Y   A S I T   K   S K I F

         190                 200                 210
(1)  G G Y E I S I V D A S D G R Q V I D L I P P G P E L L V S E G E
(2)            S   V   A       R Q   I   I I P P         L V
(3)            N   A   A       R E   V   I I P R         L V
(4)            T   V   A       S E   I   I I P P         L V
(5)            T   V   A       R E   I   I I P P         L V
(6)            T   T   A       R Q   V   I I P P         L V
(7)            T   D   I       H K   V   I S A A         I I
```

```
      220               230                   240                   250
(1)   S I K L D Q P L T S N P N V G G F G Q G D A E I V L Q D P L R V Q
(2)   S I   L           S                           I               V
(3)   S I   L           S                           V               I
(4)   S I   L           S                           I               V
(5)   S I   L           S                           I               V
(6)   S I   F           S                           I               V
(7)   L V   V           N                           V               I

                  260               270                   280
(1)   G L L F F F A S V I L A Q V F L V L K K K Q F E K V Q L Y E M N F
(2)         F   F A   V         V F                         Y
(3)         F   F A   V         I F                         S
(4)         L   L A   I         I L                         S
(5)         L   L A   I         I F                         S
(6)         F   L A   V         I F                         A
(7)         L   F G   V         I F                         A
```

Fig. 3.4 The amino acid sequences of the cytochromes *f*

	1. Rice	2. Wheat	3. Spinach	4. Pea	5. *Vicia faba*	6. Tobacco
2. Wheat	98					
3. Spinach	92	92				
4. Pea	90	89	90			
5. *Vicia faba*	91	90	90	96		
6. Tobacco	93	93	94	90	91	
7. Liverwort	83	84	84	80	81	84

Table 3.3 The matrix of % identities of the cytochromes *f*

3.4 Amino Acid Sequences of Cytochromes c_2

The cytochrome c_2 sequences are aligned in Fig. 3.5. The organisms are (**1**) *Rhodopseudomonas acidophila* (ATCC 25092), (**2**) *Rhodopila* (f. Rhodopseudomonas) *globiformis* (strain 7950, DSM 161), (**3**) *Rhodopseudomonas viridis* (NTHC 133), (**4**) *Nitrobacter agilis* (ATCC 25391), (**5**) *Rhodomicrobium vannielii* (ATCC 17100), (**6, 8**) *Rhodospirillum molischianum* (ATCC 14031), (**7, 9**) *Rhodospirillum fulvum* (ATCC 15798), (**10**) *Rhodobacter* (f. Rhodopseudomonas) *capsulatus,* strain Saint Louis (ATCC 23782), (**11**) *Rhodopseudomonas* TJ12 isolated by P Weaver and described as a marine *capsulata* (Ambler 1980), (**12**) *Erythrobacter* sp., strain OCh 114

(ATCC 33942), (**13**) *Paracoccus denitrificans* (ATCC 13543), (**14**) *Rhodobacter* (f. Rhodopseudomonas) *sphaeroides* (ATCC 17023), (**15**) *Rhodospirillum rubrum* (ATCC 11170), (**16**) *Aquaspirillum itersonii* (ATCC 12639), (**17**) *Rhodospirillum photometricum* strain SP113, (**18**) *Agrobacterium tumefaciens* (IIChrys), (**19**) *Rhodopseudomonas palustris* strain 2.1.6 (ATCC 17001), (**20**) *Rhodopseudomonas palustris* 2.1.37 (ATCC 17007).

Amino acid sequences (**1, 3, 5, 6, 7, 8, 9, 10, 14, 17, 19, 20**) are from Ambler *et al.* (1979a); (**2, 11**) are from Ambler (1980) and Ambler *et al.* (1987a); (**13**) is from Ambler *et al.* (1981b) which is a revision of Timkovich *et al.* (1976); (**15**) is from Dus *et al.* (1968); (**4**) is from Tanaka *et al.* (1982); (**12**) is from Okamura *et al.* (1987); (**16**) is from Woolley (1987); (**18**) is from van Beeumen (1980).

Residue numbers are for the complete alignment and in some cases will include gaps. Some sequence information is tentative and is shown in parentheses. In *Rps. acidophila* cytochrome c_2 the amide assignments in the region 116-119 are uncertain. Some evidence suggests an additional glycine between residues 68 and 69 in *R. photometricum* cytochrome c_2. The sequence shown for *Rb. capsulatus* is for the Saint Louis strain; the cytochrome c_2 of strain 2.3.1 contains threonine not leucine at position 92 and serine not threonine at position 105. In the iso-2 cytochromes c_2 of *R. molischianum* and *R. fulvum,* position 80 contains phenylalanine in about 30% of the molecules. The DNA sequences for the cytochromes c_2 genes of *Rb. sphaeroides* (Donohue *et al.*, 1986) and *Rb. capsulatus* (Daldal *et al.*, 1986) are in agreement with the protein sequences and in addition show the presence of N-terminal signal peptides. The alignment is similar to that of Ambler *et al.* (1979a) but differs in minimising the number of gaps placed in the regions 1 - 10, 90 - 105, 110 - 122.

The amino acid sequences of cytochrome c_2 from *Pseudomonas* AM1 (cytochrome c_H), *Rhodospirillum salexigens*, *Rhodobacter sphaeroides* ('suppressor' iso-c_2) and *Agrobacterium tumefaciens,* strain B2a have been determined but not published (R.P.Ambler, T.E.Meyer, J.van Beeumen, personal communication).

The matrix of identities is shown in Table 3.4.

Table 3.4 The matrix of % identities for the cytochromes c_2

	1	2	3	4	5	6	7	8	9	10	11	12	13	14	15	16	17	18	19
1. *R. molischianum* iso-1																			
2. *R. fulvum* iso-1	86																		
3. *R. molischianum* iso-2	62	70																	
4. *R. fulvum* iso-2	66	66	88																
5. *Rps. viridis*	43	45	49	49															
6. *N. agilis*	43	44	41	42	63														
7. *Rps. acidophila*	47	43	49	50	59	60													
8. *Rm. vannielii*	49	49	54	59	56	49	58												
9. *Rp. globiformis*	42	38	43	41	47	44	45	47											
10. *Ag. tumefaciens*	40	42	41	42	47	50	43	45	41										
11. *R. photometricum*	40	35	41	41	42	38	36	38	38	37									
12. *R. rubrum*	38	38	39	38	41	40	42	42	37	40	58								
13. *Aq. itersonii*	35	36	37	37	41	37	39	40	36	35	48	55							
14. *Rb. capsulatus*	35	36	36	37	42	39	33	36	33	33	38	47	47						
15. *Rhodopseudomonas* TJ12	32	32	35	35	38	32	33	34	32	36	37	44	45	70					
16. *Erythrobacter* sp.	31	30	29	33	34	33	39	35	31	37	35	39	39	60	53				
17. *P. denitrificans*	36	34	37	39	38	37	40	38	31	29	35	44	40	57	54	46			
18. *Rb. sphaeroides*	29	29	31	34	34	35	43	32	30	34	39	45	43	53	45	51	45		
19. *Rps. palustris* 6	37	36	43	40	49	36	42	45	40	37	43	44	47	49	42	42	43	39	
20. *Rps. palustris* 37	38	36	42	39	46	38	42	44	41	35	39	44	47	47	43	45	43	41	89

		1 … 10	HEME
(1)	*Rps. acidophila*	A G D P D A G Q K V F L K	C A A C H
(2)	*Rp. globiformis*	G S A P P G D P V E G K H L F H T I	C I L C H
(3)	*Rps. viridis*	<Q D A A S G E Q V F K Q	C L V C H
(4)	*N. agilis*	G D V F A G K A A F N K	C K A C H
(5)	*Rm. vannielii*	A G D P V K G E Q V F K Q	C K I C H
(6)	*R. molischianum* (iso-1)	A D A P P P A F N Q	C K A C H
(7)	*R. fulvum* (iso-1)	A D A P T A F N Q	C K A C H
(8)	*R. molischianum* (iso-2)	A D A P A G F T L	C K A C H
(9)	*R. fulvum* (iso-2)	A D A P P A F G M	C K A C H
(10)	*Rb. capsulatus*	G D A A K G E K E F N K	C K T C H
(11)	*Rhodopseudomonas* TJ12	E G D A A V G E K E F T K	C K T C H
(12)	*Erythrobacter* sp	E G D I E A G E K A F N K	C K S C H
(13)	*P. denitrificans*	<Q D G D A A K G E K E F N K	C K A C H
(14)	*Rb. sphaeroides*	<Q E G D P E A G A K A F N Q	C Q T C H
(15)	*R. rubrum*	E G D A A A G E K V S K K	C L A C H
(16)	*Aqu. itersonii*	G D A A K G A N V A K S	C G T C H
(17)	*R. photometricum*	A G D A A V G E K I A K A K	C T A C H
(18)	*Agr. tumefaciens*	E G D V A K G E A A F K R	C S A C H
(19)	*Rps. palustris* (2.1.6)	<Q D A A K G E A V F K Q	C M T C H
(20)	*Rps. palustris* (2.1.37)	<Q D A K A G E A V F K Q	C M T C H

	20 … 30	40 … 50
(1)	K I G P G A	K N G V G P S L N G V A N R K A G Q A
(2)	T D I K G	R N K V G P S L Y G V V G R H S G I E
(3)	S I G P G A	K N K V G P V L N G L F G R H S G T I
(4)	E I G E S A	K N K V G P E L N G L D G R H S G A V
(5)	Q V G P T A	K N G V G P E Q N D V F G Q K A G A R
(6)	S I D A G	K N G V G P S L S G A Y G R K V G L A
(7)	S I E A G	K N G V G P S L S G A Y G R K V G L A
(8)	S V E A G	K N G V G P S L A G V Y G R K A G T I
(9)	S V E A G	K N G V G P S(L A G V Y)G R K A G T L
(10)	S I I A P D G T E I V	K G A K T G P N L Y G V V G R T A G T Y
(11)	S I T A P D G T D I V	K G G K V G P N L Y G V I G R Q A G S Y
(12)	Q I V S D A G E E I V	K G G R T G P N L Y G V L G R Q A G T A
(13)	M I Q A P D G T D I I	K G G K T G P N L Y G V V G R K I A S E
(14)	V I V D D S G T T I A G R	N A K T G P N L Y G V V G R T A G T Q
(15)	T F D Q G G	A N K V G P N L F G V V F E N T A A H K
(16)	S F E Q G G	A K K Q G P N L F G I T T R G P G K A
(17)	D L N K G G	P I K V G P P L F G V F G R T T G T F
(18)	A I G E G A	K N K V G P Q L N G I I G R T A A G D
(19)	R A D	K N M V G P A L G G V V G R K A G T A
(20)	R A D	K N M V G P A L G G V V G R K A G T A

	60	70 … 80
(1)	E G F A Y S D A N K N	S G L T W D E A T F K E Y I T
(2)	P G Y N Y S E A N I K	S G I V W T P D V L F K Y I E
(3)	E G F S Y S D A N K N	S G I T W T E E V F R E Y I R
(4)	E G Y A Y S P A N K A	S G I T W T E A E F K E Y I K
(5)	P G F N Y S D A M K N	S G L T W D E A T L D K Y L E
(6)	P N Y K Y S P A H L A	S G M T I D D A M L T K Y L A
(7)	P N Y K Y S A A H L A	S G M T I D E A M L T N Y L A
(8)	S G F K F S D P H I K	S G L T W D E P T L T K Y L A
(9)	A G F K F S D P H A K	S G L T W D E P T L T K Y L A
(10)	P E F K Y K D S I V A L G A	S G F A W T E E D I A T Y V K
(11)	P D F R Y K D S I V E L G E	S G F V W T A E E I A V Y M Q
(12)	D F R Y G D D L V A A G E	A G L V W D A D N F V E Y V T
(13)	E G F K Y G E G I L E V A E K N	P D L T W T E A D L I E Y V T
(14)	A D F K G Y G E G M K E A G A	K G L A W D E E H F V Q Y V Q
(15)	D N Y A Y S E S Y T E M K A	K G L T W T E A N L A A Y V K
(16)	E G F N Y S P S Y K A A A A	K G F A W D A A T L Q D Y I T
(17)	A G Y S Y S P G Y V T M G Q	K(G)H T W D D N A L K A Y L L
(18)	P D Y N Y S N A M K K A G G	E G L V W T P Q E L R D F L S
(19)	A G F T Y S P L N H N S G E	A G L V W T Q E N I I A Y L P
(20)	A G F T Y S P L N H N S G E	A G L V W T A D N I I N Y L N

```
                   90                  100                 110
(1)   A P Q K K V P G                             T K M     T F P G L
(2)   H P Q K I V P G                             T K M     G Y P G Q
(3)   D P K A K I P G                             T K M     I F A G I
(4)   D P K A K V P G                             T K M     V F A G I
(5)   N P K A V V P G                             T K M     V F V G L
(6)   N P K E T I P G                             N K M G A A F G G L
(7)   N P K A T I P G                             N K M G A S F G G L
(8)   D P K T V I P G                             N K M     V F A G L
(9)   D P K G V I P G                             N K M     V F A G L
(10)  D P G A F L K E K L D D K K             A K T G M     A F K L A
(11)  D P T A F L K E K T G N K R             A R S G M     T H K Q N
(12)  D P R A F L R A Y L D D S K             A K S K M     A Y K L R
(13)  D P K P W L V K M T D D K G             A K T K M     T F K M G
(14)  D P T K F L K E Y T G D A K             A K G K M     T F K L K
(15)  D P K A F V L E K S G D P K             A K S K M     T F K L T
(16)  D P T A F L S N K T G D A A             A R D K M     T F K L A
(17)  D P K G Y V Q A K S G D P K             A N S K M     I F R L E
(18)  A P K K K I P G                             N K M     A L A G I
(19)  D P N A Y L K K F L T D K G Q A D K A T G S T K M     T F K L A
(20)  D P N A F L K K F L T D K G K A D Q A V G V T K M     T F K L A

               120                 130
(1)   P(N E A D)R D N I W A Y L S Q F K A D G S K
(2)   P D P Q K R A D I I A Y L E T L K
(3)   K D E Q K V S D L I A Y L K Q F N A D G S K K
(4)   K K D S E L D N L W A Y V S Q F D K D G K V K A K
(5)   K N P Q D R A D V I A Y L K Q L S G K
(6)   K N P A D V A A V I A Y L K T V K
(7)   K K P E D V K A V I E Y L K T V K
(8)   K N P D D V K A V I E Y L K T L K
(9)   K N P A D V A(A V I A Y)L K S L
(10)  K G G E       D V A A Y L A S V V K
(11)  K N Q E       D V I A F L A S V V T E
(12)  S G G E       D I A A Y L A S V S G S S S
(13)  K N Q A       D V V A F L A Q N S P D A G G D G G E A A
(14)  K E A D   A H N I W A Y L Q Q V A V R P
(15)  K D D E   I E N V I A Y L K T L K
(16)  K P D E   R A D V I A Y L A T L K
(17)  K D D D   V A N V I A Y L H T M K
(18)  S K P E E L D N L I A Y L I F S A S S K P A E
(19)  N D Q Q   R K D V A A Y L A T L K
(20)  N E Q Q   R K D V V A Y L A T L K
```

Fig. 3.5 The amino acid sequences of the cytochromes c_2

3.5 Amino Acid Sequences of Cytochromes c_3

The cytochrome c_3 sequences are aligned in Fig. 3.6. The organisms are (**1**) *Desulfovibrio desulfuricans* (Norway 4), (**2**) *Desulfovibrio desulfuricans* (El Agheila Z, NCIB 8380), (**3**) *Desulfovibrio salexigens* (NCIB 8403), (**4**) *Desulfovibrio vulgaris* (Hildenborough, NCIB 8303), (**5**) *Desulfovibrio vulgaris* (Miyazaki), (**6**) *Desulfovibrio gigas* (NCIB 9332), (**7**) *Desulfuromonas acetoxidans*. The last can live in syntrophic association with a green sulfur bacterium, this syntrophy originally being called '*Chloropseudomonas ethylica* 2K' (Pfennig and Biebl, 1976).

Amino acid sequence (**1**) is from Bruschi (1981); (**2**) is from Ambler *et al.* (1971); (**3**) is from Ambler (1973a) and contains regions of uncertainty indicated by the parentheses; (**4**) is from Ambler (1968); (**5**) is from Shinkai *et al.* (1980); (**6**) is from Ambler *et al.* (1969); (**7**) is from Ambler (1971). Sequence (**4**) was also determined by Trousil and Campbell (1974) and the gene sequence is known (Voordouw and Brenner, 1986).

Heme nomenclature and the pattern of histidine ligation (boxed) is from Haser *et al.* (1979a) based on the x-ray crystallographic structure of *D. desulfuricans* (Norway 4) (Chapter 5.3). The sequence alignment is similar to that proposed from the comparison of the x-ray crystallographic structures of *D. desulfuricans* (Norway) and *D. vulgaris* (Miyazaki) (Higuchi *et al.*,1981a). This differs from the alignment of Bruschi (1981) mainly at the N-terminus.

The matrix of identities is shown in Table 3.5.

	1. *D. desulfuricans* N4	2. *D. desulfuricans* EA Z	3. *D. salexigens*	4. *D. vulgaris* H	5. *D. vulgaris* M	6. *D. gigas*
2. *D. desulfuricans* EA Z	45					
3. *D. salexigens*	46	66				
4. *D. vulgaris* H	35	41	43			
5. *D. vulgaris* M	37	43	46	88		
6. *D. gigas*	36	39	36	52	53	
7. *Dsm. acetoxidans*	32	36	40	22	23	32

Table 3.5 Matrix of % identities for cytochromes c_3

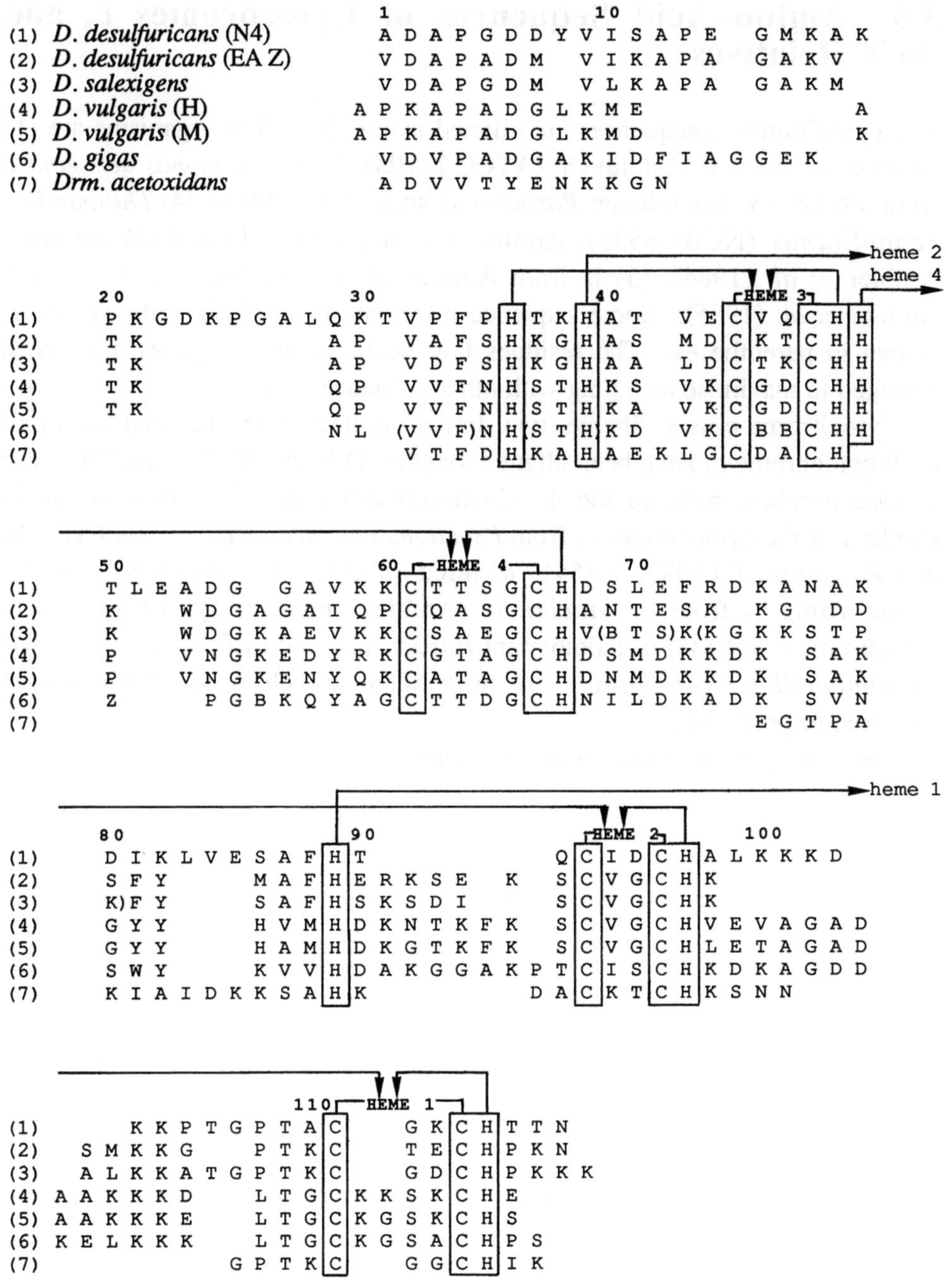

Fig. 3.6 The amino acid sequences of the cytochromes c_3

3.6 Amino Acid Sequences of Cytochromes c_4 and their Relatives

The cytochrome c_4 sequences are aligned in Fig. 3.7. The organisms are (**1**) *Azotobacter vinelandii* strain OP (ATCC 13705), (**2**) *Pseudomonas aeruginosa* strain P6009, (**3**) halotolerant *Paracoccus* sp.(ATCC 12084), (**4**) *Thiobacillus neapolitanus* (NCIB 8539). Amino acid sequences (**1**) and (**2**) are from Ambler *et al.* (1984); (**3**) is from Ambler *et al.* (1987b); and (**4**) is from Ambler *et al.* (1985). Some sequence information for (**2**) is tentative and is shown in parentheses. The residues that occur in these regions have been arranged to maximise matching with other sequences.

Cytochrome c_4 is a diheme, two domain molecule and the two halves of each cytochrome c_4 have been aligned together (1A and 1B; 2A and 2B). The residue numbers refer to the *A. vinelandii* sequence. The N-terminal 18 residues of the cytochromes c_4 from *Pseudomonas mendocina* (strain CH-110) and *Ps. stutzeri* (ATCC 17641) differed in only two positions from the cytochrome c_4 from *A. vinelandii* (Ambler and Murray, 1973). Both cytochrome *c*-554(548) of halotolerant *Paracoccus* and cytochrome *c*-554(547) of *Thiobacillus neopolatinus* are monoheme cytochromes of 83 and 91 residues, respectively.

The matrix of identities is shown in Table 3.6.

1A. *A. vinelandii* (1-105)					
83	2A. *Ps. aeruginosa* (1-96)				
47	52	3. Halotolerant *Paracoccus*			
31	37	33	4. *T. neapolitanus*		
24	31	31	24	1B. *A. vinelandii*	
27	33	31	27	77	2B. *Ps. aeruginosa*

Table 3.6 The matrix of % identities for the cytochromes c_4 and their relatives

```
                                  1                 10          HEME
(1A) A. vinelandii (1-105)        A G D A A A G Q G K    A A V C G A C H
(2A) Ps. aeruginosa (1-96)        A G D A A A G Q A K    A A V C G A C H
(3)  Halotolerant Paracoccus      A G D A A A G E D K    I G T C V A C H
(4)  T. neapolitanus              A G D A A A G K T L Y D A S C A S C H
                                        110                  120
(1B) A. vinelandii (106-190)   ...L Y R G G K L A D G    M P A C T G C H
(2B) Ps. aeruginosa (97-181)   ...L F R G G K I A E G    M P A C T G C H

          20               30               40
(1A)      G P D G N S A A P  N F P K L A G Q G E R Y L L K Q M Q D I K
(2A)      G(P B G B S A A P  B F P)K L A G Q G E R Y L L K Q M H D I K
(3)       G T D G Q G L A P  I Y P N L T G Q S A T Y L E S S I K A Y R
(4)       G M Q A Q G Q G M    F P K L A G L T S E R I K T T L V A F K
                    130              140               150
(1B)      S P N G E G N T P A A Y P R L S G Q H A Q Y V A K Q L T D F R
(2B)      G(P S S V G I A T A G F P H)L G G Q H A T Y V A K Q L T D F R

          50                60                      70
(1A)      A G T K P G A P E G S G R     K V L E M T G M L D N F S D Q D
(2A)      D G K R T                       V L E M T G L L T B(L S B Z D)
(3)       D G Q R K G                 G N A A L M T P M A Q G L S D E D
(4)       S G D T A T L K K E G L G G P N S A I M A P N A A G L S E Q D
                        160                     170
(1B)      E G A R T N D               G D N M I M R S I A A K L S N K D
(2B)      E G T R T N(D               G D T)K I M Q S I A A K L S N K D

            80                  90                100
(1A)      L A D L A A Y F T S Q K P T V G A A D P Q L V E A G E T... → 1B
(2A)      I A D L A A Y F A S Q K M S(V G M A B P B)L(V Z A G)E A... → 2B
(3)       I A D I A A Y Y S S Q E
(4)       M D N L S A Y I A T L K
          180                 190
(1B)      I A A I S S Y I Q G L H
(2B)      I A A I S S Y I Q G L H
```

Fig. 3.7 The amino acid sequences of the cytochromes c_4 and relatives

3.7 Amino Acid Sequences of the Cytochromes c_5

The cytochrome c_5 sequences are aligned in Fig. 3.8. The organisms are (**1**) *Pseudomonas mendocina* CH-110 (sequence from Ambler and Taylor, 1973) and (**2**) *Azotobacter vinelandii* (sequence from Carter *et al.*, 1985).

The sequences show 75% identity.

```
                                   1                 10
(1)  Pseudomonas mendocina         A A S A G G G A R S A D D I I
(2)  Azotobacter vinelandii                G G G A R S G D D V V

           HEME                  30             40
(1)  A K H C N A C H G A G V L G A P K I G D T A A W K E R A D H Q
(2)  A K Y C N A C H G T G L L N A P K V G D S A A W K T R A D A K

         50                60                70
(1)  G G L D G I L A K A I S G I N A M P P K G T C A D C S D D E L
(2)  G G L D G L L A Q S L S G L N A M P P K G T C A D C S D D E L

         80
(1)  R E A I Q K M S G L
(2)  K A A I G K M S G L
```

Fig. 3.8 The amino acid sequences of the cytochromes c_5

3.8 Amino Acid Sequences of Cytochromes *c*-555

The cytochrome *c*-555 sequences are aligned in Fig. 3.9. The organisms are (**1**) *Chlorobium limicola* (f. thiosulfatophilum), PM (NCIB 8346), (**2**) *Prosthecochloris aestuarii* (this is the green sulfur partner in the syntrophic mixture originally called '*Chloropseudomonas ethylica* 2K')

Amino acid sequences (**1**) and (**2**) are from van Beeumen *et al.*(1976) (strain L of *Chl. limicola* probably contains an identical cytochrome *c*-555).

The sequences show 53% identity.

cytochrome c_6 from *Porphorydium cruentium* (not shown) and *Microcystis aeruginosa* (Cohn *et al.*, 1989a), and the gene sequence of cytochrome c_6 from *Chlamydomonas reinhardtii* (not shown; Merchant & Bogorad, 1987) have been reported recently. Some amides in the sequence of *Monochrysis lutherii* cytochrome *c*-553 were not characterised in Laycock (1972) but were described in Laycock (1975).

X is monomethyllysine.

The matrix of identities is shown in Table 3.7.

1. *Alaria esculenta*													
2. *Petalonia fascia*	81												
3. *Bumilleriopsis filiformis*	80	80											
4. *Porphyra tenera*	70	76	80										
5. *Synechococcus sp.*	60	63	67	67									
6. *Synechococcus lividus*	56	59	64	69	83								
7. *Anacystsis nidulans*	55	57	58	60	61	56							
8. *Microcystis aeruginosa*	55	54	60	58	59	58	64						
9. *Spirulina maxima*	51	56	56	56	62	61	54	58					
10. *Plectonema boryanum*	53	54	62	61	58	60	59	59	59				
11. *Anabaena variabilis*	54	52	60	60	62	59	56	62	60	59			
12. *Aphanizomenon flos-aquae*	52	52	56	54	65	64	58	62	53	57	67		
13. *Monochrysis lutheri*	46	48	55	51	48	49	43	43	45	51	47	41	
14. *Euglena gracilis*	41	45	43	40	48	45	44	42	45	44	41	41	39

Table 3.7 Matrix of % identities of the algal cytochromes *c*-553

```
                                       1                 10
(1)  Alaria esculenta                  I D I N N G E N I F
(2)  Petalonia fascia                  V D I N N G E S V F
(3)  Bumilleriopsis filiformis         A D I E N G E R I F
(4)  Porphyra tenera                   A D L D N G E K V F
(5)  Synechococcus sp.                 A D I A D G A K V F
(6)  Synechococcus lividus             A D L A N G A K V F
(7)  Anacystis nidulans                A D L A H G G Q V F
(8)  Microcystis aeruginosa                    D G A S I F
(9)  Spirulina maxima                  G D V A A G A S V F
(10) Plectonema boryanum               A D A A A G G K V F
(11) Anabaena variabilis               A D S V N G A K I F
(12) Aphanizomenon flos-aquae          A D T V S G A A L F
(13) Monochrysis lutheri               G D I A N G E Q V F
(14) Euglena gracilis                          G G A D V F

              HEME
                         20                  30                  40
(1)   T A N C S A C H A G G N N V I M P E K T L K K   D A L A D N
(2)   T A N C S A C H A G G N N V I M P E K T L K K   D A L E E N
(3)   T A N C A A C H A G G N N V I M P E K T L K K   D A L E A N
(4)   S A N C A A C H A G G N N A I M P D K T L K K   D V L E A N
(5)   S A N C A A C H M G G G N V V M A N K T L K K   E A L E Q F
(6)   S G N C A A C H M G G G N V V M A N K T L K K   E A L E Q F
(7)   S A N C A S C H L G G R N V V N P A K T L E K   A D L D E Y
(8)   S A N C A S C H M G G K N V V N A A K T L K K   E D L V G G
(9)   S A N C A A C H M G G R N V I V A N K T L S K   S D L A K Y
(10)  N A N C A A C H A S G G G Q I N G A K T L K K   N A L T A N
(11)  S A N C A S C H A G G K N L G V A Q K T L K K   A D L E K Y
(12)  K A N C A Q C H V G G G N L V N R A K T L K K   E A L E K Y
(13)  T G N C A A C H S V Z Z Z X T L E L S S L W K A K S Y L A N
(14)  A D N C S T C H V N G G N V I S A G K V L S K   T A I E E Y

                           50                  60                  70
(1)   K M V S       V N A I T   Y Q V T N G K N A M P A F G S R L
(2)   E M N N       I K S I T   Y Q V T N G K N A M P A F G G R L
(3)   G M N A       V S A I T   Y Q V T N G K N A M P A F G G R L
(4)   S M N T       I D A I T   Y Q V Q N G K N A M P A F G G R L
(5)   G M N S       A D A I M   Y Q V Q N G K N A M P A F G G R L
(6)   G M N S       E D A I I   Y Q V Q H G K N A M P A F A G R L
(7)   G M A S       I E A I T T   Q V T N G K G A M P A F G A K L
(8)   K N A         V E A I V T   Q V T K G K G A M P A F G G R L
(9)   L K G F D D D A V A A V A Y Q V T N G K N A M P G F N G R L
(10)  G K D T       V E A I V A   Q V T N G K G A M P A F K G R L
(11)  G A Y S       A M A I G A   Q V T N G K N A M P A F K G R L
(12)  N M Y S       A K A I I A   Q V T H G K G A M P A F G I R L
(13)  F N G D       E S A I V   Y Q V T N G K N A M P A F G G R L
(14)  L D G G Y T   K E A I E   Y Q V R N G K G P M P A W E G V L

                         80                  90
(1)   A E T D I E D V A N F V L T Q S D K G W D
(2)   S E T D I E D V A N F V I S Q S Q K G W '
(3)   S D S D I E D V A N Y V L S Q S E Q G W D
(4)   V D E D I E D A A N Y V L S Q S E K G W
(5)   S E A Q I E N V A A Y V L D Q S S N K W A G
(6)   T D E Q I Q D V A A Y V L D Q A A K G W A G
(7)   S A D D I E G V A S Y A L D Q S G K E W
(8)   G A E D I E A V A N Y V L A E A E K G W
(9)   S P K Q I E D V A A Y V V D Q A E K G W
(10)  S D D Q I Q S V A L Y V L D K A E K G W
(11)  K P E E I Z B V A A Y V L G K A E A E W K
(12)  K A E Q I E N V A A Y V L E Q A D N G W K K
(13)  E D D E I A N V A S Y V L S K A G
(14)  S E D E I V A V T D Y V Y T Q A G G A W A N V S
```

Fig. 3.10 The amino acid sequences of the algal cytochromes *c*-553 (c_6)

3.10 Amino Acid Sequences of Cytochromes *c*-551 (Cytochromes c_7)

The cytochrome *c*-551 sequences are aligned in Fig. 3.11. The organisms are (**1**) *Pseudomonas stutzeri* strain 221, (**2**) *Pseudomonas mendocina* strain CH110, (**3**) *Pseudomonas fluorescens* strain C18, (**4**) *Pseudomonas aeruginosa* strain P6009, (**5**) *Azotobacter vinelandii* (NCIB 8789), (**6**) *Rhodocyclus* (f. Rhodospirillum) *tenue* strain 3761, (**7**) *Rhodocyclus* (f. Rhodopseudomonas) *gelatinosa* (ATCC 17011), (**8**) *Nitrosomonas europaea* (ATCC 25978).

Amino acid sequences (**1**, **2**, **3**) are from Ambler and Wynn (1973); (**4**) is from Ambler (1963); (**5**, **6**, **7**) are from Ambler *et al.* (1979b); (**8**) is from Miller and Nicholas (1986). Only the N-terminal region of the latter sequence has been determined. Residue numbers are for the complete alignment and in individual cases will include gaps. The DNA sequence for the gene of *Ps. aeruginosa* cytochrome *c*-551 is in agreement with protein sequence and, in addition, shows the presence of an N-terminal signal peptide (Nording *et al.*,1990; Arai *et al.*, 1990). The gene is located 50 base pairs downstream of the cytochrome cd_1 gene.

In some cases further information has been obtained within a species. Of the cytochromes c_7 isolated from 9 strains of *Ps. aeruginosa* only one substitution was observed in one strain (Ambler, 1974). Within *Ps. fluorescens* biotype C, 0 - 4 substitutions were observed in pairwise comparisons of the cytochromes c_7 from 6 strains (Ambler, 1974). Examination of several strains of *Ps. stutzeri* led to the general conclusion that their cytochrome c_7 sequences fell into two distinct groups which corresponded to two clusters on the basis of GC content (Ambler 1978). The high GC group (64 - 66% G + C) contains strains 221, 222, 226, 227, 273, 275 and 321, which differ by less than 5% in their cytochrome c_7 sequences. The low GC group (62 - 63% G + C) contains strains 220, 224, 228, 320 and *Ps. perfectomarinus* (J.van Beeumen and R.P.Ambler, personal communication) which has been recently reclassified as a *stutzeri* (Dohler *et al.*,1987). These low GC strains are described as *'stanierii'* by R.P.Ambler and their cytochrome c_7 sequences differ by less than 3%. The cytochrome c_7 sequences of members of the two separate clusters differ by about 10%.

Senn and Wüthrich (1983) examined strain 29-1 of *Rc. gelatinosa* and found evidence for the presence of two isocytochromes which could not be separated. Comparison of the N-terminal 27 residues with those of the *Rc. gelatinosa* ATCC 17011 sequence reveals only 2 differences for the iso-1 cytochrome and 15 differences for the iso-2 cytochrome.

The amino acid sequences of cytochrome *c*-551 from *Rhodocyclus purpureus*, *Rc. tenue* 2761, *Ps. stutzeri* 320 and *Methylophilus methylotrophus* have been determined but not published (R.P.Ambler, personal communication)

The matrix of identities is shown in Table 3.8.

```
                         1                 10      HEME        20
(1) Ps. stutzeri         Q D G E A L F K S K P C A A C H S I D A K L V G
(2) Ps. mendocina        A S G E E L F K S K P C G A C H S V Q A K L V G
(3) Ps. fluorescens      E D G A A L F K S K P C A A C H T I D S K M V G
(4) Ps. aeruginosa       E D P E V L F K N K G C V A C H A I D T K M V G
(5) A. vinelandii        E T G E E L Y K T K G C T V C H A I D S K L V G
(6) Rc. tenue            A D E S A L A Q T K G C L A C H N P E K K V V G
(7) Rc. gelatinosa       A T P A E L A T K A G C A V C H Q P T A K G L G
(8) Nm. europaea             N A D L A K K N N C I A C H Q V E T K V V G

                 30                  40                  50
(1)     P A F K E V A A K Y A G Q D G A A D L L A G H I K N G S Q G V
(2)     P A L K D V A A K N A G V D G A A D V L A G H I K N G S T G V
(3)     P A L K E V A A K N A G V K D A D K T L A G H I K N G T Q G N
(4)     P A Y K D V A A K F A G Q A G A E A E L A Q R I K N G S Q G V
(5)     P S F K E V T A K Y A G Q A G I A D T L A A K I K A G G S G N
(6)     P A Y G W V A K K Y A G Q A G A E A K L V A K V M A G G Q G V
(7)     P S Y Q E I A K K Y K G Q A G A P A L M A E R V R K G S V G I
(8)     P A L L D I A A K Y A D N N S A A T Y L A G K I K...

                60                  70                  80
(1)     W           G P I P M       P P N P V T E E E A K I L A E W I
(2)     W           G A M P M       P P N P V T E E E A K T L A E W V
(3)     W           G P I P M       P P N Q V T D A E A L T L A Q W V
(4)     W           G P I P M       P P N A V S D D E A Q T L A K W V
(5)     W           G Q I P M       P P N P V S E A E A K T L A E W V
(6)     W A K Q L G A E I P M       P A N N V T K E E A T R L V K W V
(7)     F           G K L P M T P T P P A R I S D A D L K L V I D W I
(8)

              90
(1)     L S Q K
(2)     L T L K
(3)     L S L K
(4)     L S Q K
(5)     L T H K
(6)     L S L K Q I D Y K
(7)     L K T P
(8)
```

Fig. 3.11 The amino acid sequences of the cytochromes *c*-551 (cytochromes c_7)

	1	2	3	4	5	6	7
1. *Ps. mendocina*							
2. *Ps. stutzeri*	79						
3. *Ps. fluorescens*	69	73					
4. *Ps. aeruginosa*	63	70	69				
5. *A. vinelandii*	64	67	62	63			
6. *Rc. tenue*	52	48	49	56	48		
7. *Rc. gelatinosa*	40	35	34	40	41	36	
8. *N. europaea*	48	52	48	46	40	42	35

Table 3.8 Matrix of % identities for the cytochromes *c*-551

3.11 Amino Acid Sequences of Class II Cytochromes *c* (Cytochrome c_8)

The cytochrome c_8 sequences are aligned in Fig. 3.12. The organisms are (**1**) halotolerant *Paracoccus* sp. (ATCC 12084), (**2**) *Alcaligenes* sp. (f. *'Pseudomonas denitrificans'*, NCIB 11015), (3) *Rhodocyclus* (f. Rhodopseudomonas) *gelatinosa* (ATCC 17011), (**4**) *Rhodobacter* (f. Rhodopseudomonas) *sphaeroides* (ATCC 17023), (**5**) *Rhodospirillum photometricum* (strain SP113), (**6**) *Rhodospirillum rubrum* S1 (ATCC 11170), (**7**) *Rhodobacter* (f. *Rhodopseudomonas*) *capsulatus* (strain SP7), (**8**) *Rhodopseudomonas* TJ12 (P F Weaver isolate), (**9**) *Chromatium vinosum* (strain D - ATCC 17899), (**10**) *Rhodocyclus* (f. Rhodospirillum) *tenue* strain 3761, (**11**) *Rhodospirillum molischianum* (ATCC 14031), (**12**) *Rhodospirillum fulvum* (ATCC 15798), (**13**) *Rhodopseudomonas palustris* 2.1.37 (ATCC 17007), (**14**) *Rhodopseudomonas palustris* 2.1.37 (ATCC 17007), (**15**) *Agrobacterium tumefaciens* (strain B2a), (**16**) *Agrobacterium tumefaciens* (strain Apple).

Amino acid sequences (**1** - **15**) appear in Ambler *et al.* (1981a) and some have been published individually: (**1**) is from Ambler *et al.* (1987b); (**2**) is from Ambler (1973b); (**3**) and (**10**) are from Ambler *et al.* (1979b); (**6**) is from Meyer *et al.* (1975); (**9**) is from Ambler *et al.* (1979c); (**16**) is from Tempst and van Beeumen (1983). The sequence of cytochrome *c'* from *Rb. capsulatus* M110 (which is a carotenoid-less mutant of Saint Louis (ATCC

23782)) differs from that of strain SP7 by approximately 11 residues (Ambler *et al.*, 1981b).

The alignment is similar to those of Ambler *et al.* (1981b) and Meyer and Kamen (1982) but differs in minimising the number of gaps placed in the regions 50 - 60, 68 - 78 and 107 - 112. Residue numbers are for the complete alignment and in individual cases will include gaps.

Cytochromes (**14** - **16**) are low-spin (for example, *Rps. palustris* *c*-556; see Chapter 5.2.4).

The matrix of identities is shown in Table 3.9.

```
                                                10                  20
(1)   Paracoccus sp.                A E P E D A I H Y R Q S A L S V M G W Q M G
(2)   Alcaligenes sp.           Q F A K P E D A V K Y R Q S A L T L M A S H F G
(3)   Rc. gelatinosa            Q F Q K P G D A I E Y R Q S A F T L I A N H F G
(4)   Rb. sphaeroides               A D A E H V V E A R K G Y F S L V A L E F G
(5)   R. photometricum              A S P E A Y V E Y R K Q A L K A S G D H M K
(6)   R. rubrum                     A D P A A Y V E Y R K S V L S A T S N Y M K
(7)   Rb. capsulatus                A D T K E V L E A R E A Y F K S L G K S M K
(8)   Rps. TJ12                       D G M E T V K A R Q D Y F K S L G G A M K
(9)   Chr. vinosum              A G L S P E E Q I E T R Q A G Y E F M G W N M G
(10)  Rc. tenue                 E P A K S E D L I K W R Q S A Y Q V L H W N M D
(11)  R. molischianum           Q Q S K P E D L L K L R Q G L M Q T L K S Q W V
(12)  R. fulvum                 Q Q S K P E E L L K L R Q G L M Q T L K S Q W A
(13)  Rps. palustris                    Q T D V I A Q R K A I L K Q M G E A T K

(14)  Rps. palustris                    Q Q D L V D K T Q K L[M]K D N G R N M M
(15)  Ag. tumefaciens B2a                 A G E V E K R E G M[M]K Q I G G A M G
(16)  Ag. tumefaciens apple           A D G G T H D A R I A L[M]K K I G G A T G


                  30                  40                  50
(1)     P M G A M A Q G   D I E Y D A D E F A T R A N N L A A V A H L P   W E
(2)     R M T P V V K G   Q A P Y D A A Q I K A N V E V L K T L S A L P   W A
(3)     R V A A M A Q G   K A P F D A K V A A E N I A L V S T L S K L P   L T
(4)     P L A A M A K G   E M P Y D A A A A K A H A S D L V T L T K Y D P S D
(5)     A L S A I V K G   Q L P L N A E A A K H A E A I A A I M E S L P     A
(6)     A I G I T L K E   D L A V P N Q T A D H A K A I A S I M E T L P     A
(7)     A M T G V A K     S F D A E A A K A E A A A L E K I L A T D V A     P
(8)     A L S G V A K     N Y D A E A A K A E A A K L E A I L A T D I K     P
(9)     K I K A N L E G   E Y N A A Q V E A A A N V I A A I A N S G M G     A
(10)    R L K A N I D S P Q Y N K D D G I K A A N T I A A I A N S G M G     S
(11)    P I A G F A A G K A D L P A D A A Q R A E N M A M V A K L A P I
(12)    P I A G F A A G K A D L P A D A A Q R A E N M V L V A K L A P I
(13)    P I A A M L K G   E A K F D Q A V V Q K S L A A I A D D S K K L P   A

(14)    V L G A I A K G   E K P Y D Q A A V D A A L K Q F D E T A K D L P   K
(15)    S L A A I S K G   E K P F D A D T V K A A V T T I G T N A K A F P   E
(16)    A L G A I A K G   E K P Y D A E I V K A S L T T I A E T A K A F P   D


        60                  70                  80                  90
(1)     G F T E G T L Q G D D H G V E T D A L A D I G D D W E G F E E R Q E T
(2)     A F G P G T E G G D               A R P E I W S D A A S F K Q K Q Q A
(3)     A F G P G T D K G H G         T E A K P A V W S D A A G F K A A A D K
(4)     L Y A P G T S A D D V K G     T A A K A A I W Q D A D G F Q A K G M A
(5)     A F P E G T A G I A K         T E A K A V V W S K A D E F K A D A V K
(6)     A F P E G T A G I A K         T E A K A A I W K D F E A F K V A S K K
(7)     L F P A G T S S T D L P G   Q T E A K A A I W T N M A D F G A K G K A
(8)     L F A P G T S D A D F P G   E S E A K A S I W E N M E D F G A K G Q A
(9)     L Y G P G T D K N V G D V   K T R V K P E F F Q N M E D V G K I A R E
(10)    L F A A G T E T G K G W H   P T S V K P A F F T D G K K V G E V A V A
(11)    G W A K G T E A L P N G       E T K P E A F G S K S A E F L E G W K A
(12)    G W A K G T E A L P N S       E T K A E A F G A K S A Q F M E G W K A
(13)    L F P A D S K T G G D         T A A L P K I W E D K A K F D D L F A K

(14)    L F P D S V K G L K P F D S K Y S S S P K I W A E R A K F D T E I A D
(15)    Q F P A G T E T G             S A A A P A I W E N F E D F K A K A A K
(16)    Q F N P K D S T D             A E V N P K I W D N L D D F K A K A A K
```

```
                100                    110                    120
(1)     F K Q E A A T L A Q M V D D G E E F S A L R R Q V G A V G K S
(2)     F Q D N I V K L S A A A D A G   D L D K L R A A F G D V G A S
(3)     F A A A V D K L D A A G K T G   D F A Q I K A A V G E T G G A
(4)     F F E A V A A L E P A A G A G     Q K E L A A A V G K V G T G
(5)     S A D A A K A L A Q A A T A G   D T A Q M G K A L A A L G G T
(6)     S Q D A A L E L A S A A E T G   D K A A I G A K L Q A L G G T
(7)     M N D A G A E V I A A A N A G   D A T A F G A A L Q K L G G T
(8)     M H E A G M E L I A A A N T G   E A S A F G P A L K K L G G T
(9)     F V G A A N T L A E V A A T G   E A E A V K T A F G D V G A A
(10)    F N K E A N E L A K V A A T G   D A A A V K A Q F G K V G Q T
(11)    L A T E S T K L A A A A K A G     P D A L K A Q A A A T G K V
(12)    M A A E S T K L A A A A K A G     P D A L K A Q A A A T G K V
(13)    L A A A A T A A Q G T I K       D E A S L K A N I G G V L G N

(14)    F A K A V D G A K G K I K       D V D T L K A A M Q P I G K A
(15)    L G T D A D I V L A N L P G     D Q A G V A T A M K T L G A D
(16)    L S T D A E T A L A Q L P A     D Q A G V G N T L K T L G G N
```

```
         HEME 130
(1)     C K G C H D D F R A E
(2)     C K A C H D A Y R K K K
(3)     C K G C H D K F K E K
(4)     C K S C H D D F R V K R
(5)     C K G C H E T F R E
(6)     C K A C H K E F K A D
(7)     C K A C H D D Y R E E D
(8)     C K A C H D D Y R A E H
(9)     C K S C H E K Y R A K
(10)    C K A C H D D F R R K D
(11)    C K A C H E E F K Q D
(12)    C K A C H E E F K Q D
(13)    C K S C H D D F R A K K S

(14)    C G N C H E N F R D K E G
(15)    C G T C H Q T Y R L K K
(16)    C G A C H Q A Y R I K K D
```

Fig. 3.12 The amino acid sequences of the class II cytochromes *c* (cytochrome c_8)

Table 3.9 Matrix of % identities for the class II cytochromes *c*

	1	2	3	4	5	6	7	8	9	10	11	12	13	14	15
1. *Paracoccus* sp.															
2. *Alcaligenes* sp.	38														
3. *Rc. gelatinosa*	35	51													
4. *Rb. sphaeroides*	32	41	38												
5. *R. photometricum*	29	36	40	32											
6. *R. rubrum*	30	29	36	26	56										
7. *Rb. capsulatus*	23	30	26	39	36	37									
8. *Rhodopseudomonas* TJ12	24	28	24	37	33	34	70								
9. *Chr. vinosum*	28	29	32	23	26	24	26	31							
10. *Rc. tenue*	30	33	32	26	25	25	30	30	47						
11. *R. molischianum*	24	29	26	24	21	21	21	19	20	29					
12. *R. fulvum*	25	29	25	25	19	21	22	21	20	27	92				
13. *Rps. palustrus* (*c'*)	27	27	34	30	27	25	23	19	24	25	17	16			
14. *Rps. palustris* (*c*-556)	18	24	24	25	23	17	19	18	18	20	20	20	38		
15. *Ag. tumefaciens* (B2a)	21	30	29	31	30	24	29	29	25	22	15	14	33	32	
16. *Ag. tumefaciens* (apple)	19	25	24	28	29	19	26	30	20	17	13	11	33	34	59

3.12 Cytochrome *c* Peroxidase

A possible class I region of *Ps. aeruginosa* cytochrome *c* peroxidase is shown in Fig. 3.15. This is a cyanogen bromide peptide labelled II by Ronnberg (1987). Ronnberg *et al.* (1989) have determined the complete amino acid sequence of the enzyme (Fig. 3.13) in which this cyanogen bromide peptide is labelled CB4. This region of the molecule binds a high potential heme at cysteines 177 and 180 with His 181 and Met 254 the presumed axial ligands. The second heme is bound at cysteines 51 and 54 and is pentacoordinate at room temperature with His 55 as the fifth ligand. At low temperature, Ronnberg *et al.* propose that His 240 becomes the sixth ligand. It is this heme which is proposed to be the peroxidatic centre (Pettigrew & Moore, 1987: Chapter 3.5).

↓ indicates cleavage point of *Pseudomonas* elastase.

Some sequence positions showed evidence of microheterogeneity.

```
1                 10                  20                  30
D A L H D Q A S A L F K P I P E Q V T E L R G Q P I S E Q Q R E L G

        40                  50  HEME            60
K K L F F D P R L S R S H V L S C N T C H N V D T D G A G N V P T S
                                                G   G

  70                  80                  90                  100
V H G G Q K G P R A K D L G E Q A K G P I Q N S V Q M H S T P Q L V

            110                 120                 130
E Q T L G S I P E Y V D A F R K A F A K A G K P V S F D N M A L A I
            V                       P

      140                 150                 160                 170
E A Y E A T L V T P D S P F D L Y L I G D D K A L D A Q Q K K G L K
                                    K

                HEME                    190                 200  ↓
A F M D S G C S A C H N G I L G G Q A Y F P P F G L V K K P D A S V L

        210                 220                 230
P S G D K G R F A V T K T Q S D E Y V F R A A P L R N V A L T A P Y

  240                 250                 260                 270
F H S Q G V W Q L K D A V A I M G N A Q L G K Q L A P D D V E N I V

            280                 290                 300
A F L H S L S G K Q P R V E Y P L L P A S T E T T P R P A E
                                  W
```

Fig. 3.13 The amino acid sequence of cytochrome *c* peroxidase from *Pseudomonas aeruginosa*

3.13 The Reaction Centre Cytochrome from *Rhodopseudomonas viridis*

The photosynthetic reaction centre of *Rhodopseudomonas viridis* contains a tightly bound tetraheme cytochrome *c* (Chapter 5.4), the sequence of which has been determined by Weyer *et al.* (1987a) as part of a complete structural study of the reaction centre (Deisenhofer *et al.*, 1985). The amino acid sequence shows four Cys-?-?-Cys-His heme binding sites (Fig. 3.14) but that is the limit of the sequence similarity to the known cytochrome *c* classes. There is some internal sequence homology between the four domains of the protein, consistent with remote gene duplication events.

* indicates the N-terminus of the protein, which is a cysteine substituted at the sulfydryl group by a glycerol carrying two fatty acids (Weyer *et al.*, 1987a).

```
                  10                  20                  30
*C F E P P P A T T T Q T G F R G L S M G E V L H P A T V K A K K E R

       40                  50                  60
 D A Q Y P P A L A A V K A E G P P V S Q V Y K N V K V L G N L T E A

  70                  80          HEME 1                      100
 E F L R T M T A I T E W V S P Q E G[C]T Y[C H]D E N N L A S E A K Y

           110                 120                 130  HEME 2
 P Y V V A R R M L E M T R A I N T N W T Q H V A Q T G V T[C]Y T[C H]

     140                 150                 160                 170
 R G T P L P P Y V R Y L E P T L P L N N R E T P T H V E R V E T R S

                  180                 190                 200
 G Y V V R L A K Y T A Y S A L N Y D P F T M F L A N D K R Q V R V V

        210                 220                 230
 P Q T A L P L V G V S R G K E R R P L S D A Y A T F A L M M S I S D

  240      HEME 3     250                 260                 270
 S L G T N[C]T F[C H]N A Q T F E S W G K K S T P Q R A I A W W G I R

           280                 290                 300
 M V R D L N M N Y L A P L N A S L P A S R L G R Q G E A P Q A D

 HEME 4   310                 320                 330
[C]R T[C H]Q G V T K P L F G A S R L K D Y P E L G P I K A A A K
```

Fig. 3.14 The amino acid sequence of the reaction centre cytochrome *c* from *Rhodopseudomonas viridis*

3.14 Amino Acid Sequences of Cytochromes *c* Which Do Not Fall Into the Established Sequence Classes

Selected members of known sequence classes are compared in Fig. 3.15 with cytochromes *c* which have no characterised close relatives. Sequences (**1** - **6**) are for (**1**) *Rhodobacter sphaeroides* cytochrome c_2 (Ambler *et al.*, 1979a), (**2**) tuna cytochrome *c* (Kreil, 1963), (**3**) *Pseudomonas aeruginosa* cytochrome *c*-551 (Ambler, 1963), (**4**) *Chlorobium limicola* (f. thiosulfatophilum) cytochrome *c*-555 (Van Beeumen *et al.*, 1976), (**5**) *Azotobacter vinelandii* cytochrome c_5 (Carter *et al.*,1985), (**6**) *Porphyra tenera* cytochrome *c*-553 (Ambler and Bartsch 1975).

These sequences were chosen to represent the relationships between what has been described as the large (**L**), medium (**M**) and small (**S**) cytochromes *c*. These differ in the presence of surface loops of chain which can be identified by comparison of the 3-dimensional structures (Dickerson, 1980c). A large cytochrome, like *Rb. sphaeroides* cytochrome c_2, is distinguished from a medium cytochrome, like tuna cytochrome *c*, by the presence of loops IV, II and III positioned at residues 24, 54 and 76 respectively (tuna numbering). The small cytochromes, such as sequences (**3** - **6**), also lack these loops but, in addition, lack loop I which runs between 38 and 54 (tuna numbering) (Fig. 4.6).

Cytochromes *c* which do not readily fit into the established sequence classes are shown as sequences (**7** - **12**). These are (**7**) a cyanogen bromide fragment (II) of *Pseudomonas aeruginosa* cytochrome *c* peroxidase (Ronnberg 1987); (**8**) *Thermus thermophilus* cytochrome *c*-552 (Titani *et al.*, 1985); (**9**) *Desulfovibrio vulgaris* strain Miyazaki cytochrome *c*-553 (Nakano *et al.*, 1983) (The sequence of this cytochrome type from the Hildenborough strain has also been determined (Bruschi and Le Gall, 1972) but shows little similarity to the Miyazaki sequence. Nakano *et al.* (1983) suggested that this was due to an incorrect ordering of the tryptic peptides in the Hildenborough sequence. This has been confirmed by the gene sequence from the Hildenborough strain (van Rooijen *et al.*, 1989) which shows 63 identities in 79 comparisons with the Miyazaki sequence.); (**10**) *Methylococcus capsulatus* cytochrome *c*-555 (Ambler *et al.*, 1986); (**11**) the cytochrome *c* subunit of *Pseudomonas putida* flavocytochrome *c* (McIntire *et al.*, 1986); (**12**) a cyanogen bromide fragment of *Pseudomonas aeruginosa* cytochrome cd_1 (Kalkinnen and Ellfolk, 1978). The sequence of this fragment has been

```
                                                                    |-HEME-|
                                                 Gly     Aro
                                                  ↓       ↓
(1)  Rb. sphaeroides c2            <Q E G D P E A G A K A F     N Q C Q T C H
                                        1                 10
(2)  Tuna c                             G D V A K G K K T F   V Q K C A Q C H
(3)  Ps. aeruginosa c-551                     E D P E V L F K N K G C V A C H
(4)  Chl. limicola c-555                Y D A A A G K A T Y   D A S C A M C H
(5)  A. vinelandii c5                 G G G A R S G D D V V   A K Y C N A C H
(6)  P. tenera c-553                    A D L D N G E K V F   S A N C A A C H

(7)  Ps. aeruginosa CCP                                   ....D S G C S A C H
(8)  Th. thermophilus c-552                 <Q A D G A K I Y     A Q C A G C H
(9)  D. vulgaris c-553                        A D G A A L Y     K S C V G C H
(10) M. capsulatus c-555        A P V D Q A T Y N G F K I Y K Q Q R C E T C H
(11) Ps. putida FC                    D S Q W G S G K N L Y   D K V C G H C H
(12) Ps. aeruginosa cd1         ....S E S E F N E A K Q I Y   F Q R C A G C H


          |-------LOOP IV-------|        Pro                |-----
                                          ↓
(1)   V I V D D S G T T I A G R N A K T G P N L Y G V V G R T A G T
        20                                30                  40
(2)   T V E N G G               K H K V G P N L W G L F G R K T G Q
(3)   A I D                     T K M V G P A Y K D V A A K
(4)   K T                       G M M G A P K V G D K A A
(5)   G T                       G L L N A P K V G D S A A
(6)   A G G                     N N A I M P D K T L K K D

(7)   N G I L G                 G Q A Y F P     F G L V K K P D A S V
(8)   Q Q                       N G Q G I P G A F P P L A
(9)   G A                       D G S K Q A M G V G H A V K
(10)  G A T G E                 G S A A F P N L L
(11)  K P                       E V G V G P   V L E G R G
(12)  G V L R K                 G A T G K P


      ---------LOOP I---------||---LOOP II--|
(1)     Q A D F K G   Y G E G M K E A G     A K G L A W D E E H F V Q
                          50                            60
(2)     A E G Y S     Y T D A N K S           K G I V W N N D T L M E
(3)                                         F A G Q A G A E A E L A Q
(4)                                             W A P H I A K G M N V
(5)                                             W K T R A D A K G G
(6)                                             V L E A N S M N T I D

(7)   L P S G D K G R F A V T   K T Q S D E Y V F R A A P L R N V A L
(8)                                         G H V A E I L A K E G G R
(9)                                         G Q K A D E L F K K L K G
(10)                                          N S L K N L S K D Q
(11)                                            L P E A Y
(12)                                            L T P D I T Q Q R G Q
```

```
                                   LOOP III
(1)      Y     V Q D P T K F L K E Y T G D A K     A K G K M T F K
                  70                                         80
(2)      Y     L E N P K K Y I P                     G T K M I F A G
(3)      R     I K N G S Q G V W                   G P I P M P P N
(4)      M     V A N S I K G Y K G                 T K G M M P A K G G N
(5)            L D G L L A Q S L S                 G L N A M P P K G T C
(6)      A     I T Y Q V Q N                       G K N A M P A F G G

(7)      T   A P Y F H S Q G V W Q L K           D A V A I M . . .
(8)      E Y L I L V L L Y G L Q G Q I E V K G M K Y N G V M S S F A
(9)            Y A D G S Y G G E                   K K A V M T N L V K
(10)           F K E V V L K                       G R N A M P P F E A N
(11)           I K D I V R N                       G F R A M P A F P A
(12)     Q Y   L E A L I T Y G                     T P L G M . . .


                Acidic              Aro
(1)          L K K E A D   A H N I W A Y L Q Q V A V R P
                       90              100
(2)          I K K K G E   R Q D L V A Y L K S A T S
(3)          A V S D D E   A Q T L A K W V L S Q K
(4)        P K L T D A Q   V G N A V A Y M V G Q S K
(5)        A D C S D D E   L K A A I G K M S G L
(6)          R L V D E D   I E D A A N Y V L S Q S E K G W

(7)
(8)          Q L K D E E   I A A V L N H I A T A W G D A K K V K G F K P F T
(9)          R Y S D E E   M K A M A D Y M S K L
(10)         K K V A E G   I D D L Y T Y I K G R S D G T V P A G E L E K P Q
(11)         S V V D D E S L T Q V A E Y L S S L P A P A Q


(8)          A E E V K K L R A K K L T P Q Q V L A E R K K L G L K
```

Fig. 3.15 An alignment for cytochromes *c* which do not fall into the established sequence classes confirmed by Silvestrini *et al.* (1989) who have determined the gene sequence. Divergent sequences not shown in Fig. 3.15 include the cytochrome *c*-550 of the cyanobacterium *Microcystis aeruginosa* (Cohn *et al.*, 1989b), the heme *c* peptide of the cytochrome c_1aa_3 oxidase of *Thermus thermophilus* (Buse *et al.*, 1989), and the cytochrome c_L from *Methylobacter extorquens* (Fig. 3.16).

Some conserved positions are noted by Gly, glycine; Aro, aromatic; Pro, proline; and acidic, glutamate or aspartate. .... at the N-terminal and C-terminal in the sequences (**7**) and (**12**) indicate that these are parts of larger polypeptide chains.

The matrix of identities is shown in Table 3.10.

	1	2	3	4	5	6	7	8	9	10	11
1. *Rb. sphaeroides* c_2											
2. Tuna *c*	32										
3. *Ps. aeruginosa c*-551	20	19									
4. *Chl. limicola c*-555	22	24	19								
5. *A. vinelandii* c_5	10	14	24	32							
6. *P. tenera c*-553	17	21	26	29	21						
7. *Ps. aeruginosa* CCP	16	14	26	14	14	22					
8. *Th. thermophilus c*-552	15	16	15	25	19	21	12				
9. *D. vulgaris c*-553	15	13	26	24	23	14	17	29			
10. *M. capsulatus c*-555	16	19	15	19	18	28	20	18	19		
11. *Ps. aeruginosa* FC	13	19	25	22	27	30	16	19	21	22	
12. *Ps. aeruginosa* cd_1	13	22	13	27	20	13	15	24	16	21	20

Table 3.10 Matrix of % identities of cytochromes in Fig. 3.15

The amino acid sequence of the cytochrome c_L from *Methylobacter extorquens* (Fig. 3.16) deduced from the gene sequence (Nunn and Anthony, 1988), is another cytochrome that does not fit into an established sequence class, and nor does it align simply with the sequences of Fig. 3.15. The heme is coordinated by histidine and methionine so it is possible that the cytochrome has a long N-terminal extension to a class I domain.

```
1                 10                  20                  30
Q P Q S G P Q T G V V F R N T V T G E A L D V S Q G K E G G R D T P A V K K

 40                  50                  60          HEME        70
F L E T G E N L Y I D D K S C L R N G E S L F A T S [C] S G [C H] G H L A E G K

     80                  90                  100                 110
L G P G L N D N Y W T Y P S N T T D V G L F A T I F G G A N G M M G P H N E

         120                 130                 140                 150
N L T P D E M L Q T I A W I R H L Y T G P K Q D A V W L N D E Q K K A Y T P

             160                 170
Y K Q G E V I P K D A K G Q C K P L D E
```

Fig. 3.16 The amino acid sequence of cytochrome c_L from *Methylobacter extorquens* AM1 (Nunn and Anthony 1988). The amino acid sequence is derived from the gene sequence and contains a deduced signal peptide of 25 N-terminal amino acids (not shown). The N-terminus of the mature protein is blocked (probably by pyrrolidone carboxylic acid)

Chapter 4 The Structures of Class I Cytochromes *c*

4.1 Introduction

Knowledge of the three dimensional structures of proteins is an essential basis for understanding their chemical properties and biological functions. At present such structures can only be obtained by crystallographic methods and the work of four groups - those of Takano & Dickerson, Kraut & Salemme, Kakudo and Brayer - has made the class I cytochromes *c* one of the best-defined protein families. Structures of five mitochondrial and seven bacterial cytochromes *c* have been determined (Table 4.1; reviewed in Salemme *et al.*, 1973b; Dickerson & Timkovich, 1975; Salemme, 1977; Timkovich, 1979; Dickerson, 1980b;c; Takano & Dickerson, 1981a;b; Matsuura *et al.*, 1982; Meyer & Kamen, 1982; Ochi *et al.*, 1983; Louie *et al.*, 1988a).

Our aim in the present chapter is to describe these structures with the following three general questions in mind:

1. Do the structures help to define mechanisms of action?
2. Can the structural bases of particular physicochemical properties such as redox potential, be described?
3. Does structural similarity imply evolutionary homology?

These questions are specifically addressed in the present book as follows: question 1 in Chapters 7 and 8; question 2 in Chapters 1, 2 and 7; and question 3 in Chapter 6. We have also addressed question 1 elsewhere (Pettigrew & Moore, 1987: Chapters 2 and 3).

In this chapter we also consider the questions:

4. Is there a redox state dependent conformation change?
5. What is the significance of the buried heme propionates?

4.2 General Features of the Structure of Mitochondrial Cytochrome *c*

4.2.1 A General Mitochondrial Cytochrome *c* Structure

The crystal structures of cytochrome *c* from five eukaryotic species have been determined (Table 4.1), the most accurate of which are the structures of yeast ferrocytochrome *c* (G.Brayer, personal communication), tuna cytochrome *c* (Takano & Dickerson, 1981a;b) and rice ferricytochrome *c* (Ochi *et al.*, 1983).

Table 4.1 X-ray structures of class I cytochromes *c*

Protein [a]		Size [b]	Crystal pH	Resolution (Å)	Ref [c]	Refs.
Horse	ferricytochrome *c*	104,1	6.2	2.8	-	e
Bonito	ferricytochrome *c*	103,1	6.0	2.8	-	f
	ferrocytochrome *c*	103,1	8.0	2.3	-	g
Tuna	ferricytochrome *c* [i]	103,1	7.0	1.8	+	h
	ferrocytochrome *c*	103,1	7.5	1.5	+	i
Rice	ferricytochrome *c*	111,1	6.0	2.0	+	j
S. cerevisiae iso-1	ferrocytochrome *c* [ii]	108,1	6.7	2.8	+	k - m
R. rubrum	ferricytochrome c_2	112,1	5.8	2.0	+	n o
	ferrocytochrome c_2	112,1	5.8	2.0	+	n o
Pa. denitrificans	ferricytochrome *c*-550	134,1	7.5	2.45	-	p q
Ps. aeruginosa	ferricytochrome *c*-551	82,1	5.7	1.6	+	r
	ferrocytochrome *c*-551	82,1	5.7	1.6	+	r
Chl. limicola [iii]	ferricytochrome *c*-555	86,1	d	2.7	-	s t
An. nidulans	ferricytochrome *c*-554	85,1	8.0	3.0	-	u
Ps. aeruginosa	ferricytochrome c_4	181,2	6.0		-	v w
A. vinelandii	ferricytochrome c_5 [iv]	83,1	6.9	2.5	-	x

[a] [i] There are two independent structures of this protein.
[ii] The structures of the Ser^{82} variant at 2.8 Å resolution and the Gly^{82} variant at 2.6 Å resolution have been reported (Louie *et al.*, 1988b; Louie & Brayer, 1989).
[iii] Formerly *Chl. thiosulfatophilum*.
[iv] Monomer in solution but α_2 dimer in the crystal.
[b] The figures are the number of amino acids and heme groups.
[c] Ref = Refined; + = yes; - = no. [d] Not reported.
[e]Dickerson *et al.* (1971); [f]Matsuura *et al.* (1979); [g]Tanaka *et al.* (1975); [h]Takano & Dickerson (1981b); [i]Takano & Dickerson (1981a); [j]Ochi *et al.* (1983); [k]Sherwood & Brayer (1985); [l]Louie *et al.* (1988a); [m]Brayer pers. comm.; [n]Salemme *et al.* (1973a); [o]Bhatia *et al.* (1984); [p]Timkovich & Dickerson (1976); [q]Ambler *et al.* (1981b); [r]Matsuura *et al.* (1982); [s]Korszun & Salemme (1977); [t]Salemme (1977); [u]Ludwig *et al.* (1982); [v]Sawyer *et al.* (1981); [w]Sawyer pers. comm.; [x]Carter *et al.* (1985).

These structures are at a resolution where most of the non-hydrogen atomic positions have been determined. The structures of bonito cytochrome *c* (Tanaka *et al.*, 1975; Matsuura *et al.*, 1979) and horse ferricytochrome *c* (Dickerson *et al.*, 1971; Swanson *et al.*, 1977) show the main chain and some of the side chain positions but they have not been refined and some of the side chains are poorly defined.

Despite the relatively low sequence homology between yeast, rice and tuna cytochromes *c* (Fig. 3.1), thay have remarkably similar main chain structures and, despite the difference in crystallographic resolution (Table 4.1), it is clear that the horse and bonito cytochromes also have similar structures. Furthermore, comparison of the reduced and oxidised forms of tuna cytochrome *c* (Takano & Dickerson, 1981b) together with solution NMR data (Moore *et al.*, 1980a; Williams *et al.*, 1985c) shows that the redox-linked conformation change is small. Therefore, the general comments made in the present chapter apply to all five cytochromes and to both redox states. Also, given the extensive sequence homology (Fig. 3.1) and similar CD and NMR spectroscopic properties of a wide range of mitochondrial cytochromes (Ulmer, 1965; Zand & Vinogradov, 1968; Myer, 1970; Looze *et al.*, 1976; Moore & Williams, 1980c;d; Moore *et al.*, 1985a; Robinson *et al.*, 1983a; Williams *et al.*, 1985a; Pielak *et al.*, 1988a;b; Gao *et al.*, 1989), the x-ray structures define a general mitochondrial cytochrome *c* structure.

4.2.2 The Polypeptide Fold and Internal Structure

The polypeptide chain is organised into five α-helical segments that contain ~45% of the amino acids (Fig. 4.1). There is no regular β-strand structure but there are six type II 3_{10} bends, most of which are located at the base and right side of the structure according to the orientation of Fig. 4.1.

The presence of the type II 3_{10} bends explains the evolutionary conservation of several glycines (23, 34, 37, 45, and 77) in cytochrome *c* (Takano *et al.*, 1977). In these bends the carbonyl group of residue R_i+1 approaches so closely to the side chain of residue R_i+2 that only a hydrogen atom can be accommodated at this position (Venkatachalan, 1968).

Amino acid residues of cytochrome *c* have been classified into three groups according to the degree of exposure of their side chains (Dickerson & Timkovich, 1975): exposed, partly exposed and buried (Fig. 4.2). Naturally, most of the charged residues are exposed and most of the hydrophobic residues are buried but the full significance of these groupings can only be appreciated by considering the way in which the heme and polypeptide interact.

The polypeptide chain wraps itself around the prosthetic group (Fig. 4.1) so that only about 4% of the heme is exposed at the molecular surface. This exposure is most clearly seen in the space-filling representation of Fig. 4.3. Only about 0.06% of the total solvent accessible surface area of cytochrome *c* (5,300 Å^2: defined as the area in contact with a rolling sphere of radius 1.4 Å) is occupied by heme (Stellwagen, 1978).

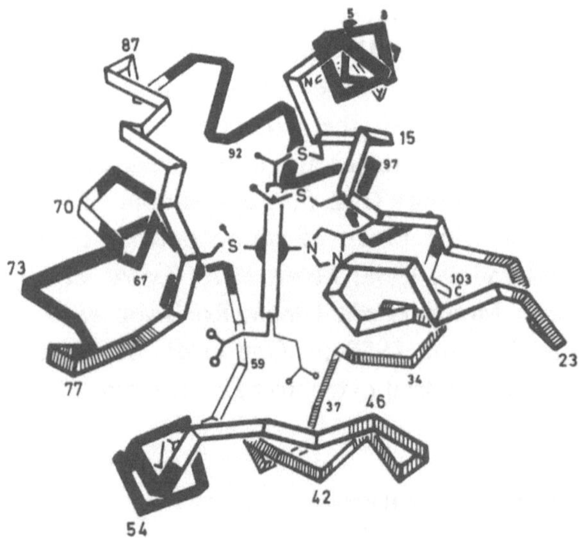

Fig. 4.1 Ribbon diagram of tuna cytochrome *c* showing the polypeptide fold and the unusual heme orientation. This is the conventional view of cytochrome *c*; the protein is viewed looking onto the face bearing the exposed heme edge. (Top, bottom, right, left, front and back refer to this orientation.) α-helices are indicated by solid ribbons between residues 3-12, 50-55, 61-69, 70-75 and 88-101. Type II 3_{10} bends are shown as striped ribbons between residues 21-24, 32-35, 35-38, 39-42, 43-46 and 75-78. The 39-42 bend is slightly deformed with the 39 carbonyl O atom hydrogen-bonded to the side chain of 42 rather than its NH (Takano & Dickerson, 1981a,b). The secondary structure of rice (Ochi *et al.*, 1983) and yeast (Louie *et al.*, 1988a) cytochromes *c* is the same as that of tuna cytochrome *c*. The stretches of polypeptide from residue 18-32, 34-43, 40-54 and 70-84 have been proposed to be examples of a recently designated category of secondary structure, an Ω-loop. These are compact loops of residues whose polypeptide chain traces the shape of a Greek omega (Leszczynski & Rose, 1986)

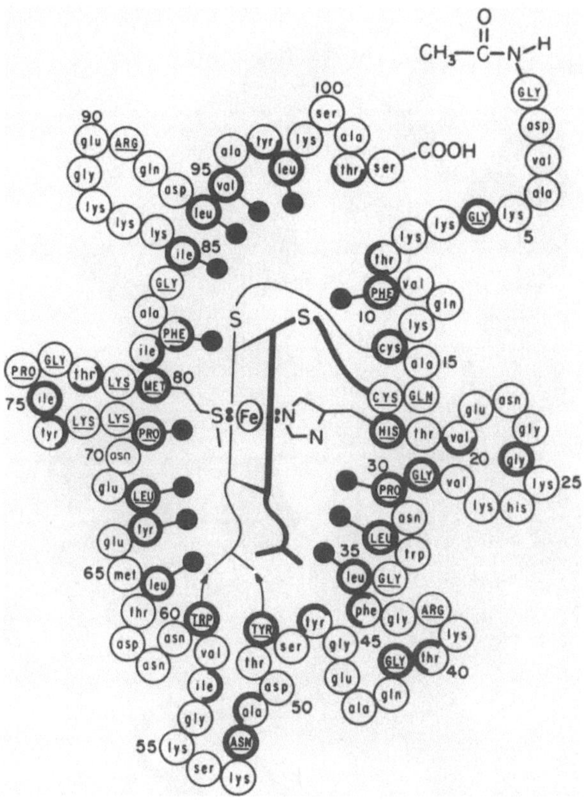

Fig. 4.2 Chain packing diagram of tuna cytochrome *c*. This illustrates the classification of side chains into the following categories: buried (heavy circles), partially buried (heavy semi-circles) and exposed (light circles). Side chains packed against the heme macrocycle are indicated by heavy black dots, and unvaried residues are underlined and upper case. (Adapted from Dickerson & Timkovich, 1975)

Covalent bonds between the heme and four amino acids - Cys 14, Cys 17, His 18 and Met 80 - anchor the heme and together with the extensive array of non-covalent side chain contacts (Fig. 4.2) these bonds help to make the internal structure around the heme macrocycle relatively rigid (Williams *et al.*, 1985c).

All of the internal residues in Van der Waals contact with the heme macrocycle - the heme contact residues - are bulky hydrophobic groups (Fig. 4.2). Only six of these thirteen residues are unvaried in the 84 sequenced mitochondrial cytochromes but the remaining seven are conservatively replaced (Fig. 3.1). These hydrophobic groups help stabilise the structure and control the level of the redox potential (Chapter 7.4.3). Further stabilisation

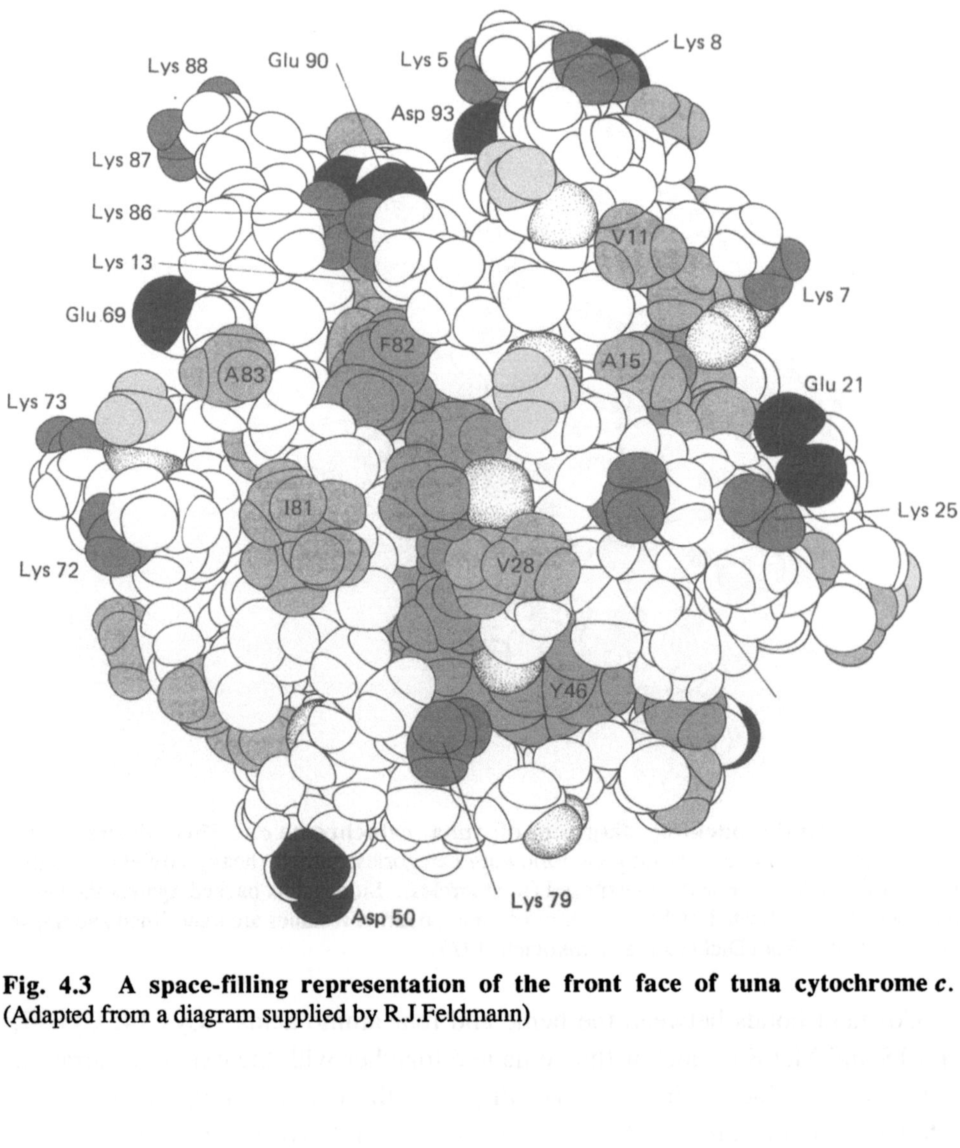

Fig. 4.3 A space-filling representation of the front face of tuna cytochrome *c*. (Adapted from a diagram supplied by R.J.Feldmann)

of the native structure is conferred by an extensive array of hydrogen-bonds. Excluding those involved in α-helices and with H_2O molecules, there are 44 hydrogen bonds in tuna cytochrome *c*. All fifteen of those involving the polypeptide backbone and sixteen of the twenty nine involving side chains are conserved in the rice protein.

The N-terminal eight-residue extension in rice cytochrome *c* is in a relatively extended conformation that runs along the back of the molecule. Hydrogen-bonds between four groups in the extension and the remainder of the protein, including one between the N-terminal acetyl and Tyr 65, help maintain the position of the extension (Ochi *et al*., 1983). Both Gln 89, whose side chain is involved in one of the other hydrogen bonds, and Tyr 65, are unvaried among plant cytochromes *c* (Fig. 3.1) consistent with the N-terminal extension in these proteins having a similar conformation to that of the rice protein.

The N-terminal five residue extension in yeast cytochrome *c* is in a different conformation to the N-terminal extension of rice cytochrome *c*. The yeast extension projects away from the remainder of the molecule, unlike the rice extension. The only non-covalent interaction between the yeast extension and the bulk of the protein occurs with the side chain of Phe -3. This is packed into a small pocket in the surface of the protein. The difference between the yeast and rice extensions reflects differences in the sequence of the N-terminal extensions themselves (Fig. 3.1) and the differences between the type of side chains in the region of the core section of the protein closest to the extension (Louie *et al*., 1988a).

A notable feature of the general mitochondrial cytochrome *c* structure is that the propionic acids of the heme are buried within the protein (Fig. 4.1 and 4.3), where they are completely shielded from the bulk solvent. We attach some importance to this structural feature and consider it more fully in section 4.5.

The most probable heme propionate interaction scheme is that given in Fig. 4.4. With the exception of Thr 49 and Thr 78, all the contact residues are unvaried. When Thr 49 is replaced it is always by serine; Thr 78 is replaced by asparagine in one protein (Fig. 3.1). In contrast to regions of the protein in contact with the heme macrocycle, some areas around the heme propionates are relatively mobile and the hydrogen-bond network involving the heme propionates is probably in a dynamic state of disruption and reformation. This is one reason why the various x-ray structures differ in this region: each of the crystals has preferentially trapped a population of molecules with a different averaged network of interacting groups. The residues forming the base of the molecule include some of the most mobile of the partly buried residues (Robinson *et al.*, 1983a; Williams *et al.*, 1985c) and, given the susceptibility of this region of the protein to proteolytic digestion (Hantgen & Taniuchi, 1977; Harris & Offord 1977; Parr *et al.*, 1978), the main chain may also be flexible.

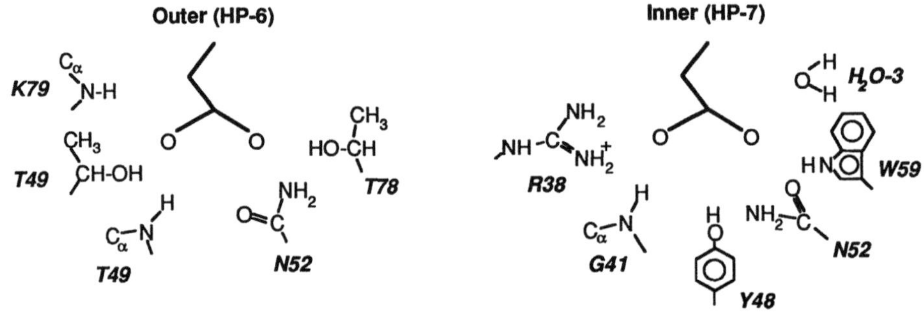

Fig. 4.4 Heme propionate interactions in cytochrome *c*. The heme propionates are called inner (**HP-7**) and outer (**HP-6**) by some authors (e.g. Takano & Dickerson, 1981a;b). ***H_2O-3*** is one of three buried water molecules. It is also hydrogen-bonded to Arg 38. ***H_2O-1*** is close to the side chains of Asn 52, Tyr 67 and Thr 78. ***H_2O-2*** is located in a different part of the structure. Many of the residues given in the diagram are sufficiently close to other groups to interact with them as well as with the propionates. The following have inter-residue atomic distances ≤3.1Å in tuna cytochrome *c* (C represents a peptide carbonyl and N a peptide NH): Gly 37-N---Trp 59-C; Arg 38-N---Leu 35-C; Arg 38---Trp 33-C; Gly 41-N---Asn 52; Asn 52-N---Thr 49; Trp 59-N---Arg 38-C; Thr 78-N---Ile 75-C. All atoms capable of forming hydrogen-bonds or salt bridges and within 3.6Å of the heme propionates are included in the diagram. Not all of these atoms can interact with a propionate at the same time, nor can some of them interact simultaneously with propionates and other groups. Probably there is a network of interacting groups in which some of the weaker bonds are continuously made and broken so that the precise interactions in any one molecule cannot be specified

If it is assumed that the crystallographic temperature factors of tuna cytochrome *c* reflect chain flexibility and side chain mobility (Takano & Dickerson, 1981a;b; 1982) then in addition to the terminal residues, the loops from Val 20 to Val 28, Leu 35 to Arg 38, Glu 42 to Tyr 46, Asn 52 to Ile 57, and Ile 85 to Leu 87 are also more mobile than other parts of the structure. However, the temperature factors of yeast cytochrome *c* have a somewhat different profile (Louie *et al.*, 1988a) which raises the question of whether this parameter reflects the intrinsic flexibility of a structure or crystallographic packing forces. The major difference between the yeast and tuna proteins is that the largest continuous segment of yeast cytochrome *c* having lower than average thermal motion is from residues 9 to 35. Residues 37 to 42, 50 to 58, 69 to 72, 84 to 88, and the N-terminal extension have above average thermal motion. The α-helical regions and the hemes of both proteins are generally of low mobility. Measurements of the rate of exchange of peptide NH protons for deuterons confirms that the N-terminal helix of horse cytochrome *c* is disorganised and flexible at its chain terminus but is more rigid further along the chain (Wand *et al.*, 1986). Side chain mobility, as indicated by the crystallographic temperature factors, generally follows the pattern of increased burial being associated with decreasing mobility (see section 4.2.3).

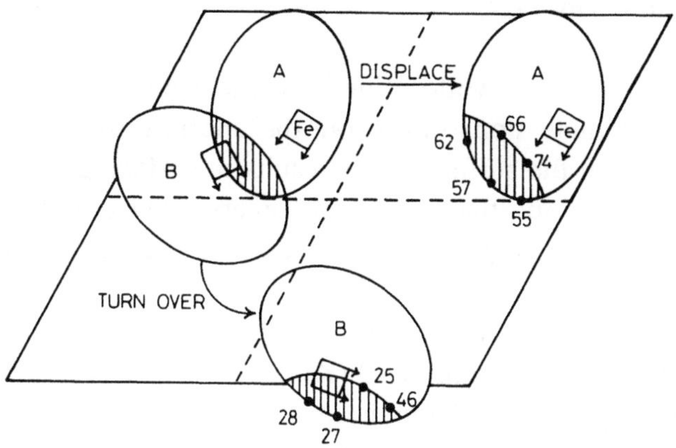

Fig. 4.5 Schematic representation of the tight intermolecular contacts between two molecules of rice ferricytochrome *c* in crystals of space group $P6_1$. The contact region between molecules **A** and **B** is indicated by the stripes. The residues involved in the tight contact are: for **A** Ala -8, Ser -7, Asn 54, Met 55, Ala 56, Val 57, Asn 62, Asp 66 and Tyr 74; for **B** Gln 16, Gly 25, His 26, Lys 27, Gln 28, Gly 45, Tyr 46 and Ser 47. All the cytochrome molecules in the rice crystals are equivalent (Ochi *et al.*, 1983). In crystals of tuna ferricytochrome *c* with space group $P6_3$ the cytochrome molecules fall into two non-equivalent groups: the inner and the outer. Nevertheless, in both sets of molecules the 50's group of residues form one of the main tight intermolecular contact areas (Swanson *et al.*, 1977; Matsuura *et al.*, 1979; M.R.Parsons, G.R.Moore & G.L.Taylor, unpublished data). Another contact region involves residues on the front face: Pro 76, Ile 81 and Ala 83. Detailed studies show the NMR properties of some residues bordering the contact regions of tuna cytochrome *c* in solution are different to those expected on the basis of the crystal structure, presumably because in solution the tight intermolecular contacts are broken and the structure can relax into its ground state (Williams *et al.*, 1985c)

The deformability of the more flexible regions of cytochrome *c* is illustrated by the structural perturbations caused by tight intermolecular packing in crystals (Williams *et al.*, 1985c). The crystal packing of rice ferricytochrome *c* is shown schematically in Fig. 4.5 (Ochi *et al.*, 1983). Only the regions of closest contact are indicated, one of which is the lower left side of the molecule and includes the mobile 50's loop. This contact region is common to crystals of tuna, bonito, horse and rice cytochromes *c* (Matsuura *et al* , 1979) suggesting that it reflects some common feature of mitochondrial cytochromes *c*. A direct result of the close contacts is that the conformations of some side chains are different in the crystal and solution states (Williams *et al.*, 1985c).

4.2.3 The Protein Surface

The protein surface is not well-defined by the standards of the protein interior, mainly because constraints on the motion of side chains are considerably smaller on the surface. This results in surface groups possessing a number of conformations that are almost energetically equivalent and it means that it is a major simplification to describe the surface as though it possessed a unique structure.

Tuna cytochrome *c*, like other mitochondrial cytochromes *c*, is a basic protein. It has an isoelectric point of 10 (Kreil 1963; 1965; Kim *et al.*, 1980), and a net charge at pH 7 of +8 for the oxidised state (section 4.4.1). Yet despite so many charged groups, about 55% of the molecular surface is formed by hydrophobic residues (Stellwagen, 1978). For example the front face of the molecule contains the heme edge and the side chains of Val 11, Val 28, Tyr 46, Ile 81 and Phe 82 (Fig. 4.3).

Most of the charged groups are located on the protein surface where they are distributed in a highly asymmetric manner (Dickerson *et al.*, 1971; Salemme *et al.*, 1973b; Koppenol & Margoliash, 1982). Lysines are clustered predominantly around the exposed heme edge (Fig. 4.3) and on the right and left sides, while carboxylates are located primarily on the top and back of the molecule. This segregation is not absolute and a few lysines are found on the back while some carboxylates are found on the left and right sides, and even on the periphery of the front face (Fig. 4.3).

The locations of some of the surface groups are blurred due to conformational heterogeneity (Northrup *et al.*, 1980, 1981; Moore *et al.*, 1982a; Takano & Dickerson, 1981a;b; 1982; Williams *et al.*, 1985b;c). This is because the surface of the protein is the most mobile region.

For nine of the sixteen lysines (positions 5, 7, 8, 25, 39, 72, 73, 87 and 88) the position of the amine cannot be reliably located (Takano & Dickerson, 1982; Williams *et al* , 1985b). The remaining seven lysines are relatively well-defined in the crystallographic map; interactions with neighbouring residues reduce the conformational flexibility of five of these. Spectroscopic and chemical studies show that for at least three of these (positions 13, 55 and 79), interactions persist in solution (Osheroff *et al.*, 1980; Bosshard & Zürrer, 1980; Robinson *et al.*, 1983a). Carboxylate groups are better defined than lysines, probably because of their smaller side chains.

Both TML 72 and TML 86 of rice cytochrome *c* and TML 72 of yeast cytochrome *c* occupy different positions in the crystals from their non-methylated counterparts in tuna cytochrome *c*. In solution, TML 72 of both plant and yeast cytochromes is relatively mobile and spends at least part of the time located close to the heme and residue 81 (G.R.Moore, unpublished data). This position is consistent with the location of TML 72 in crystals of rice cytochrome *c*.

Some aromatic residues exposed, or partly exposed, at the protein surface are rather mobile whilst others are not. Thus for Tyr 46 the rate of ring flipping about its Cβ-Cγ bond at 25°C is $<10^2$ s^{-1} with a relatively high energy of activation, while the corresponding rate for Phe 82 is $>10^4$ s^{-1} (Campbell *et al.*, 1976; Moore & Williams, 1980c; Moore *et al.*, 1982a). Mobility around the exposed heme edge is also apparent in an exchange process with a rate of approximately 10^3 s^{-1} at 25°C that affects the heme methyl-3 group of horse ferricytochrome *c* (Burns & La Mar, 1979; 1981). This implies that the protein around this methyl is switching rapidly between different conformations but the nature of the change is not known.

The aspects of conformational mobility discussed above should be borne in mind when considering the static pictures of x-ray crystallography. Functionally important aspects of the surface include both the precise positions of groups involved in interactions with other redox proteins, and the overall charge distribution, features which are discussed further in Pettigrew & Moore (1987: Chapter 2).

4.3 The Cytochrome Fold and Bacterial Class I Cytochromes

4.3.1 A General Class I Cytochrome *c* Structure

With the exception of *Desulfovibrio* cytochrome *c*-553, crystallographic structures of representatives of all the subclasses of class I cytochromes *c* listed by Pettigrew & Moore, (1987: Table 1.5), have been determined (Table 4.1), although not all to the same degree of reliability and resolution. Only low resolution structures are available for *Chlorobium* cytochrome *c*-555, *Pseudomonas* cytochrome c_4 and *Anacystis* cytochrome *c*-554, while that for *Paracoccus* cytochrome *c*-550 has not yet been refined and has been seriously criticised on the basis of a reassessment of amino acid sequence evidence (Ambler *et al.*, 1981b).

In spite of these differences in quality, the emerging picture is one of strong similarity in the basic polypeptide folding pattern among the class I bacterial and mitochondrial cytochromes *c* (Fig. 4.6). The evolutionary significance of this will be discussed in Chapter 6. The following discussion deals with the polypeptide fold, internal structure and protein surface of the bacterial class I cytochromes, and thus mirrors the preceding discussion of the mitochondrial cytochromes *c*. The dimeric structure of cytochrome c_5 and the two domain structure of cytochrome c_4 are considered in separate sections.

4.3.2 The Polypeptide Fold and Internal Structure

In all the class I cytochromes studied the polypeptide chain almost encloses the heme leaving only the edges of pyrrole rings II and III exposed. Like

Fig. 4.6 Ribbon drawings of the structures of various cytochromes *c*. A *Chlorobium limicola* cytochrome *c*-555; **B** *Pseudomonas aeruginosa* cytochrome *c*-551; **C** Tuna cytochrome *c*; **D** *Rhodospirillum rubrum* cytochrome c_2; **E** *Paracoccus denitrificans* cytochrome *c*-550. The shaded portion of the chain of mitochondrial cytochrome *c*, consisting of 16 residues, is apparently missing from the smaller cytochromes. The shaded portions of the chain of *R. rubrum* cytochrome c_2 correspond to 3- and 8-residue insertions relative to mitochondrial cytochrome *c*. *Paracoccus* cytochrome *c*-550 is the largest in its class: the shaded areas correspond to 2- and 6-residue insertions plus a 10-residue C-terminal tail. *Pseudomonas* cytochrome *c*-551 has a single residue insertion relative to the others, indicated by shading. (Reproduced from Dickerson, 1980c)

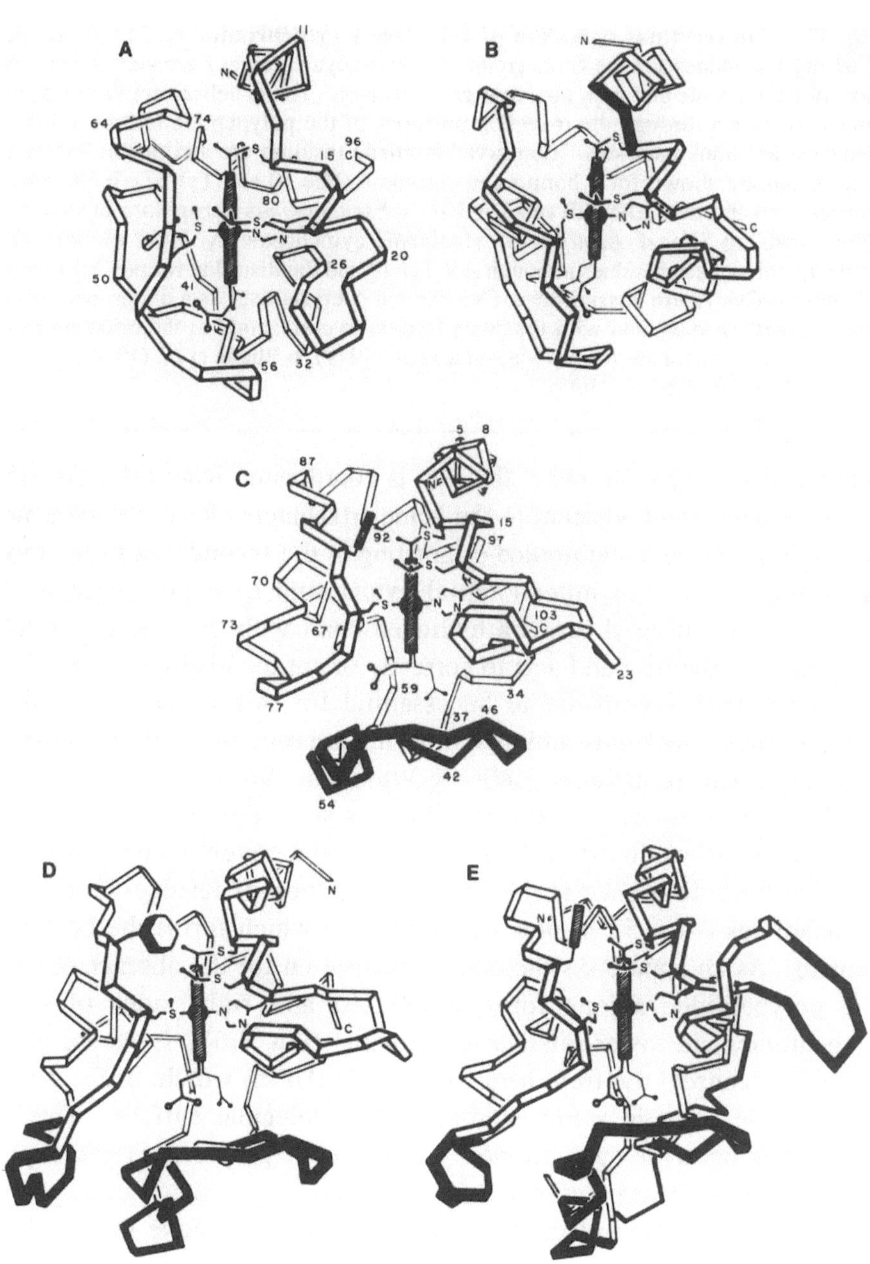
A
11
N
64
74
96
C
15
80
S
N
50
41
26
20
56
32
B
N
S
C
C
5
8
87
N
92
15
97
70
S
N
103
C
73
67
23
59
34
77
37
46
42
54
D
N
S
C
E
N
S
N

Fig. 4.7 The terminal α-helices of the class I cytochromes *c*. The N-terminal and C-terminal α-helices and the heme group of selected cytochromes *c* are viewed from the 'right' side of the molecule and with the heme group face-on. The α-helices are shown as cylinders which are drawn through the α-carbon positions of the polypeptide backbone between the sequence positions indicated. Conserved aromatic residues are a common feature of these helices and are shown for **a** bonito cytochrome *c* (Phe 10 and Tyr 97), **b** *Rhodospirillum rubrum* cytochrome c_2 (Phe 20 and Tyr 107), **c** *Pseudomonas aeruginosa* cytochrome *c*-551 (Phe 7 and Trp 77). **d**, *Azotobacter vinelandii* cytochrome c_5, lacks obvious structural counterparts to these residues although AV Tyr 18 and the disulfide formed between AV Cys 69 and AV Cys 72 are possibilities. Cytochrome c_5 crystallises as a dimer, one monomer of which (right) is orientated with the heme face-on to correspond to the orientations of **a** - **c**. Crystallographic data are from **a** Matsuura *et al.* (1979), **b** Bhatia *et al.* (1984), **c** Matsuura *et al.* (1982), **d** Carter *et al.* (1985)

mitochondrial cytochrome *c*, the iron is coordinated from the right side by a histidine positioned adjacent to the heme attachment site in the sequence, and on the left side by a methionine originating in the second half of the molecule. A proline (Pro 30 in mitochondrial cytochrome *c*) is positioned so that its peptide bond carbonyl forms a hydrogen-bond with the side of the histidine ring distal to the iron and acts to correctly orient the histidine ring. However, a proline does not appear to be essential for this purpose as some algal cytochromes *c* lack this amino acid in the appropriate position yet contain a similarly oriented histidine (Keller & Wüthrich, 1981).

Differences in size between cytochromes are mainly due to the presence or absence of surface loops of the polypeptide on a common core structure. The most striking difference is the absence of a loop of residues from the small cytochromes (41 - 55 in tuna cytochrome *c*) which forms the bottom of the protein. As an apparent structural compensation for the absence of this loop, the polypeptide regions forming the left and right sides of the small cytochromes are lower relative to the heme (Fig. 4.6). Even so, this is not enough to conceal the front heme propionate (HP-6) which, in *Ps. aeruginosa* cytochrome *c*-551 is partly exposed at the molecular surface. This is even more striking with cytochrome c_5 where both propionates are exposed to solvent (however, see section 4.3.5).

The general folding[1] of the polypeptide chains relative to their hemes is

[1]To simplify discussion of the various cytochromes the following residue designations will be used: mitochondrial cytochrome *c*, none; *Rhodospirillum* cytochrome c_2, R; *Paracoccus* cytochrome *c*-550, PA; *Pseudomonas* cytochrome *c*-551, PS; *Chlorobium* cytochrome *c*-555, CL; *Euglena* cytochrome *c*-552, E; *Azotobacter* cytochrome c_5, AV. Sequence numbering will be according to the alignments given in Chapter 3.

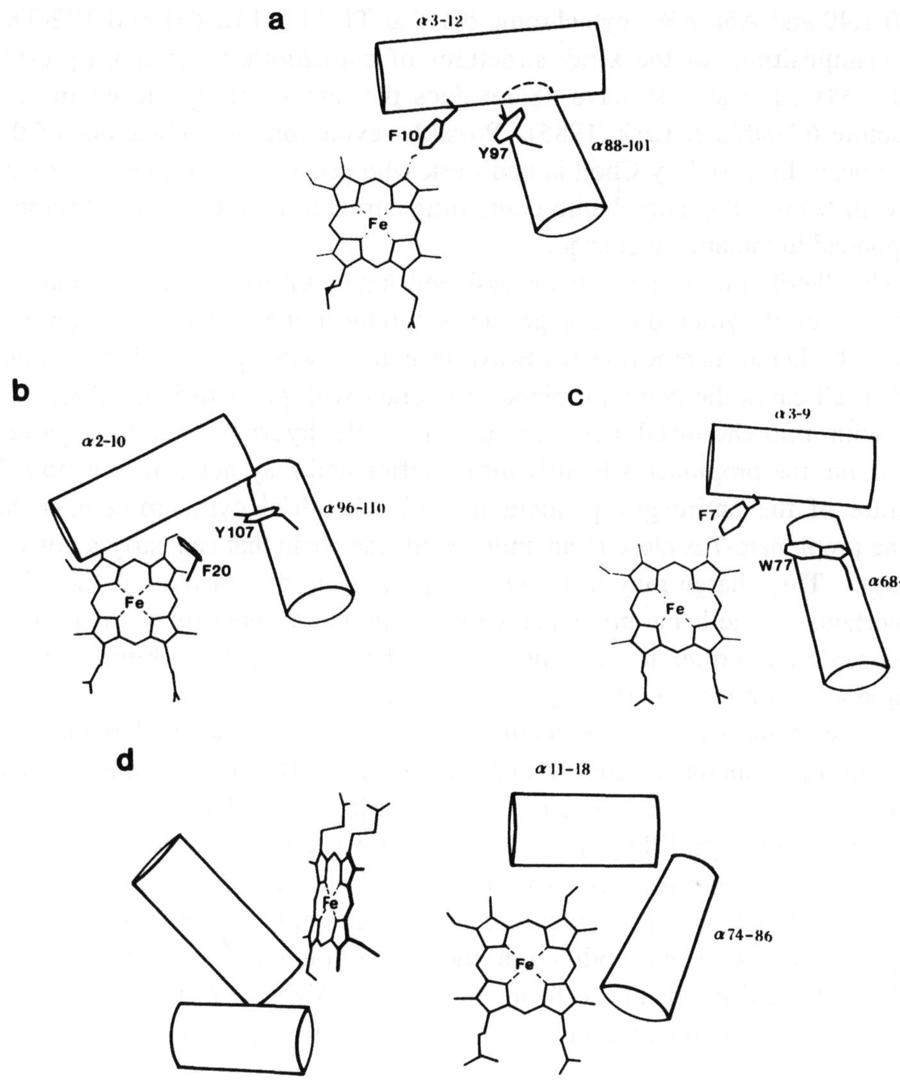

similar in all the structures of Fig. 4.6. The proteins begin and end in α-helices that interact with each other (Fig. 4.7), a feature also seen in many other proteins (Thornton & Sibanda, 1983), and the pattern of folding of the other helices in the large cytochrome structures are similar. The two smaller structures have a different distribution of α-helices (Fig. 4.6 & 4.7), partly because the extra bottom loop of the large cytochromes contains a small helix but also because the remaining polypeptide chain folds differently in some regions. Thus cytochrome *c*-551 has helices at residues A3-A9, A27-A33,

A40-A49 and A68-A80; cytochrome c-555 at T1-T12, T42-T51 and T72-T84. Superimposition of the x-ray structures of cytochromes c (tuna), c_2, c-550 and c-551 suggest they have 48 residues that are similarly placed in each structure (Chothia & Lesk, 1985). Possible evolutionary implications of this have been discussed by Chothia and Lesk who proposed a mechanism for the way in which the core heme-containing structural unit of cytochrome c responded to mutational change.

The distribution of side chains assigned important roles in the structure of mitochondrial cytochrome c is generally similar in the bacterial cytochromes. Thus, the heme macrocycle is always in contact with hydrophobic residues, and in all cases the heme propionates interact with polar residues (Fig. 4.8). As with mitochondrial cytochrome c a single hydrogen-bonding pattern involving the propionates is difficult to define and may not exist. A notable feature of the entire group, including mitochondrial cytochrome c, is that heme propionate-7 is close to an amino acid side chain that can carry a positive charge. This charge may influence the propionate pK_a (section 4.4.2). The importance of the heme propionates for the structural integrity of the proteins is considered further in sections 4.4 and 4.5, and the perturbing effect of propionate ionisation on redox potential is described in Chapter 7.6.

Many of the similar side chain features have key functional roles. The interacting aromatic residues which help lock the N and C-terminal helices together (Fig. 4.7) are a good example. Most class I cytochromes have aromatic residues fulfilling this role although it is clearly not an essential one as the sequence comparisons in Chapter 3 and chemical modification studies (section 4.6) show. In cytochrome c_5 for example, interacting aromatic residues are not found (although their function may be carried out by AV-Tyr 18 and the disulfide formed between AV-Cys 69 and AV-Cys 72). Nevertheless, the aromatic duo is important enough that compensating amino acid changes to maintain it have occurred in some proteins, such as *R. rubrum* cytochrome c_2 (Salemme *et al.*, 1973a).

Most class I cytochromes have aromatic residues clustered around the heme and axial ligands; Tyr 46, Tyr 48, Trp 59, Tyr 67 and Phe 82 in tuna cytochrome c, for example, and Tyr 27, Phe 34 and Trp 56 in *Ps. aeruginosa* cytochrome c-551. At one time these groups were assumed to play an active part in electron transfer, perhaps by formation of free radicals (Chapter 8.2.4.2), but subsequent chemical modification experiments (section 4.6) have caused this view to be abandoned. However, the reasons why these aromatic residues are unvaried or conservatively replaced are not entirely clear.

Presumably in some circumstances aromatic groups can perform particular roles that aliphatic hydrophobic groups cannot and it may be, for Tyr and Trp, that as well as their bulk and non-polar character, their hydrogen-bonding capability is important.

Trp 59 of mitochondrial cytochrome *c* is one aromatic group proposed to play more than just a structure-filling role. Like other heme propionate contact residues (Arg 38, Tyr 48 and Asn 52), Trp 59 is unvaried among the mitochondrial cytochromes *c* (Fig. 3.1) and it has its counterpart in most class I cytochromes *c*. In the structures of Fig. 4.6, the interacting Trp originates in two different regions of the polypeptide sequence for different proteins, an example of evolutionary convergence (Almassy & Dickerson, 1978). However, although the sequence and structural position of AV-Trp 39 resembles that of CL-Trp 34, it is not close enough to either heme propionate to hydrogen-bond with one of them (although see section 4.3).

Although widespread, a tryptophan in this structural position is not an absolute requirement for a class I structure because neither *R. molischianum* nor *R. fulvum* iso-1 cytochromes c_2 contain any tryptophan, unlike their iso-2 cytochrome c_2 counterparts. Tryptophan is also absent from cytochrome c_4.

Among the algal cytochromes, a tryptophan appears to be close to the heme but which residue it is has not been established. In most of these proteins there is a conserved tryptophan near the C-terminus (Fig. 3.10) which, with some manipulation of the chain, could be folded back into contact with the heme group. The x-ray crystallographic analysis of the *Anacystis* cytochrome (Ludwig *et al.*, 1982) is not secure enough to definitely decide whether this occurs but preliminary conclusions were that such an arrangement could not be fitted to the electron density. In contrast, the NMR properties of the C-terminal tryptophan in a variety of algal cytochromes suggest that these residues lie close to the heme group (Keller & Wüthrich, 1977, 1981; Ulrich *et al.*, 1982; Moore *et al.*, 1980b). However, the cytochrome from *Monochrysis lutheri* does not contain the C-terminal tryptophan although it does contain a tryptophan at position 32, that is sequentially equivalent to CL-Trp 34 and may act in an analogous way to it.

Hydrogen-bonding interactions between the axial ligand and surrounding groups are another common feature of the class I cytochromes. As mentioned previously the His ligand nearly always interacts with a Pro carbonyl and the Met-S may be hydrogen-bonded to a Tyr, as in tuna cytochrome *c* (Takano & Dickerson, 1981a, b) or Asn as in *Ps. aeruginosa* cytochrome *c*-551 (Matsuura *et al.*, 1982).

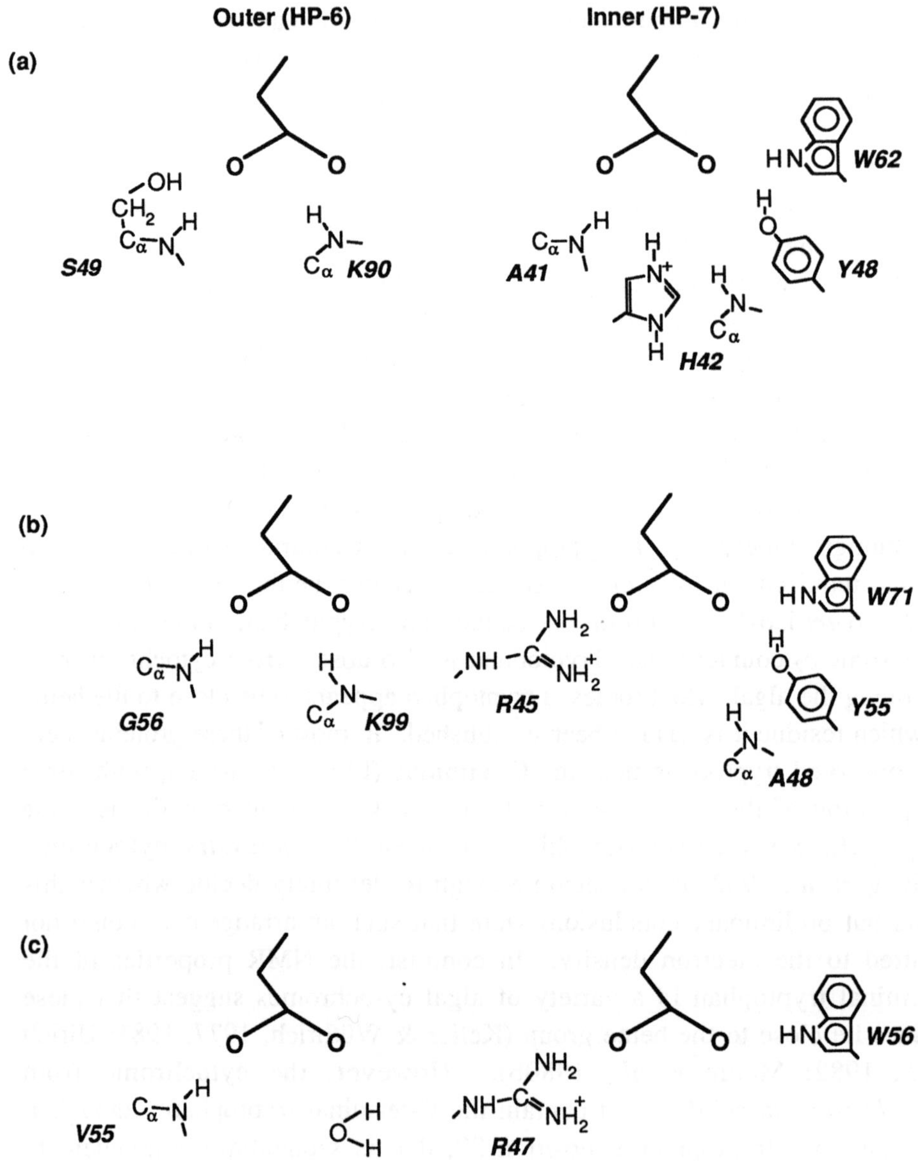
Outer (HP-6)
Inner (HP-7)
(a)
S49
K90
A41
H42
Y48
W62
(b)
G56
K99
R45
Y55
A48
W71
(c)
V55
R47
W56

(d)

(e)

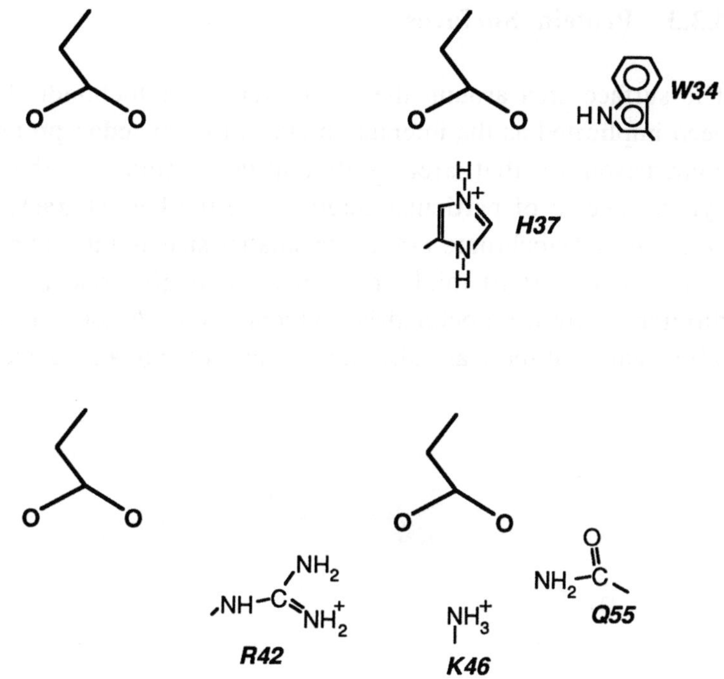

Fig. 4.8 Heme propionate interactions in bacterial cytochromes *c*. The heme propionate ionisation states are not shown (see Table 4.2). **(a)** The set of residues close to HP-7 in *R. rubrum* cytochrome c_2 are from the refined crystal structure (Bhatia *et al.*, 1984) and differ from the original set obtained from the unrefined coordinates (Salemme *et al.*, 1973b). The new set are in better agreement with the sequence alignment of mitochondrial cytochromes *c* and bacterial class I cytochromes *c*, and the x-ray structure of mitochondrial cytochrome *c*. **(b)** The proposed interactions for *Pa. denitrificans* cytochrome *c*-550 should be regarded as tentative until the structure has been refined. **(c)** *Ps. aeruginosa c*-551. Other members of the *c*-551 family have a histidine in place of Arg 47 (Leitch *et al.*, 1984). **(d)** Few atomic positions in the x-ray structure of *Chl. limicola* cytochrome *c*-555 are accurately defined and only Trp 34 has been clearly identified by XRD as a heme propionate contact residue. His 37 has been shown to interact with HP-7 on the basis of NMR studies (Moore *et al.*, 1986b). All algal cytochromes *c*-553 contain a lysine at position 29 (*E. gracilis* numbering) which may interact with the buried heme propionate. **(e)** The proposed interactions for *A. vinelandii* cytochrome c_5 are taken from the XRD structure. There is some concern that this may not represent the solution structure (section 4.3.5)

4.3.3 Protein Surfaces

The surface area around the heme edge in mitochondrial cytochrome *c* has been implicated as the interaction site for other redox proteins. This makes a comparison of that area with similar regions of the bacterial class I cytochromes *c* of particular interest. In the best characterised member, *Ps. aeruginosa* cytochrome *c*-551, the smaller size and the acidic nature (pI = 4.1, Horio *et al.*, 1960) might be expected to give rise to substantial surface differences from mitochondrial cytochrome *c*. A comparison of Figs. 4.3 and 4.9 reveals that there are differences but there are also unexpected similarities.

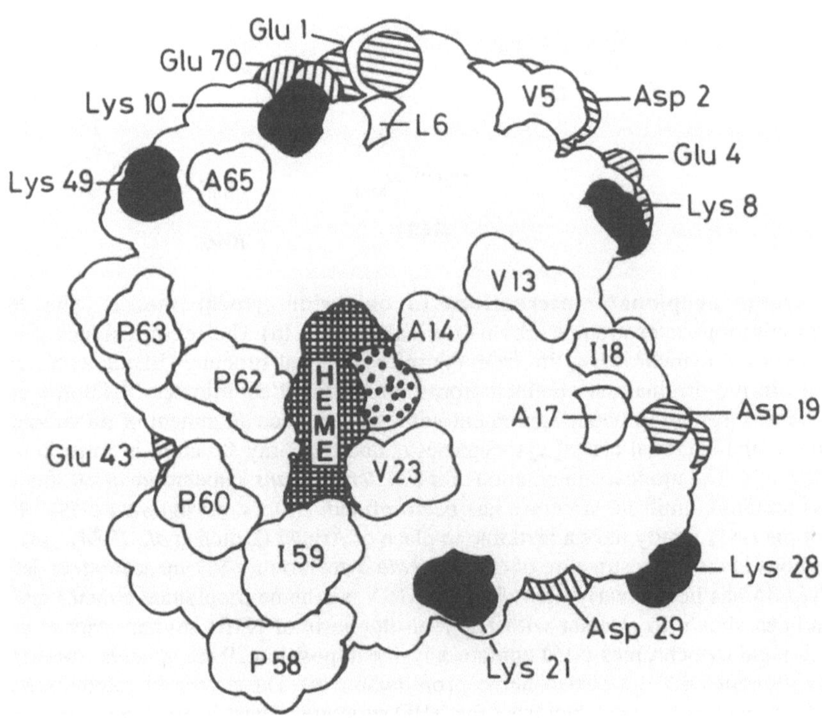

Fig. 4.9 A space-filling representation of the front face of *Pseudomonas* cytochrome *c*-551. (Adapted from a diagram supplied by R.J.Feldman)

Thus although mitochondrial cytochrome *c* contains some lysines on the front face close to the heme edge, unlike cytochrome *c*-551, both cytochromes share the feature of a ring of lysines around the periphery of the front face. There are also similarities in the distribution of negative charge at the top and right of the molecule and in the position, if not always the nature, of hydrophobic amino acids around the heme edge.

R. rubrum cytochrome c_2 is even more similar to mitochondrial cytochrome *c* in this region having both the ring of lysines around the periphery and lysines close to the heme edge. Again part of the front face is hydrophobic with residues R-Val 28, R-Tyr 46, R-Thr 92 and R-Phe 93 (equivalent to Thr 28, Tyr 46, Ile 81 and Phe 82; see Fig. 4.3). The degree of heme exposure for cytochromes *c* and c_2 is similar and less than that of the smaller cytochrome *c*-551. This feature has been proposed to have some influence upon the level of the redox potential (Chapter 7.4.3).

4.3.4 Two Domain Cytochrome c_4

Cytochrome c_4 a is predominantly membrane-attached diheme protein of M_r ~20,000 (Pettigrew & Brown, 1988). The cytochrome c_4 of *Ps. aeruginosa* is a two domain molecule, each domain containing a covalently bound heme *c* and resembling the class I cytochrome fold (L.Sawyer, personal communication). The two hemes are spectroscopically inequivalent (Chapter 2.3.2.1) and possess different redox properties (Chapter 7.5).

The two domains contain approximately 80 amino acid residues each, and are connected by a long extended polypeptide chain. They are related by an ~180° rotation so that their 'front' surfaces face in opposite directions (Fig. 4.10). The domains are arranged so that the heme groups are almost coplanar and the propionates of one heme point towards those of the second. The inner propionates (HP-7) are within hydrogen-bonding distance of each other.

Some of these features raise intriguing questions as to the nature of the interactions with the membrane and partner proteins, and the completion of the XRD structure is eagerly awaited.

4.3.5 Dimeric Cytochrome c_5

Azotobacter cytochrome c_5 has been isolated as both a dimer of M_r ~24,000 and a monomer of M_r ~12,000 (Meyer and Kamen, 1982; G.W. Pettigrew, unpublished data). The 83 residue protein used for the x-ray structure determination behaves as a monomer on Sephadex columns (T.E.Meyer, personal communication) but appears to form a dimer in the crystal (Carter *et al.*, 1985). This behaviour needs further investigation to establish whether the form studied in the crystals is a perturbed form of the protein or a biochemically relevant form.

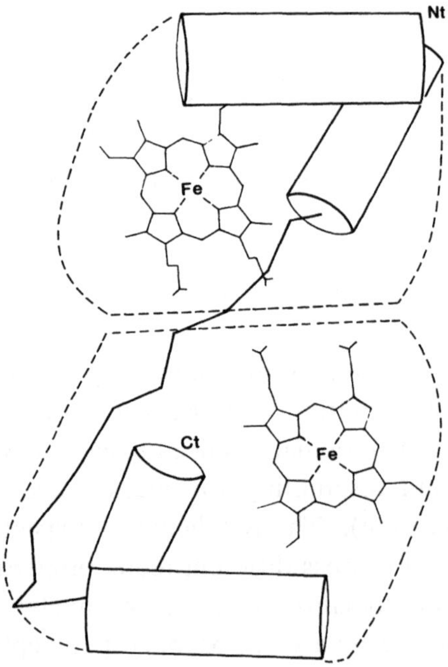

Fig. 4.10 The two domain structure of *Pseudomonas* cytochrome c_4. The N-terminal domain (top) is drawn in the orientation of Fig. 4.7 and shows the heme face-on with the N-terminal α-helix crossing a second α-helix situated near the middle of the whole protein sequence. The second domain (bottom) is connected to the first by an extended loop and has an arrangement of helices and heme group similar to the first domain but rotated through ~180°. Unpublished crystallographic data were kindly provided by L.Sawyer

Each monomer of crystalline cytochrome c_5 resembles *Pseudomonas* cytochrome *c*-551 in overall structure (Fig. 4.11). Not only is the general polypeptide fold similar but also the distribution of particular side chains. However, superimposition of the two structures and least-squares fitting of the atomic positions (Carter *et al.*, 1985) reveals that there are sizable differences between the two .

The dimer interface is quite hydrophobic and contains no charged residues (Fig. 4.11). As well as the heme edge normally exposed (Fig. 4.6), AV residues Leu 27, Leu 28, Ala 62, Pro 64 and Pro 65 form part of the interface. These prolines are equivalent to those of *Ps. aeruginosa* cytochrome *c*-551 that are close to the methionine ligand (Fig. 4.9). Thus it may be that *Azotobacter* cytochrome c_5 has a hydrophobic surface region similar to that of cytochrome *c*-551, but one that, on removal of the cytochrome from its natural redox partner, and under certain conditions - such as those prevailing in the crystal - acts as a dimerisation centre. If this is so, the properties of the molecule may be perturbed from those that obtain in the physiological environment. For example, the conformation may be altered to produce the unusual exposure of the propionates noted in section 4.3.2. Also the redox behaviour may be altered due to changes in the accessibility of the heme group. These uncertainties must be resolved before the crystallographic dimer can be taken as a guide to the physiological electron transferring properties of the molecule.

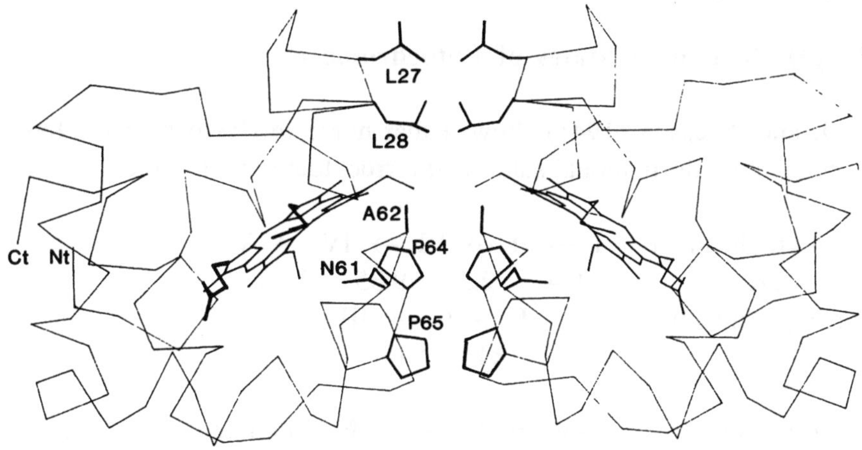

Fig. 4.11 The cytochrome c_5 dimer. The path of the polypeptide backbone is shown by joined α-carbon atoms. The heme groups and selected side chains that are present at the dimer interface are shown in bold. The left hand monomer has its heme propionate groups pointing outward, while in the right hand monomer these groups point inwards. Crystallographic data are from Carter *et al.* (1985)

The structure of the form of cytochrome c_5 that is a dimer in solution might help to resolve these questions but unfortunately the crystal structure of this form is not available. However, its NMR spectrum is very similar to that of the monomeric cytochrome c_5 (G.R.Moore & S.E.J.Rigby, unpublished data) which indicates that in contrast to the crystal dimer (Fig. 4.11), a close approach of the two heme groups does not occur in the solution dimer.

4.3.6 Class I Cytochrome *c* as Part of Larger Proteins

Although there is only limited data available, it seems probable that there is a widespread distribution of class I cytochromes *c* as domains or subunits of larger proteins in much the same way that cytochrome b_5 occurs as a domain of other proteins (Guiard & Lederer, 1979). Examples of this are the cytochrome *c* subunit of p-cresol methylhydroxylase, which shows some sequence similarity to other bacterial class I cytochrome *c* (Fig. 3.15) (McIntire *et al.*, 1986), and *Pseudomonas* cytochrome *c* peroxidase (Fig. 3.13) (Pettigrew & Moore, 1987; Chapter 3.4). Cytochrome cd_1 is another possible example (Pettigrew & Moore, 1987; Chapter 3.5).

4.4 pH Dependent Transitions of Cytochromes *c*

4.4.1 pH Dependent States of Cytochrome *c*

Theorell and Åkesson (1941) showed that horse ferricytochrome *c* has five pH dependent conformational states and ferrocytochrome *c* three:

Ferricytochrome *c*	I $\rightarrow$ II $\rightarrow$ III $\rightarrow$ IV $\rightarrow$ V
pK	0.4 2.5 9.35 12.8
Ferrocytochrome *c*	I $\rightarrow$ II $\rightarrow$ III
pK	<4 >12

Subsequent work (see review by Dickerson & Timkovich, 1975) has largely confirmed the conclusions of Theorell and Åkesson, though it is now clear that there are more conformational states at the extremes of pH.

State V ferricytochrome *c* appears to resemble state III ferrocytochrome *c* in being largely unfolded, but state IV ferricytochrome *c* is a stable, globular form (Stellwagen, 1964; Myer, 1968; Hennessey & Johnson, 1981; Provencher & Glockner, 1981) that does not have a ferrocytochrome *c* counterpart (Dickerson & Timkovich, 1975; Moore & Williams, 1980a). The differences in the stabilities of ferricytochrome *c* and ferrocytochrome *c* may be another reflection of the displaced N $\rightleftharpoons$ U equilibrium discussed in section 4.5.

The acid transition (III $\rightarrow$ II) in ferricytochrome *c* is complex, involving two or three protons and yielding a number of conformational states in equilibrium (Greenwood & Wilson, 1971; Babul & Stellwagen, 1972; Lanir &

Aviram, 1975; Stellwagen & Babul, 1975; Tsong, 1975; Drew & Dickerson, 1978; Lanir *et al.*, 1979; Dyson & Beattie, 1982; Robinson *et al.*, 1983b). Both His 18 and Met 80 are displaced from the iron. Essentially similar transitions may occur for the ferrocytochrome *c* (Drew & Dickerson, 1978) and even porphyrin cytochrome *c* (Vanderkooi & Erecińska, 1975).

Bacterial cytochromes have been less extensively studied than their mitochondrial counterparts but Drew and Dickerson (1978) showed that a range of proteins, including *Ps. aeruginosa* ferricytochrome *c*-551 and *Pa. denitrificans* ferricytochrome *c*-550, have a similar acid transition (II $\rightleftharpoons$ III) to the mitochondrial protein, and a number of bacterial proteins have been shown to have an analogous alkaline form (III $\rightleftharpoons$ IV; see section 4.4.5).

In the present section we are mainly concerned with individual group ionisation constants and with the alkaline transition of ferricytochrome *c*.

Table 4.2 Heme propionic acid pK_a values in class I cytochromes *c*

Protein		HP-6	HP-7	Refs.
mitochondrial	cytochrome *c*	≥9 [a]	<4.5	d-h
R. rubrum	cytochrome c_2	n.d. [b]	6.2 (ox), [b] 7.0 (red) [b]	f i
Rm. vannielii	cytochrome c_2	n.d.	6.3 (ox), 7.4 (red)	j k
Pa. denitrificans	cytochrome *c*-550	n.d.	6.2 (ox), n.d. (red)	l
Ps. aeruginosa	cytochrome *c*-551	<4.5	6.2 (ox), 7.3 (red)	h m
Ps. mendocina	cytochrome *c*-551	n.d.	7.2 (ox), [c] 8.0 (red) [c]	m
Ps. stutzeri (224)	cytochrome *c*-551	n.d.	7.8 (ox), [c] 8.45(red) [c]	m
Chl. limicola	cytochrome *c*-555	n.d.	6.3 (ox), [c] 7.0 (red) [c]	f

[a] See text for discussion of why the HP-6 pK_a is not given as <4.5.
[b] n.d. = not determined; ox = pK_a in oxidised protein; red = pK_a in reduced protein.
[c] These are not simple carboxylate ionisations, they are composite His-carboxylate ionisations (Chapter 7.6.2)
[d]Tanaka *et al.* (1975); [e]Moore (1983); [f]Moore *et al.* (1986b); [g]Hartshorn & Moore (1989); [h]Tonge *et al.* (1989); [i]Pettigrew *et al.* (1978); [j]Pettigrew *et al.* (1975b); [k]Moore *et al.* (1984a); [l]Timkovich & Cork (1984); [m]Leitch *et al.* (1984)

4.4.2 Ionisation of Heme Propionates

Generally, determination of pK_a values for groups in proteins can be difficult, particularly at extremes of pH where denaturation may occur. For bacterial cytochromes *c* under conditions where denaturation does not occur, spectroscopic and potentiometric methods have been used to detect heme-linked pH transitions (Table 4.2) and the identity of the ionisable groups established by

NMR and FT-IR spectroscopies (Chapter 2.3.3.4) to be heme propionates. Similar experiments with mitochondrial cytochrome *c* indicate that its heme propionates do not have pK values in the pH range over which state III ferricytochrome *c* is stable, namely pH 4.5 to 9.0.

In order to determine whether the heme propionates of native mitochondrial cytochrome *c* are ionised, four approaches have been used:

1. It has been calculated that horse ferricytochrome *c* has a net charge of +7.1 and horse ferricytochrome *c* a net charge of +6.6 at pH 7.15 based on titration data (Shaw & Hartzell, 1976), and phosphate binding data (Eden *et al.*, 1982). From the amino acid composition and histidine pK_a values (see Table 4.3), it can then be deduced that the heme propionates are charged at pH 7.15 (J.Matthew, personal communication), consistent with pK_a values of less than 4.5. The problem with this approach is that fitting pH titration and ion-binding data to a molecular model requires a number of assumptions that may lead to incorrect conclusions. In the present case, the analysis of Eden *et al.* (1982) predicts the release of 0.5 protons per molecule on oxidation of horse cytochrome *c* at pH 7 (because of shifts in histidine pK_a values), but this is not consistent with measured pK_a values (Table 4.3) or with the pH dependence of the redox potential (Chapter 7.6.2). Also, there is a hysteresis in the pH titration curve of ferricytochrome *c* that has been suggested to reflect a carboxylic acid ionising with a pK_a of 5.4 during alkalinisation, and 9.4 during acidification (Shaw & Hartzell, 1976). Whatever the explanation of this hysteresis, it introduces uncertainty into the value of the net charge obtained from the titration.
2. Theoretical investigations of the electrostatic properties of horse cytochrome *c* indicate that the environment of the propionates is polar rather than hydrophobic, which has led to the proposal that both heme propionates have pK_a values <4.5 (Churg & Warshel, 1986). However, the model used for these calculations has also been used to estimate carboxylic acid pK_a values in bovine pancreatic trypsin inhibitor protein (Russell & Warshel, 1985) and the reported error range of these calculations, ±21 KJ $mole^{-1}$ (equivalent to ±3.7 pH units at 25°C) casts some doubt on the reliability of the cytochrome proposal.
3. The pH dependence of the proton uptake of horse cytochrome *c* in the presence of urea indicates that one heme propionate has a low pK_a and the other a high pK_a (Hartshorn & Moore, 1989). The basis of this experiment is that urea denatures ferricytochrome *c* but not ferrocytochrome *c* under the conditions employed so that the difference in proton uptake (ΔH^+)

reflects the ionisation of groups in ferricytochrome *c* with abnormal pK_a values. The main problem with this approach is the assumption that ΔH^+ reflects only the ionisation of groups buried in the protein. Groups on the surface of the protein are assumed to behave the same in both oxidation states of cytochrome *c*.

4. FT-IR studies of the denaturation of horse ferricytochrome *c* indicate that a carboxylic acid ionisation occurs with a pK of approximately 9 (Tonge *et al.*, 1989). This suggests that at least one heme propionate has a high pK_a. The difficulty with this method has been discussed in Chapter 2.3.3.4.

All four studies agree that at least one heme propionate has a low pK_a; the conflict is with the second heme propionate. Our preferred scheme is the one given in Table 4.2 but this topic requires further work to resolve the conflict.

A physicochemical analysis of pK_a values is very similar to such an analysis of redox potentials (Chapter 7). Both parameters are strongly affected by electrostatic interactions and both may be affected by protein conformation changes. Thus a rigorous quantitative treatment of pK_a values of cytochrome *c* has not yet been attempted. In this, and the following section, we offer some comments on the question of why certain groups present in cytochromes *c*, including the propionate groups discussed above, appear to have unusual pH dependent behaviour.

The variation in the heme propionate pK_a values has not been adequately explained on a structural level. As we have discussed, with the exception of cytochrome c_5, HP-7 and, usually, HP-6 are buried within the protein interior. The fact that they are buried tends to raise their pK_a's but various interactions, such as the hydrogen-bonding networks shown in Fig. 4.4 and 4.8, oppose this tendency. Also, the pK_a values may be strongly affected by electrostatic interactions. These interactions may occur not only in the immediate environment but also with more remote groups within the protein or even in the solvent itself. Propionates exposed to the solvent are expected to have relatively low pK_a values (see Chapter 1.4 for intrinsic pK_a values of heme propionates). The pK_a values for buried groups depend upon the distance between the group and the protein surface, and the proximity of other buried charges and protein dipoles to the ionisable group.

The pK_a values of HP-7 of *Pseudomonas* and *Chlorobium* cytochromes have been unambiguously determined. They are higher than the pK_a values of the corresponding HP-6 groups (Table 4.2) because of the difference in their solvent accessibilities. The heme propionate pK_a values of the larger bacterial cytochromes have not been unambiguously assigned to specific propionates. The assignment in Table 4.2 is made by comparison with the smaller cytochromes: both *R. rubrum* and *Pa. denitrificans* cytochromes have similar HP-7 pK_a values to those of the smaller cytochromes, and all have similar structures around their HP-7 groups (Fig. 4.8). However, this line of reasoning is weakened by two sets of data. First, *Rm. vannielli* cytochrome c_2 lacks an obvious candidate for a positively charged group to fulfill a similar structural role to the interacting Arg or His of the other cytochromes (Fig. 4.8) yet despite this its corresponding propionate has a similar pK_a. Second, the corresponding propionate pK_a of mitochondrial cytochrome *c* is very different. It has been suggested (Moore, 1983) that HP-7 of mitochondrial cytochrome *c* has a pK_a <4.5, rather than >9, because of the proximity of Arg 38 (Fig. 4.4). However, recent site-directed mutagenesis studies of yeast cytochrome *c* in which Arg 38 was replaced by a variety of groups, including Gln, Ala and Leu, have failed to show a relationship between the pK_a of HP-7 and the presence of the Arg 38 (section 4.6.5), and this casts some doubt on the proposed substantial influence of Arg 38 on the HP-7 pK_a (Cutler *et al.*, 1989).

An explanation of why HP-6 of mitochondrial cytochrome *c* has a high pK_a must await experimental confirmation that the pK_a is indeed >9. Since HP-7 appears to retain a low pK_a when buried and not interacting with a neighbouring buried positive charge, HP-6, which has a similar environment (Fig. 4.4), might be expected to also have a low pK_a.

4.4.3 Side Chain Ionisation Constants

In this section we shall concentrate upon the ionisation constants of polypeptide functional groups over the pH range in which the protein is stable. Most, if not all, amino acid carboxyl groups on mitochondrial ferricytochrome *c* have pK_a values which fall outside this range, as do those of the tyrosines, arginines and lysines. Tyrosine ionisation has been studied by several groups using spectrophotometry (Stellwagen, 1964; Cronin *et al.*, 1985) and NMR (Boswell *et al.*, 1983). These show the lowest pK for a tyrosine of ferrocytochrome *c* is still >11.2. Ionisation of lysines of horse cytochrome *c* has been studied in an elegant series of differential chemical modification experiments by Bosshard & Zürrer (1980) and Bosshard (1981). The lowest identified pK_a, 9.3, was for Lys 39 of ferricytochrome *c*. This was raised to 9.5 in the ferrous state as a result of the redox state conformation change (section 4.4.4).

Several calorimetric and potentiometric measurements of horse cytochrome *c*, which measured the total number of protons taken up by the protein as the solution pH was changed, identify two titrating groups between pH 5.5 and 8.5 (Theorell & Åkesson, 1941; Paul, 1951; Paleus, 1954; Bull & Breese, 1966; Greenwood & Wilson, 1971; Shaw & Hartzell, 1976; Shiao & Sturtevant, 1976; Marini *et al.*, 1981). One of these has been unambiguously assigned to His 33 (Table 4.3) but the identity of the second ionisable group is unknown. It is not His 18 or His 26 because both of these have low pK_a values: the former because of coordination to the iron (Chapter 1.4); and the latter because of its non-polar location (Cohen *et al.*, 1974).

Table 4.3 pK_a values of histidine residues of cytochromes *c*

Protein		Residue	Ferri pK_a [a]	Ferro pK_a [a]	Refs.
Mitochondrial	cytochrome *c*	His 18	<3	<3	c-e
Ps. aeruginosa	cytochrome *c*-551	His 16	<3	<3	f
Mitochondrial	cytochrome *c*	His 26	<3.6	<3.2	c-e
Horse	cytochrome *c*	His 33	6.4	6.5	c-e
S. cerevisiae iso-1	cytochrome *c*	His 33	{ 6.7	6.7	d g h
		His 39	{ 6.8	7.2	d g h
C. krusei	cytochrome *c*	His 33	{ 6.6	n.d.	d g h
		His 39	{ 6.7	7.3	d g h
Rps. viridis	cytochrome c_2	His 39	6.8	7.1	i
Chl. limicola	cytochrome *c*-555	His 37 [b]	6.3	7.0	j
Ps. stutzeri (224)	cytochrome *c*-551	His 47 [b]	7.6	8.2	k

[a] The quoted pK_a values were determined by NMR. The redox state differences are significant in all cases with the possible exception of His 33 of horse cytochrome *c*; the reported difference is close to the margin of experimental error. The brackets indicate that the histidines of the ferric protein were not distinguished.

[b] See note [c] to Table 4.2.

[c]Cohen *et al.* (1974); [d]Cohen & Hayes (1974); [e]Moore & Williams (1980a,b); [f]Moore *et al.* (1980c); [g]Robinson *et al.* (1983a); [h]Cutler *et al.* (1989); [i]Moore *et al.* (1984a); [j]Moore *et al.* (1986b); [k]Leitch *et al.* (1984).

The ionisation of His 33 appears to be responsible (Ångström *et al.*, 1982) for the small perturbations of heme NMR and optical spectral properties reported by Czerlinski & Bracokova (1973), and Burns & La Mar (1981), but its pK_a is not redox-state dependent, in contrast to His 39 of yeast cytochrome *c* (Table 4.3).

Of the lysines of horse cytochrome *c* only Lys 39 and Lys 53 and/or 55 have redox state dependent pK_a values, the remaining 16 lysine pK_a's being redox state independent (Bosshard & Zürrer, 1980). These data were originally interpreted to indicate that there was not a significant electrostatic interaction between the charged form of the ionisable groups and the ferriheme charge, and that when redox dependent pK_a shifts do occur they reflect redox-state conformation changes. However, recent work with subtilisin (Russell & Fersht, 1987) suggests that charges located on the surface of a protein can have a significant electrostatic interaction with buried charges. Carrying this view over to cytochrome *c* would suggest that most of its surface charges have an interaction with the ferriheme charge but that compensating redox-state conformation changes leave the pK_a values independent of redox-state. According to this view, those groups with a redox-state dependent pK_a are not affected by redox-state conformation changes. Our preference is for the original interpretation because there is independent evidence identifying regions in cytochrome *c* where conformation changes occur (section 4.5) and because of the high degree of coincidence needed to have two opposing factors exactly cancelling each other in their contributions to so many pK_a values. Further discussion of the electrostatic influence of surface charge is given in Chapter 7.4.2

4.4.4 The Alkaline Transition of Ferricytochrome *c*

The presence of state IV ferricytochrome *c* is demonstrated by its non-reducibility by ascorbate (Greenwood & Palmer, 1965; Brandt *et al.*, 1966; Wilson & Greenwood, 1971; Lambeth *et al.*, 1973) and its much lower mid-point redox potential (Chapter 7.6). The results of investigations using a variety of spectroscopic methods (Chapter 2.3.2.4) are all consistent with a conformational change in the protein associated with replacement of the sixth ligand methionine 80. The methods include: optical spectroscopy (Theorell & Åkesson, 1941; Greenwood & Wilson, 1971; Lambeth *et al.*, 1973; Osheroff *et al.*, 1980; Gadsby *et al.*, 1987); circular dichroism (Myer, 1968; Hasumi, 1980); RR (Kitigawa *et al.*, 1977b); EPR (Lambeth *et al.*, 1973; Morton, 1973; Brautigan *et al.*, 1977; Gadsby *et al.*, 1987); and NMR (Gupta & Koenig, 1971; Morishima *et al.*, 1977; Wooten *et al.*, 1981). Under equilibrium conditions all these different methods yield a single proton ionisation with an apparent pK, for horse cytochrome *c*, of 9 - 9.4.

Stopped flow and temperature jump kinetic analysis of this transition reveals that the apparent pK is a composite of an ionisation with a pK_H of 11 coupled to a conformational equilibrium with a pK_c of -2 (Brandt *et al.*, 1966; Davis *et al.*, 1974).

$$C_o\text{-}H \underset{+H^+}{\overset{-H^+}{\rightleftharpoons}} C_o^- \underset{0.05s^{-1}}{\overset{6s^{-1}}{\rightleftharpoons}} C_o^{-*} \qquad (4.1)$$

$$pK_H=11 \quad pK_c\sim-2$$

The effect of pK_c ($K_c \sim 100$) is to displace the pK_H of 11 to produce an apparent pK of ~9 ($pK_H + pK_c$). The species C_o-H and C_o^- possess an axial methionine ligand, but C_o^{-*} (state IV ferricytochrome *c*) does not (Davis *et al.*, 1974; Kihara *et al.*, 1976; Gadsby *et al.*, 1987; Wooten *et al.*, 1981). Saigo (1981a;b) found that one or more transient forms were involved in the conformational change and that one was high spin.

Spectroscopic studies of horse cytochrome *c* at pH values of 9 - 11 reveal that the alkaline isomer is not a single molecular species (Chapter 2.3.2.4); there are at least two forms. Hong and Dixon (1989) have shown that at 45°C these different alkaline forms chemically exchange with each other. The extent of the structural differences between the different alkaline forms is not known but it appears likely that the proportions of the different forms vary with the cytochrome. For example, the position 82 variants of *S. cerevisiae* iso-1 cytochrome *c* with low pK values (Table 4.4) have different spectral properties to those of the wild-type protein (A.M.Davies, A.G.Mauk & G.R.Moore, unpublished data).

The alkaline transitions of other cytochromes and chemically modified forms of horse cytochrome *c* have not been studied in the same detail as above. Generally, optical studies centred at the Soret bands and the 605 nm and 695 nm absorbancies have been recorded but other spectroscopic techniques have not often been used. Also, most of the reported data are equilibrium studies only. Thus considerable uncertainty exists in comparing one protein with another. *E. coli* ferricytochrome b_{562} illustrates this very well. The equilibrium pH dependence of the optical spectrum revealed a transition with a pK of 8.7 apparently similar to that of mitochondrial cytochrome *c*; the 710 nm band was bleached and the spectra showed the iron retained a low-spin configuration (Myer & Bullock, 1978). These authors attributed this to the displacement of the methionine ligand in an analogous fashion to the transition of mitochondrial cytochrome *c*. However, subsequent equilibrium pH studies with NMR, EPR and MCD spectroscopies showed that the optical transition occurred with a rate of ~10^5 sec^{-1} and did not involve displacement of an iron ligand (Moore *et al.*, 1985b). In this case ionisation of the histidine ligand shifted the 710 nm band to 680 nm where it was masked by the α/β bands in the conventional optical spectrum, but resolved in the MCD spectrum.

Table 4.4 lists relevant studies for a variety of proteins. Enough work has been done to support the view that the alkaline transition of mitochondrial cytochrome *c* and bacterial cytochrome c_2 follows the general scheme in Eqn. 4.1, though Pettigrew *et al.* (1975a) noted that *Euglena* cytochrome *c*-558 did not obey the simple kinetic analysis, perhaps because more than one ionization was involved. Kinetic studies of smaller proteins such as *Ps. aeruginosa* cytochrome *c*-551, are required to characterise their alkaline transitions.

The reason why the pK values differ for related proteins is usually not known. Changes in pK_H, pK_c or both simultaneously may occur, and even when these are known the structural influences upon them can rarely be defined. This is analogous to the problem of identifying the structural basis for the altered $N \rightleftharpoons U$ equilibrium seen in some systems (section 4.5.4). A well defined series of proteins are provided by the study of Osheroff *et al.* (1980). In their work the alkaline transitions of ten primate cytochromes and nine singly modified lysine derivatives of horse cytochrome *c* were compared. The pK values varied from 9.5 to 8.1 (Table 4.4). Osheroff *et al.* convincingly demonstrated that the variation was primarily due to the disruption of two key surface interactions: the salt-bridge between Lys 13 and Glu 90; and the hydrogen-bond between Lys 79 and Ser/Thr 47. The differences in stability between human, horse and spider monkey cytochromes was attributed to the nature of the amino acid at position 50. The x-ray structures (Table 4.1) shows that the 79---47 interaction cross-links the left and right sides of the protein at the base of the heme (just as the 13---90 interaction cross-links above the heme). However, in solution Osheroff *et al.* concluded that there is a fraction of the horse cytochrome *c* with a Lys 79---Asp 50 salt bridge. A result of this interaction is that the bottom of the heme crevice is no longer closed off as before and it becomes less stable. With Ala 50 (human) there is no competition for the 79---47 hydrogen-bond, but with Asp 50 (horse) or Glu 50 (spider monkey), there is competition. Human ferricytochrome *c* is therefore the most stable (Table 4.4).

The series of site-directed mutants of *S. cerevisiae* iso-1 cytochrome *c* are also revealing (Table 4.4). The low equilibrium pK observed for some of the

Table 4.4 The alkaline transition of mitochondrial ferricytochrome *c* and bacterial cytochromes

Protein				Apparent pK[a b]	K_C[a]	pK_H[a]	Refs.
Horse		cytochrome *c*		9.0	124	11	e
Horse		cytochrome *c*	TNP-Lys 13	8.1			f
Horse		cytochrome *c*	CDNP-Lys 13	8.2			f
Horse		cytochrome *c*	CDNP-Lys 86	8.7			f
Horse		cytochrome *c*	CDNP-Lys 87	8.75			f
Horse		cytochrome *c*	CDNP-Lys 72	9.3			f
Human		cytochrome *c*		9.5			f
Human		cytochrome *c*	TNP-Lys 13	9.1			f
Spider monkey		cytochrome *c*		8.7			f
Rat		cytochrome *c*	Tyr 67	9.5			g
			Phe 67	10.7			g
C. oncopelti		cytochrome *c*-557		8.6	119	10.6	h
E. gracilis		cytochrome *c*-558		10.0	c	c	h
S. cerevisiae	iso-1	cytochrome *c*	Thr 102	8.6	244	11.0	i j
S. cerevisiae	iso-1	cytochrome *c*	Cys 102	8.3	460	10.9	j
S. cerevisiae	iso-1	cytochrome *c* [d]	Gly^{82}	7.8	22	9.1	j
S. cerevisiae	iso-1	cytochrome *c* [d]	Ser^{82}	8.0	16	9.2	j
S. cerevisiae	iso-1	cytochrome *c* [d]	Ile^{82}	7.0	200	9.4	j
S. cerevisiae	iso-1	cytochrome *c* [d]	Leu^{82}	7.0	200	9.6	j
R. rubrum		cytochrome c_2		9.1			k
Rm. vannielii		cytochrome c_2		9.1			k l
R. molischianum	iso-1	cytochrome c_2		7.3			k
R. molischianum	iso-2	cytochrome c_2		8.5			k
R. fulvum	iso-1	cytochrome c_2		6.9			k
R. fulvum	iso-2	cytochrome c_2		8.4			k
Ps. aeruginosa		cytochrome *c*-551		11.0			m n

[a] Apparent pK, K_C and pK_H refer to the scheme proposed by Davis *et al.* (1974) and described in the text.

[b] The apparent pK of mitochondrial cytochrome *c* varies with ionic strength and composition (Osheroff *et al.*, 1980).

[c] The alkaline isomerisation of *E. gracilis* cytochrome *c*-558 is more complex than that of horse cytochrome *c* and does not fit the scheme of Davis *et al.* (1974).

[d] Cys 102 replaced by threonine.

[e]Davis *et al.* (1974); [f]Osheroff *et al.* (1980); [g]Luntz *et al.* (1989); [h]Pettigrew *et al.* (1975a); [i]Aviram & Schejter (1969); [j]Pearce *et al.* (1989); [k]Pettigrew *et al.* (1978); [l]Moore *et al.* (1984a); [m]Vinogradov (1970); [n]M.C.Cox & G.R.Moore, unpublished data.

position 82 variants is largely a result of a decrease in pK_H. This has important implications for the identity of the triggering ionisation (see below).

Comparison of proteins with substitutions of internal groups is not straightforward because there are usually many other accompanying amino acid changes. However, it may be that the low pK values for *R. molischianum* and *R. fulvum* iso-1 cytochromes c_2 (Table 4.4) are because of an increased K_c resulting from the lack of Trp at a position equivalent to 59 of horse cytochrome *c*. This is not proven but is consistent with the instability of certain yeast mutant proteins lacking Trp 59 (Schweingruber *et al.*, 1978; 1979). Similarly, the high pK for *Euglena* cytochrome *c*-558 (Table 4.4) may reflect the replacement of Tyr 67 by Phe. This is supported by the effect on the alkaline transition of the replacement of Tyr 67 by phenylalanine in rat cytochrome *c* (Table 4.4). This is one area where further site directed mutagenesis (section 4.6) should resolve many outstanding questions.

The Nature of the Replacement Ligand and the Ionising Group. In principle, the ionising group with pK_H ~11 may also be the replacement ligand for methionine in state IV ferricytochrome *c*. Alternatively, the ionising group may only act as a trigger for the conformational change which leads to replacement of methionine. As we will see, no definite conclusions can be drawn regarding the nature of the ionisation.

Spectroscopic studies of the native protein are consistent with lysine as the replacement ligand (Chapter 2.3.2.4). However, many chemical studies appear to be incompatible with this. Thus although guanidation (Hettinger & Harbury, 1964; Fanger & Harbury, 1965; Morton, 1973) or trifluoroacetylation (Stellwagen *et al.*, 1975; Smith & Millet, 1980) of all the lysines of horse cytochrome *c* gives rise to an alkaline transition to a high spin species, this is not so with amidination (Pettigrew *et al.*, 1976; Wallace, 1984). Also, the individually lysine-modified cytochromes *c* that have been tested retain apparently normal alkaline transitions (Aviram & Schejter, 1969; Looze *et al.*, 1978; Osheroff *et al.*, 1980; Smith & Millet, 1980). Similarly, lysine 72 and 79, two of the lysines most frequently considered as replacement ligands because of their proximity to the heme, behave as freely accessible solvent exposed amines in differential chemical modification studies between pH 7 and 11 (Bosshard, 1981).

There is no model yet that reconciles all these apparently contradictory conclusions but recent optical and NMR studies of fully guanidated, trifluoroacetylated, and amidinated horse ferricytochrome *c* indicate that the alkaline forms of these proteins differ considerably from that of native ferricytochrome *c* (A.M.Burch, M.C.Cox, G.R.Moore & A.F.Parsons, unpublished data). The former two derivatives possess a hydroxide ligand at high pH but in the latter derivative at high pH and 25°C, a modified lysine is the sixth ligand through one of its nitrogen atoms. This accounts for the different spin-states of the two sets of derivatives since their ligand field strengths vary considerably (Chapter 1.4). However, as with the native protein, at alkaline pH there are multiple species of the modified proteins.

We propose that further spectroscopic study of singly lysine-modified proteins will establish whether their methionine ligands have really been displaced and, if so, the nature of the replacement.

For the ionisable group in the native protein, the possible candidates are tyrosine, arginine, lysine, His 18, HP-6 or protein associated H_2O. The first two are eliminated by consideration of their measured or predicted pK values (section 4.4) but each of the others remain possibilities.

Gadsby *et al.* (1987) clearly show that at pH 11, His 18 is not ionised *at equilibrium* but they propose that its ionisation is part of a kinetic pathway leading to a transitional 5-coordinate intermediate with a histidinate ligand. The subsequent coordination of a lysine leads to an increase in the His 18 pK resulting in reprotonation. This mechanism neatly accounts for the fact that lysine must be deprotonated before it can coordinate to the iron, yet the whole transition is a one-proton process.

An alternative proposal is based on FT-IR studies suggesting a carboxyl ionisation with an apparent pK near 9 (Tonge *et al.*, 1989). If this ionisation were that of HP-6 it could account for the difference in alkaline pK between *Ps. aeruginosa* cytochrome *c*-551, in which HP-6 has a low pK_a (Table 4.2), and the larger cytochromes *c* and c_2 (Table 4.4). However, there are problems in assigning the FT-IR spectral feature to HP-6 (Chapter 2.3.3.4 & section 4.4) and in any case, a model involving a triggering HP-6 ionisation and lysine ligation is difficult to reconcile with the observed proton stoicheiometry of one.

Finally, Takano & Dickerson (1981b) proposed that ionisation of the buried H_2O-1 molecule is the conformational trigger. They envisaged a transient intermediate with a OH^- ligand to the iron. The hydroxide ligation of the guanidated and trifluoroacetylated ferricytochromes at high pH is consistent with this proposal.

The alkaline transitions of site-directed mutant forms of cytochrome *c* should help identify the triggering ionisation. For example, if the replacement of Tyr 67 by phenylalanine led to the displacement of H_2O-1 then the raised pK (Table 4.4) might be taken as evidence in favour of Takano and Dickersons proposal. Also, the reduction in pK_H accompanying the substitutions at position 82 (Table 4.4) appear to us to argue against the proposal of Gadsby *et al.* (1987). It seems more likely that a destabilising mutation would lower an abnormally high pK_a rather than depress a normal or abnormally low pK_a (compare Tables 1.4 & 4.4). Further work in this area is needed.

4.5 Differences in Conformation and Dynamics between Reduced and Oxidised Cytochrome *c*

4.5.1 Differences in Properties of the Two Redox States

As long ago as 1966 Margoliash & Schejter discussed a large body of data which showed differences in the properties of the reduced and oxidised forms of cytochrome *c*. These, and more recent, data are summarised in Table 4.5. The origin of these differences must lie in the single charge change associated with the redox process but the immediate cause could be:

1. a direct effect of the charge difference at the heme;
2. different protein conformations;
3. different protein dynamic states.

These effects might contribute to each of the redox state differences, though to different extents in different cases, and a central problem is to disentangle their relative contributions.

Our view is that in most cases charge difference is important only in so far as it leads to changes in structure or dynamics (see Chapter 7 for the effect of electrostatic interactions on the redox potential). Our proposal is that the oxidation of the heme is associated with the appearance of a positive charge in an energetically unfavourable environment and that the need to stabilise this charge leads to the observed redox-state differences. Charge stabilisation comes from the interaction of the heme charge with all other charges and dipoles in the protein and surrounding solvent. However, some interactions are more important than others and the following could make major contributions:

1. interactions with small ions in solution leading to altered ion-binding properties;
2. interactions with charged surface groups leading to different conformations

Table 4.5 Redox state dependent properties of mitochondrial cytochrome *c*

Stability

Ferricytochrome *c* is less stable than ferrocytochrome *c* to the following conditions: extremes of temperature and pH; the presence of chaotrophs such as alcohols, urea and guanidinium HCl (Theorell & Åkesson, 1941; Butt & Keilin, 1962; Myer, 1968; Kaminsky *et al.*, 1973b; Knapp & Pace, 1974; Dickerson & Timkovich, 1975; McLendon & Smith, 1978; Myer *et al.*, 1979; 1981: Ridge *et al.*, 1981; and cited references).

Proteolysis

Ferricytochrome *c* is digested by subtilisin and specifically cleaved by trypsin. Ferrocytochrome *c* is completely resistant to these proteases (Nozaki *et al.*, 1958; Yamanaka *et al.*, 1959; Hantgen & Taniuchi, 1977; 1978; Harris & Offord, 1977; Endo *et al.*, 1985).

Peptide and side chain reactivities

Certain buried peptide NH's and side chains of ferrocytochrome *c* are more resistant to chemical reaction than their counterparts of ferricytochrome *c*, as revealed by the exchange of NH protons for deuterons (Kägi & Ulmer, 1968; Ulmer & Kägi, 1968; Nabedryk-Viala *et al.*, 1976; Patel & Canuel, 1976; Moore *et al.*, 1980b) and the chemical modification of side chains such as those of the tyrosine residues (Cronin *et al.*, 1985 & references therein) and Met 80 (Ando *et al.*, 1966).

Ion-binding

The binding of some ions to cytochrome *c* is redox state dependent: e.g. ferricytochrome *c* binds Cl^- more strongly than does ferrocytochrome *c* (see Pettigrew & Moore, 1987: Chapter 2.6 for futher details). Binding of cytochrome *c* to some chromatographic materials is also redox state dependent (Margoliash & Schejter, 1966; Ferguson-Miller *et al.*, 1979 & references therein).

pK_a's of surface groups

Some surface groups of ferrocytochrome *c* have higher pK_a's than their ferricytochrome *c* counterparts: e.g. Lys 39 and Lys 55 of horse cytochrome *c* and His 39 of yeast cytochrome *c* (Bosshard & Zürrer, 1980; Robinson *et al.*, 1983a).

Crystallisation

Ferricytochrome *c* and ferrocytochrome *c* have different crystal forms. Crystals cannot be completely redox cycled without shattering although crystals of ferrocytochrome *c* can be ~60 % oxidised by O_2 without damage (Margoliash *et al.*, 1968; Tsukihara *et al.*, 1973; Matsuura *et al.*, 1979; Takano & Dickerson, 1981a,b).

Antigenicity

Ferricytochrome *c* and ferrocytochrome *c* have slightly different antigenic properties as revealed by complement fixation (Reichlin *et al.*, 1966; Margoliash *et al.*, 1968).

for these groups;

3. interactions with buried solvent molecules (Takano & Dickerson, 1981a,b), and the heme propionates (Moore, 1983).

Various data indicate that the first two are not major factors for mitochondrial cytochrome *c*. Ion-binding association constants are not greatly redox-state dependent (Pettigrew & Moore, 1987; Chapter 2.6) and chemical modification of the surface charge does not greatly affect either the protein (section 4.6) or redox potential (Chapter 7.4.2).

Experimental data are not available to assess the strength of the interactions of the buried H_2O molecules but the interaction between the heme and the buried, ionised heme propionate has been measured for various bacterial cytochromes (section 4.4 and Chapter 7.6.2). This is of the order of ~6 kJ mol^{-1}. Structural mechanisms related to these two stabilisation influences have been proposed and these are described below.

4.5.2 Conformational Differences between the Two Redox States

Direct crystallographic information concerning conformational differences between reduced and oxidised cytochrome *c* has a chequered history! Dickerson and co-workers initially claimed that there were substantial structural differences between redox states of horse cytochrome *c* (Takano *et al.*,1971; Dickerson *et al.*, 1972). So certain were they that, in discussion of a scheme for the mechanism of electron transfer, they stated *"Feasibility arguments may be brought against the scheme ... but not on the issue of side chain conformational changes."* (Takano *et al.*, 1973). The largest effect was thought to be an 8Å movement of Phe 82 from within the heme crevice of the reduced form to the surface of the protein on oxidation. However, studies by other workers on related cytochromes did not reveal such an effect (Tsukihara *et al.*, 1973; Salemme *et al.*, 1973a,b) and Dickerson's group themselves found no such difference in the two redox states of tuna cytochrome *c* at 2Å resolution (Mandel *et al.*, 1977). The error had been due to the misassignment of the electron density belonging to Phe 82 in horse ferricytochrome to a postulated anion bound at the heme crevice. On the basis of the 2Å map, the claim was then made that *"... if believable differences do appear in the course of the 1.5Å refinement ... these should be no more than minor breathing of main chain or adjustment of side chains."* (Mandel *et al.*, 1977). Refinement, however, revealed rather larger differences than predicted (Takano & Dickerson, 1981b, 1982). These mainly affected the loop of residues from Ser 47 to Trp 59 at the base of the heme, and the region below Met 80 to the left of the heme. Movements included Asn 52 (1.6Å), Trp 59 (0.4Å), Tyr 67 (0.7Å), Phe 82 (0.4Å) and H_2O-1 (0.9Å). Associated with these side chain movements were small changes in the main chain and heme group.

In view of the small distances involved, the occurence of crystal packing artifacts (Williams *et al.*, 1985c), and the unreliability of previous pronouncements on the subject, it is important to remain critical of the accuracy of the x-ray crystallographic description. However, independent methods tend to support the present crystallographic conclusions. These include measurement of the reactivity of the lysines of horse cytochrome *c* (Bosshard & Zürrer, 1980) and NMR studies of native and chemically modified cytochromes *c* (Moore & Williams, 1980b;c; Moore *et al.*, 1980a; Robinson *et al.*, 1983a; Williams *et al.*, 1985c). The involvement of the Asn 52 region of the molecule in a redox state dependent movement is also consistent with the different abilities of the two redox states to fix complement with rabbit anti-horse cytochrome *c* sera (Reichlin *et al.*, 1966; Margoliash *et al.*, 1968; Moore & Williams, 1980b;d). Residues 58-60 form one of the most immunogenic regions of the molecule (Urbanski & Margoliash, 1977; Jemmerson & Margoliash, 1979)[2].

Two bacterial cytochromes *c*, *R. rubrum* c_2 and *Ps. aeruginosa* *c*-551, have been studied to high resolution (Salemme *et al.*, 1973a;b; Bhatia *et al.*, 1984; Matsuura *et al.*, 1982) and reveal little conformational difference between their redox states. Unlike most of their mitochondrial counterparts (Margoliash *et al.*, 1968), crystals of both can be oxidised and reduced without physical damage. A comparison of the root mean square difference in atomic positions of the oxidised and reduced structures of tuna cytochrome *c* (0.5 Å) and cytochrome *c*-551 (0.08 Å) is very striking (Takano & Dickerson, 1981b; Matsuura *et al.*, 1982). Thus there seems to be no conformational change associated with the redox process in the two bacterial cytochromes. Moore (1983) suggested an explanation for this difference from mitochondrial cytochrome *c*, based on the observation that studies on the bacterial proteins were performed on crystals prepared at pH values lower than the pK_a of the buried propionates (Tables 4.1 & 4.2). Because of this there can be no Fe(III)---HP-7 charge interaction, the central element of one of the conformational triggers described below. Crystallographic studies at higher pH will test this proposal.

[2] A full review of the literature concerning the antigenicity and immunogenicity of mitochondrial cytochrome *c* is beyond the scope of the present book. However, this is an active area of research. Recent work includes the identification of antigenic sites on horse and *S. cerevisiae* iso-1 cytochrome *c*, and the use of monoclonal antibodies to determine whether chemical modifications cause long-range conformational perturbations. The following articles, and references therein, should be consulted for further details: Paterson (1985); Jemmerson *et al.* (1985); Cooper *et al.* (1987); Collawn *et al.* (1988); Silvestri and Taniuchi (1988); Oertle *et al.* (1989). The XRD structure determination of a cytochrome *c*: antibody complex is underway (Mylvaganam *et al.*, 1988).

The Trigger and Transmission of the Conformation Change. The similarity in the geometry of the hemes and axial ligands of ferri- and ferro-cytochrome *c* (Chapter 1.3) rules out a heme-driven conformational change such as occurs on oxygenation of deoxyhemoglobin (Perutz, 1970). In only a few cases has a substantial redox-state conformational difference been found for the redox-centre of a class I cytochrome *c* (Chapter 2.3.1.2) but whether there are subsequent protein conformational changes has not been determined.

Two possible triggering mechanisms have been proposed for mitochondrial cytochrome *c*. That suggested by Takano & Dickerson (1981a;b) involves reorganisation of the hydrogen-bond network around Met 80 (Fig. 4.12). In this scheme the positive charge appearing on the Fe in the oxidised state is partly delocalised onto the Met 80-S resulting in a weakening of the hydrogen-bond between Met 80-S and Tyr 67-OH, and an increase in the distance between them. The movement of Tyr 67 causes the buried H_2O molecule with which it is hydrogen-bonded to move, leading to movement of other groups in the hydrogen-bond network. The subsequent movements of Asn 52, Tyr 48 and Thr 49, which are hydrogen-bonded to the heme propionates (Fig. 4.4), push the heme up and out of the crevice by 0.16Å. It is a combination of the greater solvent exposure of the heme and the movement of the internal H_2O molecule towards the iron (a movement of ~1Å) that, in Takano & Dickerson's view, act to stabilise the ferric heme. In *Ps. aeruginosa* cytochrome *c*-551 the hydrogen-bonding network around the methionine ligand S is situated in the top left rather than the bottom left of the heme crevice (according to the view shown in Fig. 4.6). Thus although a small movement (0.1Å) of an internal H_2O molecule towards the iron occurs on oxidation it does not lead to the side chain movements at the bottom of the molecule seen in tuna cytochrome *c*.

The other mechanistic proposal involves the electrostatic relationships between Fe(III), HP-7 and Arg 38 (Moore, 1983). According to this scheme, oxidation of the iron leads to a requirement for the stabilisation of the internal positive charge, and this is in part achieved by interaction with the rear heme propionate. On reduction the iron does not require electrostatic stabilisation but the propionate does, and this is achieved by a stronger interaction with its environment, in particular Arg 38. The small movements associated with this strengthened interaction lead to the observed conformational changes in the Asn 52 region. This proposal also neatly explains the redox dependent pK values of Lys/His 39 and Lys 53 and 55 (Bosshard & Zürrer, 1980; Robinson *et al.*, 1983a) by a change in their immediate chemical environments.

There are no additional data addressing the question of which (if either!) of these two mechanisms is correct. Our view is that since the stabilisation of the heme charge involves many groups it is most likely that no single mechanism is operative, although the relative importance of particular interactions will not be the same.

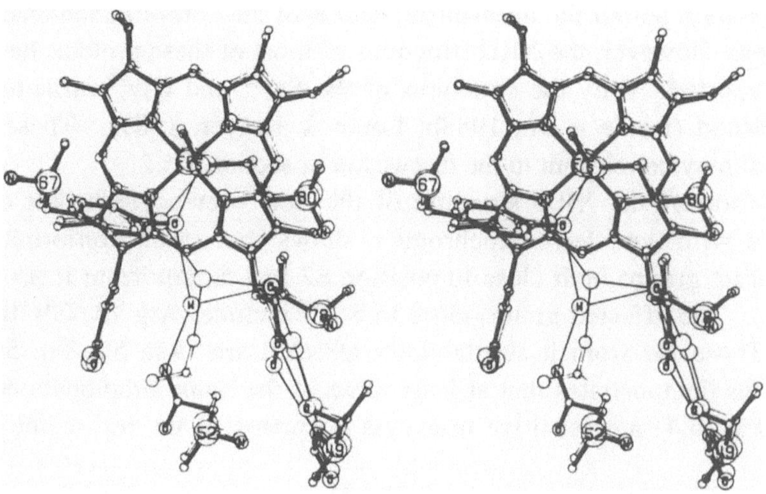

Fig. 4.12 Close-up stereo view of the heme and selected amino acids of tuna ferrocytochrome *c* (solid bonds) and tuna ferricytochrome *c* (open bonds), showing some of the main oxidation state changes observed by XRD. (Reproduced with permission from Takano & Dickerson 1981b)

NMR studies of mitochondrial cytochrome *c* have indicated that groups in the protein remote from the left side and bottom of the molecule also show redox-state dependent properties (Moore *et al.*, 1985a; Williams *et al.*, 1985b;c). Examples are the side chains of Phe 10, Val 11, Tyr 97 and Ala 101, which are clustered close together in the N- and C- terminal helices. Conformational changes at these positions were not found in the refined crystal structures of tuna cytochrome *c* (Takano & Dickerson, 1981a;b) and it is difficult to assess how large they would need to be to explain the NMR data. They may occur because of the presence of an electrostatic relay between a group in this region (perhaps Arg 91) and the heme or because the main conformation change is transmitted through the protein by minor movements of the 60's helix and the N- and C-terminal helices (Moore, 1983).

4.5.3 Conformational Differences between Site-Directed Mutants of *S. cerevisiae* iso-1 Cytochrome *c*

Site-directed mutagenesis is being used to to explore cytochrome *c* (section 4.6), and many of the variants being studied have residues proposed to be involved in the redox-state conformation change (section 4.5.2) replaced. Thus they should permit the mechanistic details of the conformation change to be described. However, the XRD structures of most of these proteins have not yet been reported. Only the structures of the Ser^{82} and Gly^{82} variants have been published (Louie *et al.*, 1988b; Louie & Brayer, 1989). These have features that may be relevant to the discussion in section 4.5.2

Comparison of the XRD structure of the Ser^{82} ferrocytochrome *c* with that of the wild-type ferrocytochrome *c* shows that small conformational changes affect groups both close to position 82 and remote from it (Louie *et al.*, 1988b). The affected groups close to Ser^{82} include: Arg 13, Gly 83 and Gly 84. Those far from it substantially affected are: Asn 52, Trp 59 and H_2O-1. This demonstrates that at least some of the heme propionate contact residues (Fig. 4.4) are sensitive to events occurring at the top of the heme crevice.

Louie *et al.* (1988b) suggest that the Ser^{82} protein is less able to adopt the normal reduced conformation of the protein (consistent with its lower redox potential - see Table 4.8) and therefore has a conformation tending towards that of the oxidised protein. They further suggest that the shift of Asn 52 and Trp 59 represents one of the stages of the redox state conformational change of native cytochrome *c*. If this view is correct then the proposal concerning the mechanism involving the heme propionates (section 4.5.2) is strengthened.

The structure of the Gly^{82} ferrocytochrome shows there is a larger conformational change at the mutation site than occurs in the Ser^{82} protein and a smaller conformational change at the base of the heme. Nevertheless, H_2O-1 is substantially moved in the Gly^{82} protein. Louie and Brayer (1989) emphasize this point in connection with the proposal of Takano and Dickerson for the involvement of H_2O-1 in the redox state conformation change (section 4.5.2).

NMR studies of the Tyr^{82} (Pielak *et al* , 1988b) and Ala^{38} (Cutler *et al* , 1989; A.G.P.Thurgood & G.R.Moore, unpublished data) proteins show that the replacements do not lead to substantial long-range structural perturbations. XRD structures of these proteins to calibrate the NMR method would be welcome.

4.5.4 Dynamic Fluctuations in the Cytochrome *c* Structure

The cytochrome *c* molecule described in previous sections is a compact and globular structure, which we signify by N (for Native state), but in solution this structure is in equilibrium with various Unfolded states (denoted by $U_1 ...... U_n$) whose degree of unfolding may range from small, localised fluctuations in main chain conformations to gross unfolding. This equilibrium:

$$N \rightleftharpoons \Sigma U \tag{4.2}$$

is a feature of globular proteins in general, not just cytochrome *c*, and the large body of physicochemical data on cytochrome *c* (Table 4.5) cannot be understood without reference to it (Linderström-Lang & Schellman, 1959; Hvidt & Nielsen, 1966; Pace, 1975). The situation is really more complex than Eqn. 4.2 expresses because there are conformational fluctuations of the N state which may not involve unfolding. Thus Eqn. 4.2 should be rewritten as:

$$\Sigma N \rightleftharpoons \Sigma U \tag{4.3}$$

Fig. 4.13 illustrates the difference between Eqns. 4.2 and 4.3. We have tried to separate time dependent effects within the N state from unfolding effects, but with our present understanding of protein structure these are difficult properties to classify. Indeed, a clear-cut distinction may not exist.

All studies of the stability of cytochrome *c* have found that the reduced protein is more stable than the oxidised protein, and this has often been taken to indicate conformational differences between the two redox states (Margoliash & Schejter, 1966; Dickerson *et al.*, 1971; Lemberg & Barrett, 1973). Even as recently as 1981, Takano & Dickerson discussed the stability differences in terms of the static conformations of cytochrome *c* determined by x-ray crystallography. However, the most probable explanation for the differences in stability is that the equilibrium constant of equation 4.3 (K^u) is different for the two redox states so that $K^u_{ferri} > K^u_{ferro}$ (Kägi & Ulmer, 1968; Salemme, 1977; Hantgen & Taniuchi, 1978; Moore & Williams, 1980b; Eden *et al.*, 1982).

The free energy diagrams of Fig. 4.14 illustrate possible origins of the different values of K^u. These diagrams are grossly simplified representations of three limiting cases and the actual situation is probably a combination of the different schemes. Thus the key structural features responsible for the different stabilities may not be easily recognised from the crystal structures.

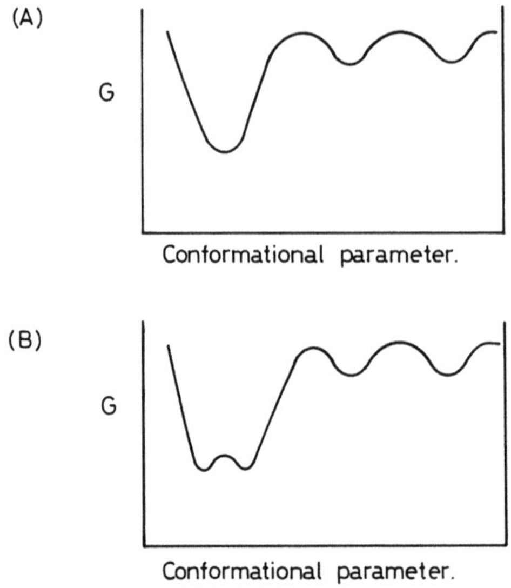

Fig. 4.13 A simplified representation of the relative values of the free energies (G) of two hypothetical proteins. The conformational parameter represents the structure of a protein. Ideally this should be a function of all the bond angles and distances of the molecule, including the solvation layer around it. Such a representation would be multi-dimensional. **(A)** A globular protein with a single free energy minimum for the native state and two minima for partially or completely unfolded states. **(B)** A globular protein with two free energy minima for the native state, corresponding to different conformations resulting from local motions (note that they need not be of equal energy, as shown, to be significant; the energy difference should be of the same order as **KT**, or less), and two minima for partially or completely unfolded states

The problem is to decide how much of the stability difference results from changes in conformation (such as the exposure of groups and the number of hydrogen-bonding and non-bonding interactions) and how much results from the oxidised cytochrome having an extra positive charge compared to the reduced cytochrome.

Although the conformational differences are triggered by the charge change, it is important to distinguish between these two causes. The burial of a charge without an accompanying conformation change can give rise to substantial stability differences because the overall free energy of the folded protein is changed (Churg & Warshel, 1986; Moore *et al.*, 1986b). Also, in the case of cytochrome *c*, there appears to be a strengthening of the Fe-S bond on reduction (Harbury *et al.*, 1965). Studies with *Ps. aeruginosa* cytochrome *c*-551 and *R. rubrum* cytochrome c_2 address directly the question of whether the redox dependence of protein stability is a direct effect of the charge

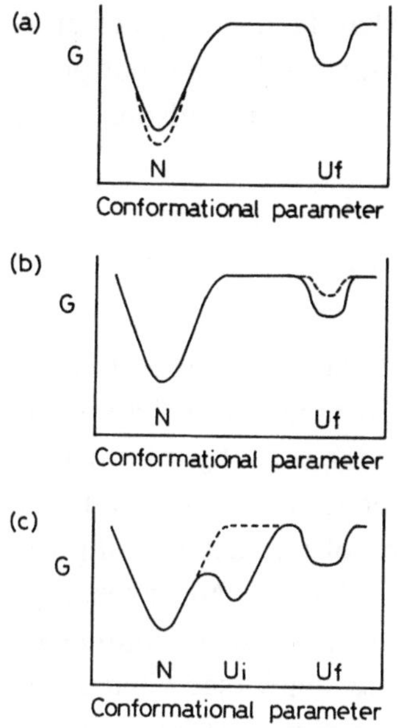

Fig. 4.14 Simplified representation of the relative values of the Gibbs free energy (G) of ferricytochrome *c* (solid line) and ferrocytochrome *c* (broken line). The stabilisation free energy (ΔG^{s}) is the energy difference between the native (**N**) and fully unfolded (**U$_f$**) states. **(a)** The **N**-state of ferrocytochrome *c* is stabilised with respect to that of ferricytochrome *c* and, since the **U$_f$** states are of equal energy, $\Delta G^{s}_{ferro} > \Delta G^{s}_{ferri}$. **(b)** The **U$_f$**-state of ferrocytochrome *c* is destabilised with respect to that of ferricytochrome *c* while the **N**-states are of equal energy. Therefore $\Delta G^{s}_{ferro} > \Delta G^{s}_{ferri}$. **(c)** Although both **N**-states and both **U$_f$** states are of equal energy, ferricytochrome *c* has a partially unfolded globular state (**U$_i$**) that has no direct ferrocytochrome *c* equivalent. In addition to these limiting cases both ferricytochrome *c* and ferrocytochrome *c* have a number of locally unfolded states ($U_{l_{1..n}}$). The difference in energy between these states and the **N**-states are also likely to be redox state dependent; thus $\Delta G^{l_{1..n}}_{ferro} > \Delta G^{l_{1..n}}_{ferri}$

difference or due to different protein conformations, because these proteins do not have a redox state conformation change. The indications are that their ferric forms are less stable that their ferrous forms (Pettigrew *et al.*, 1975b; 1978; Moore *et al.*, 1977) but further work is required to clarify this.

A number of experimental methods investigate the position of equilibrium of Eqn. 4.3, and most involve determining chemical reactivities of particular groups. The rate of exchange between buried peptide NH protons and deuterons is the most generally applicable method and it has been extensively investigated with a variety of proteins (for reviews see Tsuboi & Nakanishi, 1979; Woodward & Hilton, 1979; CIBA Foundation Symposium No. 93).

The central assumption with this method is that the exchange of an NH in the interior of a globular protein is impossible from the N-state but possible from the U-state. Thus the exchange mechanism is (Hvidt & Nielson, 1966):

$$N \underset{k_2}{\overset{k_1}{\rightleftharpoons}} U \overset{k_e}{\rightarrow} U^* \underset{k_1}{\overset{k_2}{\rightleftharpoons}} N^* \tag{4.4}$$

Since the N-state (which contains the NH) and the N*-state (which contains the ND) are monitored, the measured rate is a function of the rates of folding (k_2) and unfolding (k_1), and the rate of exchange (k_e). k_e is known from work on peptides and therefore k_1 and k_2 can be determined. The method has many unsolved problems but the central assumption of rate determining fluctuations in the protein structure seems to be valid. Thus the NH exchange data for cytochrome *c* (Table 4.5) can be rationalised, at least qualitatively, in terms of redox state changes in the protein dynamic states (see Chapter 2.3.3.2 for details of NMR studies of the NH exchange properties of horse cytochrome *c*).

Other chemical modification experiments may be interpreted with a similar reaction scheme. For example, Sutin & Yandell (1972) propose that the binding of CN^-, N_3^- and imidazole to ferricytochrome *c* at pH 7 requires a rate limiting fluctuation of the protein structure that causes the displacement of Met 80 as an axial ligand. The resulting unstable intermediate then reacts with the incoming ligand:

$$\overline{S - Fe - N} \underset{k_2}{\overset{k_1}{\rightleftharpoons}} \overline{S \quad Fe - N} \underset{k_4}{\overset{k_3}{\rightleftharpoons}} \overline{S \quad X - Fe - N} \tag{4.5}$$

where X can be CN^-, N_3^- or imidazole. At pH 7, 25°C and I = 0.1 M, k_1 was determined to be 30 - 60 s^{-1} for horse ferricytochrome *c*; k_2 has not been determined. Sutin & Yandell (1972) suggest that the open, intermediate form contains either pentacoordinate Fe(III) or hexacoordinate Fe(III) with H_2O as a sixth ligand. Apart from the expected axial perturbations around Met 80, cyanoferricytochrome *c* is similar in conformation to the native protein (Moore *et al.*, 1986b).

Another example is the carboxymethylation of Met 65 and Met 80 (Ando *et al.*, 1966). The surface residue Met 65 can be modified in both redox states but the buried Met 80 can only be modified in ferricytochrome *c*. Many of these kinds of reactivity differences have been previously interpreted solely in favour of conformational differences between the N states of cytochrome *c* - an example is the modification of Arg 38 and Arg 91 of horse cytochrome *c* (Pande & Myer, 1980) - but clearly, differences in the N $\rightleftharpoons$ U equilibria may also be important.

The N $\rightleftharpoons$ U equilibrium can be displaced by a wide range of chaotrophic agents and this has been exploited by many workers with cytochrome *c* (Table 4.5). However, despite the large body of data that has been accumulated, the regions of cytochrome *c* most susceptible to local unfolding have not been identified. Since proteolytic enzymes digest the U state much faster than the N state (Pace, 1975; Imoto *et al.*, 1974), and since ferricytochrome *c* is readily digested by some proteolytic enzymes while ferrocytochrome *c* is more resistant (Table 4.5), the use of time-course proteolysis might reveal which regions of cytochrome *c* are susceptible to redox dependent unfolding. The characterisation of non-covalent complexes of horse cytochrome *c*, such as (H1-38):(56-104) (Parr *et al.*, 1978; Westerhuis *et al.*, 1982), and the observation that acetimidylated ferricytochrome *c* is cleaved by trypsin only at Arg 38 (Harris & Offord, 1977) suggests that the region that is conformationally most sensitive to the redox change is also the region least structurally defined and most prone to local disorganisation.

Displacement of the equilibrium in Eqn. 4.3 has been used by many workers to study the unfolding pathways of cytochrome *c* (Myer, 1968; Myer *et al.*, 1979; 1980a;b; 1981; Nall & Landers, 1981; Roder *et al.*, 1988; Tsong, 1973; 1974; 1975). The basic rationale for these experiments is that the unfolding and folding pathways of cytochrome *c* may mirror one another and that insight may be gained into how cytochrome *c* folds into its native form under physiological conditions. However, there is no direct evidence bearing on this point. We have considered the folding pathways of cytochrome *c* elsewhere (Pettigrew & Moore. 1987: Chapter 4.2.7).

Comparison of the stabilities of different cytochromes may lead to information about the effect particular amino acids have on the equilibrium of Eqn. 4.3, but generally such analyses have not been as successful at identifying stabilising influences as those described in section 4.4.5. Susceptibility to proteolysis of mitochondrial ferricytochromes varies markedly with the protein studied (Fig. 4.15), and this correlates well.with thermal stability (Endo *et al.*, 1985). However, comparison with guanidinium HCl denaturation (Fig. 4.15) (Knapp & Pace, 1974; McLendon & Smith, 1978) does not show the same degree of correlation. In both sets of experiments cow ferricytochrome *c* is more stable than the horse protein but although there are only

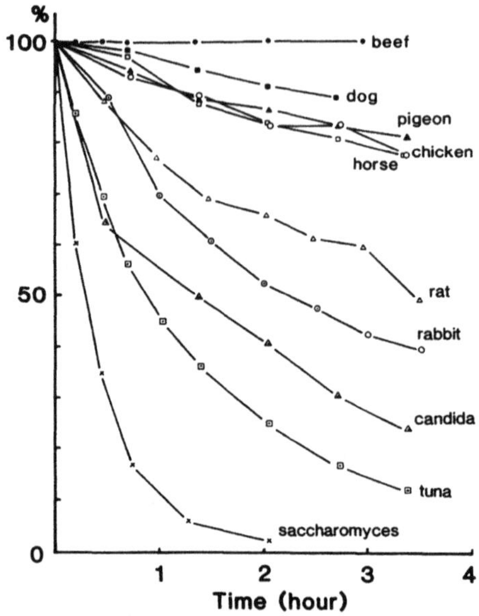

Fig. 4.15 Proteolysis of mitochondrial ferricytochromes *c* by chymotrypsin. (From Endo *et al.*, 1985). The relative order for proteolysis is the same as that for thermal denaturation monitored by CD at 222 and 416 nm, with the most rapidly digested cytochrome being the most readily denatured (Endo *et al.*, 1985). This does not correlate simply with guanidinium-HCl denaturation which shows the cytochrome stability order to be cow> horse>*Candida* (Knapp & Pace, 1974) and tuna>chicken~cow~rabbit~donkey >horse>dog (McLendon &Smith, 1978)

three amino acid differences (Fig. 3.1), the structural basis of the enhanced stability is not known. Even with horse and donkey cytochromes *c,* which differ only at position 47 (Ser(donkey) → Thr(horse)), the reason for the enhanced stability of the latter position is not clear (McLendon & Smith, 1978). It may be a reflection of the mechanisms contributing to stability suggested by Nall & Landers (1981) and Zuniga & Nall (1983). These authors proposed that the stability differences between cytochromes are due to different exposures of hydrophobic groups. On this basis they calculated the differences in stability between horse and *S. cerevisiae* iso-1 and iso-2 cytochromes and they found reasonable agreement between the calculated and observed values. However, further studies are required to verify this proposal.

Perhaps the most dramatic of the observed stability differences is shown by the comparison of horse and bullfrog cytochromes *c*, carried out by Brems *et al.* (1982). Bullfrog cytochrome *c* is the only mitochondrial cytochrome *c* known to contain an intramolecular disulfide bond (Cys 20 - Cys 102; Chan *et al.*, 1967) and this stabilises the native conformation by reducing the entropy of the completely unfolded state. Frog cytochrome *c* retains its globular structure even at extreme values of acid pH, or at high concentrations of guanidinium. The disulfide bond of cytochrome c_5 between Cys 69 and Cys 72 will not be such a stabilising interaction because it only encloses a few residues and thus does not greatly reduce the entropy of the unfolded state.

4.5.5 The Physiological Relevance of Conformational Transitions in Cytochrome *c*

In view of the evidence that native cytochrome *c* exists in equilibrium with locally or extensively unfolded forms it is important to know whether these forms can be physiologically significant. For example, ferricytochrome *c* undergoes a transition at approximately 50°C associated with the loss of the 695 nm band (Schejter & George, 1964; Kaminsky *et al.*, 1973b; Osheroff *et al.*, 1980; Ångstrom *et al.*, 1982) and there is a significant reduction in intensity of this band at pH 7 and 37°C compared to pH 7 and 25°C (Schejter & George, 1964). Similarly, at physiological pH, small amounts of state IV ferricytochrome *c* may exist (Davis *et al.*, 1974). Indeed the NMR spectrum of heated ferricytochrome *c* at pH 7 resembles that of a mixture of native state III and alkaline state IV forms (Ångstrom *et al.*, 1982). This may be due to an increase in the equilibrium constant K_c (Eqn 4.1) on warming so that the apparent pK_a is reduced, an explanation consistent with the correlation found by Osheroff *et al.* (1980) between this pK_a and the stability of the 695 nm band to temperature in a variety of native and modified cytochromes *c*.

The physiological significance of these altered forms is not clear. Our own view is that unfolded forms are present in small amounts under physiological conditions but they do not contribute significantly to cytochrome *c* function. The rate of conversion of state IV to state III ferricytochrome *c* (Brandt *et al.*, 1966) is slower than the rate of turnover of cytochrome *c* in the mitochondrion (Chance *et al.*, 1964). Also the operating environment of cytochrome *c* is the acid side of the bioenergetic membrane and at pH values between 5 and 7 the 695 nm absorption is much less sensitive to temperature (Ångstrom *et al.*, 1982).

4.6 Structural and Functional Effects of Chemical Modifications on Mitochondrial Cytochrome *c*

4.6.1 General Aspects of Chemical Modifications

Techniques now exist to systematically vary particular residues of cytochrome *c*. Also, many residues have undergone substitution during the natural course of eukaryotic evolution (Chapter 3.2), though rarely have the structural and functional consequences of this been determined. In the compilation of Fig. 3.2 only 26 residues are unvaried and many of these have been natural choices for chemical modifications. In this section we consider a number of derivatives of cytochrome *c* that have recently been described. Only modifications of the heme, the polypeptide chain and aromatic amino acids, and modifications produced by semisynthetic or genetic methods are described. Lysine modifications, which have been crucial in establishing details of the reactions of cytochrome *c* with its oxidoreductases, are described in Pettigrew & Moore (1987: Chapter 2); and previous reviews (Harbury & Marks 1973; Erecińska & Vanderkooi, 1978; Ferguson-Miller *et al.*, 1979) have thoroughly covered other areas.

Modified cytochromes should be characterised in three general ways: structure; stability; and electron transfer. Of these, investigation of stability is rarely carried out, which is unfortunate in view of the importance we have ascribed to stability differences in the previous section.

The characterisation of structure is important, although it has sometimes been disregarded. It may allow general effects of the modification on the conformation (such as breakage of the Fe-S bond when Trp 59 is formylated - see Fig. 4.18), and their resulting secondary effects on activity, to be distinguished from the specific effect of the modification on the route of electron transfer, or on the stability or orientation of the reactive complex (such as occurs with derivatives of lysines).

We have only considered in detail derivatives whose structures have been characterised and which appear to resemble native cytochrome *c*. The most widely used indicator of this is the 695 nm band of ferricytochrome *c*, but many of the derivatives have also been studied with other techniques described in Chapter 2. Site-specific methods based on semisynthetic or genetic techniques were used for most of the recent modification studies of cytochrome *c*. With these techniques there is no uncertainty concerning the site or sites of modification, or the nature of the modifying group, uncertainties which abound in much of the earlier literature on chemical modifications.

Measurement of electron transfer activity should be done in the oxidative and reductive directions along with a determination of mid-point redox potentials. However, a variety of different reductase and oxidase assays are favoured by different workers. Some use purified enzyme complexes but others use Keilin-Hartree particles and cytochrome *c* depleted mitochondria. Although the latter is an intact system (Jacobs & Sanardi, 1960) it suffers from the drawback that electron transfer through cytochrome *c* is not a rate limiting step within the respiratory chain (Pettigrew & Moore, 1987; Chapter 2.2.3.3) and thus activity differences between native and modified cytochromes may be missed. Assays based on reactions with small chemical reagents, such as ascorbate and ferricyanide, although widely used, are not particularly useful in this context.

4.6.2 Modification of the Heme

There have been no modifications made to the heme without resultant perturbations of the structure and properties of cytochrome *c*. Removal of the iron and its replacement by different metals has led to a range of derivatives that are structurally similar to the native protein and that have particularly interesting spectroscopic properties (Chapter 1.5), useful for probing the structure and interactions of cytochrome *c*, but which are generally non-reactive because the metal has different redox properties to iron. Reactions involving the thioether bonds between the heme vinyl groups and cysteines 14 and 17 result in their cleavage and loss of the heme. The major use of such reactions (reviewed by Ferguson-Miller *et al.*, 1979) is to prepare apocytochrome *c* for amino acid sequence studies.

Although direct modification of the heme propionates has been attempted, a major unsolved problem, common to all internal modifications, is to obtain selective reactions with a group that does not disrupt the native structure. Thus, acidic methanol esterifies many amino acid carboxylates, in addition to the heme propionates, and produces a derivative lacking Met 80 as a ligand (Myer *et al* , 1966; Stellwagen *et al.*, 1975); whilst treatment of ferricytochrome *c* with a water soluble carbodiimide results in a high-spin protein with decreased reactivity with cytochrome oxidase and an $E_{m,7}$ of 105 mV (Timkovich, 1980). This latter reaction is complex, and in addition to the modification of Asp 62 and a heme propionate identified by Timkovich, the heme group undergoes additional intramolecular covalent crosslinking to the protein (T.Brittain, personal communication).

Met-80 ligation and reactivity with yeast cytochrome b_2 are retained in the non-covalent complex (H1-25):(1-104) in which all the carboxylates of the heme peptide are methylated (Brems & Stellwagen, 1981). However in many such non-covalent complexes, the polypeptide chain around the base of the heme is disordered (Hantgan & Tanuichi, 1977) which complicates any assessment of the role of the propionates.

4.6.3 Modifications of the Main Chain

Modifications to the polypeptide backbone, which all take the form of peptide bond cleavage, were among the first modifications of cytochrome *c* to be made. This was partly because the resulting fragments were used in protein sequencing and partly because the heme containing peptides were used as 'model' heme complexes. A comprehensive account of the work with heme peptides is given by Lemberg & Barrett (1973) and, although they are still used to investigate the spectroscopic properties (Baumgartner *et al.*, 1974; Smith & McLendon, 1981) and chemical properties (Aron *et al.*, 1986; Baldwin *et al.*, 1986; Marques *et al.*, 1987) of heme, we will not consider them further because their relevance to cytochrome *c* is only marginal.

More recently, large peptides, some with the heme and some without it, have been prepared by the controlled action of proteolytic enzymes, BNPS-skatole and cyanogen bromide (Hantgan & Taniuchi, 1977; Harris & Offord, 1977; Harbury, 1978; Wilgus *et al.*, 1978; Juillerat *et al.*, 1980), and they have been used to prepare functional, non-covalent complexes and chemically modified cytochrome *c*. These derivatives are described below.

Non-covalent Complexes of Fragments of Cytochrome c. The nomenclature used is to represent a peptide of cytochrome *c* by the N and C terminal residue numbers in parentheses, for example (39-104); a heme-containing peptide is indicated by the inclusion of H, for example (H1-38); and a non-covalent complex is indicated by the inclusion of a colon, for example (H1-38):(39-104).

The first non-covalent complex of cytochrome *c* was prepared by Corradin & Harbury (1971), who found that when peptides (ferric H1-65) and

(66-104) were mixed, a complex was formed between them which had similar spectral properties to native ferricytochrome *c*. A similar complex was formed with (ferrous H1-65), but Corradin & Harbury (1974) subsequently discovered that in the ferrous state the peptide bond between the fragments reformed, provided sufficient time was allowed for the reaction (12 - 24 hrs.). This fragment complementation system is now almost exclusively used in semi-synthesis, although Harbury (1978) has experimented with the preparation of hybrid complexes in which the peptide bond often does not reform. Complexes were formed between fragments of the cytochromes from *C. krusei* (which contains Met-Ser at positions 64-65 instead of Leu-Met) and from horse, but they were inactive in cytochrome oxidase assays. However, the hybrid complex (horse, H1-65):(*R. rubrum*, 56-112) was active with cytochrome oxidase and, moreover, its activity was intermediate between that of horse cytochrome *c* and that of *R. rubrum* cytochrome c_2, a result attributed by Harbury (1978) to the fact that the cytochrome oxidase binding site on cytochrome *c* is composed of groups from both fragments of the protein.

A large number of non-covalent complexes have now been prepared and Table 4.6 summarises what is known about their structures and functional properties. The classification into structural types I - IV, which is illustrated in Fig. 4.16, comes from the work of Taniuchi and his colleagues (Hantgen & Taniuchi, 1978; Parr *et al.*, 1978; Juillerat *et al.*, 1980) who used trypsin to further cleave the associated peptides, thus yielding the limiting complexes.

Among the more interesting of the complexes which retain electron transfer activity are those, such as (H1-38):(59-104), which lack a large region forming the base of the heme pocket in the native protein. It is this region that constitutes the major structural difference between horse cytochrome *c* and *Pseudomonas* cytochrome *c*-551, and in view of the known differences in propionate ionisations of these two proteins (section 4.4) and the properties of the des-(40-55) semisynthetic derivative (see below), further characterisation of the complex will be of interest. Among the unvaried residues missing from the complex are Arg 38, Tyr 48 and Asn 52, all of which are implicated in propionate stabilisation (Fig. 4.4), yet the K_m and V_{max} with purified cytochrome oxidase (5.2×10^{-7} M and 1.5 μmol O_2 s^{-1}) are comparable to those of the native protein (3.7×10^{-7} M and 2.35 μmol O_2 s^{-1}) under the same conditions (Westerhuis *et al.*, 1982).

Taniuchi and coworkers have carried out an extensive study of the kinetics and thermodynamics of the fragment complementation process with the aim of characterising the folding pathways of cytochrome *c* (Poerio *et al.*, 1986).

Table 4.6 Non-covalent complexes of fragments of horse cytochrome *c*

Complex [a]		Ferri K_D [b]	Structure characterisation				$E_{m,7}$ [d]	Functional characterisation [e]					Refs.
Peptide with heme	Peptide without heme		695	CD	Trp	Type	(mV)	mit	YCCP	b_2	aa_3	mic	
1-25	(28-38):(56-104)	$<10^{-12}$	+	+	+	IV				+			f
	23-104	$\leq 3 \times 10^{-8}$	+	+	+	I				+			g
	39-104	7.5×10^{-7}		+						-			g
	56 -104	7.5×10^{-7}								-			g
1-37	38-104		+				216	+					h i
	39-104		+				160	+					h - j
	40-104		+					+					h i
1-38	1-104	$\leq 3 \times 10^{-8}$	+	+	+		186		+	+		+	g k
	39-104	$\leq 3 \times 10^{-8}$	+	+	+		172	+		+	+		g h j l - n
	40-104		+					+					l
	56-104	$\leq 3 \times 10^{-8}$	+	+	+	II				+			g
	59-104		+								+		n
	60-104		-								-		n
1-50	51-104		+				166	+					m
	56-104		+				160	+					m
1-53	39-104	$<3 \times 10^{-7}$			+					+			o
	40-104	$<3 \times 10^{-7}$			+					+			o
	54-104	$<3 \times 10^{-7}$		+	+	III				+			op
	56-104	$<3 \times 10^{-7}$			+					+			o
1-55	56-104		+				193			+			m
1-64	66-104											-	q
1-65	1-104		+	+			196		+			+	k r
	56-104		+				184	+					m
	60-104											-	q
	66-104		-	-			-18		-			-	k
	67-104						-112						k
1-80	1-104	$\sim 7 \times 10^{-4}$	+	+			189		+			+	k

11-21	1-104	-	-		-	kq
11-26	1-104	-	-		-	k
14-21	1-104	-	-	-119	-	kq

[a] Some of the peptides were acetimidylated.

[b] K_D measured at pH 7 and 25°C for all complexes other than (H1-80):(1-104) which was measured at pH 6 and 25°C. K_D is M for all complexes other than (H1-25):(28-38):(56-104) for which it is M^2.

[c] 695, CD and Trp are spectroscopic indicators of structure. + indicates the complex exhibited a 695 nm absorption band, intrinsic CD indicative of α-helix or Trp fluorescence similar to that of native ferrocytochrome *c* (Chapter 2). - indicates these spectral features were absent. NMR (Chapter 2) has been used to investigate some complexes and large spectral differences observed compared to the native protein have been found. It is not clear whether these differences reflect different structures for the formed complexes or the exchange process of peptide association and dissociation (G.R.Moore, unpublished data). The structural types I, II, III and IV are defined in Fig. 4.16.

[d] The pH dependence of the redox potential has only been reported for (H1-37):(39-104) and (H1-38):(39-104) (Moore *et al.*, 1984a). Both complexes are pH dependent over the pH range 5 - 9. Wallace and Proudfoot (1987) give the $E_{m,7}$ of (H1-38):(39-104) as 150 mV.

[e] A variety of assay systems have been used to test for activity: reconstituted cytochrome *c* depleted mitochondria (mit), yeast cytochrome *c* peroxidase (YCCP), yeast cytochrome b_2 (b_2), cytochrome *c* oxidase (aa_3) and rabbit liver microsomal NADPH-cytochrome *c* reductase (mic). Activity is indicated by + and inactivity by -.

[f]Juillerat *et al.* (1980);
[g]Parr *et al.* (1978);
[h]Harris (1979);
[i]Proudfoot *et al.* (1986);
[j]Moore *et al.* (1984a);
[k]Wilgus *et al.* (1978);
[l]Harris & Offord (1977);
[m]Wallace & Proudfoot (1987);
[n]Westerhuis *et al.* (1982);
[o]Hantgan & Taniuchi (1977);
[p]Hantgan &Taniuchi (1978);
[q]Harbury (1978);
[r]Fisher *et al.* (1973).

Like the folding of native cytochrome *c* (Pettigrew & Moore, 1987; Chapter 4.2.7) the fragment complementation process is complicated and probably includes a number of intermediates. Formation of the Fe-S bond occurs in one of the last stages.

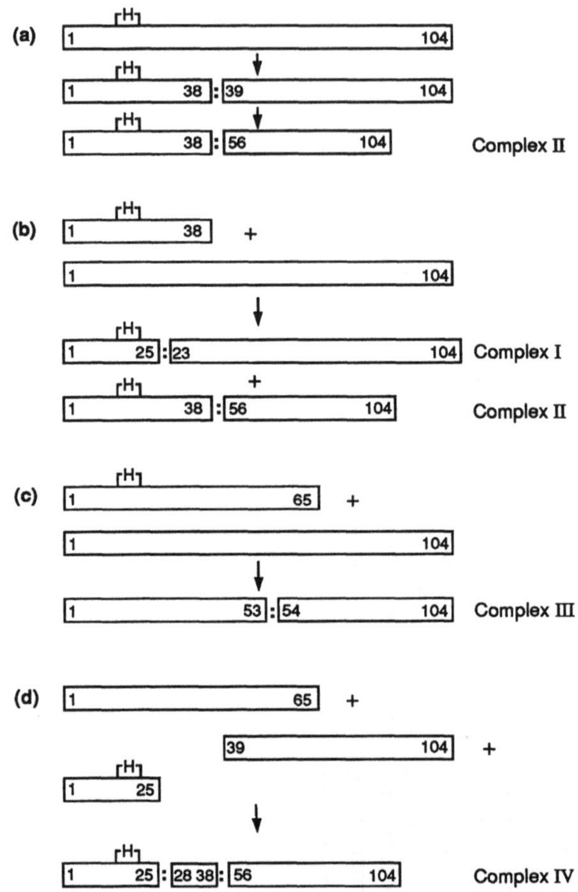

Fig. 4.16 Non-covalent complexes of horse cytochrome *c*. **(a)** Horse cytochrome *c* in which all the lysines are blocked is cleaved by trypsin at peptide bond 38-39 (Fanger & Harbury, 1965; Harris & Offord, 1977; Westerhuis *et al.*, 1979). Deblocking of the lysines followed by controlled proteolysis of the ferrous complex with trypsin produces the limiting complex II (Parr *et al.*, 1978). **(b)** Controlled proteolysis of the ferrous complex (H1-38):(1-104) with trypsin yields a mixture of the two limiting complexes I and II (Parr *et al.*, 1978). **(c)** Controlled proteolysis of the ferrous complex (H1-65):(1-104) with trypsin produces the limiting complex III (Hantgan & Taniuchi, 1978). **(d)** Controlled proteolysis of the ferrous complex (H1-25):(1-65):(39-104) with trypsin yields a mixture of the limiting complex IV and the inactive complex (H1-25):(56-104). (Juillerat *et al.*, 1980)

Semisynthetic Cytochrome c. In principle, the complete synthesis of the cytochrome *c* molecule should open the way to precisely defined side chain modification and substitution work. In practice the cytochrome *c* synthesised by Sano and Kurihara (cited in Lemberg & Barrett, 1973: p154) proved to be inactive and no further progress in this area has been reported. A more fruitful area has been the use of fragments of cytochrome *c* to obtain selective modification and replacement of amino acids (Barstow *et al.*, 1977; Boon *et al.*, 1979a; Koul *et al.*, 1979; Nix & Warme, 1979; Wallace & Offord, 1979), followed by coupling the fragments to obtain modified cytochrome *c*.

These semisynthetic methods make use of the spontaneous coupling reaction discovered by Corradin and Harbury (1974) and illustrated in Fig. 4.17. Cyanogen bromide cleaves horse cytochrome *c* at Met 65 and Met 80 with the formation of C-terminal homoserine-lactones. When peptide **A** is mixed with peptide **BC** in the presence of a reducing agent, the peptide bond between residues 65 and 66 reforms to make intact cytochrome *c* in which Met 65 has

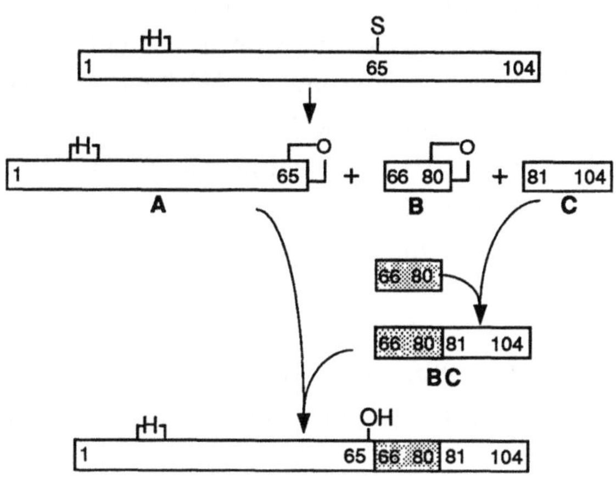

Fig. 4.17 Strategy for the semi-synthesis of HSe⁶⁵-cytochrome *c* (Boon *et. al.*, 1979a). Native cytochrome *c* is cleaved by CNBr at the two methionines and the three peptides separated. In the diagram peptide **B** is shown as the synthetic peptide, though any of the peptides may be modified. The coupling of **B** + **C** to produce **BC** involves blocking the lysines, adding Met 80 to peptide **C**, removing the C-terminal homoserine lactone from peptide **B** and converting the terminus to an azide. The spontaneous coupling reaction then produces **BC**, which is linked to peptide **A** by the conformationally directed spontaneous coupling reaction of Corradin & Harbury (1974) after the lysines have been unblocked. Wallace and coworkers (Wallace & Rose, 1983; Wallace & Corthesy, 1986) use the same scheme but couple the peptides together without deprotecting the lysines. They favour N-acetimidylation as the blocking method

been converted into homoserine 65. Neither the conformation of cytochrome *c* (Barstow *et al.*, 1977; Boswell *et al.*, 1981) nor its oxidation by cytochrome oxidase (Boon *et al.*, 1979a) are significantly affected by the modification. This coupling reaction permits modification to be carried out in any of the three peptides by the use of group specific reagents or by synthetic methods. Thus, Tyr 48 may be specifically modified since the other three tyrosines of horse cytochrome *c* are in peptides **B** and **C**. Tyr 67 and Tyr 74 can be replaced with leucine by synthesis of peptide **B** (Boon *et al.*, 1979b; Boon, 1981) and Phe 82 can be replaced with leucine by semisynthesis of peptide **C** (Ten Kortenaar, 1983). Results of the studies with substituted proteins are given in Tables 4.7 and discussed in the following sections.

Semisynthesis using the tryptic peptides, (H1-38) and (39-104), opens up the possibility of a further range of derivatives, and the demonstration by Westerhuis *et al.*, (1980) of the enzymatic synthesis of a peptide bond between (H1-38) and a dipeptide suggests that resynthesis of the complete protein may be possible.

Spontaneous coupling between peptides (H1-39) and (56-104) when the C-terminus of (H1-39) has been activated by conversion to a dichlorophenyl ester gives a high yield of des-(40-55) cytochrome *c*. Wallace (1987) has shown that this interesting derivative retains some biological activity (Table 4.7) but has a much reduced $E_{m,7}$ (130 mV) and alkaline pK: (7.4). Wallace points out that the suggestion of Leszczynski and Rose (1986), that the loop of residues 40-54 constitutes an Ω-loop,is supported by the formation of the des-(40-55) derivative.

A related variant of *S. cerevisiae* iso-1 cytochrome *c* is present in the CYC1-453 strain. Hampsey *et al.* (1988) have shown that this has loop deletions between residues -3 and 2, and between residues 40 and 53. Since the CYC1-453 variant is non-functional and the -3 to 2 loop removal alone produces a functional protein, Hampsey *et al.* claim the 40 to 53 loop deletion is reponsible for the loss of activity. This runs counter to the suggestion of Wallace (1987) who worked with the horse des-(40-55) cytochrome.

Further characterisation of the structure and properties of both these variants are needed. Since many of the heme propionate contact residues have been removed (Fig. 4.4), an altered pH dependence to the redox potential might occur. Also the relationship between the structures of the variants and those of the small cytochromes (Fig. 4.6) will be interesting.

4.6.4 Modifications of Aromatic Amino Acids

As a result of their evolutionary conservatism (Fig. 3.2) and their structural location (sections 4.2 & 4.3), the aromatic amino acids have been subject to many chemical modifications. However, it is only recently that unambiguous results have been obtained for some of the tyrosines and phenylalanines, and the sole tryptophan of horse cytochrome *c*.

Tyrosine and phenylalanine. A variety of approaches have demonstrated that Tyrosines 67, 74 and 97, and Phe 82 are not essential for the function of cytochrome *c* in *in vitro* assays. Thus 3-iodo-Tyr 74 cytochrome *c* is indistinguishable from native cytochrome in its $E_{m,7}$, 695 nm absorption and electron transfer activity in Keilin Hartree particles (Osheroff *et al.*, 1977). Also substitution of Tyr 67, 74 and 97 and Phe 82 by leucine in semisynthetic experiments yields derivatives which retain activity with cytochrome *c* oxidase (Table 4.7). In fact in most cases V_m is raised. Further characterisation with cytochrome *c* reductase is required. Finally, site-directed mutagenesis of *S. cerevisiae* iso-1 cytochrome *c* (Table 4.8) confirms that Phe 82 is not essential for electron transfer.

The replacement of Tyr 48 by phenylalanine leads to a protein with a structure and redox potential similar to that of the wild-type protein (Table 4.8). However, its electron-transfer activity has not yet been measured.

These experiments raise the question of why these aromatic residues should be conserved, and in the case of Phe 82 unvaried, in evolution. One possibility is that they are important in the protein folding process, but are not critical for the functioning of the folded molecule. In this respect it is interesting that the semisynthetic leucine derivatives varied greatly in the efficiency and kinetics of refolding. Only the 67 and 74 derivatives folded quickly and in good yield, while the 97 derivative folded only slowly and in poor yield (Harbury, 1978; Boon, 1981; Ten Kortenaar *et al.*, 1985).

Tryptophan. Trp 59 is present in all naturally occurring eukaryotic cytochromes *c* (Fig. 3.1). The fragment complementation study of Westerhuis *et al.* (1982), which shows that (H1-38):(59-104) retains native properties while removal of the tryptophan to form (H1-38):(60-104) abolishes them (Table 4.6), suggests that Trp 59 has a key role. Additional modification studies do not identify satisfactorily what it is.

Modification of Trp 59 by N-bromosuccinimide results in retention of the hydrogen-bonding activity of the indole ring, and of biological activity (Fig.

Table 4.7 Semisynthetic derivatives of horse cytochrome *c*

Derivative [a]	Structural characterisation [b] 695	NMR	CD	$E_{m,7}$ (mV)	Functional characterisation [c] *aa*$_3$	mit	Refs.
—	+	+	+	262	+	+	e - j
(—)	+	d		245		+	k
(Gln^{66})	+			261		+	k
(Lys 66)	+			269		+	k
(Nor^{66})	-					-	k
Leu^{67}	+	+		201	+		l m
F-Phe^{67}	+			199	+		n
Phe 67	+			225		+	r
Leu^{74}	+	+			+		l m o
Lys^{75}	-			<100	-		p
Val^{78}	-			170	-		p
Asn 78	+			260		+	r
Asn 78,Pro 83	+			240		+	r
Leu^{81}	+			257	+		p
Val 81	+			253	+		p
Ala 81	+			256	+		p
Leu^{82}	+			265	+		n
Pro 83	+			265	+		r
Leu^{97}	+			319	+		n
des-(40-55)	+			130		+	q

[a] All derivatives contain homoserine in place of Met 65. Nor = norvaline and F-Phe = 2,3,5,6-fluoroPhe. ( ) indicates all the lysines are acetimidylated. A superscript (e.g. Leu^{81}) indicates the substitution does not occur naturally and normal type (e.g.Val 81) indicates it does occur naturally (Fig. 3.1).

[b] 695, NMR and CD are spectroscopic indicators of structure (see note to Table 4.6). + or - indicates that the derivative had similar or significantly different spectroscopic properties to the native protein, respectively.

[c] Activity with purified cytochrome oxidase (*aa*$_3$) and cytochrome *c* depleted mitochondria (mit) is indicated by +. Low, or zero, activity is indicated by -.

[d] Blank spaces indicate that the measurement has not been reported.

[e]Barstow *et al.* (1977); [f]Harbury (1978); [g]Boon *et al.* (1979a); [h]Wallace & Offord (1979); [i]Nix & Warme (1979); [j]Boswell *et al.* (1981); [k]Wallace & Corthésy (1986); [l]Boon (1981); [m]Eley *et al.* (1982a); [n]Ten Kortenaar *et al.* (1985); [o]Boon *et al.* (1979b); [p]Boots (1986); [q]Wallace(1987); [r]Wallace *et al.* (1989)

4.18) (O'Hern *et al.*, 1975; Myer *et al.*, 1980c). In contrast, both hydrogen-bonding capacity and biological activity are lost if the indole ring is formylated (Fig. 4.18) (Aviram & Schejter, 1971; Erecińska, 1975). These results suggest that the hydrogen-bond to the heme propionate (Fig. 4.4) is important. However, this conclusion is not supported by studies on mutants of

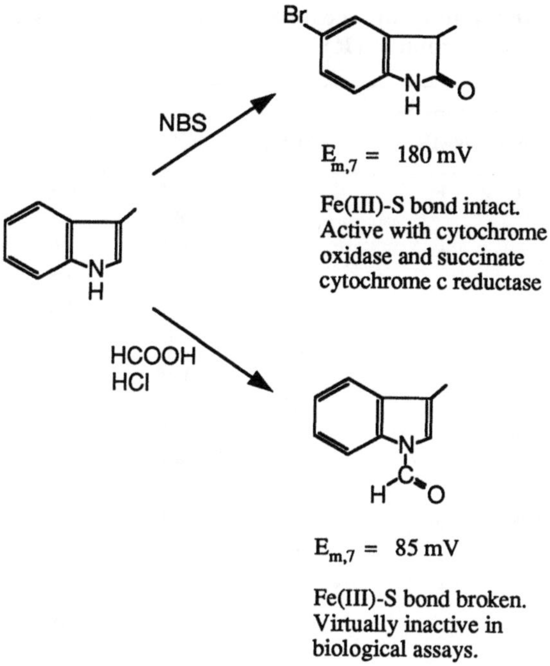

Fig. 4.18 Chemical modification of Trp 59. Data taken from O'Hern *et al.* (1975) and Myer *et al.* (1980c) for the N-bromosuccinimide modification, and from Aviram & Schejter (1971) and Erecińska (1975) for the formyl derivative. The NBS modification is accompanied by the conversion of Met 65 to a sulfoxide

iso-1 cytochrome *c* of *S. cerevisiae* (Table 4.8) (Schweingruber *et al.*, 1977; 1978; 1979). When Trp 59 is replaced by a cysteine as a consequence of a missense mutation, all biological activity is lost, but this is recovered by reversion at the primary site to give Phe 59, or at a secondary site to give Phe 40 in place of Ser 40. Both revertants have phenylalanine occupying the space normally occupied by Trp 59 in the native structure. Similarly, the structural and functional integrity of the Phe^{48} and Phe^{59} variant of *S. cerevisiae* iso-1 cytochrome *c* (Table 4.8) indicates the heme propionate hydrogen-bond is not essential. These results suggest that the hydrogen-bonding ability of Trp 59 is less important than its hydrophobic bulk in the maintenance of the native state, and pose the question: why is Trp 59 not replaced by phenylalanine in naturally occurring cytochromes *c*? Again, it could be a matter of the stability of the protein.

Table 4.8 Genetic replacements of residues in *S. cerevisiae* iso-1 cytochrome *c* (S) and rat cytochrome *c* (R)

Position	Amino acid [a]	Structure [b] 695	NMR	$E_{m,7}$ [c] (mV)	Function [d] YCCP [e]	Growth	Refs.
S27	Lys (natural)	+	+		100	100	f
	Leu				80	85	f
	Gln		+		110	85	f
	Trp					65	f
	Tyr					65	f
S38	Arg (natural)	+	+	272	100		g
	Lys	+		249	104		g
	His	+	+	245	97		g
	Asn	+		242	95		g
	Gln	+	+	238	97		g
	Leu	+		231	98		g
	Ala	+	+	225	99		g
S48	Tyr (natural)	+	+	265			h
	Phe	+	+	252			h
S59	Trp (natural)	+	+			100	i j
	Cys					0	i j
	Cys (revertant with Phe[40])					100	i j
	Phe					100	i j
	Tyr					<25	i j
S48/S59	Phe	+	+	202			h
R67	Tyr (natural)	+	+	259			k
	Phe	+	+	224			k
S72	TML (natural)	+	+		100		l
	Arg		+		100		l
S76	Pro (natural)	+	+			100	m
	Val					90	m
	Thr					60	m
	Ser					30	m
	Ile					20	m
82	Phe (natural)	+	+	270	100		n
	Tyr	+	+	270	30		n,o
	Ser	+	+	220	70		n
	Gly	+	+	220	20		n

[a] The position 38, 48, 67, 72 and 82 mutants were obtained by site-directed mutagenesis. Residue 102 was Thr for the position 38 and 48 mutants and Cys for the position 72 and 82 mutants. The S48/S59 variant contains Phe at both positions. It was obtained by site-directed mutagenesis. The position 27, 59 and 76 replacements were carried out by classical genetic methods and the mutant proteins were not always isolated.

[b] 695 and NMR are spectroscopic indicators of structure (Chapter 2). + indicates that the

Histidine. His 18 is the sole invariant histidine of mitochondrial cytochrome *c* (Fig. 3.2). Sorrell *et al.*(1989) report that it can be replaced by site-directed mutagenesis of *S. cerevisiae* iso-2 cytochrome *c* by arginine and that the mutant protein is capable of supporting growth, albeit at a lower rate than the wild-type protein. Futhermore, the purified protein possesses a 695 nm absorption band and unaltered redox potential. Given that His 18 is a ligand to the iron, these properties are surprising and therefore a structural study of the Arg^{18} cytochrome is desirable to ascertain what its axial ligation is. Hampsey *et al.* (1988) report that replacement of His 18 by tyrosine or asparagine produces a non-functional protein.

4.6.5 Modification of Aliphatic Amino Acids

Lysine Residues. Two of the unvaried lysines (Fig. 3.2) proposed to play key roles in the formation of cytochrome *c*: oxidoreductase complexes - TML 72 and Lys 27 of *S. cerevisiae* iso-1 cytochrome *c* (Table 4.8) - have been replaced by other residues.. The former was replaced by arginine, which had no discernable affect on the *in vitro* activity of the cytochrome (Holzschu *et al.*, 1987), and the latter was replaced by leucine, glutamine, tryptophan and tyrosine (Das *et al.*, 1988). All of the residue-27 variants retained a high level of biological activity, with the Glu^{27} protein binding to YCCP considerably more strongly than to the wild-type cytochrome *c*. These results suggest either that the proposed model of the cytochrome *c*:YCCP complex (Poulos & Kraut, 1980; Poulos & Finzel, 1984) is incorrect, or that an alternative structure, strongly favoured by the Glu^{27} protein, is possible (see Pettigrew & Moore, 1987: Chapter 2.7).

---

spectra are similar to those of the native protein. The XRD structures of the Ser^{82} and Gly^{82} variants have been reported (section 4.5.3).

[c] The position 38 mutants had pH dependent redox potentials with profiles similar to that of the wild-type protein.

[d] Functional studies were carried out with purified yeast cytochrome *c* peroxidase (YCCP) and the rate of growth of yeast cells on lactate media corrected for the % normal amount of cytochrome *c* present (Growth). The data are presented as relative specific activities.

[e] YCCP is yeast cytochrome *c* peroxidase.

[f]Das *et al.* (1988); [g]Cutler *et al.* (1989); [h]G.Guillemette, A.M.Davies, A.G.Mauk, G.R.Moore, A.G.P.Thurgood & M.Smith, unpublished data; [i]Schweingruber *et al.* (1978); [j]Schweingruber *et al.* (1979); [k]Luntz *et al.* (1989); [l]Holzschu *et al.* (1987); [m]Ernst *et al.* (1985); [n]Pielak *et al.* (1985); [o]Pielak *et al.* (1988b).

Arginine Residues. Mitochondrial cytochromes *c* have two unvaried arginine residues at positions 38 and 91. Their role has been investigated using direct chemical modification, formation of non-covalent complexes (section 4.6.3), and site-directed mutagenesis, but with conflicting results.

By preparing the two cyanogen bromide fragments of horse cytochrome *c* resulting from cleavage at residue 65, Wallace & Rose (1983) were able to reconstruct cytochrome *c* molecules selectively modified at one or other of the conserved arginines (Fig. 4.19). Whereas modification of Arg 38 caused a major change in the structure and activity of the protein, modification of Arg 91 had little effect.

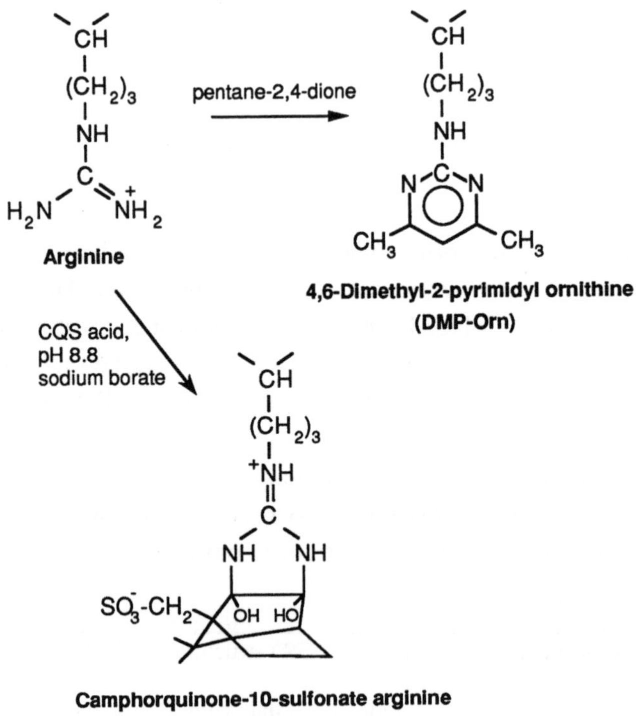

Fig. 4.19 Chemical modification of Arg 38 and Arg 91. The derivatives contained HSe^{65} and were fully acetimidylated. $E_{m,7}$ values: CQS-Arg^{91}, 258 mV; DMP-Orn^{91}, 248 mV; CQS-Arg^{38} and DMP-Orn^{38}, ~170 mV. The Arg 91 derivatives both possessed 695 nm absorptions and had >90% activity in cytochrome *c* depleted mitochondria but the Arg 38 derivatives had no 695 nm band and considerably diminished activities. (Wallace & Rose, 1983). Pande & Myer (1980) have described the reaction of intact horse cytochrome *c* with butan-2,3-dione but they did not characterise the products in detail except to show that both Arg 38 and Arg 91 were modified

In contrast, the (1-37):(39-104) complex formed by selective tryptic digestion and carboxypeptidase B action retained electron transfer activity and the 695 nm band (Table 4.6). The mid point redox potential was, however, lowered to 160 mV; although a similar lowering was associated simply with cleavage of the Arg 38-Lys 39 bond (Moore *et al.*, 1984a). Interestingly, the complex with a break at 37-38 has a redox potential closer to the normal value than does the complex cleaved at 38-39, and the former is also more resistant to trypsin. To account for these results Proudfoot *et al.*(1986) suggest that the role of Arg 38 is to shield the heme from the solvent by providing a significant amount of the binding energy needed to locate the 36-59 loop of residues.

Replacement of Arg 38 in iso-1-cytochrome *c* of *S. cerevisiae* by site-directed mutagenesis always lowers the mid point potential (Table 4.8). However, the profile of the pH dependence of the redox potential is the same as that of the normal protein (Cutler *et al.*, 1989), indicating that a redox dependent heme propionate ionisation does not occur (Chapter 7.6.2). NMR studies are consistent with this. In view of this it appears that the positive charge of Arg 38 is not as important an influence on the redox potential as has been suggested (Moore, 1983).

Proline Residues. The role of Pro 30 has been investigated with the fragment complementation procedure described earlier (Polerio *et al.*, 1986). Replacement by glycine in the complex (H1-25):(28-38):(39-104) yielded a product active with yeast cytochrome b_2. The replacement of Pro 30 with leucine in *S. cerevisiae* iso-1 cytochrome *c* by a similar procedure to that described for mutations of Trp 59 (section 4.6.4), and by threonine in *S. cerevisiae* iso-2 cytochrome *c* by site-directed mutagenesis, yields active proteins (Hampsey *et al.*, 1986; Wood *et al.*, 1988).

The replacement of the unvaried Pro 76 in yeast iso-1 cytochrome *c* by a variety of residues yields proteins of differing activity (Table 4.8). Like the Phe 82 mutants considered earlier, these activities appear at first sight to be somewhat suprising with no obvious reason for the marked differences. Ernst *et al.* (1985) rationalised the trend in terms of particular side chain structural parameters. They concluded that functional residues at position 76 induce the native back-bone conformation and have small side chains. The replacement of Pro 76 by glycine in *S. cerevisiae* iso-2 cytochrome *c* yields a reduced protein whose conformation is almost indistinguishable from that of the wild-type protein (Wood *et al.*, 1988). However, the oxidised Gly protein is less stable than the wild-type oxidised protein.

Methyl-containing Amino Acids. The replacement by classical genetic methods of various unvaried methyl containing residues in *S. cerevisiae* iso-1 cytochrome *c* has been reported (Hampsey *et al.*, 1988). None of the proteins have been isolated but both the replacement of Leu 68 by serine and, surprisingly, Met 80 by isoleucine, have been shown to support growth on lactate medium, albeit at a greatly reduced level compared to the wild-type protein. The replacement of Leu 32 by serine and tryptophan yielded non-functional proteins, but the replacement of Leu 9, Leu 94 or Leu 99 by serine yielded functional proteins. Detailed characterisation of the isolated proteins, particularly that containing Ile 80, will be interesting.

Replacement of various methyl containing amino acids by semi-synthetic methods has been achieved (Table 4.7). Not all of these are unvaried residues, for example; Ile 81 is often replaced by alanine, Ala 83 by proline and Met 65 by tyrosine (Fig. 3.1). This is reflected in the similarity of these modified proteins to native cytochrome *c*.

Both the Lys^{75} and Val^{78} derivatives are much perturbed by the replacements; the former perhaps unsurprisingly since an internal isoleucine is replaced by lysine. Attempts to prepare the Glu^{75} derivative failed because the peptides did not recombine. The Val^{78} derivative is more interesting. This is an isosteric substitution that leads to the replacement of a hydroxyl by a methyl. Thr 78 is hydrogen-bonded to a heme propionate (Fig. 4.4) and it is tempting to suggest that the removal of this bond is the cause of the altered properties. However, it may simply be the insertion of a hydrophobic group into a relatively polar region of the protein that causes the native structure to be destabilised. The retention of activity on the replacement of Thr 78 by the polar asparagine is consistent with both views.

Taborsky & McCollum (1979) have described an unusual reaction in which electron transfer from Fe^{2+} to ferricytochrome *c* in the presence of phosphate is accompanied by oxidation of Thr 89 to its keto derivative. This reaction is specific for Thr 89, which is not an unvaried residue (Fig. 3.2), and is caused by the presence of a phosphate binding site close to position 89; this is proposed to act as a Fe^{2+} binding site once phosphate has bound.

Met 65 of horse and tuna cytochromes *c* can be modified by a variety of reagents without materially affecting the protein structure. However, carboxymethylation, unlike conversion to homoserine, diminishes the stability of cytochrome *c* (Stellwagen, 1968; Boswell *et al.*, 1981) and causes small differences in NMR spectra for groups far from the site of modification, thus implying minor conformation changes.

Other Amino Acids. Cys 102 of *S. cerevisiae* iso-1 cytochrome *c* has been replaced by threonine to obtain a more suitable protein for study (Pielak *et al.*, 1988a). The Thr 102 protein is less prone to aggregation and has cleaner electrochemical properties than the native protein, changes expected from studies of other proteins containing Thr 102 and from chemically modified yeast cytochrome *c* in which Cys 102 has been blocked (Motonaga *et al.*, 1965; Drott *et al.*, 1970).

Wallace and Corthésy (1986) have replaced Glu 66 of cytochrome *c* by glutamine, lysine and norvaline semi-synthetically. The first two derivatives had almost normal redox potentials and activities but the latter derivative had strongly changed properties (Table 4.7). Wallace and Corthésy suggest that this is because hydrophobicity cannot be accomodated at this part of the 60's helix although in some cytochromes residue 66 is hydrophobic (Fig. 3.1). Possibly in these cytochromes changes in other parts of the helix may compensate for hydrophobicity at position 66.

Das *et al.* (1989) have reported second-site revertants of *S. cerevisiae* iso-1 cytochrome *c* variants. The primary sites contained Ser^{29} (instead of glycine) and Pro^{33} (instead of histidine), and the secondary sites contained Ile^{52} (instead of asparagine). In both cases the replacement at position 52 enhanced stability considerably. A variant containing only the position 52 replacement is needed, particularly in view of the location of this residue at the base of the heme (Fig. 4.4).

Hampsey *et al.* (1988) have reported the replacement of Gly 6, Cys 14, Cys 17, Gly 29, His 33, Asn 70, and Lys 79 to yield singly substituted mutants of yeast iso-1 cytochrome *c* (only replacements of Gly 6 and His 33 were functional). Bowler *et al.* (1989) have replaced Asn 62 by histidine in order to attach ruthenium pentaammine to cytochrome *c* at position 62 (see Chapter 8.5.2). They report that the His protein has identical UV-visible spectra to those of the wild-type protein and a redox potential of 268 mV at pH 7.

Chapter 5 The Structures of Class II, Class III and Class IV Cytochromes *c*

5.1 Introduction

Judged spectrally, a wide variety of bacterial cytochromes *c* are known (Meyer & Kamen, 1982; Pettigrew & Moore, 1987: Chapter 3) but only a relatively small number are sufficiently characterised to be placed within the classification scheme described in section 1.3 of Pettigrew & Moore (1987). These are the class I cytochromes *c*, which are described in the previous chapter, and the class II and class III cytochromes *c* which are described in the present chapter. Photosynthetic reaction centre (PRC) cytochrome *c* from *Rhodopseudomonas viridis* is also described in the present chapter. This protein has only recently been characterised structurally, but it is clear that it does not fall into any of the currently defined classes (I, II or III) (Ambler, 1980). Accordingly, we have proposed that this protein be the prototype of a new group - class IV (Pettigrew & Moore, 1987: Table 1.4).

5.2 Class II Cytochromes *c*

5.2.1 Introduction

Class II cytochromes *c* are monoheme proteins whose heme attachment sites are located at their C-terminii (Pettigrew & Moore, 1987: Table 1.4). There is a natural sub-classification into class IIa, the predominately high spin cytochromes *c*′, and the low spin cytochromes of class IIb. However, while cytochromes *c*′ are readily distinguished by their optical spectra (Fig. 2.29), class IIb cytochromes *c* are not because they have spectra very similar to those of class I cytochromes *c* (Pettigrew & Moore, 1987: Fig. 1.7). Nevertheless, Meyer & Kamen (1982) have proposed a classification scheme for the IIb proteins based upon the following spectral features; red-shifted

Soret and α-bands for the ferro protein (~419 nm and 554 - 556 nm respectively) and a broad δ-band for the ferri protein at ~350 nm, a shorter wavelength than for class I cytochromes *c*. Using these criteria Meyer & Kamen grouped together four cytochromes *c*-554 and five cytochromes *c*-556, and although only two of these have been sequenced, both are class IIb cytochromes *c*: the cytochromes *c*-556 from *Rps. palustris* (Ambler *et al.*, 1981a) and *Agrobacterium tumefaciens* (van Beeumen *et al.*, 1980).

Cytochrome *c'*, for which an x-ray structure is available, is the main subject of this section. The probable structural relationship between class IIa and IIb cytochromes *c* is also discussed. The similarity between cytochrome *c'* and *E. coli* cytochrome b_{562} is discussed in Chapter 6.10.2.

5.2.2 Cytochromes *c'*

Cytochromes *c'* form a homogeneous class with respect to optical spectra (Pettigrew & Moore, 1987: Fig. 1.7), mid-point potential and monomer molecular weight (Table 5.1). However there are substantial differences within the group with respect to the state of aggregation (most, but not all, are dimers when isolated - Table 5.1) and the value of the pK of the heme-linked ionisation (discussed in Chapter 2.3.2.5 and below). The amino acid sequences also reveal that although the proteins are closely related there is a high degree of variation within the group (Fig. 3.12). Cytochrome *c'* was first isolated by Vernon & Kamen (1954), who named it pseudohemoglobin, and it went under many names before its present one was finally adopted. A full description of the discovery and characterisation of the proteins is given by Lemberg & Barrett (1973) and Bartsch (1978).

A type of cytochrome with similar optical spectra, mid-point potential and molecular weight to those of cytochrome *c'* is found in a few members of the Rhodospirillaceae (Bartsch, 1978; Meyer & Cusanovich, 1985). This protein has been named Sphaeroides heme protein (SHP). It can be distinguished from cytochrome *c'* because the extinction coefficient of the Soret peak of ferro-SHP is lower than that of ferri-SHP, whilst for cytochrome *c'* the situation is reversed. The full amino acid sequence of SHP has not been determined but its amino acid composition and partial sequence shows it is not an altered form of cytochrome *c'* (Meyer & Cusanovich, 1985). However, it may be a class II cytochrome *c* because its heme attachment site is not located in the first 39 residues. SHP has not been further characterized.

Table 5.1 Properties of class II cytochromes *c* and a related cytochrome *b*

Cytochrome		m/d [a]	pI	$E_{m,7}$	pK	A → B method [b]	Refs.
Chr. vinosum	*c'*	d	4.6	-5	9	optical	c - e
				20	8.4	E_m	f
					9 - 10	EPR, MCD	g h
Alc. denitrificans	*c'*	d		132	7.1	optical	e
R. rubrum	*c'*	d	5.6	-8	8.2	optical	c d l
				10	8.1	E_m	f
					8.5 - 9.0	EPR, NMR, MCD RR, Mössbauer	h j m
R. molischianum	*c'*	d			8.8	optical, NMR	n
Rc. gelatinosa	*c'*	d	9.6				d
Rps. palustris	*c'*	m	9.4	102	7.8	optical, MCD	d h o - t
				100	8.0	E_m	f
Rps. palustris	*c*-556	m		230			d
Agr. tumifaciens (B2a)	*c*-556	m	5.7	195	~9	E_m, NMR	u - w
E. coli	b_{562}	m		180	9.0	E_m, NMR, optical	x y

[a] m = monomer and d = dimer. Meyer & Kamen (1982) point out that there may be a rapid equilibrium between monomer and dimer forms of *Rps. palustris* cytochrome *c'*. This is consistent with the crystallisation of both monomeric and dimeric forms (Salemme, 1974).
[b] A → B represents the ionisation of His, the axial histidine ligand, to a histidinate (see Chapter 2.3.2.5).
[c]Bartsch & Kamen (1960); [d]Bartsch (1978); [e]Cusanovich *et al.* (1970); [f]Barakat & Strekas (1982); [g]Maltempo *et al.* (1974); [h]Rawlings *et al.* (1977); [i]Horio & Kamen (1961); [j]Emptage *et al.* (1977); [k]Emptage *et al.* (1981); [l]Kitigawa *et al.* (1977a); [m]La Mar *et al.* (1981); [n]Moore *et al.* (1982b); [o]DeKlerk *et al.* (1965); [p]Henderson & Nankiville (1966); [q]Dus *et al.* (1967); [r]Cusanovich (1971); [s]Salemme (1974); [t]Meyer & Kamen (1982); [u]van den Branden *et al.* (1975); [v]van Beeuman *et al.* (1980); [w]J.van Beeumen, G.R.Moore & G.W.Pettigrew, unpublished data; [x]Myer & Bullock (1978); [y]Moore *et al.* (1985b).

One peculiarity of cytochrome *c'*, discussed more fully in Chapter 3.5.5 of Pettigrew & Moore (1987), is that it has not been detected *in situ*. Despite having characteristic spectroscopic properties, cytochrome *c'* has not been detected spectrally in whole bacterial cells (Corker & Sharpe, 1975; Prince *et al.*, 1974), nor have more biochemical methods identified it in chromatophores. This has led to the view that cytochrome *c'* is modified upon isolation so that its properties are altered. Indeed, Kakuno *et al.* (1971) have suggested that *in situ* cytochrome *c'* may have a cytochrome *b*-type optical spectrum, and although this proposal is based on slender evidence it is noteworthy that the low-spin structural homologues of cytochrome *c′* have red-shifted absorption bands.

5.2.3 Structure of Cytochrome *c′*

The amino acid sequences of 13 cytochromes *c′* have been determined and, unlike the different types of class I cytochromes *c*, there are few unvaried residues; according to the alignment of Fig. 3.12 only the Cys-?-?-Cys-His heme attachment residues and an arginine close to the N-terminus are unvaried. Some residues appear to be conservative substitutions, such as Phe/Tyr 125 and Lys/Arg 126 (*R. molischianum* numbering), but a proper search for structural homology must await the completion of several x-ray structure determinations. To date only the x-ray structure of *R. molischianum* ferricytochrome *c′* has been determined though suitable crystals of others have been reported (Salemme, 1974; Norris *et al.*, 1979; Yasui *et al.*, 1985).

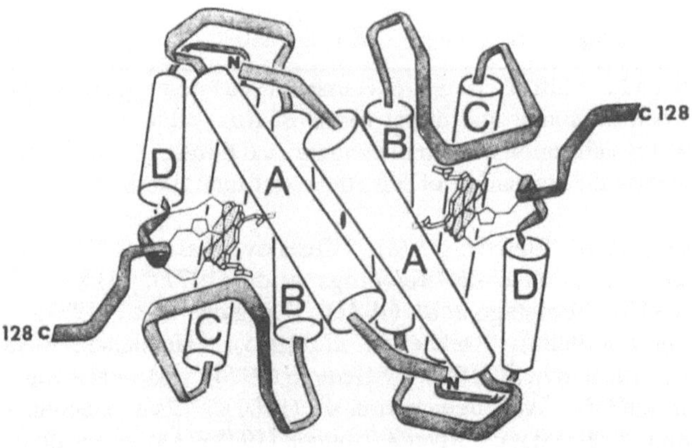

Fig. 5.1 Schematic representation of the *R. molischianum* cytochrome *c′* dimer. The monomers are principally composed of four roughly parallel α-helices, shown as the cylinders **A** - **D**. The left-twisted 4-α-helical arrangement of the monomers is repeated at the subunit interface due to the pairwise interaction of two helices from each subunit. Regions of essentially extended polypeptide chain (shaded) occur at the **N** and **C** terminii of the helices. The heme groups are situated at one end of each monomer and are oriented roughly parallel to the subunit interface with their propionate groups pointed towards the molecular surface. The chain folding pattern is: 1-4 (extended), 5-30 (helix **A**), 31-38 (**AB** loop), 40-53 (helix **B**), 54-78 (**BC** loop), 79-102 (helix **C**), 103 (**CD** connection), 104-125 (helix **D**) and 126-128 (extended). The following hairpin turns occur in the loops: 29-32 and 37-40 (**AB** loop); 56-59, 59-62, 60-63, 62-65, 65-68, 73-76 and 76-79 (**BC** loop). (Fig. reproduced from Weber *et al.*, 1980; the secondary structure description taken from Finzel *et al.*, 1985)

R. molischianum **Ferricytochrome *c'*.** The x-ray structure of *R. molischianum* ferricytochrome *c'* has been determined to 2.5Å resolution (Weber *et al.*, 1981a) and subsequently refined to 1.67Å (Finzel *et al.*, 1985). The crystals were obtained from a solution of undefined pH but they probably contain protein in the state A form (Weber, 1982; Finzel *et al.*, 1985). However, it requires single-crystal spectroscopic studies to clearly establish this and none have been reported.

The structure (Fig. 5.1) consists of two identical subunits each composed of four α-helices packed sequentially into a left-twisted anti-parallel bundle; i.e. with the N-terminus of one helix packed against the C-terminus of another helix. There are also two short 3_{10} helices. 73% of the amino acids are contained in the helical stretches, in good agreement with the CD measurements of Imai *et al.* (1969a,b) on *R. rubrum* and *Rps. palustris* cytochromes *c'*, who found a helix content of 63%. Imai *et al.* ascribed this high value to the large number of alanine residues, which are helix-formers, and, consistent with this, the x-ray structure shows that 23 of the 28 alanines of *R. molischianum* cytochromes *c'* are incorporated into helices. The inter-helical loops are not devoid of secondary structure; there are nine hair-pin turns (Fig. 5.1). The sharp change in direction between helices **C** and **D** is achieved with only one residue in the interconnecting peptide, Gly 103 (109 in Fig. 3.12 numbering. Finzel *et al.* (1985) suggest that the 180° turn may be possible because of the high percentage of alanines (with their small side chains) involved in the helical contacts for helices **C** and **D**. However, they also point out that only *R. molischianum* cytochrome *c'* has alanines at all the key contact points, which are residues 99, 102, and 106. Interestingly Gly 103 is unvaried in the dimeric cytochromes *c'* while in the monomeric cytochrome *c'* and class IIb cytochromes *c*-556 there is a deletion at this point of the sequence which removes the glycine in at least three of the four proteins sequenced (Fig. 3.12).

The two monomer subunits are almost identical and only small structural differences occur. Ignoring the two pairs of N-teminal and C-terminal residues as well as Gln 2, all of which are disordered, side chains of eight residues have significantly different conformations in the two subunits, with differences in average atomic positions being greater than 1Å. Three of these disordered residues (lysines 60, 116 and 119) extend into the solvent and the remaining five (glutamates 81, 84 and 124; Lys 86; and Thr 91) have different conformations as a result of different crystal contacts.

Helix Packing and Subunit Interactions. The dimer interface is an antiparallel 4-α-helical AA'BB' bundle, where A and B come from one subunit and A' and B' from the other (Fig. 5.1). Most residues at the interface are hydrophobic though none are aromatic. Hydrophobic packing interactions at the AA' interface are Leu 15-Leu 15' and Leu 11-Gln 17'/Thr 18', and at the BB' interface they are Met 49-Met 49' and Val 50-Leu 53'. The only intersubunit ionic or hydrogen-bonded interactions are those between HP-7 and Lys 10', and Lys 52 and Glu 45'. The point of closest approach of the two subunits, which is 6.1Å disregarding side chain interactions, is near the centre of the helix bundle in the AA' interface by the two Gly 14 residues.

It is not apparent from the structure (Fig. 5.1) why some of the class II cytochromes *c* are dimeric and some monomeric (Table 5.1), although it is probable that the aggregation is controlled by the disposition of the helices as well as by the nature of the amino acids on the subunit interface. Comparison of the amino acid sequences (Fig. 3.12) does not identify which of the interhelical interactions are important in controlling dimerisation. Gly 61, Thr 62, Gly 103 and Gly 115 (positions 64, 65, 109 and 123 in the numbering of Fig. 3.12) are conserved in the dimeric cytochromes *c'* and all are involved in helix contacts. However, these residues are also present in the monomeric *Agr. tumefaciens* (B2a) cytochrome *c*-556 although this protein, in common with the other monomeric proteins, has a deletion in the CD connection which includes Gly 103.

Steric interactions play a large part in determining the twist of the helical bundle. Weber *et al.* (1980) observe an interhelical axis angle of 18°, a value predicted by Crick (1953) who described twisted helical bundles from a theoretical stance and showed that with undistorted helices the twist would require an angle of about 20° to minimise unfavourable side chain interactions.

The antiparallel nature of the helix packing - with the N-terminus of one helix packed against the C-terminus of a second helix - has led to the suggestion that electrostatic interactions involving the helical dipoles are an important stabilization force (Sheridan *et al.*, 1982). However, Rogers & Sternberg (1984) suggest that the α-helix dipole is not as large as Sheridan *et al.* (1982) calculate, and Gilson and Honig (1989) indicate it may even be a destabilizing influence on the formation of α-helix bundles. This important point regarding the α-helix dipole is difficult to address because it relies on theoretical calculations of interaction energies and these require a physically reasonable model of the interaction to be constructed. At present there is a wide range of models (Chapter 7.4) which yield a wide range of results. Nevertheless, the experimental observation remains that the antiparallel 4-α-helical bundle structure is a stable structure, not just for cytochrome *c′* but for other proteins as well (Chapter 6.10.2).

The Heme and its Immediate Environment. The heme environments of the dimer are identical, with both irons being pentacoordinate and having an axial histidine ligand each, thus confirming the spectroscopic studies described in Chapter 2.3.2.5. As expected for a pentacoordinate iron (Chapter 1.3), the iron is displaced from the mean pyrrole nitrogen plane toward the histidine by 0.2Å, and as with other cytochromes *c* the heme is saddle-shaped. The geometry around the cysteine minimises unfavourable histidine-heme interactions. However, there are small differences between cytochrome *c′* and other proteins in these respects and these are described in Chapter 1.3.

The two hemes of the dimer are oriented at an angle of 42° between their planes with an Fe-Fe distance of 24Å. Each heme lies in a plane roughly parallel to that containing the **A** and **C** helix axes (Fig. 5.1) with one face largely exposed at the protein surface, and the other oriented toward the protein interior where it is bound in a pocket lined with bulky hydrophobic groups, including five of the seven aromatic side chains (Fig. 5.2). The amino acid sequence alignment of Fig. 3.12 indicates that few of these groups packed around the heme face are conservatively replaced. However, given the strong tendency for the heme pockets of different proteins of the same class to be constructed of similar side chains, we expect that when further x-ray structures of cytochrome *c′* are available it will be seen that the protein fold adapts itself to allow maintainence of a similar heme environment.

The heme propionates are partially buried within the protein. The heme is oriented so that the propionate bearing edge is close to the molecular surface where a variety of side chains interact with them and partially shield them from the solvent (Fig. 5.2). These interactions, which include a bridging H_2O molecule, form an extensive network that introduces a positively charged residue close to each heme propionate. The invariant Arg 12 interacts with the more buried propionate. NMR studies have revealed that a heme propionate of *R. rubrum*, *R. molischianum* and *Rps. palustris* ferricytochromes *c′* has a pK_a in the range 5.8 - 6.4 while the other propionate probably has a $pK_a < 4.5$ since its ionisation has not been observed over the pH range 4.5 - 10.5 (Emptage *et al.*, 1981; Moore *et al.*, 1982b; Jackson *et al.*, 1983). The assignment of the pK's to particular propionates is not known.

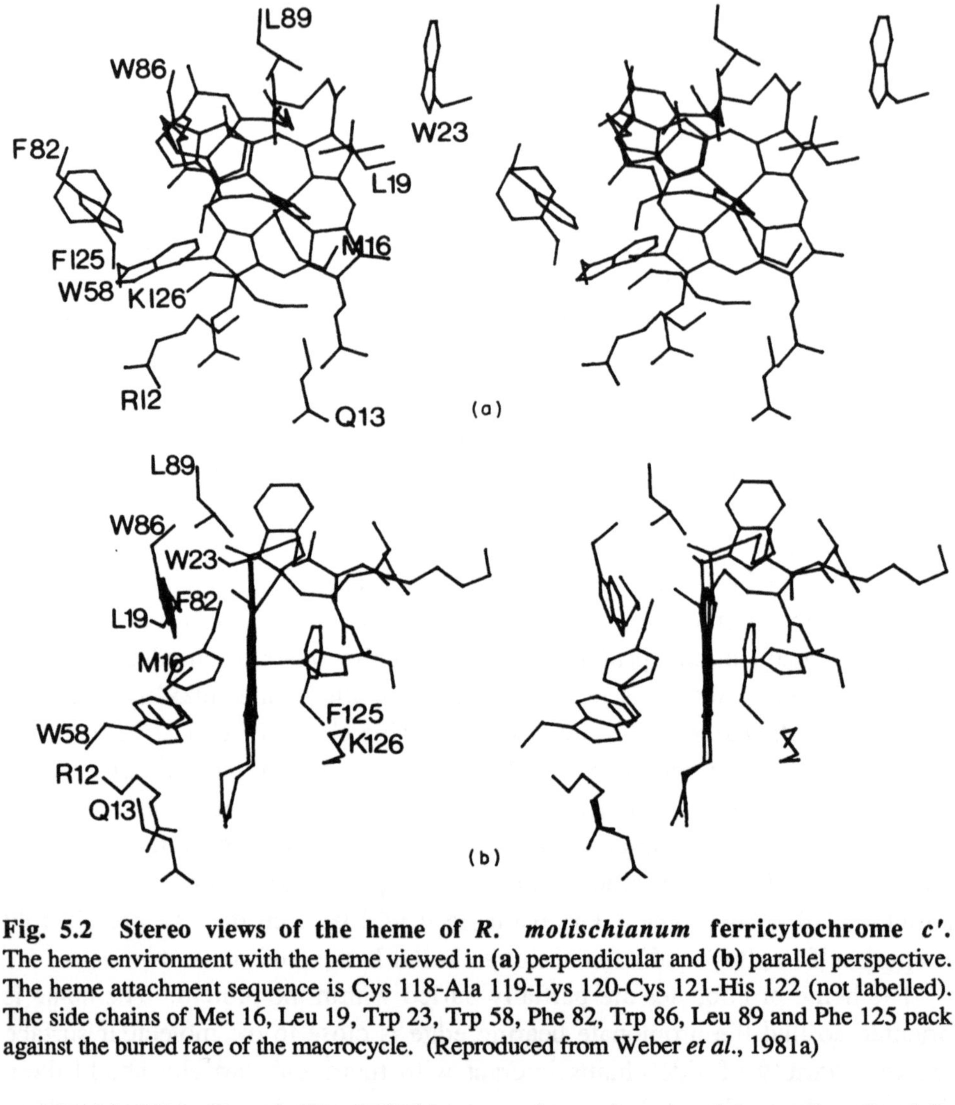

Fig. 5.2 Stereo views of the heme of *R. molischianum* ferricytochrome *c*′. The heme environment with the heme viewed in **(a)** perpendicular and **(b)** parallel perspective. The heme attachment sequence is Cys 118-Ala 119-Lys 120-Cys 121-His 122 (not labelled). The side chains of Met 16, Leu 19, Trp 23, Trp 58, Phe 82, Trp 86, Leu 89 and Phe 125 pack against the buried face of the macrocycle. (Reproduced from Weber *et al.*, 1981a)

The histidine ligand, His 122, is exposed to solvent and is not involved in hydrogen-bonding through its N1 hydrogen (Weber, 1982). The implications of this structural feature for the determination of the iron spin-state, and the correlation of the ionisation of His 122 with the A → B spectroscopic transition (Table 5.1), are given in Chapter 2.3.2.5. It should be noted that the pK observed by equilibrium methods for the A → B transition is probably the real pK_a and not one perturbed by conformation changes (as with the alkaline denaturation of class I ferricytochromes *c* (Chapter 4.4)) because NMR studies indicate the ionisation occurs with a proton exchange rate $\geq 10^4$ sec^{-1} (Emptage *et al.*, 1981; Moore *et al.*, 1985b). The wide variation in pK_a of the transition (Table 5.1) probably reflects differences in distribution of charged residues around the histidine. As described in Chapter 1.4, the normal pK_a for a Fe(III) bound histidine ionising to histidinate is ~10.5 and in the cytochromes *c'* it is proposed that it is reduced from this value by specific electrostatic interactions involving lysine or arginine residues that stabilize the histidinate.

An interesting feature of the heme environment is the location of Met 16 close to the vacant axial coordination site (Fig. 5.2). The Fe-S distance is 3.75Å which is too long for a bonding interaction; typical Fe-S bond lengths are ~2.3Å (Table 1.2). The implication of this methionine location for the His-Met ligated class IIb cytochromes *c* is discussed in section 5.2.4.

The exposed face of the heme is surrounded by charged residues, which are preferentially clustered here compared to the remainder of the protein. However, there are no large regions of clearly delineated positive or negative charge, as there are with mitochondrial cytochrome *c*.

Ferrocytochrome c'. The structure of ferrocytochrome *c'* has not been determined. However, it is unlikely to be substantially different from that of ferricytochrome *c*; certainly there is no change in the amount of helical structure (Imai *et al.*, 1969a; Yang and King, 1970) although the helix packing might change somewhat. A small conformational difference could account for the lower solubility of ferrocytochrome *c'* compared to ferricytochrome *c'* (Taniguchi and Kamen, 1963).

Ligand Binding to Cytochrome c'. The literature concerning ligand binding to cytochrome *c'* contains a number of conflicting results. The protein has been reported to have unusual ligand binding properties with the ferric ion not binding CN^-, N_3^- or F^- and the ferrous ion binding CO but not O_2 (Taniguchi & Kamen, 1963; Gibson & Kamen, 1966). Furthermore CO binding has been described as a complex reaction with a stoicheiometry of 2CO per subunit and involving a pre-binding step for activation of the cytochrome (Cusanovich & Gibson, 1973). However, *Chr. vinosum* ferricytochrome *c'* binds CO with a stoicheiometry of 1 CO per subunit (Kassner & Kykta, 1985) as does the *R. molischianum* protein (Doyle & Gill, 1985). The behaviour reported by Cusanovich and Gibson (1973) has been re-analysed with a model in which ligand binding causes dimer dissociation (Doyle *et al.*, 1986). Ethylisocyanide also binds to a variety of ferrocytochromes *c'* with a stoicheiometry of 1:1 (Rubinow & Kassner, 1984).

The x-ray structure clearly indicates that ligand binding to cytochrome *c'* will involve protein conformational changes because the ligand will have to penetrate the protein and displace side chains shielding the vacant iron ligation site. Thus it is reasonable to search for correlations between the kinetic and thermodynamic parameters governing ligand binding for different proteins and ligands. However, so far convincing correlations have not been found.

Denaturation of Cytochrome c' and Dissociation of the Dimer. Many studies of the pH dependence of ferricytochrome *c'* have been made and two alkaline states have been observed: B and C. State B cytochrome *c'* is described in Chapter 2.3.2.5 and the state C protein is described below.

The pK for the conversion of state B ferricytochrome *c'* to state C is ~12. According to CD (Imai *et al.*, 1969a,b; Yong & King, 1970) and NMR (Akutsu *et al.*, 1983) measurements, state C is practically random coil. Its axial ligands are indicated by RR to be two nitrogenous groups (Kitigawa *et al.*, 1977a). This form is a monomer judged by light scattering and ultracentrifugation (Akutsu *et al.*, 1983). No monomeric form with retained secondary structure has been detected in these studies of *R. rubrum* cytochrome *c'* and it should be noted that, at least with the *Chr. vinosum* protein, the B to C conversion is not a simple two-stage process because three separate low-spin species have been detected by EPR for this protein between pH 11.7 and 12.4 (Maltempo *et al.*, 1974).

The addition of denaturing agents such as urea and alcohols at pH 7 cause the denaturation of the *R. rubrum* protein to a random- coil state but this can be controlled and a low-spin intermediate produced that retains the normal high degree of secondary structure (Imai *et al.*, 1969a;b). Cusanovich *et al.* (1970) have also shown that addition of alcohols to *Alcaligenes* spp. ferrocytochrome *c'* produces a mixture of low-spin proteins, one of which had a 550 nm absorbance and another a 553 nm absorbance.

Akutsu *et al.* (1983) have studied the low-spin forms of *R. rubrum* ferricytochrome *c'* produced by the addition of 30% 2-propanol or 0.06% SDS. Both forms are monomeric and, according to their NMR spectra, retain some degree of structure, though the spectra, and hence structures, are different in the two cases. Imai *et al.* (1969b) had previously shown that with 25% 2-propanol the helix content of *R. rubrum* ferricytochrome *c'* was unchanged. The axial ligation of these low-spin forms has not been studied directly but EPR studies carried out by Emptage (1978) suggest that methionine ligation may occur. Emptage (1978) found that when *R. rubrum* ferricytochrome *c'* at pH 7 was treated with phenol, a mixture of high-spin (~20%) and two low-spin (~40% each) species was produced. One of the low-spin species had a g_z value of ~3.2, which is in the range for histidine-methionine ligation (Table 2.4) and comparable to the g_z value of 3.23 for *Rps. palustris* ferricytochrome *c*-556 (Moore, 1985). In view of the discussion of the class IIb cytochromes *c* in section 5.2.4, the problem of the dissociation of cytochrome *c'* ought to be reinvestigated.

5.2.4 Class IIb Cytochromes *c*

The class IIb cytochromes *c* are predominantly low-spin monomeric proteins. NMR and EPR studies of the cytochromes *c*-556 from *Agr. tumefaciens* (B2a) and *Rps. palustris* show that the axial ligands are histidine and methionine (Moore *et al.*, 1982b; Moore, 1985). The optical spectrum of *Agr. tumefaciens* ferricytochrome *c*-556 contains a 695 nm band consistent with methionine ligation (Van den Branden *et al.*, 1975). The methionine ligand is Met 13, which is the only methionine in the sequence of *Agr. tumefaciens* (Apple 185) cytochrome *c*-556 (Fig. 3.12).

NMR measurements show that the histidine ligand of *Agr. tumefaciens* ferricytochrome *c*-556 ionises to a histidinate with a pK_a of about 9 (J.von Beeumen, G.R.Moore & G.W.Pettigrew, unpublished results). The histidine ionisation is a common feature between this and the cytochromes *c'* (Chapter 2.3.2.5) and, as with cytochrome *c'*, it requires a specific interaction between a lysine or arginine and the histidinate to reduce the pK to about 9. Again, as with the cytochromes *c'*, there is a lysine or arginine four residues from the histidine towards the C-terminus (Fig. 3.12) that we propose provides this interaction.

Although the tertiary structure of cytochrome *c*-556 has not been investigated further it seems very likely that it is similar to the structure of the cytochrome *c'* monomer. This follows from the sequence similarity between them (Fig. 3.12). Even though the only unvaried amino acids between the class IIb and IIa proteins are the heme attachment residues, Cys-?-?-Cys His, there are many residues apparently conservatively replaced, for example: Phe/Tyr 133 and Lys/Arg 134. Assuming the structural identity, the properties of the ligated histidine can be rationalised, as can the appearance of a methionine ligand.

R. molischianum ferricytochrome *c'* contains Met 16 situated close to the vacant iron coordination site, but too far from it for an Fe-S bond to form (Fig. 5.2). Since the ligand for the class IIb cytochromes is Met 13 (Apple 185 numbering) it seems likely that these two residues are sequential homologues with the structural difference that in one protein there are constraints on the formation of an Fe-S bond. All class IIa cytochromes *c* have at least one methionine in the first 29 residues, though they are not in a constant position according to the alignment of Fig. 3.12. This prompts the question: does methionine approach close to the Fe in all these proteins? Recalling that cytochrome *c'* has not been detected *in situ* and may be modified on isolation (Pettigrew & Moore, 1987: Chapter 3.5.5) prompts two further questions: do all class II cytochromes *c* possess a methionine ligand *in situ* and, if so, why does the Fe-S bond break upon isolation? These questions are open to experimental study, as is a related question - can the postulated Fe-S bond be reformed *in vitro*? - and we offer the following comments concerning them.

Met 16 of *R. molischianum* cytochrome *c'* is near the point of closest approach of the subunits and is in the centre of helix A, which has a number of important roles in the structure (Weber *et al.*, 1981a): it provides residues that interact with the heme propionates; and it forms part of the subunit interface as well as covering the interior face of the heme. If the cytochromes *c'* were monomeric low-spin proteins *in situ* and became dimeric on isolation, displacement of the methionine ligand could occur as a result of a change in the position of helix A with respect to the heme (Moore *et al.*, 1982b). *Rps. palustris* cytochrome *c'* is important because it is the only monomeric cytochrome *c'* - although there are indications that it may be subject to a monomer to dimer equilibrium (Table 5.1) - and thus it may be that a rearrangement of the helix packing, without dimerisation, is enough to cause displacement of the proposed methionine ligand. The attempts to obtain structured monomers of dimeric cytochrome *c' in situ* have not been particularly successful, probably because it is difficult to disturb the 4-α-helical packing at the subunit interface without disrupting the helix packing of each subunit. Nevertheless, there are indications that when the dimer is dissociated under conditions which do not unwind the helices, a low-spin form is produced that may have histidine-methionine coordination (section 5.2.3).

5.3 Class III Cytochromes *c*

5.3.1 Introduction

Class III cytochromes *c* are multiheme proteins with bishistidinyl coordination and 30 - 40 residues per heme (Pettigrew & Moore, 1987: Table 1.4). The group includes triheme, tetraheme and octaheme proteins from sulfate- and sulfur-reducing bacteria (Pettigrew & Moore, 1987: Chapter 3.7). The best studied group are the cytochromes c_3 (4 heme)) which have 102 - 118 amino acids and markedly divergent sequences (Fig. 3.6). Only 25 residues are unvaried between the six sequenced proteins and 16 of these are heme binding residues, leaving a total level of identity not far removed from that expected on a random basis.

Cytochrome c_3 (3 heme) is clearly related to the tetraheme cytochromes (Fig. 3.6) but although the deletion of a heme binding region is obvious, the positions of other deletions are not, and a more precise alignment must await completion of the x-ray structure (Haser *et al.*, 1979b). The loss of polypeptide chain is proportionately greater than the loss of the single heme and the protein is only 68 residues long.

Amino acid sequence data for the cytochromes c_3 (8 heme) have not been reported but their amino acid compositions indicate that they are not dimers of the characterised tetraheme proteins (Guerlesquin *et al.*, 1982). Nevertheless, they may be dimers of tetraheme cytochromes because heme removal results in a M_r of 13,500 compared with 26,000 for the holocytochrome *c* (Guerlesquin *et al.*, 1982). This result also suggests that an unusual covalent attachment of two polypeptide chains via a heme group may be present. Potentiometric titrations by EPR indicate that the 26,000 protein contains four pairs of identical hemes (Gayda *et al.*, 1985) consistent with the presence of two identical tetraheme polypeptide chains.

5.3.2 Cytochromes c_3 (4 heme)

The x-ray structures of two tetraheme ferricytochromes c_3 have been determined; those of the 118 residue protein from *D. desulfuricans* (Norway 4), at a resolution of 2.5Å (Haser *et al.*, 1979a; Pierrot *et al.*, 1982), and the 107 residue protein from *D. vulgaris* (Miyazaki), at a resolution of 1.8Å (Higuchi *et al.*, 1981a,b; 1984). For the sake of convenience, these two

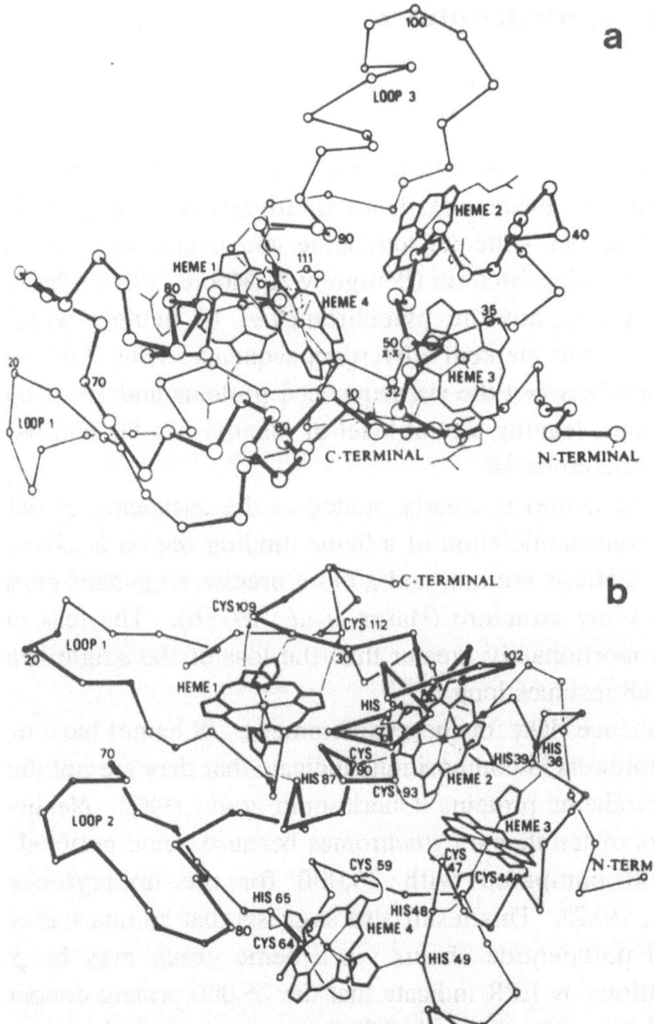

Fig. 5.3 The structure of *D desulfuricans* (Norway 4) ferricytochrome c_3 (4 heme). Only the α-carbon positions and the four heme groups are illustrated. The size of the circles indicate their relative depth. **x-x** at position 50 (view **a**) refers to residues Thr 50, Leu 51 and Glu 52. At the time the x-ray structure was reported the sequence determination was incomplete and only one residue was fitted into the space occupied by three. This does not significantly affect any of the conclusions made about the structure (as view **b** shows, the extra residues are added to a small loop of residues on the molecular surface,which is indicated by →) but it does mean that two amino acids have been left out of the illustrations and that the residue numbering after His 49 needs to be incremented by two. The two views of the molecule were selected to show the arrangement of the four hemes and the positions of the three large loops. (Reproduced from Haser *et al.*, 1979a)

proteins will be referred to as DdN and DvM respectively throughout the description of their structures.

General features of the structure. The approximate overall dimensions of the two molecules are 45Å x 28Å x 32Å for DdN and 39Å x 33Å x 34Å for DvM. There is a high degree of three-dimensional structural similarity between them. The structure of DdN is illustrated in Fig. 5.3.

The DdN structure consists of a core formed from the four heme groups and their interconnecting peptides, around which the remainder of the chain is wrapped. The chain folds in such a way as to create a cleft into which hemes 2 and 4 fit. Despite the constraints imposed by the four hemes, about 50% of the amino acids are organised into recognisable secondary structure elements. These consist of a long α-helix (residues 84 to 101) and a large number of β-turns that include about 35% of the residues. Most of the charged residues (13 of the 17 lysines and 8 of the aspartates) are located on the three large loops.

The DvM structure is very similar to the DdN structure. It too is based upon a core that is formed from the four hemes, whose relative dispositions resemble those of DdN (Table 5.2). However, differences are apparent in the location of the peptide chain at the surface and in place of the three large loops, DvM has four small loops. Also it contains three small α-helices; that from Ser 78 to Leu 84 corresponds to the larger α-helix of DdN, and there are additional helices from Tyr 65 to His 70 and from Lys 93 to Thr 98. Like DdN, ~35% of the residues of DvM are contained in β-turns. The large amount of organised secondary structure is consistent with CD studies of *D. vulgaris* (Hildenborough) cytochrome c_3 (4 heme) (Drucker *et al.*, 1970).

The Environments of the Heme Groups. The heme cluster of DdN is shown in Fig. 5.3. None of the hemes are co-planar, in contradiction to a previous suggestion (Dobson *et al.*, 1974); and each has a unique environment, in agreement with the results of previous spectroscopic studies on a variety of tetraheme cytochromes c_3, including DdN and DvM (Le Gall *et al.*, 1971; DerVartanian & Le Gall, 1974; Dobson *et al.*, 1974; DerVartanian *et al.*, 1978; Xavier *et al.*, 1979). All four hemes have large areas exposed to solvent and the order of exposure is (figures in brackets represent the exposed surface area): heme 4>2>1>3 for DdN, and heme 4(168Å^2)>2=3(136Å^2)>1(127Å^2) for DvM.

The heme propionates are either largely exposed to solvent (hemes 1, 2 and 4) or located at the molecular surface and partly shielded by hydrogen-bonded groups. NMR studies of DdN and *D. vulgaris* (Hildenborough) and *D. gigas* cytochromes c_3 (4 heme) in their fully oxidised and partly reduced states reveal that some heme propionates have pK_a's in the range 5 - 7 although the ionisations have not been identified with specific propionates (Moura *et al.*, 1982; Santos *et al.*, 1984; G.R.Moore, J.J.D.Moura & A.V.Xavier, unpublished data).

Table 5.2 Iron-iron distances and heme-heme angles in class III cytochromes *c*.

	DdN [a]	DvM [b]	Drm.a [c]
Iron-iron distances (Å)			
Fe 1 - Fe 2	12.7	12.0	11
Fe 1 - Fe 3	17.3	17.8	17
Fe 2 - Fe 3	10.9	11.0	10
Fe 1 - Fe 4	16.3	16.4	—
Fe 2 - Fe 4	16.8	15.8	—
Fe 3 - Fe 4	12.8	12.2	—
Heme-heme angles (°)			
Heme 1 - Heme 2	89	80	—
Heme 1 - Heme 3	35	22	—
Heme 2 - Heme 3	90	80	—
Heme 1 - Heme 4	55	73	—
Heme 2 - Heme 4	62	59	—
Heme 3 - Heme 4	84	89	—

[a] *D. desulfuricans* (Norway 4) cytochrome c_3 (4 heme) from Pierrot *et al.* (1982).
[b] *D. vulgaris* (Miyazaki) cytochrome c_3 (4 heme) from Higuchi *et al.* (1981a).
[c] *Drm. acetoxidans* cytochrome c_3 (3 heme) from Haser (personal communication).

The similarities between the heme cores of DdN and DvM are illustrated by the Fe-Fe distances and the inclination angles between hemes (Table 5.2); and by the similar positions of the heme substituents. Only the conformations of the thioether linkages to hemes 1 and 3 are different, and in the case of heme 1 this is because of the different number of inter-cysteine residues (Fig. 3.6). Fe-Fe distances as close as 10Å had been suggested previously from spectroscopic studies (Drucker *et al.*, 1970; Dobson *et al.*, 1974; McDonald *et al.*, 1974; Ono *et al.*, 1975), and these suggestions are in reasonable agreement with the structures. Table 5.2 does not give an immediate indication of how far the hemes are apart but that can be seen from Fig. 5.3. The closest approach of pyrrole rings from neighbouring hemes is about 5Å. The puckered conformations of the hemes are discussed in Chapter 1.4.

The orientations of the axial ligands with respect to the heme planes are different for different hemes, and also differ between corresponding hemes of the two structures. The His-heme orientations of DvM are given in Table 1.5. These differences probably reflect the different hydrogen-bonding contacts involving the histidines (Table 5.3).

Table 5.3 Hydrogen-bonding involving histidine ligands of cytochromes c_3 (4 heme) [a]

Heme	*D. sulfuricans* (Norway 4) Histidine			*D. vulgaris* (Miyazaki) Histidine		
1	89	Ser 86	Oγ	70	Tyr 66	CO
	115	Gln 29	Oε1	106	H_2O [c]	(Pro 17 CO)
2	39	His 36	CO [b]	25	Asn 21	CO
	96	Pro 108	CO	83	Leu 97	CO
3	36	Pro 34	CO	22	H_2O	(Ala 6 CO & Phe 20 CO)
	48	Lys 59	CO	34	H_2O	(Asn 42 CO & Gln 44 CO)
4	49	Thr 50	Oγ	35	Pro 36	CO
	67	Leu 83	CO	52	Ala 62	CO

[a] From Pierrot *et al.* (1982) for DdN and Higuchi *et al.* (1984) for DvM.
[b] CO ≡ carbonyl.
[c] The H_2O molecules of DvM are hydrogen-bonded to various carbonyl groups as well as to the histidines. The carbonyls are given in brackets.

Haser *et al.* (1979a) suggest that the heme arrangement of Fig. 5.3 is representative of all cytochromes c_3 (4 heme) and that the extensive variation in amino acid sequences manifests itself largely by insertions and deletions in surface loops, much as it does for class I cytochromes *c* (Fig. 4.6). However, as described in Chapter 7.5, there are substantial differences in redox potentials amongst cytochromes c_3 (4 heme), and thus it is not only the disposition of the heme groups that is important, but also the nature of the amino acid residues packed around them. Higuchi *et al.* (1984) emphasise that many carbonyl groups are <4Å from the hemes of DvM which should lead to a stabilization of the ferriheme: 8 carbonyls are close to heme 1; 3 close to heme 2; 4 close to heme 3; and 5 close to heme 4. Haser (1981) has emphasised the positions of the tyrosine and phenylalanine residues of DdN (particularly Tyr 8 which is close to heme 3 and hydrogen-bonded to one of its propionates). This residue is not present in other cytochromes c_3 (4 heme) and this might result in different redox properties of their heme 3. The only aromatic residue other than the histidine ligands that appears to be unvaried is Phe 34 (Fig. 3.6) which is located in the tertiary structure between hemes 2 and 3. The equivalent residue in DvM (Phe 20) is similarly located. However, Tyr 66 of DvM (equivalent to Lys 82 of DdN according to Fig. 3.6) occupies the same relative position as does Phe 88 of DdN. Higuchi *et al.* (1984) also point out that the side chains of 4 of the 5 benzenoid residues of DvM are approximately parallel to the imidazole rings of histidine ligands: Tyr 66 and His 70 (heme 1), Phe 20 and His 25 (heme 2), Tyr 43 and His 34 (heme 3) and Phe 76 and His 35 (heme 4). They further suggest that this feature has importance for intramolecular electron transfer.

The identification of particular hemes with specific spectroscopic signals has not been reported. EPR and optical studies of crystalline cytochrome c_3, analogous to the comparable studies of membraneous PRC cytochrome *c* (section 5.4.2), should enable identification to be carried out. This would then allow the assignment of particular redox potentials to specific hemes to be made firm. The redox potential assignments proposed by Bruschi *et al.* (1984) are speculative.

The Protein Surface. Like other *c*-type cytochromes, cytochrome c_3 (4 heme) is presumed to bind to redox proteins via specific surface interactions. Little is known about the protein complexes of class III cytochromes *c* compared to those of mitochondrial cytochrome *c* but it is clear that electrostatic interactions are important in stabilising the complex. This is not surprising given the high density of charged groups: for example, assuming all heme propionates are ionised, DdN has 24 negatively charged groups and 19 positively charged groups; while DvM has 19 and 23 respectively. These charges are not clustered into large domains of positive or negative charge. Presumably there is some pattern to the charge distribution that is functionally important, but at present it cannot be discerned. Interestingly, in the crystalline state both DdN and DvM are involved in close intermolecular contacts. Whether these lead to minor structural perturbations, as with mitochondrial cytochrome *c* (Chapter 4.3) is not known, but Pierrot *et al.* (1982) consider that they might give an indication of how cytochrome c_3 (4 heme) interacts with its oxidoreductases. In DdN the crystal packing involves neighbouring molecules interacting, with the propionates of heme 4 from one molecule binding to lysines of a second molecule (Fig. 5.4). The loop of residues bearing the lysines has a high density of positive charges. The packing arrangement leads to a close approach between heme 1 of molecule I and heme 3 of molecule II (Fe-Fe distance is 19.2Å) and places the heme planes approximately parallel. The closest approach between heme edges is 14Å. Whether this has any biological relevance remains to be determined but such an intermolecular structure is similar in some respects to the proposed cytochrome *c*:oxidoreductase complexes determined by computer-modelling studies (Pettigrew & Moore, 1987: Fig. 2.20). The packing arrangement in crystals of DvM is somewhat different, although the close intermolecular contacts again involve hydrogen-bonding and salt-bridging groups. None of the heme propionates are involved in these interactions and the closest intermolecular Fe-Fe distance is again for hemes 1 and 3 (21.8Å), although in this case the hemes are not parallel.

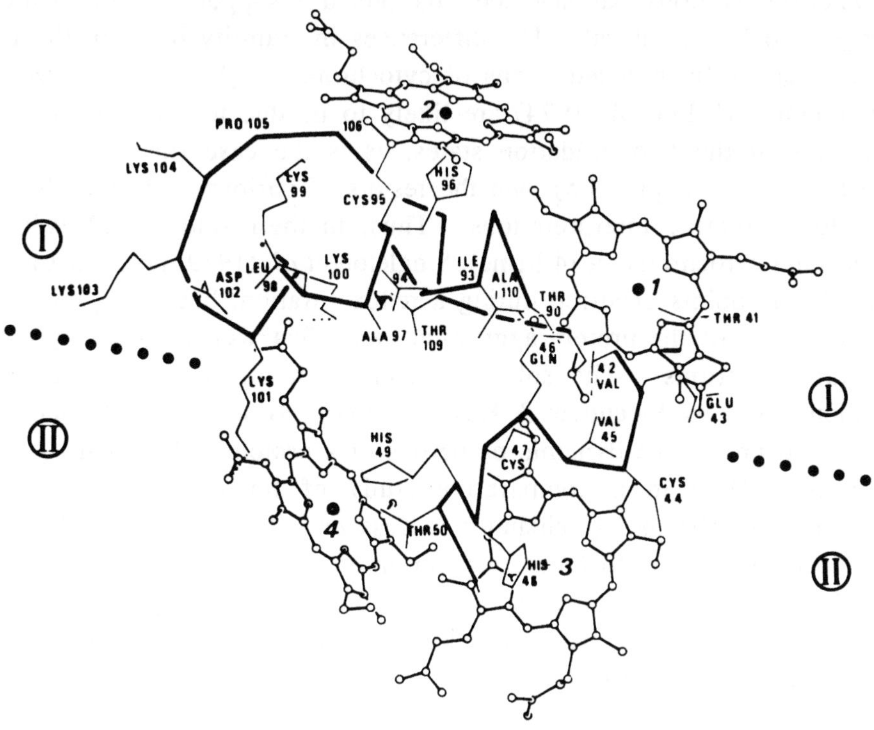

Fig. 5.4 Intermolecular interaction in crystalline *D. desulfuricans* (Norway 4) cytochrome c_3 (4 heme). The interface region is formed by molecule **I** (with hemes ***1*** and ***2***) and the symmetry-related molecule II (with hemes ***3*** and ***4***). The heme propionate interactions are indicated by dashed lines. Intermolecular iron-iron distances are: Fe ***1***-Fe ***2***, 19.2Å; Fe ***2***-Fe ***4***, 21.7Å; Fe ***4***-Fe ***1***, 18.9Å. (Reproduced from Pierrot *et al.*, 1982)

The Structure of Ferrocytochromes c_3 (4 heme). The structures of the fully reduced cytochromes and their intermediate oxidation states have not been determined. There is a general expectation that redox linked conformation changes occur, but the magnitude of these is not known.

The clearest indicator of redox linked conformation changes is the observation that the interaction energies between certain of the hemes are positive (i.e. the reduction of one heme favours the reduction of a second heme), a result that is inconsistent with simple electrostatic considerations (Niki *et al.*, 1984; Santos *et al.*, 1984; Rogers & Moore, 1988). Another possible indicator of conformational changes is that crystals of DdN crack when exposed to dithionite (Pierrot *et al.*, 1982). However, mitochondrial cytochrome *c* crystals also crack on reduction and we now know that this is due to a minor conformational change only (Chapter 4.5).

Other observations that have been interpreted in support of conformational changes are less clear-cut. The differences in stability between the fully oxidised and fully reduced forms of cytochrome c_3 (4 heme) observed by DerVartanian & LeGall (1974) are likely to be due to differing dynamic properties of the two oxidation states, as is the case for mitochondrial cytochromes *c* (Chapter 4.5), and the results of various optical studies are open to alternative interpretations. Thus, in their study of *D.vulgaris* (Africanus) cytochrome c_3 (4 heme), Singleton *et al.* (1979) noted that a shift in isosbestic points occurred during a redox titration, but during the same titration some of the protein aggregated, an effect well known to produce spectral distortions. The solvent perturbation difference spectroscopy technique used by Fiechtner & Kassner (1979) to study cytochrome c_3 (3 heme) also probes the dynamic properties of a protein rather than its static structure. Thus in a comparative study of horse cytochrome *c* and *R. rubrum* cytochrome c_2 Schlauder & Kassner (1979) observed redox state differences analogous to those they observed for cytochrome c_3 (3 heme), even though x-ray data shows that the heme exposure of cytochrome *c* is only marginally affected by a change in redox state whilst the exposure of the heme of cytochrome c_2 is unaffected.

Denaturation and Unfolding of Cytochromes c_3. A remarkable feature of the cytochrome c_3 (4 heme) structure is the complex nature of the heme binding (Fig. 5.3). With four heme attachment sites and four histidines as sixth ligands there is a theoretical maximum of twenty four different ways of attaching the four hemes. Even allowing for the fact that, because of steric constraints, His 49 cannot be the sixth ligand for the heme attached to Cys 44 and Cys 47 (DdN numbering) only reduces the number of possible structures to eight. This raises the question of how one structure is selected from the range of possible structures? This interesting problem in protein folding has not been explored but, as with mitochondrial cytochrome *c* (Pettigrew & Moore, 1987: Chapter 4.2.7), insight into the pathways of folding for cytochrome c_3 may come from denaturation-refolding experiments.

On decreasing the pH to ~2.5, cytochrome c_3 (3 heme) becomes high-spin as a result of dissociation of the histidine ligands, but with increasing pH the native protein is regenerated. However, at pH 7, increasing the temperature to 70°C causes cytochrome c_3 (3 heme) to irreversibly denature. The temperature denatured form is highly structured and has a low-spin state but it is not the native form (Moura *et al.*, 1984). Further work is required to establish whether the difference between these two experiments is because of the binding properties of histidine (at low pH histidine will not bind to Fe(III) but at pH 7 and 70°C it will), or the conformational behaviour of the polypeptide.

5.3.3 Cytochrome c_3 (3 heme)

The x-ray structure of cytochrome c_3 (3 heme) is in progress (Haser *et al.*, 1979b) but an outline structure of the protein fold is not yet available. Nevertheless, the crystallographic data that is available, and the amino acid sequence of cytochrome c_3 (3 heme), points to a high degree of structural homology with cytochrome c_3 (4 heme).

The sequence alignment given in Fig. 3.6, suggests that heme 4 is missing, along with residues 49 to 74 (DdN numbering) which form part of loop 2. If both heme 4 and part of loop 2 are removed from the DdN structure and the lesion in the chain repaired by linking the small 50's loop (at the bottom right of Fig. 5.3) with the remnant of loop 2, the remainder of the core structure might be left essentially unchanged. Another major deletion occurs at the N terminus in loop 1 but if the alignment is made at His 17, which is probably the sixth ligand of heme 3 in cytochrome c_3 (3 heme), then Val 32 and Phe 34 (DdN numbering) have their homologs in the triheme protein. The preliminary x-ray data of Haser (personal communication) showing that the Fe-Fe distances of cytochrome c_3 (3 heme) are similar to those of hemes 1, 2 and 3 of DdM and DvM (Table 5.2) are consistent with the idea that the heme core of the triheme protein resembles that of hemes 1, 2 and 3 of DdN and DvM.

5.4 Class IV Cytochromes *c*

5.4.1 Introduction

Class IV cytochromes *c* are tetraheme proteins containing both bis-His and His-Met coordinated heme, and with a tertiary structure exemplified by that of the photosynthetic reaction centre (PRC) cytochrome *c* of *Rps. viridis*. To date such proteins have been identified only in the reaction centres of certain purple non-sulfur bacteria.

In his original classification scheme, Ambler (1980) used the designation class IV to include cytochromes *c* containing heme *c* and additional prosthetic groups. We have made an alternative proposal for these proteins (Pettigrew & Moore, 1987: Table 1.4) and have suggested that class IV cytochromes *c* be defined as above to form a structurally homogeneous family.

Photosynthetic Reaction Centre Cytochrome c. A membrane associated class IV cytochrome *c* is a constituent of the *Rps. viridis* PRC (Thornber *et al.*, 1969; 1980; Prince *et al.*, 1976; Deisenhofer *et al.*, 1985). *Chr. vinosum* PRC's also contain a tetraheme cytochrome *c* that is probably a member of the same class (Tiede *et al.*, 1978). However, not all bacterial PRC's contain an intrinsic cytochrome. The *Rb. sphaeriodes* PRC's do not, even though the core unit responsible for the primary photosynthetic charge separation is similar to that of *Rps. viridis* (Michel & Deisenhofer, 1988; Feher *et al.*, 1989).

The *Rps. viridis* cytochrome, for which an x-ray structure is available, is the main subject of the present section. We also consider various sets of spectroscopic data that give some insight into the functional aspects of this cytochrome, and its relationship to the *Chr. vinosum* PRC cytochrome.

5.4.2 *Rps. viridis* PRC Cytochrome *c*

This protein is sometimes described as cytochrome *c*-553/*c*-558 after the appearance of its ferrocytochrome optical spectrum (Thornber *et al.*, 1969; 1980). However, the heme α-band absorbance maxima have now been resolved and none are at 553 nm or 558 nm (Table 5.4). We return to the spectroscopic properties of the hemes below.

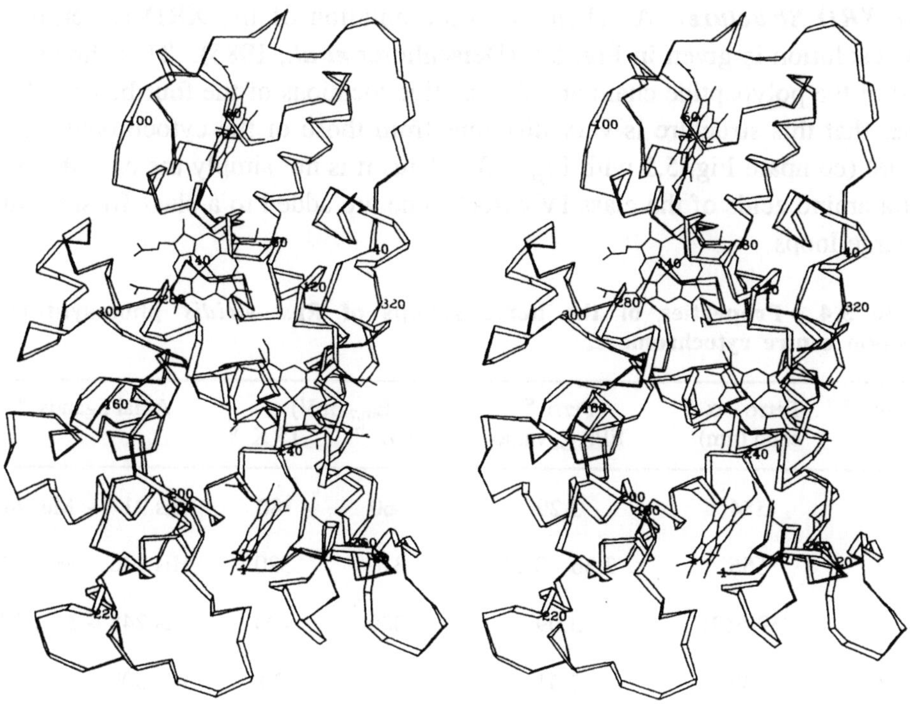

Fig. 5.5 Stereo view of the 3Å resolution XRD structure of the class IV cytochrome *c* present in *Rps. viridis* photosynthetic reaction centres. The polypeptide chain is represented by a ribbon. The heme groups in descending order through the structure as depicted are groups 1, 2, 4 and 3. Heme 4 is coordinated by two histidines, and the other three hemes by histidine and methionine. Interaction with the reaction centre occurs at the base of the cytochrome. (Reproduced from Deisenhofer *et al.*, 1985)

Rps. viridis PRC cytochrome *c* is a lipoprotein consisting of 337 amino acids. Its amino acid sequence is given in Fig. 3.15. The lipid containing group is the N-terminal cysteine which is linked to a diglyceride (Weyer *et al.*, 1987a). This helps attach the cytochrome to the surface of the membraneous PRC; most of the cytochrome is located in the aqueous periplasmic compartment. Weyer *et al.* (1987a) point out that there are extensive polar interactions between the cytochrome and the core PRC subunits and, therefore, they propose that the function of the attached diglyceride is to help fix the cytochrome to the photosynthetic membrane prior to its association with the remainder of the complex.

The XRD Structure. A schematic representation of the XRD structure at 3Å resolution is given in Fig. 5.5 (Deisenhofer *et al.*, 1985). This shows the fold of the polypeptide chain and the relative locations of the four hemes. It is clear that this structure is very different from those of the cytochromes c_3 (4 heme) (compare Fig. 5.5 with Fig. 5.3). Thus it is not simply the case that the extra amino acids of the class IV cytochrome are added to a class III structure as large loops.

Table 5.4 Properties of the heme groups of *Rps. viridis* photosynthetic reaction centre cytochrome *c*.

Heme [a]	Ferro - α [b] band (nm)	Ferri [c] EPR g_z value	$E_{m,7}$ (mV) 298 K [b]	15 K [c]	Axial ligands [a]
1	554	3.29	-60	-80	His 91 - Met 74
2	556	3.36 - 3.32	310	320	His 136 - Met 110
3	559(552)	3.09	380	400	His 248 - Met 233
4	552	3.31	20	20	His 309 - His 124

[a] The heme numbering and ligand assignments are taken from the XRD structure (Fig. 5.5). The sequence numbers come from Deisenhofer *et al.* (1985) and the amino acid identification from Weyer *et al.* (1985b). For heme 4, His 309 is part of the Cys-Arg-Thr-Cys-His region.
[b] The absorbance bands were resolved in difference spectra poised at different potentials. The α-band of heme 3 is split into major (559 nm) and minor (552 nm) components (Dracheva *et al.*, 1988; Shinkarev *et al.*, 1990).
[c] The g_z values were the same for membrane bound and isolated PRC's (Nitschke & Rutherford, 1989).

At its present stage of refinement the most striking feature of the structure is the relative positions of the 4 hemes. These do not form a close packed core, as with cytochrome c_3 (Fig. 5.3); rather they form an approximately linear array running through the long axis of the protein. The shortest Fe-Fe distances are: hemes 1 and 2, ~14Å; hemes 3 and 4, ~14Å; and hemes 2 and 4, ~16Å (Deisenhofer *et al.*, 1984). This disposition of the hemes was not anticipated prior to the XRD structure (section 5.4.3).

A detailed description of the protein fold must await refinement of the structure and fitting of all amino acid side-chains. However, even from Fig. 5.5 it is clear that at least 93 of the 337 amino acids form five α-helices, and each of the heme attachment sites is at the carboxyl terminus of a separate helix. The 3Å structure, together with the amino acid sequence, also reveals that three of the hemes are His-Met coordinated and the fourth is bis-His coordinated (Table 5.4).

Spectroscopic and Redox Properties. A central problem with all multiheme proteins is the identification of particular spectroscopic and redox properties with specific hemes in the structure. This can be achieved by using EPR and optical spectroscopy to monitor oriented samples of the protein (Chapter 2.3.3.1) at different redox potentials.

The identification scheme given in Table 5.4 for the *Rps. viridis* PRC cytochrome *c* is the most favoured scheme based on recent optical (Dracheva *et al.*, 1988; Fritzsch *et al.*, 1989; Vermeglio *et al.*, 1989) and EPR (Nitschke & Rutherford, 1989) studies. However, it is not accepted by all workers in the field. Hubbard & Evans (1989), who used EPR to monitor PRC photo-oxidation, prefer a model in which the hemes are arranged in redox potential order, i.e.: 1 (*c*-554, -80 mV); 2 (*c*-552, 20 mV); 4 (*c*-556, 320 mV); and 3 (*c*-559, 400 mV). They also claim that reduction of the low potential hemes alters the electron transfer processes involving the high potential hemes. Whichever of the schemes is accepted, it is clear that the immediate donor to the reaction centre is the high-potential heme *c*-559.

5.4.3 *Chr. vinosum* PRC Cytochrome *c*

The tetraheme cytochrome *c*-553/*c*-555 from the PRC's of *Chr. vinosum* has been assumed to be similar to the *Rps. viridis* class IV cytochrome *c*. However, the relative positions of the *Chromatium* hemes proposed from spectroscopic studies are very different from those of the *Rps. viridis* cytochrome (Fig. 5.5). Tiede *et al.* (1978) proposed that the shortest Fe-Fe separation was ~8Å; this was for a low-potential heme and high-potential heme pair. And they suggested that the four hemes formed an approximately oblong arrangement with an Fe at each corner of the oblong. In view of the XRD structure and spectroscopic studies of the *Rps. viridis* cytochrome, the *Chromatium* protein should be reinvestigated.

Chapter 6 Evolution

6.1 Introduction

The structural information described in Chapters 3 and 4 not only furthers our understanding of the chemical and biochemical properties of cytochrome *c*, but also provides a rich store of information of evolutionary significance. The evolutionary implications of the structural data is one of the themes of the present chapter in which we discuss the role that cytochrome *c* has played in the development of the relatively new field of molecular evolution. We start by defining certain general taxonomic and phylogenetic concepts and follow this with discussions of a number of topics, including: evolution of eukaryotes and the origin of eukaryotic organelles; the evidence for lateral gene transfer and gene duplication in molecular evolution; and the possible occurrence of convergence in the evolution of cytochromes.

6.2 Aspects of Taxonomy

Taxonomy has the dual function of defining species and ordering them into groups. In order to do this, the degree of similarity across a range of characters must be assessed. Classifications based on these similarities in present day features of organisms, whether they be morphological or molecular, are described as phenetic. This is to be contrasted with phylogenetic classification which is concerned with evolutionary relationships. Phylogenetic trees may include the components of anagenesis (evolutionary change) and cladogenesis (pattern of branching on a scale of elapsed time). In whole organisms, the time scale can be long with little phenotypic change (for example in the frogs and 'living fossils' such as the coelocanth) or can be short with much phenotypic change (for example during adaptive radiation of the mammals). Because the rate of evolutionary change is so variable, phenetic and phylogenetic classification will not usually coincide (Fig. 6.1). The problem of phenetics is that closely related species which have undergone rapid morphological change are placed in separate groups (e.g. **B** and **C**, Fig. 6.1(b)) while species that happen to resemble each other but are only remotely related (**C** and **D** or **A** and **B**, Fig. 6.1(b)) are grouped together.

Traditionally the component of elapsed time in phylogenetic trees was derived from the fossil record. This record is, at best, difficult to interpret in

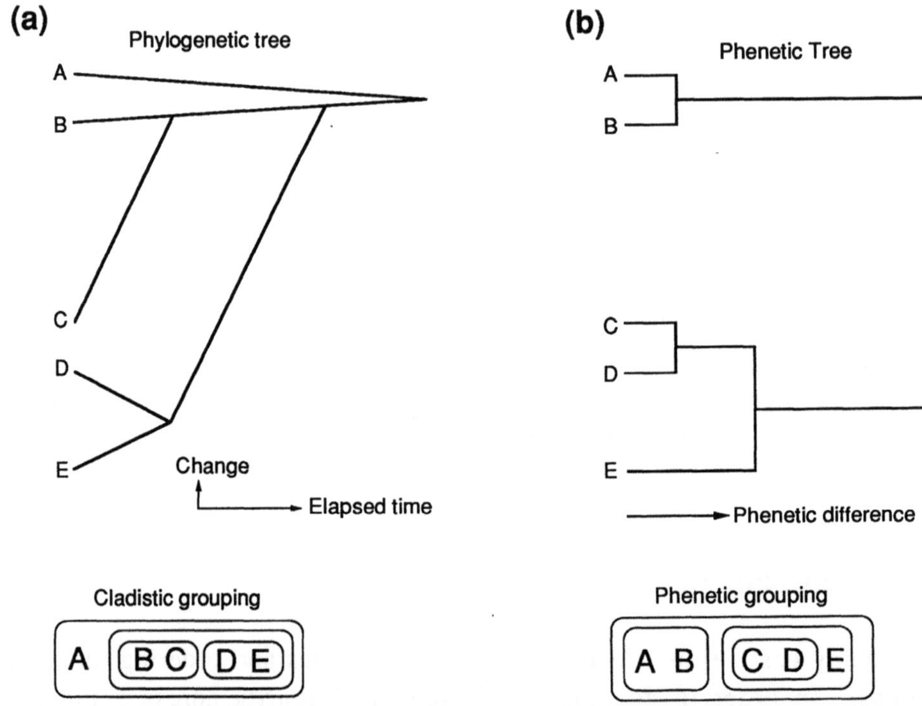

Fig. 6.1 Phylogenetic and phenetic trees. A hypothetical phylogenetic tree is shown in **(a)** established by a good fossil record (horizontal axis of elapsed time) and assessment of degree of change or anagenesis (vertical axis). The cladistic pattern is based on the order of branching only. In **(b)**, a tree is drawn using the differences observed between present-day organisms (**A** - **E**) based on the vertical axis of (**a**). The organisms (**A** - **E**) are drawn to superimpose on the vertical axes of **(a)** and **(b)** but in the case of **(b)**, this vertical axis has no dimension or significance and the degree of change is expressed in the horizontal axis. The phylogenetic tree **(a)** differs from the phenetic tree **(b)** and the cladistic pattern of **(a)** differs from the phenetic pattern of **(b)**. Only if the evolutionary rate of change were constant would the phenetic tree correspond to the phylogenetic tree. This is the claim made for molecular evolution. According to this claim, the phenetic tree of amino acid or nucleic acid sequences corresponds to the cladistic pattern of phylogenetic history of the organisms of which they are a part

terms of accurate divergence times and at worst, non-existent. With the appearance of amino acid sequence information, it became apparent that proteins changed during evolution at a much more even rate than morphological or physiological characteristics: indeed proteins seemed to have the characteristics of a molecular clock (Zuckerkandl and Pauling, 1965) and therefore the phenetic matrix of sequence similarity was also a measure of elapsed time (Fig. 6.1, legend). There was initial euphoria that amino acid sequences held the history of evolutionary change, which, when deciphered, would allow deduction of phylogenetic relationships. Such information would be available for areas in which the fossil record is poor or absent. While this view seems to many naively optimistic and a much more sceptical atmosphere now prevails, there is no doubt that the comparative study of amino acid and nucleic acid sequences has revolutionised thinking on the pathway and mechanisms of evolution.

6.3 The Construction of Phylogenetic Trees

The first step in the comparison of amino acid or gene sequences is the production of an optimum alignment in which similarity is maximised. In many cases (e.g. the mitochondrial cytochromes *c*, Fig. 3.1) the best alignment may be self-evident: in others (e.g. the cytochromes *c'*, Fig. 3.12) the number and placement of gaps required may not be so obvious. The insertion of gaps incurs a statistical cost (Needleman and Wunsch, 1970) and they should be used sparingly at positions best determined on the basis of x-ray crystallographic structures (Meyer *et al.*, 1986).

The tree or dendrogram is the basic means of presentation of phylogenetic analysis. In its simplest form it is drawn using the methods of numerical taxonomy (Sneath and Sokal, 1973) using the amino acid similarities matrix (or the derived minimum matrix of nucleotide replacement) to identify a closely related nuclear pair of sequences, followed by sequential addition of more remote sequences. This simple matrix method can also be used to deal with the T1 oligonucleotide catalogues of 16S rRNA (Doolittle and Bonen, 1981). These contain ribonuclease T1 fragments of the RNA of between 1 - 20 nucleotides. For a molecule of this size, about 10 - 12 pentamers and 2 - 4 hexamers are predicted to be identical by chance while chance identity of higher oligomers is vanishingly small. Catalogues are compared using the coefficient S_{AB} (Eqn. 6.1).

$$S_{AB} = \frac{2N_{AB}}{(N_A + N_B)} \qquad (6.1)$$

where N_{AB} is the number of common nucleotides and $(N_A + N_B)$ is the total number of nucleotides. S_{AB} is 1 if two catalogues are identical and 0.05 if they are unrelated but of the same size and composition.

It is this simple matrix method that is used in the construction of most of the dendrograms in this chapter. An illustration of the method applied to the cytochrome *c*-551 sequences is shown in Fig. 6.2(a). These dendrograms assume equal rates of change in the lineages descendent from a node and are foreshortened due to the increasing contribution of undetected multiple changes at remote divergences. This foreshortening can be corrected empirically. For example Ochman and Wilson (1987) established the relationship:

Fig. 6.2 Dendrograms and matrices for the cytochrome *c*-551 (c_7) sequences. The experimental data on which the construction of the dendrograms is based are the sequence alignment of Fig. 3.11 and the matrix of sequence identities of Table 3.8. Two methods of tree construction are examined - the simple averaged matrix method **(a)** and the Fitch and Margoliash matrix method **(b)**. In each case the derived matrix (normal type, bottom left figures) is compared with the actual experimental matrix (**bold type**, upper right figures). **(a) The simple averaged matrix.** The dendrogram is based on the matrix of sequence identities in pairwise comparisons and is assembled using closely related pairs first, and adding progressively more remote connections. When adding a new sequence to the pattern, the average % identity with the members of the dendrogram is used to place the depth of the divergence (node). The range of % identities at a node is indicated by a bar. The solid lines show unambiguous patterns of relationship in the matrix. For example, the sequence of *Rc. tenue* cytochrome is 48 - 56% identical to members of the group - *stutzeri, mendocina, fluorescens, aeruginosa* and *vinelandii* - while within that group, pairwise comparisons cover the range 62 - 79%. The more remote *Rc. gelatinosa* sequence, on the other hand, shows a range of 34 - 41% identity. Thus the divergence of the *Rc. tenue* sequence is placed at a point more remote than that of the group named above but less remote than that of the *Rc. gelatinosa* sequence. The broken lines indicate uncertainty in constructing the branch points of the dendrogram. For example, there is overlap in the range of pairwise comparisons between the *Ps. stutzeri* - *Ps. mendocina* couple and *Ps. fluorescens* on the one hand (73, 69%) and *Ps. aeruginosa* on the other (70, 63%). The position of the *Nitrosomonas* sequence in the dendrogram is uncertain because only the N-terminal part of the sequence has been determined. Thus the pairwise comparisons will be comparable to the figures for other cytochromes only if the N-terminal region has changed at a similar rate to the overall sequence. For other members of the matrix, figures for % identity based on only the first 49 residues tend to be lower than those for the entire sequence. This suggests that the *Nitrosomonas* figures may be underestimates. The sequences of the cytochromes from *Ps. stutzeri* strains have not been published but a dendrogram of their degree of identity appeared in Ambler (1978) and has been added here to the pattern of relationships for the published sequences. The strain numbers are those of the Stanier collection.

$$\% \text{ identity} = 100 \times S_{AB}^{0.21} \tag{6.2}$$

by comparing the results of oligonucleotide catalogues with whole RNA sequences. Similar empirical corrections are possible for amino acid sequence matrices (Dayhoff 1972; Dickerson 1971; Holmquist *et al.*, 1972, Moore *et al.*, 1976: see section 6.4).

(a)

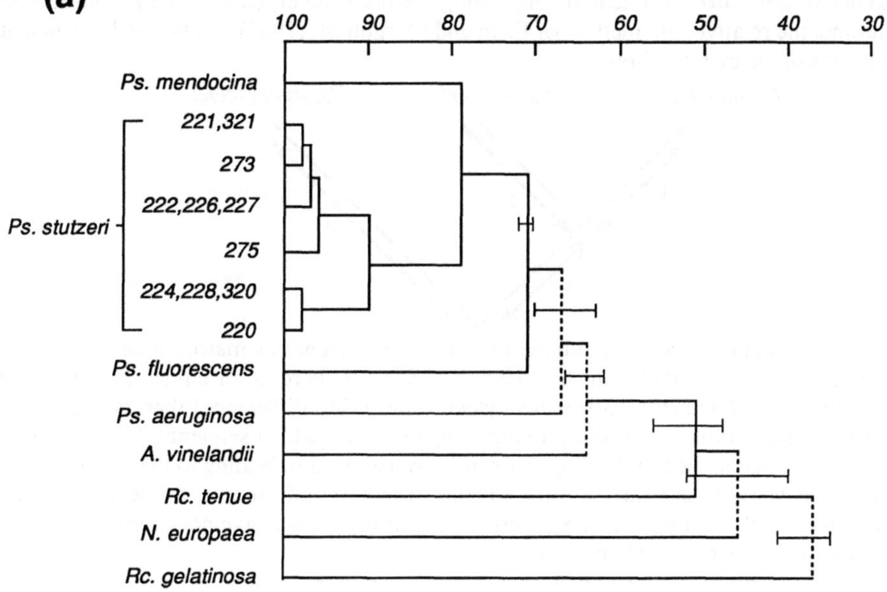

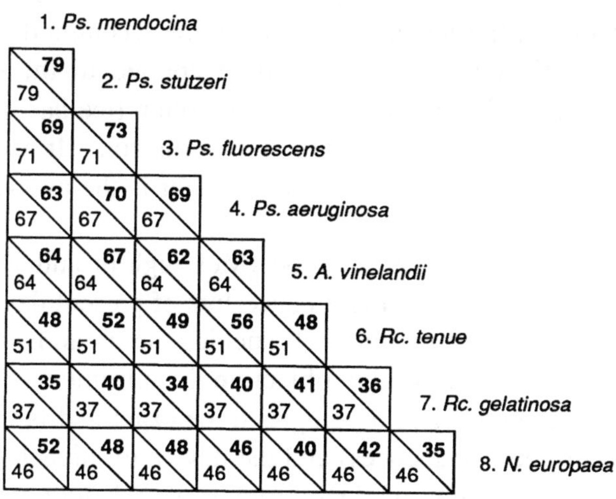

Fig. 6.2 (b) The Fitch and Margoliash matrix. Among the full sequences, pairwise comparisons to *Ps. mendocina* are regularly less than to *Ps. stutzeri* (columns 1 and 2 of the matrix). The average difference is 4.6%. This could mean either that the *Ps. mendocina* sequence changed more rapidly than the overall rate of change or that the *Ps. stutzeri* sequence changed more slowly. However if we look at the % identities of the outlying sequence, *Ps. aeruginosa,* with the group *stutzeri, mendocina and fluorescens,* we find that the *Ps. mendocina* shows a greater divergence (63% identity) than either *stutzeri* or *fluorescens.* This suggests that the *Ps. mendocina* sequence has evolved faster than the others (notice that the matrix is not unambiguous in this respect - the *A. vinelandii* comparisons suggest a slower divergence of the *stutzeri* sequence). Assuming the faster divergence of the *Ps. mendocina* cytochrome, the relative rate method of Fitch and Margoliash (1967) can be used to calculate the first node of the tree as follows:

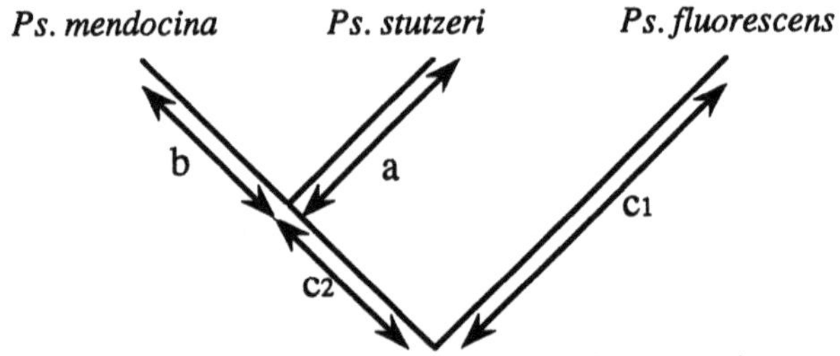

Using % differences obtained by subtraction from the % identities matrix, then (b - a) = 4.6 (average), (b + a) = 21, therefore b = 12.8, a = 8.2. If there is an equal rate of change between *Ps. stutzeri* and *Ps. fluorescens* cytochromes (27% difference) then $c_1 = 13.5$, $c_2 = 5.3$. In this simple illustration, more remote sequences are added sequentially assuming an equal rate of change in the arm leading to *Ps. stutzeri* and the arm leading to the new sequence. The Fitch and Margoliash matrix affords a better fit (bottom left figures) to the actual matrix than does the simple averaged matrix described in part **(a)** but we have not attempted to make further small adjustments to obtain a best fit

An extension of the simple matrix is the method of Fitch and Margoliash (1967) in which different rates of change are accommodated along different lineages. An illustration of this method applied to the cytochrome *c*-551 sequences is shown in Fig. 6.2(b). The optimum tree has the best fit between the branch lengths and the actual input matrix. It is represented as a dendrogram in which the ordinate node height is the average of all distances descendent from it.

In the ancestral sequence method (McLaughlin and Dayhoff, 1973) the nodes are populated by the inferred ancestors of present day sequences. There is simultaneous determination of branching order and number of changes within a lineage. Again the objective is minimum change (maximum parsimony). The number of topologies to be examined is very great (5×10^{15} for 70 cytochrome *c* sequences) and the device of 'branch swapping' must be introduced into the search to allow a computational short-cut.

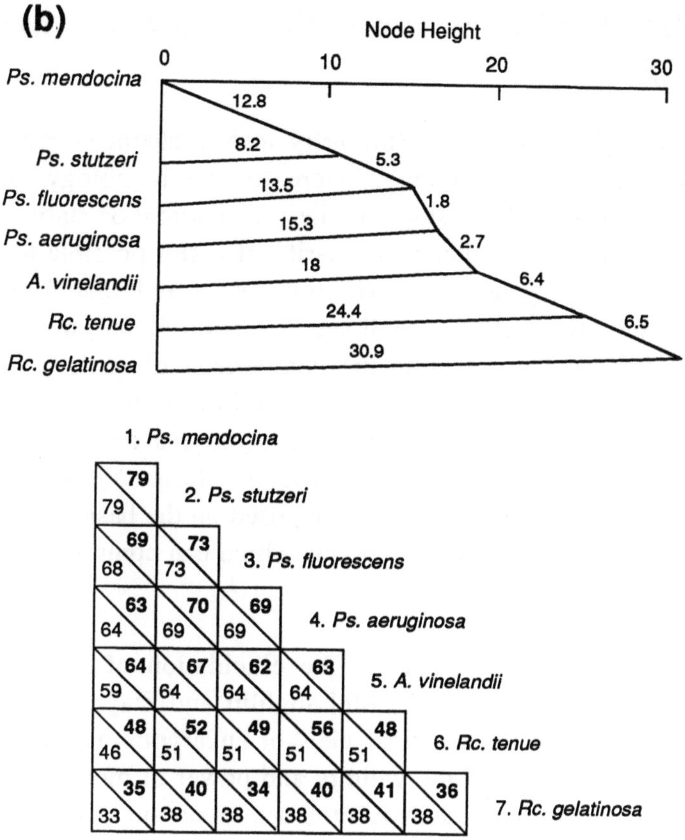

An alternative way to solve the computational problem of the ancestral sequence method is to use a matrix to determine the branching order and then select the minimum tree by constructing ancestral codons - the so-called 'maximum parsimony-ancestral sequence' method (Baba *et al.*, 1981).

Several other methods can be used to estimate the differences between proteins or genes. We mention them only briefly here because they have been little used in the comparison of cytochromes *c*. They include microcomplement fixation by antibody-antigen complexes, electrophoretic mobility of proteins, and DNA hybridisation and melting. All tend to be of most use in defining patterns of close relationship (for example within a genus) rather than in the investigation of the more remote divergences that characterise the cytochrome *c* work (Wilson *et al.*, 1977).

6.4 The Problems and Successes of Molecular Phylogeny

The Question of Homology. Similarity in any character may be due to chance, convergence or homology. Here we use homology to indicate a divergent evolutionary relationship. While it is possible on statistical grounds to assess the likelihood of chance similarity, it is not possible to distinguish homology and convergence on a rigorous basis and therefore homology remains an assumption. As we shall see in section 6.10.2, the possibility of convergence is a real problem in the comparison of tertiary structures of proteins. With sequences, overall structural and functional constraints leading to convergence seem unlikely and there is no reason to doubt that the individual families of class I cytochromes *c* (e.g. cytochromes c_2) are homologous internally. However, superimposed on the basic homology, there can be local convergent or parallel effects which can complicate construction of phylogenetic trees (see below).

Relationship to Classical Trees. The elucidation of the cytochrome *c* tree (Fitch and Margoliash 1967, McLaughlin and Dayhoff 1973) and its general agreement with the classical view of eukaryotic evolution (Fig. 6.3) was considered a measure of the validity and power of molecular phylogeny. However on close inspection, and with the determination of further sequences, a number of disagreements between the molecular and classical trees appeared. These indicate that either the classical tree or the molecular tree was wrong in these respects (or that both were wrong!). The initial response of the early proponents of the molecular methods was that both trees should stand until independent tests could distinguish them. To quote Fitch in the discussion of his paper (Fitch and Margoliash 1969): "I am sorry about putting the [flightless penguin] so close to the [chicken] ... but ... I would rather present the tree which the data dictate ... than try to adjust the data to make it fit preconceived evolutionary notions." This initial confidence in the molecular approach and desire to retain its independent *a priori* status has to some extent become diluted. Many workers now find it acceptable to increase the mutational size of the minimum molecular tree to allow agreement with the classical viewpoint.

For example, Fitch himself (1976) adjusted his tree (Fig. 6.4) to remove 'anomalous' joins between kangaroo and non-primate placentals; turtles and birds; and bony fish, shark and lamprey. Lyddiatt *et al.* (1978) found that an increase of 10 in their minimum tree of the invertebrates (225 replacements, Fig. 6.5) allowed a more classically acceptable divergence of the molluscs, annelids and arthropods on a prostomial line of descent and the echinoderms and chordates on a deuterostomial lineage (compare Fig. 6.3 and Fig. 6.5). The most parsimonious tree of Baba *et al.* (1981) contained the divergence of the prawn from the horse-donkey lineage and a close relationship between rattlesnake and humans! These oddities were removed by an additional small number of substitutions.

The view of the classical taxonomist tends to be that the molecular trees should be used selectively. Cronquist (1976) criticised the inability of the

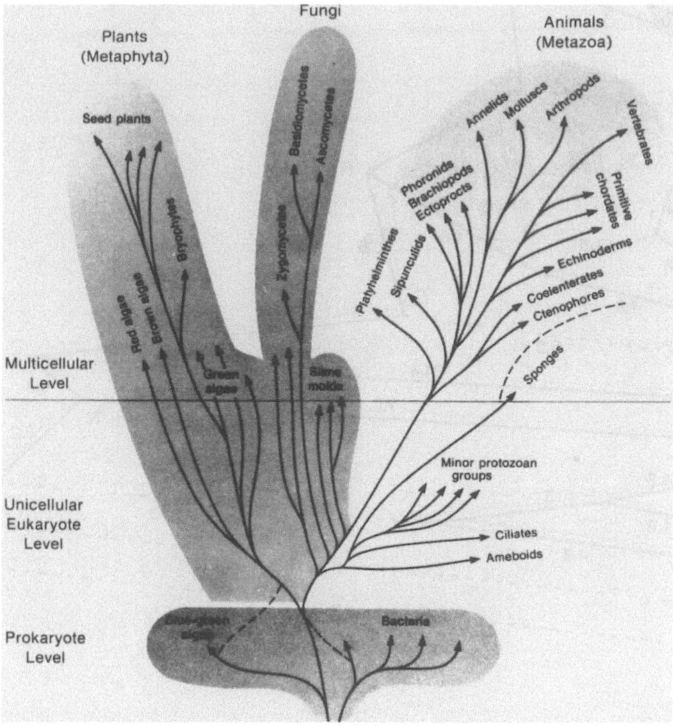

Fig. 6.3 Phylogenetic chart of living organisms. Three major grades of advancement are recognised: prokaryote, unicellular eukaryote, and multicellular. Of the five multicellular groups that have advanced to the grade that includes tissue differentiation and elaboration of distinctive form, three (the red and brown algae and the green algal - seed plant line) are photosynthetic with expansion of surface area. One line, the fungi, consists of heterotrophs that absorb rather than ingest food. The final line includes the Metazoa which are heterotrophic ingestors with usually compact body form and internal absorptive membranes. (From Dobzhansky *et al.*, 1977)

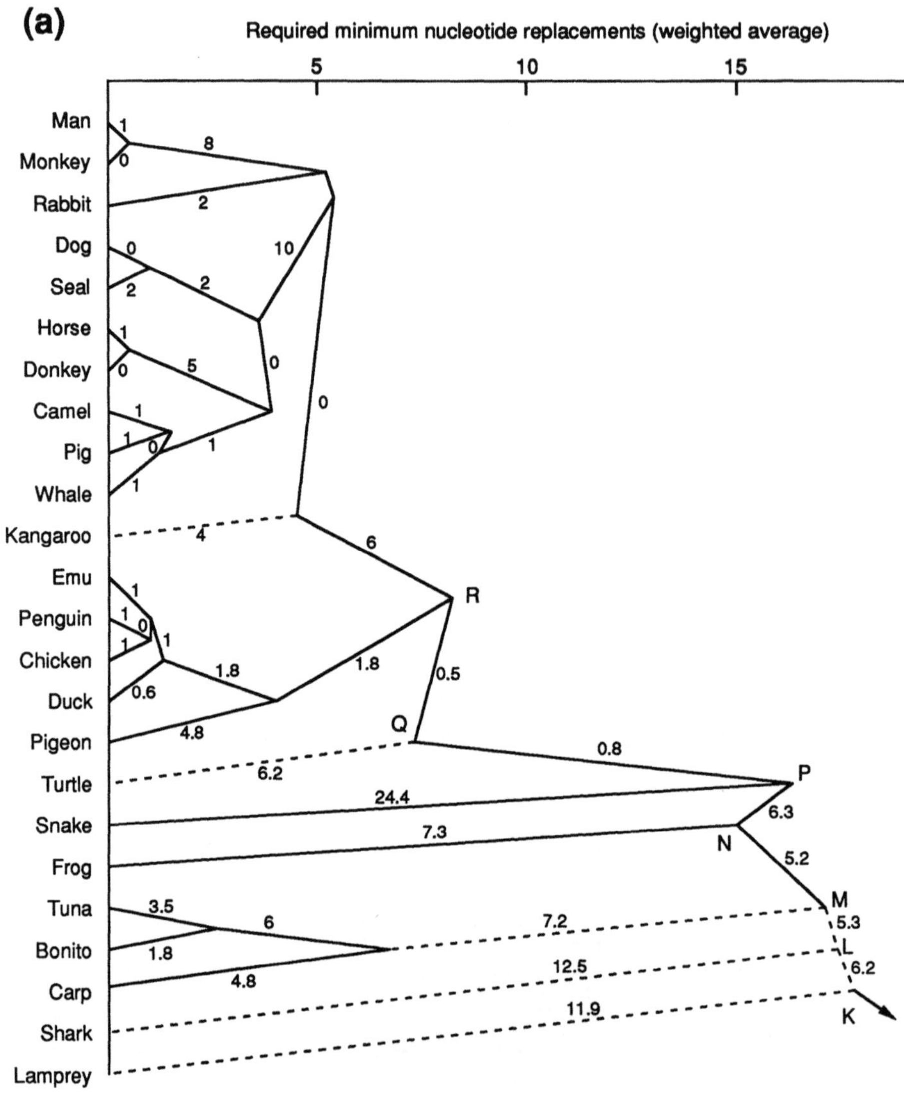

Fig. 6.4 The phylogeny of cytochrome *c*. Phyletic distances shown on the arms of the dendrogram are obtained by the method of Fitch (1971) and are the minimum number of mutations required to account for the difference between a common ancestor and descendents. Nodes are at a height equal to the average of the descendent arms. The arrow at the lower right by node **K** connects the vertebrate tree **(a)** to the arrow at the upper left by node **E** in **(b)**, with a change in horizontal scale. Selected nodes are lettered (**A** - **R**) and are used in the analysis of covarions (Fig. 6.9). (After Fitch, 1976)

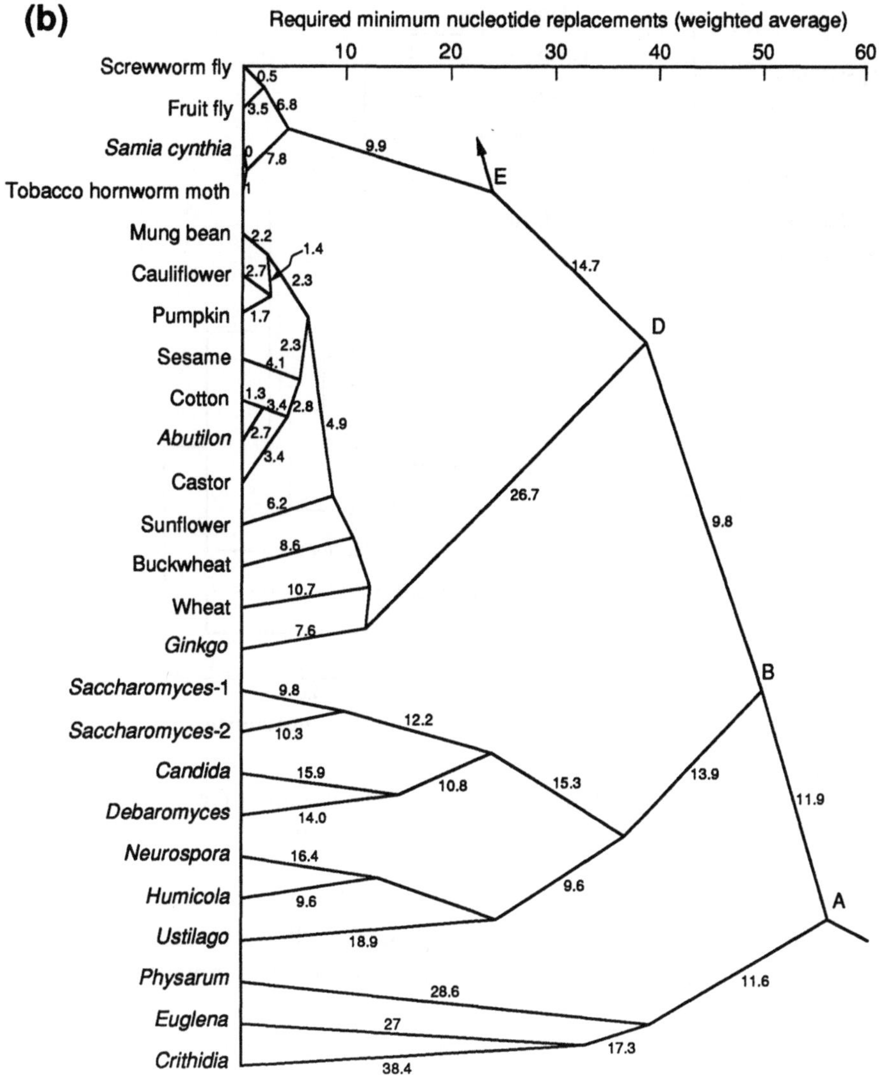

molecular trees of Boulter and co-workers to reflect reliably the basic monocotyledon-dicotyledon division of the Angiospermae and he proposed that 9 further substitutions to the group of 68 would achieve a pattern of relationship acceptable to the classical view. Some taxonomists feel that the application of the objective numerical approach to sequence data misses the opportunity to study the individual character changes and conservation which may be useful in defining taxa. For example residues 2, 5 and 8 in cytochrome *c* are constant in the vertebrates but vary in the insects; 7, 21, 55,

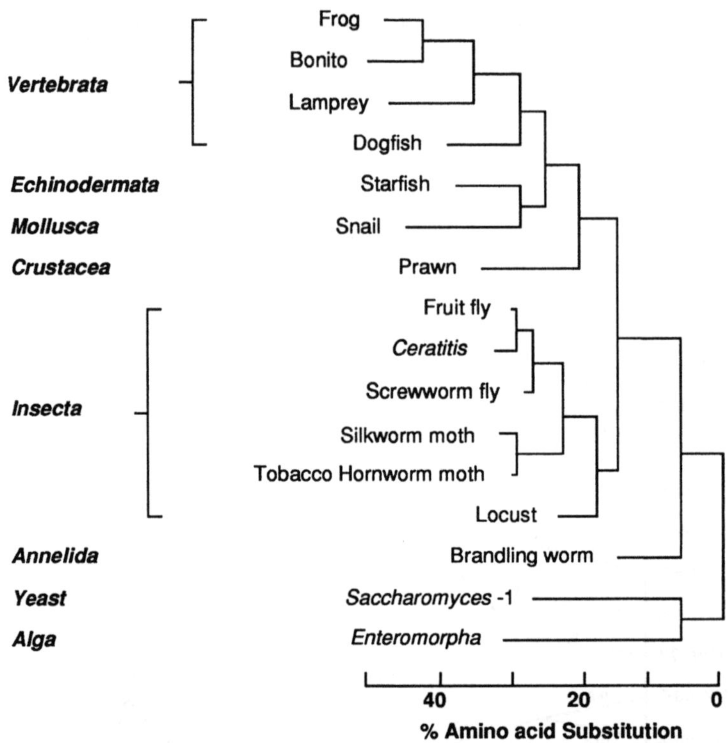

Fig. 6.5 Cytochromes *c* of the invertebrates. Relationships were calculated using a modified ancestral sequence method. (After Lyddiatt *et al.*, 1978)

69 and 90 are constant in animals but vary in the Angiospermae (Crowson, 1972). Rather than using the ancestral sequence method objectively, Crowson would prefer to guide the choice of ancestral codons by consideration of the classical tree.

Our view is that this *a posteriori* use of sequence data and molecular trees is not in the spirit of the original aim. In this modified form, the molecular data no longer stand as an independent body of evidence; if they fit the pre-existing scheme they are acceptable, if not they are discarded or adjusted. We believe a more scientifically useful role would be the simple presentation of the uncorrected data. Within that data there will be anomalies due to defects in our ability to map historical molecular events and others due to defects in the classical tree. This form of presentation will encourage independent tests to resolve the anomalies rather than neutralise them. One such test is the congruency or otherwise of two molecular trees for the same set of species.

Multiple, Back and Parallel (Convergent) Changes. Phyletic distance is the total change that occurred in bringing about the observed differences between taxa. For close relatives, the observed changes are a good estimate of phyletic distance but this estimate deteriorates as the divergence deepens due to the undetected presence of multiple mutations. For this reason dendrograms based on simple matrices are foreshortened due to the underestimate of remote internodal distances. Several corrections to allow for these undetected changes have been proposed and although none have a strict theoretical foundation they all give rise to distances that are reasonably linear with time. An example is the augmentation correction of Moore *et al.* (1976) which is based on the assumption that the most populated parts of a tree allow detection of most mutational events and their analysis permits correction (augmentation) of the sparser regions. The augmentation correction can be massive. In the cytochrome *c* tree of Moore *et al.* (1976) there are 521 observed nucleotide replacements but 1047 after augmentation. A further illustration of the extent to which observed change underestimates phyletic distance comes from studies of superoxide dismutase. Only 26% of the substitutions predicted from closely related sequences are observed at the divergence level of the mammalian and yeast enzymes (Ayala, 1986).

The accuracy of a tree depends on how much convergent change has occurred. This was elegantly demonstrated by Peacock and Boulter (1975) who simulated a pattern of substitutions in a set of sequences diverging from a single start. All changes (including convergent, back and multiple mutations) were known and these were compared with those deduced by the ancestral sequence method working from the tips of the simulated tree.

The simulation showed that undetected convergent changes were the main source of error in the the tree construction by the ancestral sequence method and that this problem increased with the depth of the tree. In the actual plant cytochrome *c* tree there are 33% detected convergent changes (27% parallel and 6% back mutations) but there is no way of knowing the extent of undetected convergent change. In the case of the minimum cytochrome *c* tree of Baba *et al.* (1981), the placement of the prawn on a lineage diverging to the horse and donkey was proposed to be result of an unusual number of convergent changes.

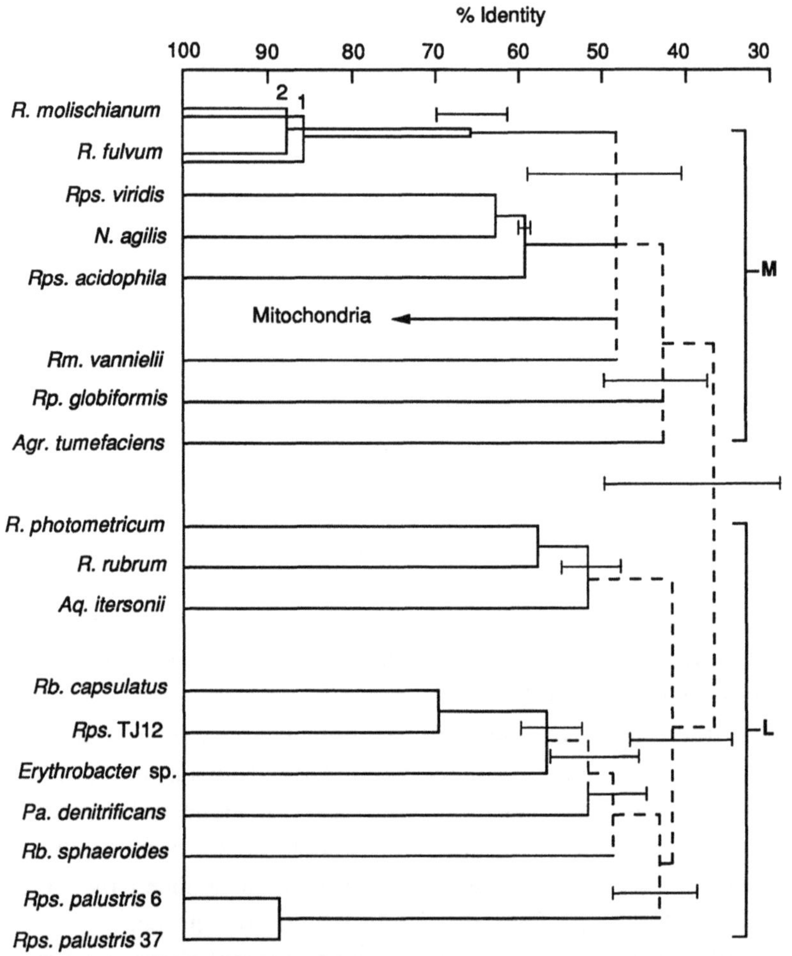

Fig. 6.6 Dendrogram of the cytochrome c_2 sequences. The dendrogram is based on the matrix of sequence identities (Table 3.4) using the simple averaged matrix method described in Fig. 6.2a. The range of % identities at a node is indicated by a bar. The solid lines are unambiguous patterns of relationship in the matrix. For example, the cytochromes c_2 of *Rps. viridis* and *N. agilis* are 63% identical and each is 59 - 60% identical to *Rps. acidophila* cytochrome c_2. All other pairwise comparisons are more remote. Ambiguous regions of the dendrogram are shown as broken lines. A striking example is the case of *Paracoccus denitrificans* cytochrome c_2 which is 54 - 57% identical to the *Rb. capsulatus/Rhodopseudomonas* TJ12 pair, but only 46% identical to the *Erythrobacter* sequence which is a member of the same group. The double pattern connecting *R. molischianum* and *R. fulvum* reflects the existence of two isocytochromes c_2 in each organism. One fork at 87% is a comparison of the iso-1 cytochromes c_2 while the other, at the same value, is a comparison of the iso-2 cytochromes. This pair of forks represents the species divergence. The gene duplication event is shown by the more remote fork at 66% identity. The groups of cytochrome c_2 of large and medium size are denoted by **L** and **M** respectively. Mitochondrial cytochromes *c* show remote affinity for

Saturation. Related to the problem of multiple changes is the problem of saturation (Meyer *et al.*, 1986). As evolutionary divergence between sequences deepens, the proportion of multiple, back and convergent changes will increase until a point may be reached where further evolutionary change is not reflected in an increase in observed sequence changes and all divergences that have reached that limit will appear equal in depth. Meyer *et al.* (1986) argue that the saturation limit will vary with different proteins depending on functional and structural constraints. Valid trees can only be built using a sequence family with conserved function and less sequence variation than the limit for that protein.

An illustration of the problem is the cytochrome c_2 dendrogram (Fig. 6.6) and its relationship to that of the mitochondrial cytochromes *c*. Many of the comparisons of cytochromes c_2 with each other and with mitochondrial cytochromes *c* fall within the region 35 - 50% identity. This could mean either that these sequences diverged at a similarly remote time or that saturation has been reached. The fact that the level of identity at the *Rb. sphaeroides*/*Pa. denitrificans* branching point in the cytochrome c_2 tree is much less than the same divergence in the cytochrome c_1 tree (Fig. 6.7) tends to support the second possibility. Also, the broken lines that appear as dominant features at the base of the cytochrome c_2 tree indicate uncertainty in assigning branch points and this is consistent with a region of change in which undetected substitutions are becoming dominant.

Thus the problem of saturation should be borne in mind when constructing trees with deep divergences. In other respects, however, we find the analysis of Meyer *et al.* unduly pessimistic. They conclude that all macromolecules will have reached saturation at divergences more remote than those of the Rhodospirillaceae and the mitochondria, and therefore that a molecular phylogeny of the bacteria is not a realistic goal. Yet the tree of rRNA sequences does not show saturation at this level (Fox *et al.*, 1980; Fig. 6.8). Meyer *et al.* (1986) propose that a plateau of variability is nevertheless present for rRNA in the Rhodospirillaceae at a level compatible with aspects of function not yet understood. rRNA is therefore seen as having evolved functionally between different plateaux of variablity. While this proposal may

---

the **M** group of c_2 sequences. % identities fall within the range 41 - 56% for horse cytochrome *c*. Within this group, horse cytochrome *c* is most similar to *Rps. viridis* c_2 (56%) and *Rm. vannielii* c_2 (52%) and the mitochondrial divergence is therefore placed at the node from which these two cytochromes descend be true, there is as yet no evidence to support it and it remains an arbitrary assumption. It also becomes entangled in the knotty question of whether functional change does not, in any case, occur within families. We will tackle this problem more fully in section 6.5.4.

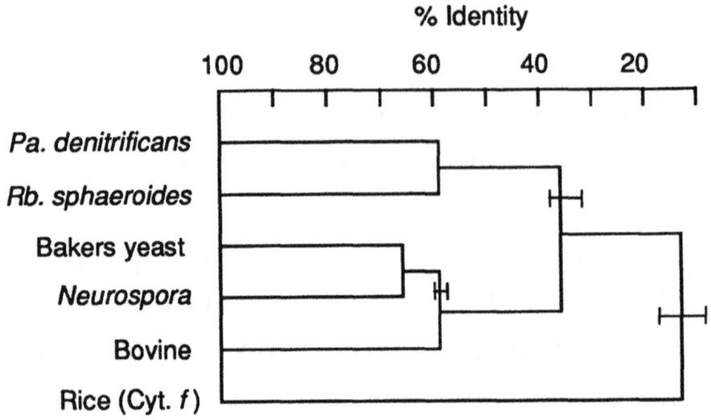

Fig. 6.7 Dendrogram of the cytochromes c_1 (and rice cytochrome *f*) sequences. The dendrogram is based on the matrix of sequence identities in pairwise comparisons (Table 3.2) and was constructed as described in Fig. 6.2a. The range of identities at a node is indicated by a bar. The eukaryotic cytochromes c_1 form a clear group, with the yeast and *Neurospora* cytochromes more closely related than either is to bovine. The bacterial cytochromes c_1 show a strong similarity. The eukaryotic/prokaryotic divergence is considerably more remote with no suggestion of a special relationship between *Paracoccus* and the eukaryotes. The divergence to rice cytochrome *f* is very remote and this could reflect either a truly distant divergence in time or a rapid change along the cytochrome *f* line of descent

Equally Parsimonious Solutions and the Pattern of Branching. Sometimes several trees are almost equally parsimonious. This happens commonly in poorly populated regions of the dendrogram where branch lengths are long and internodal distances are short (Peacock and Boulter, 1975; Demoulin, 1979). In the cytochrome *c* tree this pattern occurs at the divergence of the eukaryotic kingdoms (Fig. 6.4), at the divergence of the Angiospermae (Fig. 6.4) and at the divergence of the invertebrates (Fig. 6.5). In our dendrograms of the cytochrome *c* families (e.g. Figs. 6.2a and 6.6) we indicate these doubtful areas of branching by broken lines. These occur when there is overlap in the % identity range of the two adjacent nodes, these ranges being given as bars on the dendrograms.

The Rate of Change. Initial observations that proteins change at a roughly constant rate have not withstood detailed analysis and it is now clear that very different rates of change can exist within a molecular tree. This important

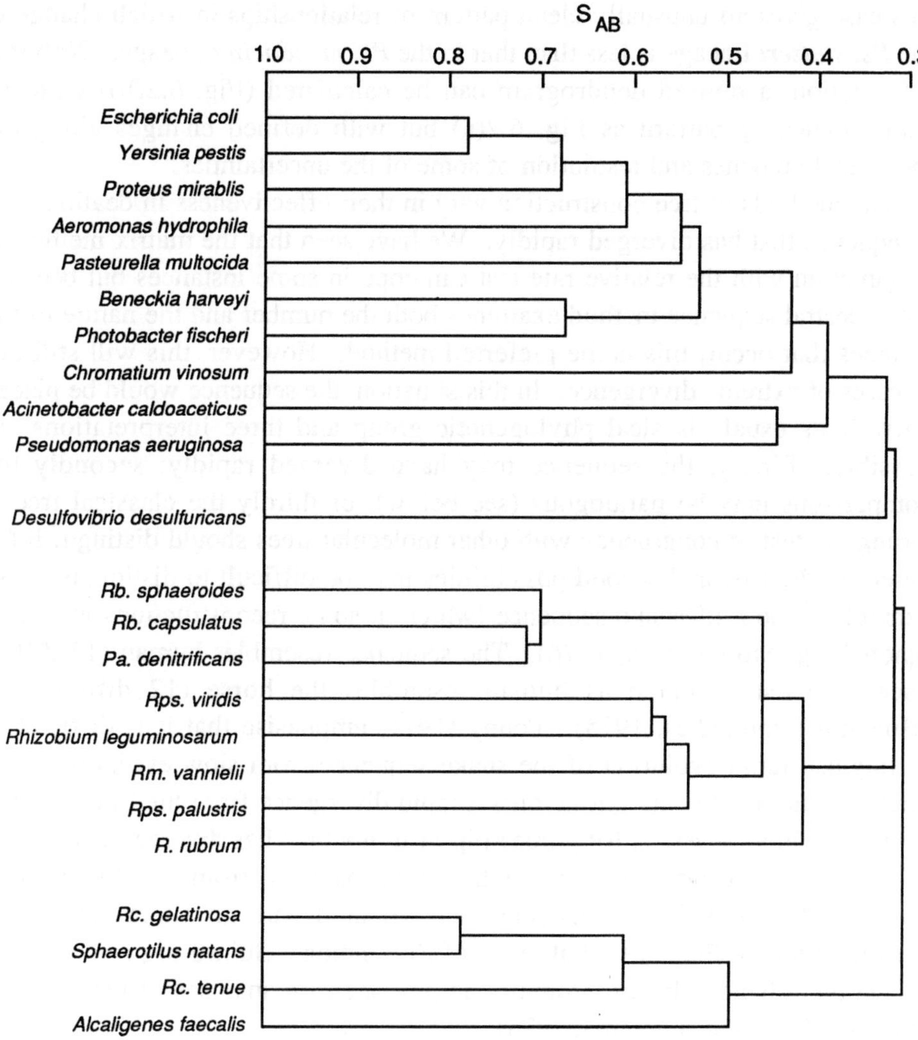

Fig. 6.8 The phylogenetic relationships among the purple photosynthetic bacteria and their relatives. The dendrogram is based on T1 oligonucleotide catalogues of 16S rRNA as described in section 6.2. (After Fox *et al.*, 1980)

area is of relevance to the mechanism of change in proteins and will be considered fully in the following section.

Differences in the rate of change in divergent lines of descent can be detected by the relative rate test (Fitch and Margoliash, 1967) which is applied to the cytochrome *c*-551 sequences in Fig. 6.2(b). The matrix of identities in this case gives an unusually clean pattern of relationships in which change in the *Ps. stutzeri* lineage is less than that in the *Ps. mendocina* lineage. With that information, a revised dendrogram can be calculated (Fig. 6.2(b)) with the same branching pattern as Fig. 6.2(a) but with defined changes along the individual branches and resolution of some of the uncertainties.

The methods of tree construction vary in their effectiveness in dealing with a sequence that has diverged rapidly. We have seen that the matrix method in conjunction with the relative rate test can cope in some instances but because the ancestral sequence method examines both the number and the nature of the changes that occur, this is the preferred method. However, this will still fail in cases of extreme divergence. In this situation, the sequence would be placed outwith its usual classical phylogenetic group and three interpretations are possible. Firstly, the sequence may have diverged rapidly; secondly the comparisons may be paralogous (see below); or thirdly the classical tree is wrong. A test of congruency with other molecular trees should distinguish the latter but the first and second possibilities may be difficult to distinguish. An example is the rattlesnake sequence (which in some reconstructions is simply ignored e.g. Moore *et al.*, 1976). The sequence resembles human (13 differences) almost as much as human resembles the horse (12 differences) (Holmquist and Jukes, 1975). Penny (1974) emphasise that it is difficult to distinguish rapid evolution of the snake sequence from slow evolution of its phylogenetic neighbours but favours a rapid divergence from the bird descent. A similar situation exists for guinea pig insulin which has diverged from other insulins to an extent much greater than that expected from its phylogenetic position (Smith, 1972). This may be correlated with an altered structural constraint due to the unusual absence of Zn binding. Guinea pig ribonuclease on the other hand behaves 'normally' in phylogenetic analyses of that enzyme (Van den Berg and Beintema, 1975).

Paralogy. Orthologous genes are direct descendants of a gene present in the most remote common ancestor of a tree whereas paralogous genes are the products of gene duplication which may have taken place within the tree or prior to the base of the tree. A phylogenetic tree of orthologous genes is a tree of divergence of the species of which they are a part. On the other hand a phylogenetic tree of paralogous genes gives information on the remoteness of the duplication event. The presence of paralogous sequences in an otherwise orthologous tree introduces aspects of gene as well as species divergence. Wilson *et al.* (1977) use the example of the globins to illustrate how confusing it would be to attempt construction of a mammalian tree using myoglobin from some species and hemoglobin α or β chains from others.

Problems may arise when paralogy is not recognised. Thus the original bird lysozyme sequence studies showed the duck and chick enzymes as more similar to each other than either was to the goose enzyme in contradiciton to classical avian phylogeny (Wilson *et al.*, 1977). However the swan was found to have two lysozymes one resembling the goose and the other the duck and it therefore seems likely that the original duck - goose comparison was for paralogous proteins and the divergence reflected the gene duplication event.

No case of paralogous comparisons has been definitely identified in the mitochondrial cytochrome *c* tree but the potential for such confusion certainly exists. Thus two functional gene products have been identified in yeast, *Drosophila*, house fly, mouse and rat (Pettigrew & Moore, 1987: Chapter 4; this volume: section 6.6.2). It is possible that an oddity like the rattlesnake cytochrome *c* sequence is the product of a paralogous gene.

The degree of disruption a paralogous gene causes to a tree, depends on its position within the branching pattern. For example, a gene duplication event in an unbranched lineage (see e.g. Fig. 6.6) would have no anomalous effect since its ancestry is shared by all comparisons.

6.5 Neutral Mutation and Natural Selection in the Evolution of Cytochrome *c*

6.5.1 The Molecular Clock and the Theory of Random Genetic Drift.

In 1962, Zuckerkandl and Pauling proposed the revolutionary idea that an individual protein changed at a constant rate on the palaeontological time scale and that the extent of change was therefore a measure of elapsed time. The proposal implied a radical departure from Darwinian evolution. This is illustrated by the case of a 'living fossil', such as the coelocanth, which has remained unchanged morphologically across millions of years but the proteins of which are predicted to have changed at the same rate as those of other species.

An apparent constancy of rate for individual proteins was established by plots of the extent of change in sequence against the time elapsed from the palaeontological divergence of the species to the present (Dickerson, 1971).

Table 6.1. The rate of change in proteins (Wilson *et al.*, 1977)

protein	UEP (MY)
Histone H4	400
Collagen	36
Glucagon	43
Cytochrome *c*	15
Insulin	14
Hemoglobin α	3.7
Fibrinopeptide	1.7
Insulin C-peptide	1.9

Rates for different proteins are shown in Table 6.1 (Wilson *et al.*, 1977) expressed in terms of the time in millions of years required for 1% difference in sequence (the Unit Evolutionary Period - UEP).

This observation of rate constancy and the finding of massive genetic polymorphism in populations (see e.g. Dobzhansky *et al.*, 1977) were the two main sources of the theory of Neutral Mutation and random genetic drift (King and Jukes, 1969; Kimura, 1969). According to this theory, evolutionary change is a stochastic process based on accidents in sampling the total potential gene pool in populations. A relative of the theory is the Founder principle which involves physical separation (for example, by geographical isolation) of a small portion of the population during the formation of a new species.

Kimura (1987) showed (Scheme 6.1, section a) that the rate of fixation of neutral mutation (k_g) is equal to the rate of mutation itself (v_n). If the latter were constant and neutral mutations were the dominant source of change in proteins, then proteins would change at a constant rate. In contrast, the equation for fixation of selectively advantageous alleles (Scheme 6.1, Eqn. 6.6) contains the factors - population size (N) and selection coefficient (s). Balancing of these variables to produce a constant rate seems highly improbable.

The different rates of change observed in different proteins are due to different functional constraints. For example, cytochrome *c* carries out a complex electron transfer between two redox partners, themselves proteins. Therefore unworkable molecules will be produced frequently and will be eliminated by natural selection. In contrast, the fibrinopeptides are parts of the fibrinogen molecule that are released proteolytically during the activation process and probably have no further role. That lack of functional constraint gives rise to rapid evolutionary change.

Scheme 6.1. The fixation of neutral mutations

A neutral allele is one for which the difference in fitness is so small that its frequency changes more due to accidents of sampling than natural selection.

(a) Chance of appearance of neutral mutation = $2Nv_n$
(where v_n is the rate of neutral mutation per gamete per generation and N is the number of breeding individuals)
Chance of random fixation of neutral mutation = 1/2N
Therefore rate of fixation of neutral mutation (k_g) = $2Nv_n/2N = v_n$ (6.3)

(b) Neutral theory states that a certain fraction (f_n) of total mutations are neutral (i.e. (1-f_n) are deleterious).
Therefore $k_g = f_n v_t$ (where v_t is the overall mutation rate) (6.4)
The rate of fixation of neutral mutations per year (k_n) = $f_n v_t/g$ (6.5)
(where g is generation time).

(c) The corresponding equation for selectively advantageous alleles is:
$k_a = 4Nsf_a v_t/g$ (6.6)
(where s is the selection coefficient and f_a is the fraction of advantageous mutations).

Thus a protein family is seen as a set of functionally equivalent sequences which are interconvertible by neutral mutation with natural selection eliminating those that fall outside the set. In choosing a protein to study phylogeny, the rate of change must be fast enough to allow resolution of the fine structure at the tips of the tree but slow enough to avoid confusion of remote evolutionary patterns by the noise of back, multiple and parallel mutations. Thus cytochrome *c* is a good choice for studying the broad sweep of eukaryotic evolution but fibrinopeptides are better for the detailed mapping of evolutionary change in the mammals.

6.5.2 Covarions

Since about 75% of cytochrome *c* residues have been observed to change during eukaryotic evolution, a simple stochastic model would involve random hits within this large set (e.g. Holmquist, 1972). However Fitch (1976) found that the constraint on the cytochrome *c* molecule is much greater than implied by this simple model. He used the method of Fitch and Markowitz (1970) to estimate the fraction of the gene that is theoretically invariable at a particular node of the tree. A plot of that invariable fraction against species range is shown in Fig. 6.9. Fitch concluded that in an individual species about 90% of the gene is invariable, the remainder being the ten or so **concomittantly variable codons** or **covarions**. Thus although 48 positions have accommodated 5 or more amino acids this does not mean that they have no structural or functional significance (Ferguson-Miller *et al.*, 1979). In an individual species most of these positions will be unable to accept change. Rapidly changing molecules like the fibrinopeptides contain a larger proportion of covarions. When the rate of change per covarion is calculated for different proteins similar values are obtained (Table 6.2) in agreement with neutral mutation theory.

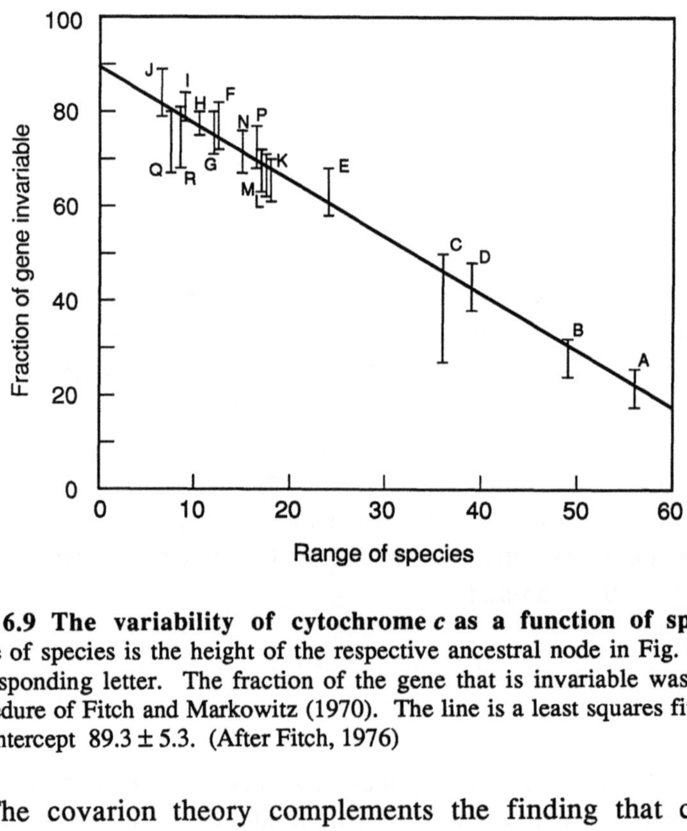

Fig. 6.9 The variability of cytochrome *c* as a function of species range. The range of species is the height of the respective ancestral node in Fig. 6.3, labelled with a corresponding letter. The fraction of the gene that is invariable was determined by the procedure of Fitch and Markowitz (1970). The line is a least squares fit with slope -0.0123 and intercept 89.3 ± 5.3. (After Fitch, 1976)

The covarion theory complements the finding that change within a phylogenetic group is clustered in the 3-dimensional structure (Margoliash *et al.*, 1972). For example, Lyddiatt *et al.* (1978) found that the pattern of variable residues in cytochrome *c* within an invertebrate taxon differed from the pattern of another taxon. Fitch and Markowitz (1970) found that change at one covarion could cause both creation of new covarions and removal of old.

This is quite close to the view of Zuckerkandl (1976) that a protein cannot optimise all key properties at once but occupies slightly suboptimal states which differ in patterns of amino acid substitution. It also resembles the balance model for diploid population structure in which fitness is due to combinations of alleles and several combinations confer similar levels of fitness (Dobzhansky *et al.*, 1977).

Table 6.2 The rate of change in proteins (Fitch 1976)

Pig-horse divergence	Fixations	Codons	$Rate_o$	Covarions	$Rate_c$(i)	$Rate_c$(ii)
Cytochrome *c*	5	104	0.05	12	0.42	3.3
Fibrinopeptides	10	19	0.53	18	0.56	4.4

$Rate_o$ is the overall rate of change per codon. $Rate_c$(i) is the rate of change per covarion. $Rate_c$(ii) is the rate of change per covarion per 10^9 years in one line of descent. The pig-horse divergence is taken to be 64 MY.

This model of a small set of variable codons which shifts its membership across the phylogenetic tree is markedly different from the simple model of random hits on a large fixed codon set but it can be accommodated in the theory of random genetic drift (Scheme 6.1). Eqn. 6.4 shows that the rate of neutral mutation can be expressed as a fraction (f_n) of the total mutation rate. f_n can be considered to be the proportion of covarions in the molecule.

6.5.3 The Dimension of Time in Neutral Mutation Theory.

There remains uncertainty as to whether the molecular clock operates on calendar time or generational time. The former is favoured by the results of the relative rate test at nodes from which short and long generation species diverge (Wilson *et al.*, 1977). A dimension of calendar time is also favoured by the inability of Ochman and Wilson (1987) to detect a significant difference in the rate of change in bacterial rRNA and its counterpart in higher organisms. On the other hand some evidence suggests a generation time effect exists (Wu and Li, 1985).

The distinction between the two is an important one because it influences our view of the origin and fixation of mutation in DNA. If the clock ran on calendar time it would suggest that gradual accumulation of mutational damage to the DNA occurred; if the time scale is measured in generations, then errors during DNA replication would be a dominant feature.

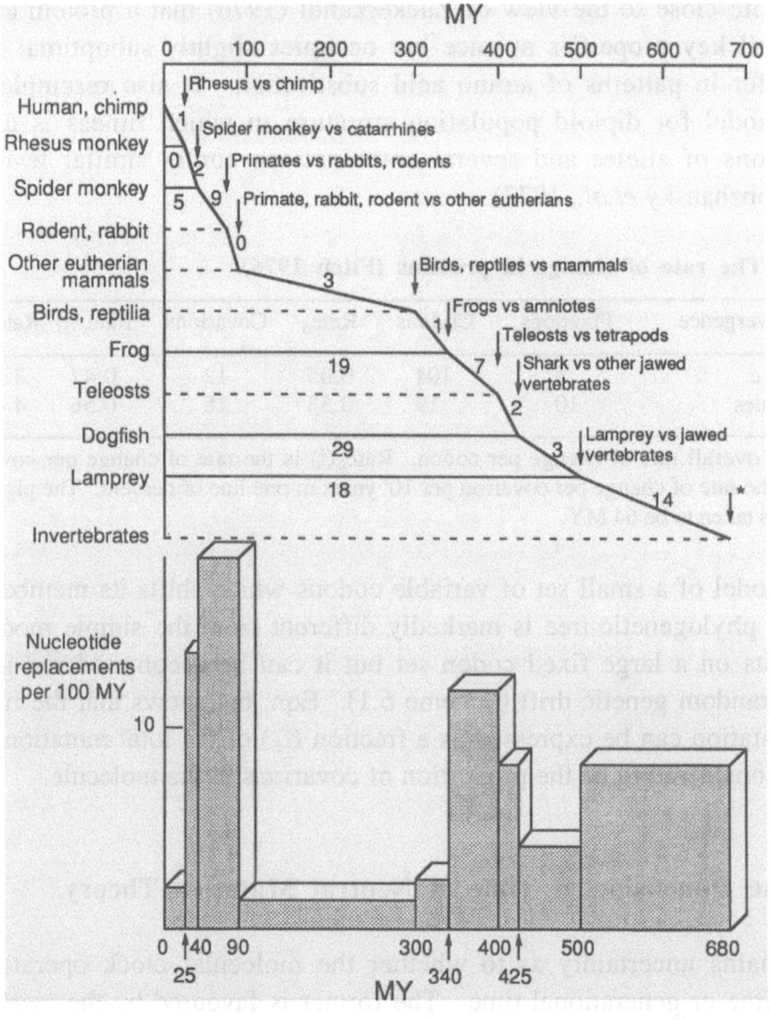

Fig. 6.10 The rate of change of cytochrome *c*. The top part of the diagram is a dendrogram obtained by the maximum parsimony method with augmentation adjustments to the number of nucleotide replacements detected (see text). The most parsimonious solution contained 1273 nucleotide replacements (NR) but violated some accepted phylogenetic relationships. A solution which satisfied these relationships contained 1303 NR and is shown here. The value of NR for each branch is shown. Some descents (e.g. teleosts) lead to several sequences with different nucleotide replacement distances. These figures are not shown. Regions of the tree which differed in branching pattern in the 1273 and the 1303 solutions are shown as broken lines. The horizontal time scale is based on palaeontological evidence (Young, 1962; Romer, 1966; McKenna, 1969). The bottom part of the diagram is a representation of the rate of evolution of cytochrome c during the internodal intervals shown in the dendrogram. (Adapted from Baba *et al.*, 1981)

The problem is accommodated within neutral mutation theory as shown in Scheme 6.1. The generation time appears in the final equation for the rate of fixation of neutral mutations per year (Eqn. 6.5) and is a possible source of variation in observed rates (see below).

6.5.4 Changes in the Rate of Evolution - Are They Due to Natural Selection or Neutral Mutation?

A stochastic clock would be expected to be accurate over long periods but erratic over short. To some extent this is observed in cytochrome *c* with main divergences correlating quite well with the fossil record but with fine details showing variable rates. However the variation in rate is 2 - 3 times greater than expected for a simple stochastic process (Fitch and Langley, 1976; Ohta and Kimura, 1971). Coates and Stone (1981) devised a computer simulation to test the degree to which neutral mutation of an ancestral sequence through an accepted phylogeny could give rise to the observed extent of change in the cytochrome *c* molecule. If a random hit produces a synonomous codon or a new allowed codon it is fixed. The matrix for the simulated sequences at the branch tips is in fair agreement with the actual matrix, and the fit is improved if the set of allowed amino acids is altered during the simulation. The authors conclude that the evolution of cytochrome *c* has largely been by random fixation of neutral mutation punctuated with periods of directional selection and adaptation, during which the covarion group may alter in size and composition.

The detailed analysis by Baba *et al.* (1981) of their phylogenetic tree tends to support this model. They find a wide range of rates depending on the evolutionary region of the tree (Fig. 6.10). It should be emphasised that divergence times based on the fossil record are often subject to much uncertainty but this could hardly be the sole cause of the striking differences that are observed. In general, the bursts of change coincide with similar bursts in other proteins and occur at times of rapid speciation into new ecological zones exemplified by the basal tetrapod and early eutherian radiations (Baba *et al.*, 1981). The evolutionary lineage of the primates leading to the anthropoid ancestor involved rapid change and was followed by a much slower rate in the human line of descent. Thus patterns are consistent with adaptive change driven by natural selection, followed by stabilising selection once a new function has been established.

Is there any evidence to suggest that cytochrome *c* has evolved in function in the way suggested by these studies? This is a problematic area that may not be amenable to definitive experimental investigation. In early studies, cytochromes *c* were shown to be identical in their reactivity with cytochrome *c* oxidase (Smith *et al.*, 1973) but closer examination revealed differences (Margoliash *et al.*, 1976b; Pettigrew & Moore, 1987: Chapter 2.2). However it is difficult to assess the significance of non-equivalence in the *in vitro* test just as it is difficult to see how functional equivalence could be exhaustively proven.

Although the results should therefore be viewed with caution, primate cytochrome *c* is very poorly reactive with beef cytochrome *c* oxidase yet highly reactive with the primate enzyme (Osheroff *et al.*, 1983). Baba *et al.* (1981) showed that during the rapid rate of change in the primate descent over a period of approximately 35 million years (MY) there were four changes in the surface of cytochrome *c* involved in interaction with cytochrome *c* oxidase. This is to be contrasted with the zero change observed in the same region during 425 MY of evolution between the arthropod-vertebrate divergence and the primate-lagomorph origin.

In conclusion, cytochrome *c* evolution may incorporate both neutral mutation within local regions of the tree, limited by the nature and size of the covarion group, and adaptive change with subtle modification of function and involving alteration of the covarion group.

6.6 Gene Duplication, Gene Fusion and Pseudogenes

6.6.1 Gene Duplication in Mitochondrial Cytochromes *c*

After a gene duplication event, one of the copies may remain under the original genetic control and functional constraint while the other may be open to structural change and functional adaptation (Wilson *et al.*, 1977). Gene doubling to produce isocytochromes *c* has occurred in bakers yeast, house fly, rat and mouse. In the two mammalian species, one of the isocytochromes is found in germinal tissue (Pettigrew & Moore, 1987: Chapter 4). Two cytochrome *c* genes (but so far only one of the corresponding proteins) have been detected in Drosophila. They differ in 32 out of 104 encoded amino acids. The relative rate test (Fig. 6.2) indicates that the DC3 gene, which is expressed

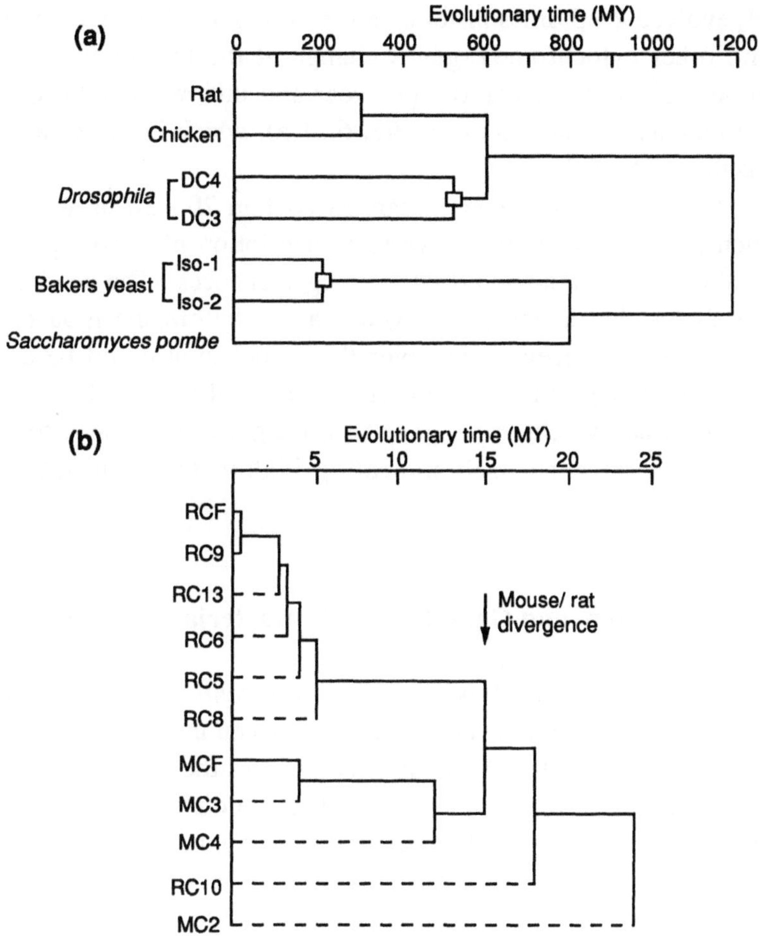

Fig. 6.11 Phylogeny of functional genes and pseudogenes for mitochondrial cytochrome c. **(a)** The analysis is based on nucleotide substitutions at nonsynonomous sites in the gene sequences. The rate of change in the rat, chick and DC4 lineages is 0.08 - 0.1 x 10^{-9} nonsynonomous substitutions per site per year based on a 300 MY divergence for the chick and the rat and a 600 MY divergence for the vertebrates and insects. DC3 has evolved 2 - 3 times faster than DC4 based on the relative rate test (Fitch and Margoliash, 1967) with the *Saccharomyces pombe* gene as the outlying sequence. Points of gene duplication of approximately 520 MY for DC3/DC4 and 200 MY for yeast iso-1/iso-2 can be estimated. **(b)** There are no nonsynonomous differences between the functional rat (RCF) and mouse (MCF) genes. The dendrogram is based on synonomous substitutions in the RCF/MCF lineages and the pseudogene descents. If a pseudogene arose before the mouse/rat divergence, then it should be equally distant from both and the distance should be greater than between the RCF and MCF genes. The time scale is based on a pseudogene rate of change of 4.9 x 10^{-9} substitutions per site per year (Li *et al.*, 1985). (After Wu *et al.*, 1986b)

at a low level, evolved 2 - 3 times faster than the DC4 gene (Wu *et al.*, 1986b). The latter and other cytochrome *c* genes change at the rate of 0.09 x 10^{-9} non-synonomous substitutions per site per year and this rate can be used to calculate the time-based dendrogram of Fig. 6.11(a), which indicates a yeast-animal divergence of 1.17 BY.

In addition to the isocytochrome, mammals contain 20 - 30 pseudogenes, many of which seem to arise from reverse transcription of messengers and therefore contain a poly A tail and no intervening sequences. The dendrogram of Fig. 6.11b shows that a few pseudogenes arose before the mouse - rat divergence but most are recent. However this conclusion could be a consequence of the stringency of the probe used. Thus, of the eight clones cross-reacting with the yeast cytochrome *c* gene in a rat genomic library, four did not hybridise to the rat gene itself indicating highly divergent sequences (Sylvanen, 1987).

6.6.2 Gene Duplication and Gene Fusion in Bacterial Cytochrome *c*

Isocytochromes c_2 are found in *Rhodospirillum molischianum* and *R. fulvum* (Fig. 6.6). The gene duplication event occurred between the divergence of the two species from the rest of the Rhodospirillaceae and their divergence from each other. Application of the relative rate test to the distance matrix (Table 3.4) suggests that 2 - 3 more substitutions occurred in the iso-1 lineage compared to the iso-2. No functional distinctions have yet been made although the iso-1 cytochromes have more positive midpoint potentials (Pettigrew & Moore, 1987: Chapter 3.6.2) and greater stability to alkaline pH (Table 4.5).

A much more remote gene duplication is observed in the class II cytochromes (cytochromes c_8) (Fig. 6.12). This group contains both high and low spin members (Chapter 5.2) and both types are found in *Rhodopseudomonas palustris* strain 2.1.37. Taken at face value, the dendrogram indicates that the gene duplication took place in the *Rps. palustris* lineage from a low spin ancestor and therefore that the high spin character evolved twice.

Some of the different subclasses of class I cytochromes *c* may have arisen by remote gene duplication events. Cytochromes c_4, c_5, *c*-551 (c_7) and the class I domains of cytochrome *c* peroxidase and cytochrome cd_1 for example, are commonly found in the same organisms. However, these events are so remote that they are now impossible to distinguish from lateral gene transfers (section 6.6) or even convergence from separate origins (section 6.10.2).

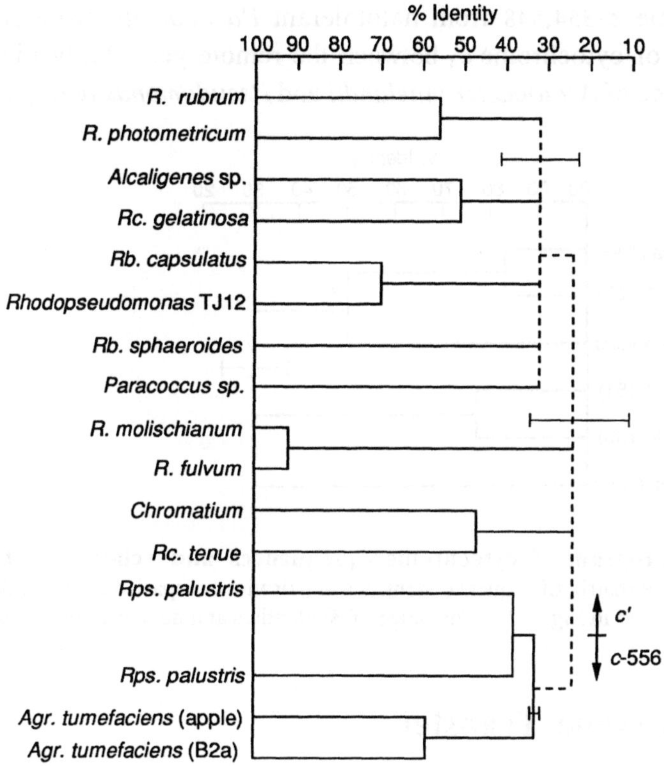

Fig. 6.12 Dendrogram of the class II cytochromes *c* (cytochrome c_8) sequences. The dendrogram is based on the matrix of sequence identities in pairwise comparisons (Table 3.9) and constructed using the method described in Fig. 6.2a. The range of % identities at a node is indicated by a bar. The solid lines indicate unambiguous patterns of relationship in the matrix. For example, cytochrome *c*-556 from the two strains of *Agrobacterium tumefaciens* are 59% identical and each is 32 - 34% identical to the *Rps. palustris* cytochrome *c*-556. All other pairwise comparisons are more remote. Ambiguous regions of the dendrogram are shown as broken lines. For example, the connectivity within the top group of sequences cannot be established. This is illustrated by the *Rb. sphaeroides* sequence which can be placed closest to the *Rb. capsulatus/Rps.* TJ12 nucleus in one pattern, closest to the *Alcaligenes sp./Rc. gelatinosa* nucleus in a second pattern, or distant from both these pairs in a pattern in which the nucleus is provided by the *R. rubrum* and *R. photometricum* sequences

Gene duplication accompanied by gene fusion has probably occurred in the diheme cytochrome c_4 (Fig. 6.13). The two halves of the molecule are class I structures and show a remote relationship. Gene fusion may have accompanied gene duplication or have occurred after divergence of the monoheme cytochromes. The dendrogram is consistent with the latter since the monoheme cytochrome *c*-554,548 from halotolerant *Paracoccus* arose from the N-terminal half of cytochrome c_4 between the remote gene duplication event and the divergence of *Azotobacter vinelandii* and *Pseudomonas aeruginosa.*

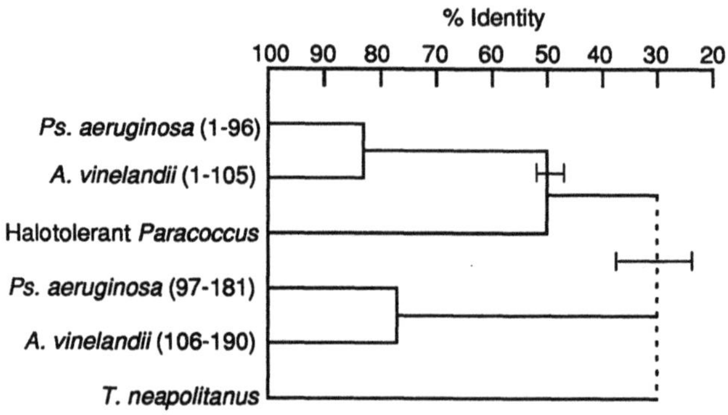

Fig. 6.13 A dendrogram of cytochrome c_4 sequences and relatives. The dendrogram is based on the matrix of sequence identities in pairwise comparisons (Table 3.6) and is constructed as described in Fig. 6.2a. The range of % identites at a node is indicated by a bar

6.7 Lateral Gene Transfer

In this section we discuss the part played in molecular evolution by gene transfer between species. Mechanisms for gene transfer are well-known in bacteria and include transformation involving extrachromosomal DNA (plasmids), transduction by bacteriophages and conjugation. These mechanisms are not available to all bacteria and all three are much more common within species or between closely related species than across deeper divisions. At the same time, the evolutionary time scale is great enough for even rare events to play a significant role.

A pessimistic conclusion is that lateral gene transfer may have complicated bacterial phylogeny to a degree which makes it impossible to decide patterns of species divergence. According to this view, prokaryotic evolution is highly reticulate giving rise to species which are evolutionary chimaera. Even a single gene may have regions of separate evolutionary lineage due to intracistronic recombination.

A less extreme view would be that gene transfer has occurred but is constrained in its effect by two factors. Firstly, it is much more probable between closely related species. This would lead to a restricted divergence of a group but would not much affect its broad pattern of relationships. Secondly, for gene transfer to be selectively advantageous, the gene product must be able to function in the new cellular environment. A gene product which works by multiple interactions with others (for example, a cytochrome in an electron transport chain or an RNA molecule in the ribosome) is unlikely to be capable of integration and will be successfully transferred only very rarely. This problem may be overcome if blocks of functionally related genes were transferred. Such blocks do exist (for example the *E. coli UNC*operon encoding the subunits of the proton translocating ATPase (Walker *et al.*, 1984) and the *pet* operon encoding the bc_1 complex in *Rhodobacter capsulatus* (Davidson and Daldal, 1987)).

Some plausible instances of gene transfer have been described. The enzyme penicillinase does not require special integration into a new cellular environment in order to be effective in conferring resistance to the antibiotic and penicillinase can be transferred between bacteria via plasmids. Sylvanen (1987) argues that superoxide dismutase has a similar independence of action. *Photobacter leiognathi* contains a Cu-Zn SOD characteristic of eukaryotes in addition to its typical prokaryotic Fe SOD. This organism is symbiotic in the ponyfish and it is proposed that gene transfer from the host to the bacterium has occurred (Martin and Fridovich, 1981). Gene transfer across equally deep phylogenetic divisions has also been observed for *Agrobacterium* in the formation of the crown gall tumour in host plants (Drummond, 1979) and may have occurred in the organism *Progenitor cryptoides* which is found associated with certain human tumours and secretes human chorionic gonadotrophin (Cohen and Strampp, 1976).

In the cytochrome *c* area, it is ironic that Ambler, who has determined the great majority of sequences from prokaryotic sources, should also be the one most deeply pessimistic about their phylogenetic usefulness. His work has defined several instances of similar sequences in bacterial species considered on classical grounds to be unrelated. In particular, in the Rhodospirillaceae there is no clustering of cytochrome *c* sequences according to genus and there are several instances of close similarities between cytochrome *c* from photosynthetic and non photosynthetic sources (Ambler, 1978; Ambler *et al.*, 1981a). However, to accept these as evidence for lateral gene transfer requires the assumption that current bacterial classification is an accurate reflection of evolutionary relationships. Much evidence now suggests that this is not so and this is discussed in sections 6.8 and 6.10.

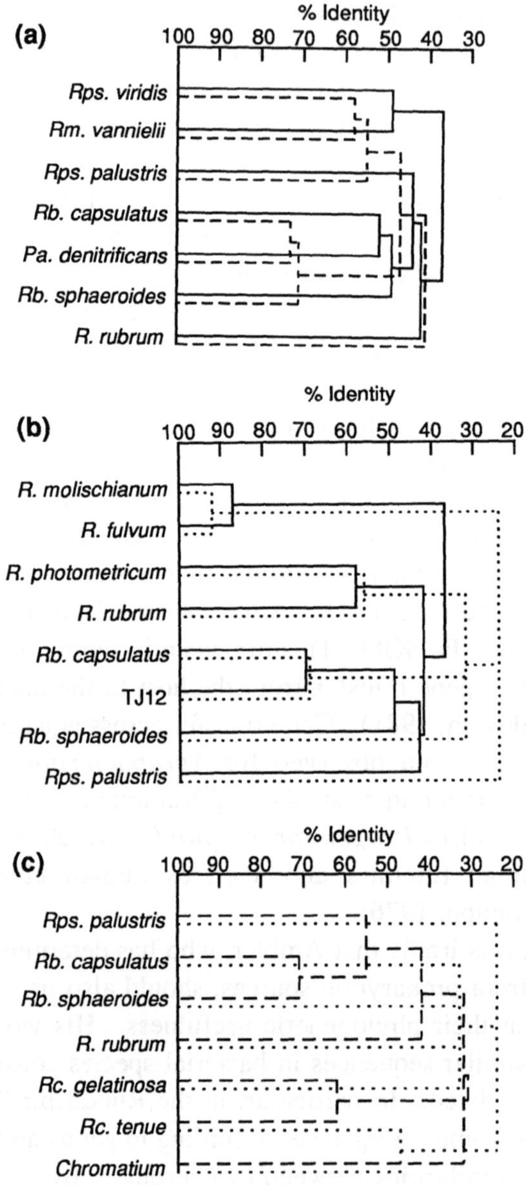

Fig. 6.14 The test of congruency for cytochrome c_2, cytochrome c' and 16S rRNA. Regions of Figs. 6.6, 6.8 and 6.12 were overlaid to test the congruence of dendrograms of cytochrome c_2, cytochrome c' and 16S rRNA sequences. **(a)** Cytochrome c_2 (—) and 16S rRNA (-----); **(b)** cytochrome c_2 (—) and cytochrome c' (······); **(c)** cytochrome c' (······) and 16S rRNA (-----). The nodes were taken directly from the full dendrograms without recalculation. They therefore represent the average depths for the full set of comparisons rather than for the selected sequences that appear here

The test of lateral transfer is the lack of phylogenetic congruence, preferably between genes with unrelated functions to minimise the possibility of linkage. This condition is met by the cytochromes c_2 and 16S rRNA which are compared in Fig. 6.14 using the group of organisms common to the individual dendrograms (Figs. 6.6 and 6.8). The dendrograms match reasonably well giving consistent grouping of *Rps. viridis* and *Rm. vannielii* and of *Rb. capsulatus*, *P. denitrificans* and *Rb. sphaeroides*. Only the position of *Rps. palustris* is ambiguous with an attachment to the *Rps. viridis* - *Rm. vannielii* pair in the case of the rRNA but not in the case of the cytochromes c_2.

Which (if either) is the more reliable dendrogram? Two arguments favour that for rRNA. Firstly cytochrome *c* may have undergone functional adaptation and periods of rapid change while the rRNA remained functionally conserved (Woese *et al.*, 1980). Secondly the cytochrome c_2 dendrogram shows signs of saturation at remote divergence while the rRNA dendrogram does not (at least for this species range) (section 6.4).

Much more severe incongruency is found however with cytochrome c' (Ambler *et al.*, 1979b, 1981). Although the matching of the cytochrome c_2 and cytochrome c' dendrogram is reasonably good (Fig. 6.14b) considering the uncertainty in the pattern of remote divergence in each (Figs. 6.6 and 6.12), the comparison of cytochrome c' with rRNA reveals one striking anomaly. In the rRNA dendogram *Rc. tenue* and *gelatinosa* are paired while in the cytochrome c' dendrogram the most recent divergence of *Rc. tenue* is from *Chromatium*. Meyer *et al.* (1986) consider this anomaly to be a consequence of the uncertainty in the dendrograms due to saturation at remote divergence but the divergence points at issue seem well short of those saturation levels and we regard the incongruency of the dendrograms as strong but circumstantial evidence for lateral gene transfer.

The function of cytochrome c' remains obscure (Pettigrew & Moore, 1987: Chapter 3) and it may be that it has a peripheral non-essential role that does not require integration with other proteins. This would be consistent with its surprising absence from just one strain of *Rps. palustris* (Ambler *et al.*, 1981). Thus it may be susceptible to lateral transfer and therefore unreliable in the mapping of species divergence.

In conclusion, there is some evidence that the caution of Ambler regarding lateral gene transfer is justifed. However, as we will discuss in the following section, many of his examples may be viewed as arising from the limitations of bacterial classification. The influence of lateral transfer on molecular phylogeny may be quite limited in scope and may be minimised by judicious choice of conserved and central macromolecular markers.

Gene Conversion. The lateral transfer of just part of a gene can be termed gene conversion and would involve intracistronic recombination. That such recombination is possible is shown by the reversion of a yeast mutant containing a nonsense codon at the position of Pro 71 in iso-1 cytochrome *c* (Ernst *et al.*, 1981). The reversion involves recombination with the non-allelic iso-2 cytochrome *c* gene leading to replacement of the segment 69 - 83.

Several instances of possible gene conversions have been identified. Mitochondrial cytochrome *c* genes show different rates of incorporation of silent third base substitutions in different regions of the molecule suggesting that the genes did not evolve as single units (Pettigrew & Moore, 1987: Chapter 4). Ambler (1980) has noted a strong similarity between the N-terminal region of *Rhodopila globiformis* cytochrome *c* and that of plants. *Rhodocyclus tenue* cytochrome *c*-553 is identical to that of *Pseudomonas aeruginosa* in the sequence 35 - 42, a region in which other Pseudomonad cytochromes *c*-551 differ (Ambler, 1978). The significance of these scattered examples is hard to assess. No systematic search for possible gene conversions has been performed and the congruency test may often fail to detect them if only small parts of the gene are involved.

6.8 The Concepts of Genus and Species in Bacteria

In sexual organisms the species is defined as a population of interbreeding individuals which are reproductively isolated. The process of formation of new species (speciation) thus involves a stage in which populations follow divergent descents and no longer interbreed. The prevalent balance model of diploid populations proposes that fitness is conferred almost equally by several gene combinations and that most individuals are heterozygous at most loci. The massive polymorphism that is a feature of this model is a reservoir for evolutionary exploration. Wright (1932) proposed that the natural world resembles a contour map of adaptive peaks and valleys (Fig. 6.15). The adaptive peaks represent fitting responses to particular ecospaces and are populated by certain gene combinations, while unfit combinations perish in the adaptive valleys. Not all peaks on the map are occupied and, within the range of genetic combinations possible in a population, there may be those fitted to a neighbouring adaptive peak. This potential for exploration is incurred at a cost in terms of the fitness of the overall population to the original niche. Thus species are a compromise between conservation and innovation.

As Stanier *et al.* (1970) have pointed out, this definition of species involving reproductive isolation cannot usefully be applied to asexual bacteria because each daughter cell then becomes a new species. Although bacterial populations do have stores of polymorphism (Milkman, 1973) which may allow occupation of new adaptive peaks, the genetic variation is much more limited in extent because there can be no shuffling by means of the recombination that accompanies the sexual process.

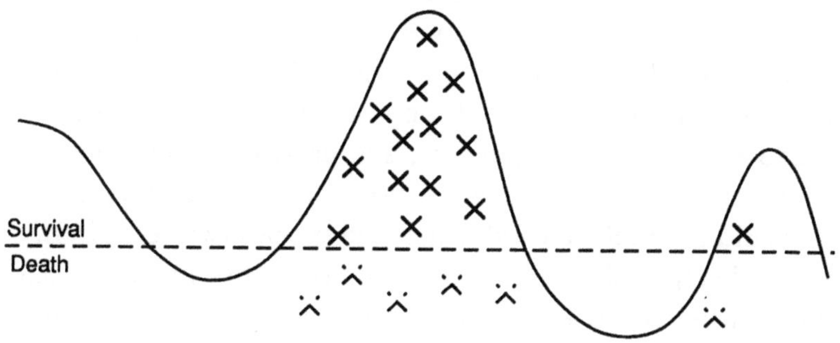

Fig. 6.15 The concept of adaptive peaks (Wright, 1932). The environment is seen as a contour map of adaptive peaks (comparable to ecological niches) and valleys. Genotypes are represented by crosses: in the case of those that do not survive, the crosses lack centres. The central adaptive peak is occupied by a population which occasionally may produce a genotype 'pre-adapted' to a neighbouring unoccupied peak

Because a definition based on reproductive isolation is inapplicable, bacterial taxonomists are forced to define a species pragmatically as a collection of individuals or clonal populations that share many features and differ considerably from the individuals of other species (Staley and Krieg 1984). Such a definition is dependent on phenetic differences being clustered rather than continuous. This is indeed observed but the degree of phenetic similarity found within one bacterial species may be very different from that used to accommodate another.

In general, the amino acid sequence work on cytochromes *c* is consistent with species identified by classical taxonomy. Of nine strains of *Pseudomonas aeruginosa* isolated from a wide range of locations and habitats, eight had identical cytochromes *c*-551 (c_7) while the ninth differed in one position (Ambler, 1974). The cytochromes *c*-551 from six strains of *Ps. fluorescens* contained 0 - 4 substitutions in pairwise comparisons (Ambler, 1974). These figures are to be compared with the 18 differences (21% of comparisons) found between the two most closely related species of Pseudomonas

(Fig. 6.2a). The strains of *Ps. stutzeri* cluster into two groups on the basis of cytochrome *c*-551 sequences. There is less than 5% difference within a cluster but 10% between the clusters (Fig 6.2a). Since the clusters are also distinguished on the basis of GC content, division into two separate species may be contemplated.

The cytochrome c_2 from two strains of *Rb. capsulatus* differ in two postions (Fig. 3.5 legend) while those from two strains of *Rps. palustris* differ by 11% (Fig. 6.5). Almost all other pairwise comparisons exceed 30% difference. The exception is the pair *R. molischianium* and *R. fulvum*, the cytochromes c_2 of which differ by only 12 - 14%. The status of these two organisms as separate species depends on just three microbiological characters: p-aminobenzoate requirement, cell size, and growth on benzoate. It may be that these have been overvalued as taxonomic characters and the two organisms should be combined within a single species (Meyer, 1980).

An extraordinary finding is the presence of a mix of phenylalanine and tyrosine at position 67 in the sequences of both *R. molischianum* and *R. fulvum* iso-2 cytochromes c_2 (Ambler *et al.*, 1979a). Since bacteria are haploid this cannot be due to alleles within an individual cell, nor can it be due to alleles in the population since the cultures are clonal. Gene doubling is a possibility but it is strange that the duplicated gene differs in just one and the same position in both lines of descent. Alternatively, the particular codon (67) may not distinguish tyrosyl and phenylalanyl tRNAs.

Although the new molecular taxonomy is in broad agreement with classical identification of species, it conflicts with the classical view in the identification of genera. This is best illustrated in the Rhodospirillaceae.

Several molecular and morphological characteristics of the Rhodospirillaceae are shown in Table 6.3 (based on Dickerson, 1980a). It is clear that groups based on similarities in cytochrome *c* are very different from groups based on the three genera, *Rhodospirillum, Rhodopseudomonas* and *Rhodomicrobium*. For example, '*Rhodospirillum tenue*' and '*Rhodopseudomonas gelatinosa*' are the only two species to contain a small class 1 cytochrome *c* which is related to Pseudomonad cytochromes *c*-551 (Fig. 6.2a). The two are also distinctive in the presence of tubular photosynthetic membranes and in the depth of divergence of their 16S rRNA (Fig. 6.8). Cellular morphology is clearly not a sound basis for assignment to genera and this problem has been recognised in a revised classification which combines *tenue* and *gelatinosa* in the genus *Rhodocyclus* (Imhoff *et al.*, 1984).

Table 6.3. The classification of the Rhodospirillaceae

Present genus [a]	Former genus	Species	rRNA[b]	Cyt.c[c]	Insertions[d] I	II	III	IV	Membrane[e]	Cell Division[f]	Morphology
Rhodocyclus	*Rhodopseudomonas*	*gelatinosa*	II	S	-	-	-	-	T	F	Small rod
Rhodocyclus	*Rhodopseudomonas*	*tenue*	II	S	-	-	-	-	T	F	Thin spiral
Rhodospirillum	*Rhodospirillum*	*molischianum*		M	+	-	-	-	L	F	Short spiral
Rhodospirillum	*Rhodospirillum*	*fulvum*		M	+	-	-	-	L	F	Spiral
Rhodopseudomonas	*Rhodopseudomonas*	*viridis*	I	M	+	-	-	-	L	F	Sphere
Rhodopseudomonas	*Rhodopseudomonas*	*acidophila*		M	+	-	-	-	L	F	Stalked ovoid
Rhodomicrobium	*Rhodomicrobium*	*vannielii*	I	M	+	-	-	-	L	B	Rod
Rhodopila	*Rhodopseudomonas*	*globiformis*		M	+	-	-	-	V	F	Large rod
Rhodospirillum	*Rhodospirillum*	*photometricum*		L	+	+	+	-	L	F	Large spiral
Rhodospirillum	*Rhodospirillum*	*rubrum*	I	L	+	+	+	-	V	F	Spiral
Rhodobacter	*Rhodopseudomonas*	*capsulatus*	I	L	+	+	+	+	V	F	Rod
Rhodobacter	*Rhodopseudomonas*	*sphaeroides*	I	L	+	+	+	+	V	F	Sphere
Rhodopseudomonas	*Rhodopseudomonas*	*palustris*	I	L	+	+	+	-	L	B	Rod

a The Rhodospirillaceae were reclassified by Imhoff et al. (1984)
b 16S rRNA types are according to the dendrogram of Woese et al. (1980)
c S M L denote small, medium and large classes of cytochrome *c*. Boxing indicates cladistic relationships on the dendrogram of Fig. 6.6.
d Insertions are between 21 and 32 (IV), 62 and 68 (II), 90 and 105 (III) and 44 and 62 (I) based on the cytochrome c_2 numbering of Fig. 3.4
e Photosynthetic membranes can be tubular (T), lamellar (L) or vesicular (V).
f Cell division can be by fusion (F) or budding (B)

The most striking and unexpected feature of both the cytochrome and RNA dendrograms is the association of non-photosynthetic bacteria with photosynthetic taxa. *Paracoccus denitrificans, Nitrobacter agilis, Aquaspirillum itersonii* and *Agrobacterium tumefaciens* are all non-photosynthetic offshoots from the Rhodospirillaceae, branching from points wthin the cytochrome c_2 dendrogram rather than at its base. In the case of the 16S rRNA dendrogram, the divergence of *Pa. denitrificans* is the most recent within the Rhodospirillaceae investigated.

We propose that this congruency between the cytochrome c_2 and rRNA dendrograms makes the possibility of lateral transfer a remote one and suggests a pattern of loss of photosynthetic activity in different lines of descent. *Aquaspirillum itersonii* and *Rhodospirillum rubrum* for example, have a similar shape and GC content and use similar carbon sources but differ in the absence of photosynthetic ability in the former.

Although we have suggested small changes in the status of certain species and some rethinking of the higher taxa in the Rhodospirillaceae, we should emphasise that sequence information produces a cladistic taxonomy with grouping solely according to branching pattern. Since anagenetic change is not measured, adaptively incongruous forms may be grouped together while congruous groups may be dismembered. An example in higher organisms would be the breakup of the reptiles to group the birds with one reptile lineage and the mammals with another. Also, the ranking of taxa in a cladistic taxonomy can only be done on the basis of node depth (or time). This can lead to static species being ranked higher than entire classes.

Classical bacterial taxonomy, for all its associated problems of identifying and weighting taxonomically useful characters, does at least incorporate some measure of anagenetic change. We are sympathetic to the view of Murray (1984) that it may be necessary to have a stable working classification scheme for bacteria based on classical methods in addition to the new information on evolutionary relationships deriving from molecular studies. Thus although the proximity of photosynthetic and non-photosynthetic species in the dendrograms is a fascinating evolutionary discovery it need not necessarily lead to placement of *Paracoccus* in the same genus as *Rhodobacter capsulatus*. Equally, the deep division between the two strains of *Agrobacterium tumefaciens* reflected in their cytochrome *c*-556 sequences (Fig. 6.12), need not necessarily require subdivision into separate species if anagenesis has been limited.

6.9 Endosymbiotic Origin for Eukaryotic Organelles

6.9.1 The Endosymbiotic and Autogenic Theories.

According to the endosymbiotic theory, eukaryotic cells are genetic chimaera arising from association of a host cell with an oxygen-respiring bacterium and, in the case of algae and plants, with an oxygenic photosynthetic bacterium (Margulis, 1970).

The host cell may have been thin-walled or wall-less with a scavenging form of nutrition. The invagination of the cell membrane associated with endocytosis allowed the development of the cytoplasmic membrane structures seen in the eukaryotic cell and also permitted the entry of free-living bacteria destined to become endosymbionts. Interest in the origins of the eukaryotic host has centred around the archaeobacteria (Woese, 1981). These organisms often inhabit extreme 'primordial' environments such as high salt (halobacteria), hot springs at low pH (thermoacidophiles) and anaerobic methane atmospheres (methanogens). They are distinguished from the eubacteria in several ways which include the absence of peptidoglycan in their cell walls and the presence of ether-linked lipids. Other features appear eukaryotic. Thus introns are present in some tRNA and rRNA genes, the ribosomes show a eukaryotic pattern of sensitivity to some antibiotics and the presence of actin is inferred from the inhibitory action of cytochalasins. *Thermoplasma acidophilum* is an eocyte, a subdivision proposed for the archaeobacteria (Lake, 1988) the basis of which is discussed below. It lacks a cell wall and cannot carry out oxidative phosphorylation but uses oxygen in a cyanide-insensitive manner consistent with a peroxisomal style of respiration.

The blue green algae and relatives capable of oxygenic photosynthesis are considered the probable ancestors of chloroplasts. The similarities in the genetic material and its translation are striking. Thus, both blue green algae and chloroplasts have histone-free DNA with no introns, 16S and 23S rRNA and ribosomes of the same size and similar sensitivity to antibiotics (Gray and Doolittle, 1982). They share the feature of oxygenic photosynthesis although detailed scrutiny of the photosynthetic pigments and associated proteins reveals that chloroplasts show considerable variation and may be polyphyletic (Fig. 6.16). The puzzling differences observed between the chloroplasts of the Rhodophyta (red algae) and the Chlorophyta (green algae) have been partly clarified by the discovery by Lewin (1981) of the green oxygenic photosynthetic bacterium *Prochloron* which is a likely ancestor to the chlorophyte chloroplast (Fig. 6.16).

Circumstantial evidence in favour of an endosymbiotic origin for the chloroplast is the existence of extant symbioses involving blue green algae. The *Nostoc* species which is symbiotic in the fungus *Geosiphon* can be cultured separately and the algae *Glaucocystis* and *Cyanidium caldarium* contain blue green cyanelles instead of chloroplasts (Raven, 1970). *Cyanophora paradoxa* is a protozoan with simple chloroplasts that resemble blue green algae but which cannot be cultured separately (Aitken, 1978).

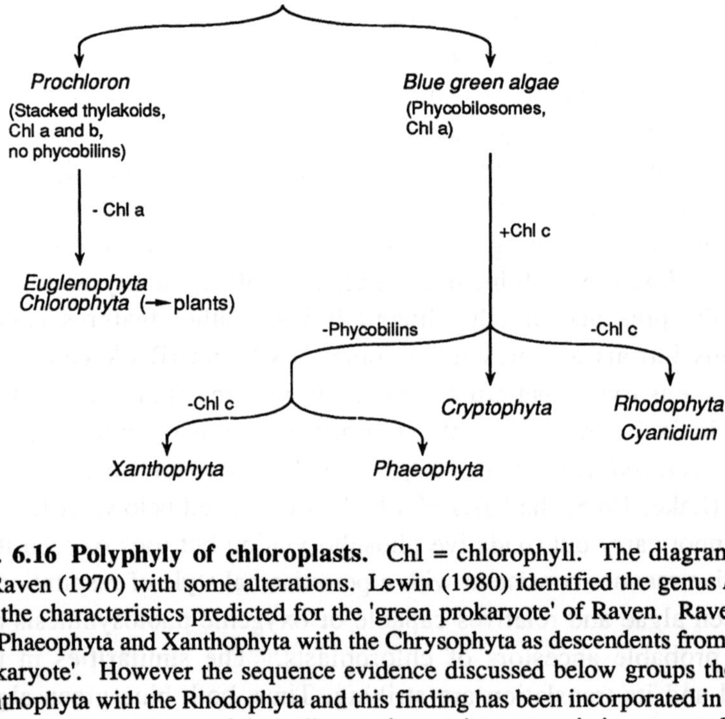

Fig. 6.16 Polyphyly of chloroplasts. Chl = chlorophyll. The diagram is based on that of Raven (1970) with some alterations. Lewin (1980) identified the genus *Prochloron* which has the characteristics predicted for the 'green prokaryote' of Raven. Raven (1970) grouped the Phaeophyta and Xanthophyta with the Chrysophyta as descendents from a putative 'yellow prokaryote'. However the sequence evidence discussed below groups the Phaeophyta and Xanthophyta with the Rhodophyta and this finding has been incorporated in the diagram. This leaves the Chrysophyta and the 'yellow prokaryote' as uncertain in status and relationship

These examples not only show blue green algae at different stages of intimate incorporation into their hosts; they also show that several cell types can act as hosts. A polyphyletic descent for the chloroplast may therefore involve different oxygenic photosynthetic bacteria and different eukaryotic hosts. Whatley (1981) has argued that, superimposed on this polyphyly, there is the possibility that some chloroplasts (those bounded by more than two membranes) may have arisen from other chloroplasts rather than blue green algae.

Mitochondria show considerable diversity in certain aspects of their genetic material and its translation. There are differences in DNA size, in ribosome structure and even in the genetic code (Gray and Doolittle, 1982). The DNA of all mitochondria seems to encode apocytochrome *b*, subunits I - III of cytochrome *c* oxidase and subunits of the ATP synthase. The proteolipid subunit of the latter enzyme is encoded in mitochondrial DNA in yeast but in the nucleus in others. This indicates the existence of gene traffic between the mitochondrion and nucleus, and many proteins (for example cytochrome *c*) that function in the mitochondrion are now nuclear-encoded, presumably as a result of such transfer. Obar and Green (1985) propose that the gene flow was largely unidirectional because of the ability of the nucleus, but not the mitochondrion, to process intervening sequences. It is interesting in this respect that, although the yeast apocytochrome *b* gene does contain intervening sequences (unlike its mammalian counterpart), these have open reading frames that may code for processing activities.

The prokaryotic origins of mitochondria are considered to lie among the purple non-sulfur bacteria and their aerobic non-photosynthetic relatives. John and Whatley (1976) chose *Paracoccus denitrificans* as a putative mitochondrial ancestor on the basis of strong similarities in the electron transport systems but it has become clear that several Rhodospirillaceae are equally good candidates on this basis if it is proposed that photosynthesis was subsequently lost. The question of polyphyly has been raised but not resolved. Stewart and Mattox (1984) argue that a major division exists between those mitochondria with tubular and those with lamellar cristae. We recall (section 6.8) that these patterns of membrane folding were also correlated with subdivision of the Rhodospirillaceae and it is possible that the two classes of mitochondria arose from organisms in these two subdivisions. The bearing of cytochrome *c* and rRNA sequences on these questions is discussed below.

As with the chloroplast, extant endosymbioses provide at least circumstantial evidence that the mitochondrion originated as an endosymbiotic bacterium. Some of the non-photosynthetic bacteria now grouped with Rhodospirillaceae are involved in intimate association with eukaryotic hosts. Examples are *Agrobacterium* in the formation of the crown gall tumour in host plants and *Rhizobium* in the nodulation of legumes. *Pelomyxa palustris* is an amoeboid cell which lacks mitochondria but contains two species of bacteria which are segregated into daughters cells at division (Whatley, 1981).

In opposition to endosymbiosis is the autogenic theory in which eukaryotic organelles arose by segregation of existing membranes and DNA fragments

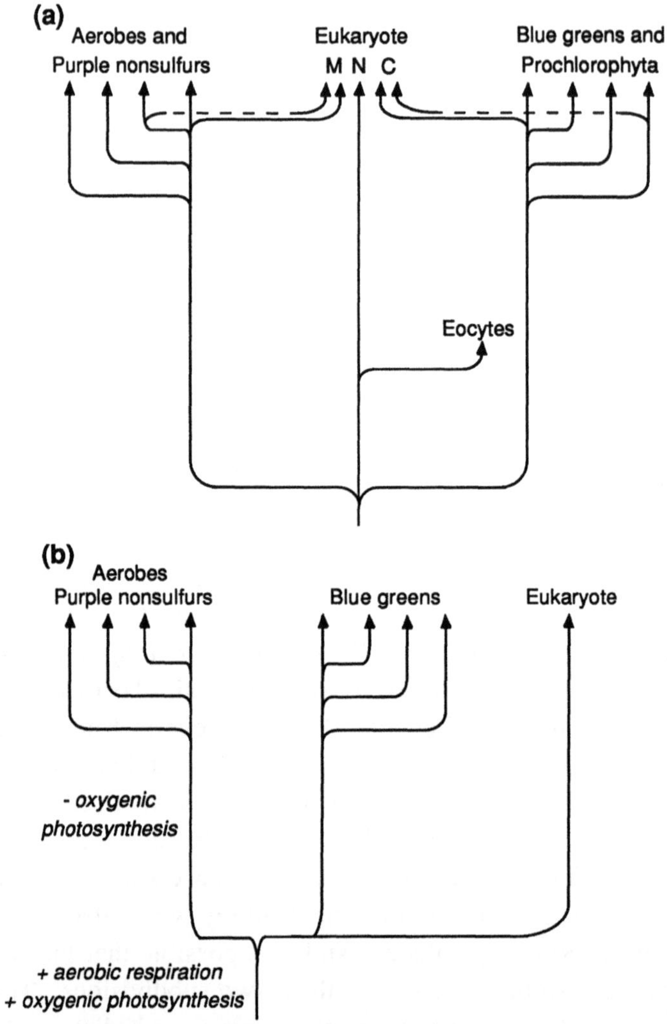

Fig. 6.17 Endosymbiotic (a) and autogenic (b) origins for mitochondria and chloroplasts. M mitochondria; **N** nucleus; **C** chloroplast. Possible polyphyly of mitochondria and chloroplasts is indicated by broken lines in **(a)**

(e.g. Raff and Mahler, 1972). According to this view, organelles retain some prokaryotic features because they have evolved slowly relative to the nucleus.

The two theories are compared in Fig. 6.17 and lead to quite different predictions as to the evolutionary relationships of present-day cellular components. Consider a macromolecule which is encoded in all three genetic compartments - mitochondria, chloroplast and nucleus - and is also found universally in bacteria. The endosymbiotic theory (Fig. 6.17(a)) predicts a relatively recent divergence of that macromolecule isolated from mitochondria and from the Rhodospirillaceae. A similarly recent divergence is predicted for the same macromolecule isolated from chloroplasts and blue green algae.

In contrast, the divergence between the two organellar forms of this molecule will be as deep as that between the prokaryotic ancestors of the organelles. The nuclear-encoded form of the molecule is predicted to show a relationship to the eocytes but a deep divergence from the organellar forms. The autogenic theory on the other hand predicts no special relationship between the forms of macromolecule isolated from organelles and from free-living bacteria. All divergences are expected to be equally deep with no intervening prokaryotic branches on the eukaryotic tree (Fig. 6.17(b)).

6.9.2 Cytochrome *c*-553 (c_6) and the Origins of the Chloroplast

A pattern of relationships of the cytochromes *c*-553 from eukaryotic algae and blue green algae is shown in Fig. 6.18. In the dendrogram, the red and brown algae cluster together and derive from the blue green algae. We regard this as strong evidence in favour of a blue green ancestry for the chloroplasts of these eukaryotes. *Euglena* and *Monochrysis* have a more remote origin consistent with the proposal that their chloroplasts derive from a relative of the blue greens such as *Prochloron*. Sequences from a green eukaryotic alga and from *Prochloron* itself would be helpful in confirming this view.

Ferredoxin sequences tend to support the close relationship of the Rhodophyta and the blue green algae and the pattern of polyphyletic divergence (Schwartz and Dayhoff, 1978; 1981). Demoulin (1979) notes however that the gene duplication that occurred in the ferredoxins introduces the risk of paralogous comparisons in the matrix.

A quite unexpected, and still poorly explained, feature of the dendrogram is the shallow divergence of the blue green algae themselves. The prevalent view is that blue green algae appeared about 3 billion years (BY) ago to produce the stromatolite formations of that period. A divergence into filamentous and unicellular forms is apparent from the fossils of Gunflint Chert (1.6 - 2 BY) (Schopf, 1970) long before the appearance of the eukaryotes about 1.1 BY ago. It may be that stromatolites are the products of other photosynthetic bacteria and that filamentous forms arose several times and more recently than Gunflint Chert. According to this view the later divergence gave rise to the ancestors of present-day blue greens and of the chloroplast while earlier forms died out. Another possibility is that the blue green algal proteins have evolved more slowly than their eukaryotic counterparts. This however should be detectable by the relative rate test with an outlying sequence and no sign of a slower rate is apparent from the matrix.

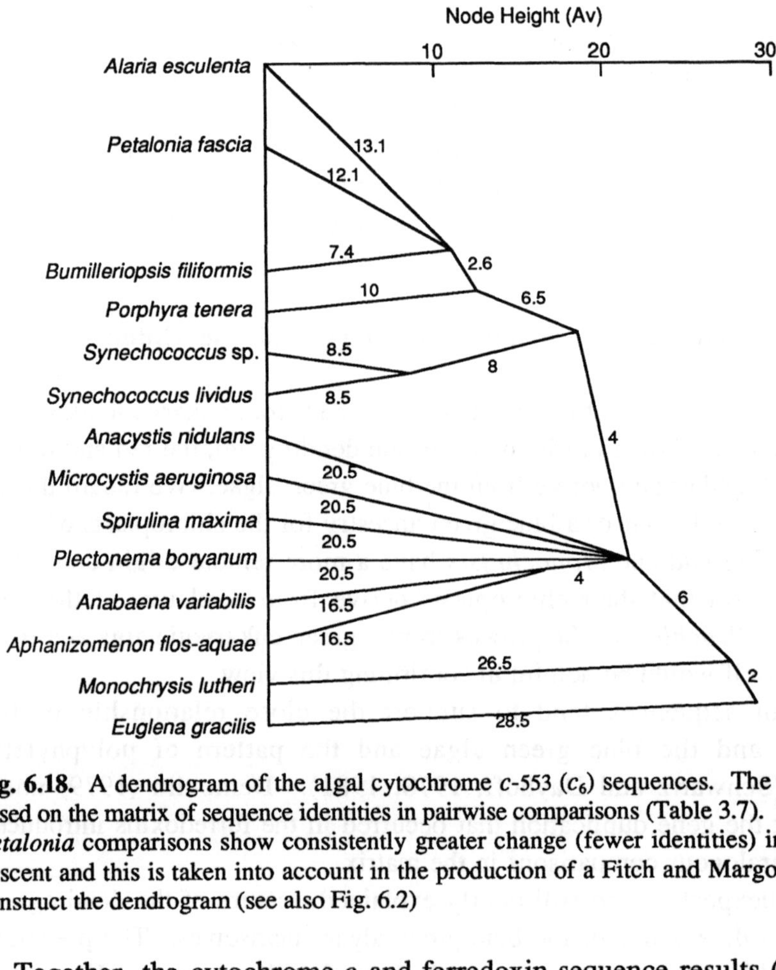

Fig. 6.18. A dendrogram of the algal cytochrome *c*-553 (c_6) sequences. The dendrogram is based on the matrix of sequence identities in pairwise comparisons (Table 3.7). The *Alaria* and *Petalonia* comparisons show consistently greater change (fewer identities) in their lines of descent and this is taken into account in the production of a Fitch and Margoliash matrix to construct the dendrogram (see also Fig. 6.2)

Together, the cytochrome *c* and ferredoxin sequence results (and the 16S rRNA catalogues - Doolittle and Bonen, 1981) reveal a strong relationship between the chloroplast and the blue green algae, a prediction of endosymbiotic theory. They do not however eliminate the possibility that the blue green algae are ancestors of the whole eukaryotic cell. That distinction requires further examination of the predictions of the endosymbiotic theory with regard to the mitochondrion.

6.9.3 Cytochrome *c* and the Origins of the Mitochondrion

In section 6.9.1 we described a hypothetical macromolecule which showed deep divergences in comparisons between chloroplast, mitochondrial and cytoplasmic forms. Cytochrome *c*-553 is not such a molecule because it is not found in the cytoplasm and it is not clear whether it is orthologous (or even homologous) to mitochondrial cytochrome *c*. Composite dendrograms containing different cytochrome *c* classes have been strongly criticised on this basis (Demoulin, 1979; Uzzell and Spolsky, 1981). An alternative test of the endosymbiotic theory would be the demonstration of a recent divergence between mitochondria and a prokaryotic group distinct from the blue green algae.

Mitochondrial cytochromes *c* are the same size as the cytochromes c_2 of the M class defined in Table 6.3. Indeed, Ambler *et al.* (1976) found that the cytochromes c_2 of *Rm. vannielii* and *Rps. viridis* are more similar to a selected group of mitochondrial cytochromes *c* than they are to other cytochromes c_2 (Fig. 6.6). These organisms would be placed only slightly beyond the most remote divergence of the cytochrome *c* tree (Fig. 6.4). We have no reason to suppose that the true ancestor of the mitochondrion yet appears in the cytochrome c_2 tree (nor even that the ancestor survives in a free-living form). It may be that future sequences from aerobes related to *Nitrobacter agilis* and *Agrobacterium tumefaciens,* which are associated with this group of the Rhodospirillaceae may reveal an extant prokaryote that falls *within* the cytochrome *c* tree. This would give rise to a pattern resembling that seen in the cytochrome *c*-553 dendrogram (Fig. 6.18) indicating a polyphyletic origin for mitochondria.

To some extent such a situation has already arisen. *Tetrahymena* cytochrome *c,* which does not appear in Fig. 6.4, is the most divergent of all the mitochondrial cytochromes *c*. The simple percentage identity is not a true reflection of this divergence. The *Tetrahymena* cytochrome *c* is unique in the mitochondrial family in having substitutions at the otherwise conserved residues *Lys/Arg 13*, Gln 16, Gly 29, Gly 45, Asn 52, *Lys 72*, Tyr/Phe 74, and *Lys 86*. Three of these (italics) are implicated in the interaction with cytochrome *c* oxidase (Pettigrew & Moore, 1987: Chapter 2). In contrast, *Euglena gracilis* cytochrome *c* is unique at only one position (Phe 67), *Crithidia oncopelti* at none and *Neurospora crassa* at one (Phe 97).

In the matrix (Table 6.4), *Rps. viridis* cytochrome c_2 intervenes between *Tetrahymena* cytochrome *c* and the other mitochondrial sequences. This diphyletic pattern is consistent with the major division proposed between mitochondria with tubular cristae and those with lamellar cristae (Stewart and Mattox, 1984). Organisms with tubular cristae are represented in the cytochrome *c* sequences only by *Physarum polycephalum* and *Tetrahymena pyriformis*.

Table 6.4 % Identities in selected cytochromes *c* and *Rps. viridis* c_2

1. Horse					
65	2. Wheat				
65	59	3. *Neurospora crassa*			
57	52	48	4. *Crithidia oncopelti*		
56	49	47	54	5. *Rhodopseudomonas viridis*	
47	41	45	43	44	6. *Tetrahymena pyriformis*

Sequences were aligned and compared as described in Chapter 3.

However, although the tubular cristae and the distinctive cytochrome *c* sequence suggest a separate origin for *Tetrahymena* mitochondria in the Rhodospirillaceae, the nature of that origin is not yet clear. *Rhodocyclus tenue* and *gelatinosa* have tubular photosynthetic membranes but contain a small cytochrome *c*-551, quite different from the cytochrome c/c_2 group. It would be useful to examine cytochromes *c* from a greater range of Rhodospirillaceae with tubular membranes, and mitochondria with tubular cristae.

We have commented on the problem of saturation in the cytochromes c_2 in section 6.4, and Meyer *et al.* (1986) consider that this problem makes the comparison of cytochrome *c* and cytochrome c_2 sequences invalid and misleading. We think this problem has been overstated and propose that cytochrome c_2 and mitochondrial cytochrome *c* are certainly close enough to support an endosymbiotic origin for mitochondria within the Rhodospirillaceae.

Meyer *et al.* (1986) are correct to the extent that the steadily failing correspondence of sequence change with elapsed time near the limits of variability and the increased noise of multiple, parallel and back mutations diminish our ability to map any fine details of the phylogenetic relationships. We need the evidence of a macromolecule that is well within its variable limits. On the basis of Fig. 6.7, cytochrome c_1 looks a possible candidate and although it is a large protein, gene sequencing may be a feasible approach to its phylogeny.

The contribution of rRNA work in this area has been slightly disappointing by comparison with its major contributions elsewhere. The problem seems to be the very rapid divergence of mitochondrial 16S rRNA. It is only with application of the relative rate test and judicious choice of species on each side of the eukaryotic-prokaryotic divide that the divergence of the mitochondrial rRNA can be placed close to *Agrobacterium* (Yang *et al.*, 1985).

6.9.4 The Present Status of the Endosymbiotic Theory

We have shown that the cytochrome *c* work places the chloroplast close to the blue green algae and the mitochondrion close to the Rhodospirillaceae. We also noted that, because of the question of paralogy, this work cannot measure the depth of divergence between these two groups. Indeed Uzzell and Spolsky (1981) have proposed an autogenic model in which duplicated cytochrome *c* genes encoding cytochromes *c*-553 and c_2 are present in the common ancestor of the blue greens, the Rhodospirillaceae and the whole eukaryotic cell, and are selectively lost in particular free-living descendants and eukaryotic organellar compartments.

The 16S rRNA catalogues and sequences can overcome this problem since this molecule is not thought to be subject to paralogy and is universal in its distribution. According to these data, the divergence between the blue greens and the chloroplasts on the one hand and the mitochondria and Rhodospirillaceae on the other is indeed deep as predicted by the endosymbiotic theory (Woese, 1981).

This means that both mitochondria and chloroplasts could not be autogenous but still allows the possibility of one being so. However the degree of similarity of cytoplasmic 18S rRNA and prokaryotic 16S rRNA is not much greater than expected for random sequences. A detailed analysis detects a relationship between cytoplasmic 18S rRNA and 16S rRNA of eocytes, a subgroup of the archaeobacteria (Lake, 1988).

Taken together, these molecular studies go far towards acceptable proof of the endosymbiosis of both chloroplasts and mitochondria. Given that extremely remote historical events are being examined using imprecise and erratic chronometers, we consider it remarkable that such a degree of success has been achieved.

6.10 Remote Relationships

6.10.1 The Evolution of Energy Metabolism

The atmosphere of the earth has altered from reducing to oxidising over the 3BY or so of biological evolution. It is therefore a reasonable hypothesis that the first organisms were fermentative anaerobes which then gave rise to anaerobic photosynthetic bacteria (Olson, 1981). Gest (1980) proposed that electron transport might have arisen from fermentation in the manner outlined in Fig. 6.19. According to this scenario, fermentation to succinate spares one pyruvate for biosynthetic purposes but is energetically wasteful at the stage of fumarate reduction. Gest proposed that fumarate reductase evolved from a simple soluble enzyme to a transmembrane system capable of producing an electrochemical potential using the energy of NADH oxidation by fumarate. The addition of a photosynthetic reaction centre converted the linear electron transport system into a cycle driven by the absorption of light energy. Non-cyclic photosynthesis in its early evolution almost certainly used H_2S as the electron donor, a feature still seen in present-day Chlorobiaceae and Chromatiaceae. The 16S rRNA catalogue suggests a very deep divergence between these taxa and indicate that most of the further bioenergetic developments occurred on the Chromatiaceae lineage. One of these developments was the use of water rather than H_2S as the source of electrons for non-cyclic photosynthesis. This oxygenic photosynthesis is found in the cyanobacteria and their descendants, the chloroplasts. A second line of photosynthetic evolution involved the replacement of H_2S by organic reductants and gave rise to the Rhodospirillaceae.

According to this scheme, photosynthesis is a central strand running through bacterial evolution and respiratory systems are offshoots (Fig. 6.20). The anaerobic respiration of *Desulfovibrio* involves reduction of sulfate produced by the green and purple sulfur bacteria and *Desulfovibrio* is related to *Chromatium* on the basis of rRNA sequences. It is proposed that aerobic respiration arose several times. While this may seem unlikely, the extant adaptability of the electron transport system of *Rhodobacter sphaeroides* (Fig. 6.21) shows that the switch between photosynthesis and respiration involves the addition of just cytochrome oxidase to the modular set of components. It will be of great interest to compare the detailed structures of the oxidases of aerobes, proposed on the basis of Fig. 6.20 to derive from independent adaptations to oxygen.

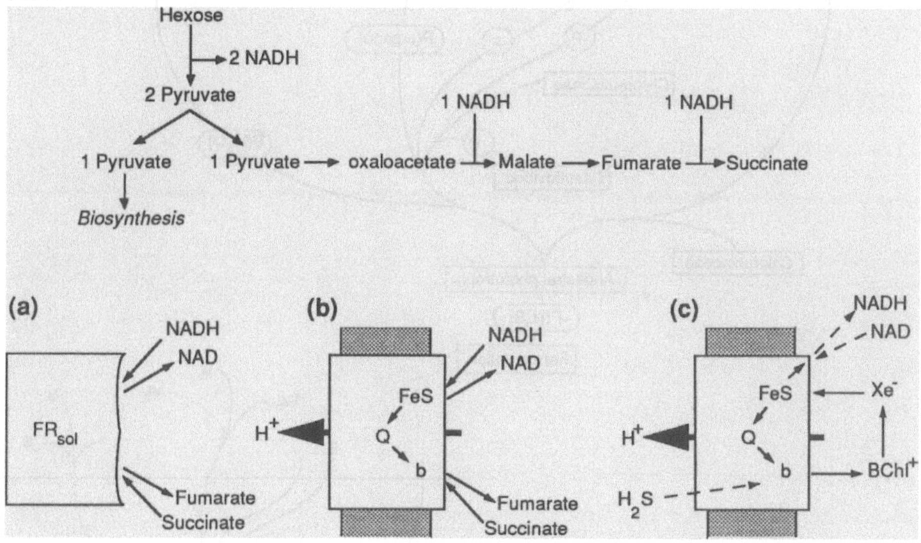

Fig. 6.19 A possible evolution of electron transport from fermentation. Gest (1980) proposed that fermentation to succinate spared a pyruvate for biosynthesis and fumarate reductase could then evolve from a soluble enzyme **(a)** to a proton translocating respiratory complex **(b)** containing iron-sulfur protein (Fe-S), Coenzyme Q (Q) and cytochrome *b*. Photosynthesis **(c)** may have evolved by addition of a bacteriochlorophyll system (BChl) that could be photooxidised to yield a reductant Xe^- capable of reducing the iron-sulfur protein of the membrane electron transport chain. (After Gest, 1980)

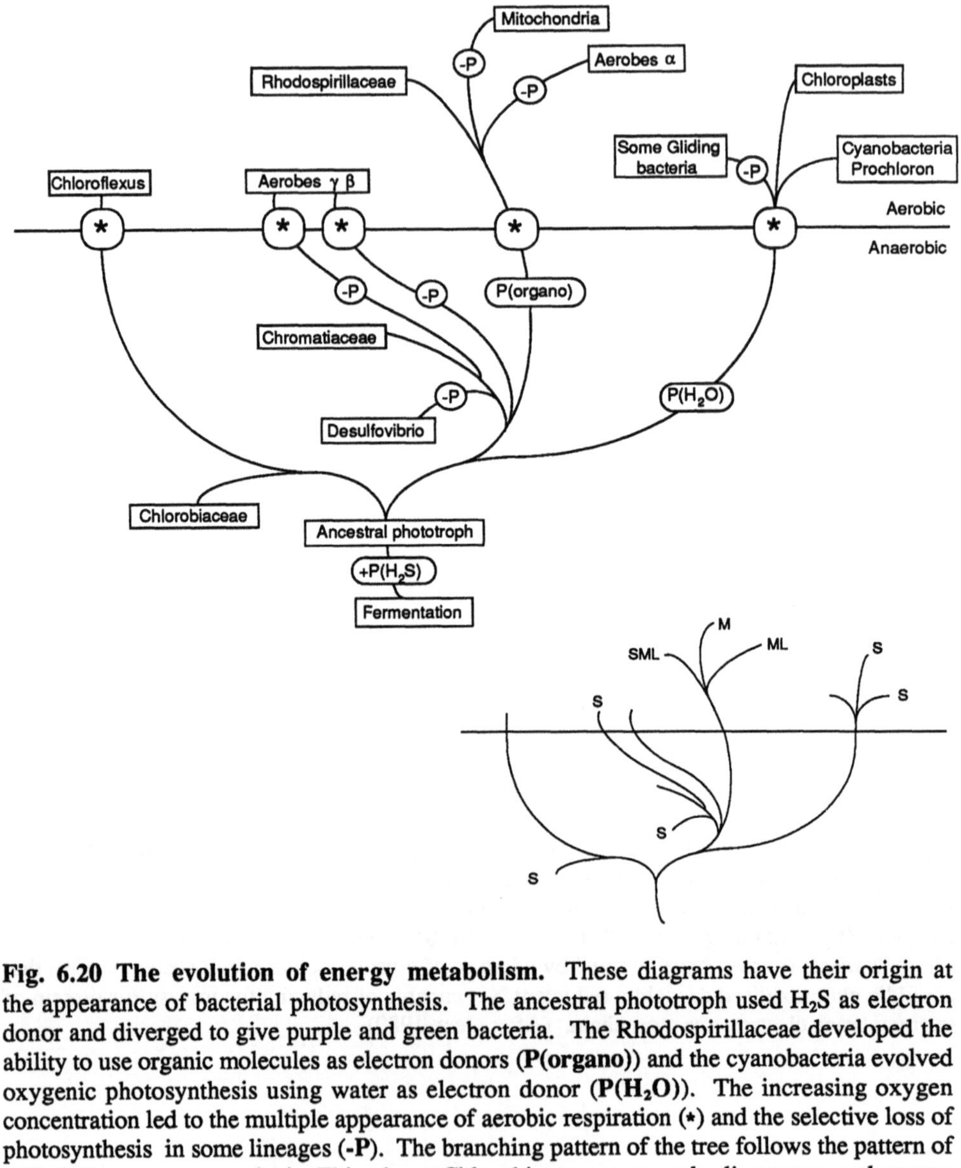

Fig. 6.20 The evolution of energy metabolism. These diagrams have their origin at the appearance of bacterial photosynthesis. The ancestral phototroph used H_2S as electron donor and diverged to give purple and green bacteria. The Rhodospirillaceae developed the ability to use organic molecules as electron donors (**P(organo)**) and the cyanobacteria evolved oxygenic photosynthesis using water as electron donor (**P(H_2O)**). The increasing oxygen concentration led to the multiple appearance of aerobic respiration (*) and the selective loss of photosynthesis in some lineages (**-P**). The branching pattern of the tree follows the pattern of 16S rRNA sequence analysis. This places Chlorobiaceae as an early divergence and groups *Desulfovibrio* with *Chromatium*. The α-group of aerobes (Woese *et al.*, 1985) derive from the Rhodospirillaceae lineage, the β-group from an early purple bacterial ancestor and the γ-group from the *Chromatium* lineage. The occurrence of small (S), medium (M) and large (L) cytochromes *c* in the evolutionary pattern is shown in the lower diagram. (Adapted from Dickerson, 1980b)

6.10.2 The Relationships between Cytochrome *c* Subclasses and the Problem of Convergence

The class I cytochromes present in the different bioenergetic contexts of Fig. 6.20 follow a logical progress from an ancestral small cytochrome to the range of class I types, some having insertions of polypeptide chain at specific points in the molecule. However, the cytochrome *c* sequences themselves could certainly not be used to construct a reliable phylogeny. This failure is due firstly to the question of possible paralogy, and secondly to the fact that sequence divergence has followed pathways of functional adaptation leaving a degree of similarity close to that expected from chance alone. We believe these criticisms invalidate the composite cytochrome *c* trees of Dayhoff (Schwartz and Dayhoff, 1978; Dayhoff and Schwartz, 1981).

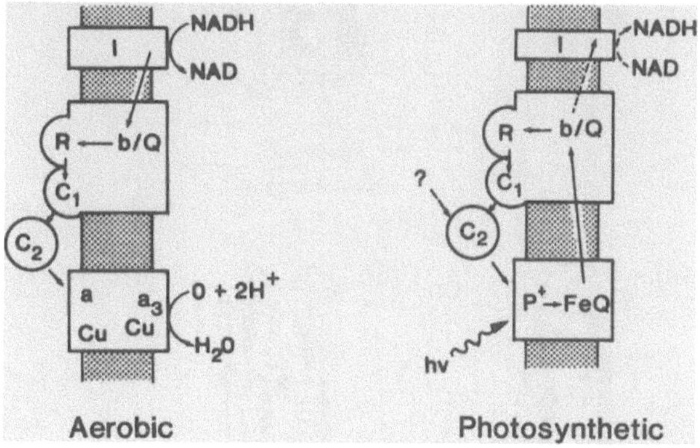

Fig. 6.21 Aerobic and photosynthetic electron transport systems of *Rb. sphaeroides*. The bc_1 complex contains the Rieske iron sulfur centre (**R**), cytochrome c_1 (**c_1**), *b*-type cytochromes (**b**) and coenzyme Q (**Q**). It is the central module that is retained in both photosynthetic and respiratory modes. Aerobic cells contain a cytochrome *c* oxidase of the mitochondrial 2Cu, aa_3 type; photosynthetic cells contain a reaction centre with a photooxidisable bacteriochlorophyll (**P^+**) and an iron quinone (**Fe-Q**). — cyclic photosynthetic electron transport. —— non-cyclic photosynthetic electron transport.

However if we examine the 3-dimensional structures we find striking similarities in the different class I cytochromes (Chapter 4). It is often stated that similarities in tertiary structures are retained in diverging proteins beyond the point where similarities in sequence become random (see e.g. Dickerson, 1978). Could these structural similarities be used to map phylogenetic relationships? Tertiary structure similarity can be quantified by rotating

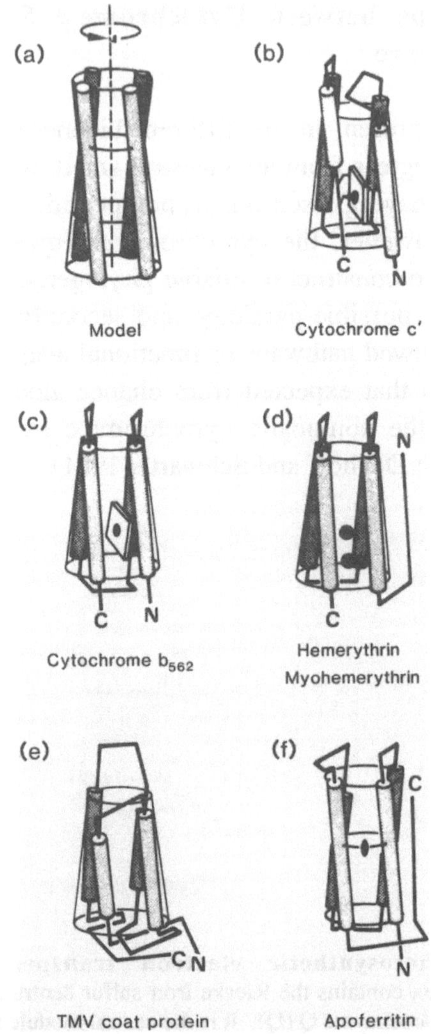

Fig. 6.22 The 4-α-helical proteins. α-helices are shown as cylinders. The 4-α-helical bundle contains heme in the case of cytochrome *c′* and cytochrome b_{562} and iron in the case of myohemerythrin and hemerythrin (black spheres). (After Weber and Salemme, 1980)

molecules in space to achieve a least squares fit of the α-carbon distances. However, we have noted in section 6.4 that the problem of convergence is more severe with tertiary structures and cannot be objectively evaluated. There may be strong structural constraints in the efficient packing of a small polypeptide chain around the heme group and these, together with the requirements for heme attachments and axial ligation, may result in the independent appearance of similar structures several times in evolution.

Similar arguments cast doubt on proposals of homology in other proteins also. The common topology of connections in the supersecondary structure of twisted β-sheet flanked by α-helices that is a feature of the dehydrogenases and kinases, was originally seen as strong evidence for the divergence of homologous structures. However the most stable conformation of β-strand has a tilted arrangement of peptide bond carbonyls and amides relative to the axis of the chain. Thus when two such strands form hydrogen-bonds, they do so at an angle of about 25° to each other which, when repeated over a six strand sheet, gives the typical 125° left handed twist. What is more, the simplest and shortest connections between neighbouring strands are right-handed giving rise to the observed topology of connections in the kinases and dehydrogenases. Thus at least some of the similarities seen in β-sheet proteins may be the consequence of structural constraint rather than kinship (Sternberg and Thornton, 1978).

Another probable example of convergence is the family of 4-α-helical proteins (Weber and Salemme, 1980). α-helices can only make effective packing contact if they are tilted at an angle equal to twice the helix pitch angle (9°) to form a left handed supercoil. This is the structure seen, for example, in myosin and keratin. A small protein can be made from 4 short α-helices packed against each other. Because of the sense of the coiling, the most straightforward connection is right-handed between helices. Thus the similarities seen in the family of 4-α-helical structures (Fig. 6.22) may be due to convergent structural effects rather than evolutionary divergence. This seems all the more likely in this case since the group is functionally diverse and includes cytochrome *c′*, ferritin and tobacco mosaic virus coat protein. However in the specific case of cytochromes *c′* and b_{562} Weber *et al.* (1981b) suggest that the structural similarity is so marked that it supports an evolutionary divergence from a common structure. The similarity extends beyond the common 4-α-helical units to include the position and orientation of the hemes, and their exposure to solvent, as well as the location of the axial histidine ligand and the locations, close to the heme, of one methionine and three aromatic sidechains.

Chapter 7 Redox Potentials

7.1 Introduction

The tendency of a redox couple to donate or accept electrons is given by the redox potential (E) and is measured as a reduction potential with reference to the standard hydrogen electrode (0 V) (Scheme 7.1). Practical aspects of the determination of redox potential and the use of the Nernst equation have been dealt with in Pettigrew & Moore (1987: Chapter 1).

The value of E exhibited by an individual couple is a reflection of the relative stability of the reduced and oxidised states. Any factor which acts to

Scheme 7.1. The Nernst equation and symbols used in this chapter

The fundamental relationship governing redox potentials is the Nernst equation. For the reaction - ($ox + e^- \rightarrow red$) - this is:

$$E = E^\circ + \frac{RT}{nF} \ln \frac{[ox]}{[red]} \qquad (7.1)$$

where E is the redox potential, E° is the redox potential for components in their standard states at pH 0, R is the gas constant (8.3144 J K^{-1} mol), F is the Faraday (9.65×10^4 J V^{-1} mol), n is the number of electrons transferred, T is the absolute temperature.

In biology, this equation is usually converted to the form:

$$E_h = E_{m,x} + \frac{0.06}{n} \log \frac{[ox]}{[red]} \qquad (7.2)$$

where E_h is the redox potential with reference to the standard hydrogen electrode, $E_{m,x}$ is the midpoint redox potential (when [ox] = [red] ≠ standard state) at a defined pH of x and 2.303 RT/F = 0.06 for 29.4°C.

Some different forms of $E_{m,x}$ will be used in particular circumstances. $E_{m,L}$ and $E_{m,NL}$ are the midpoint potentials in the presence and absence of a ligand respectively (section 7.2). If the ligand binds with different affinities to the two redox states these two values will differ. $E_{m,I=0}$ is a hypothetical midpoint potential obtained by extrapolation of a linear region of a plot of E_m vs f(I) to zero ionic strength (section 7.4.1). $\tilde{E}$ is a hypothetical midpoint potential obtained by extrapolation of a linear region of a plot of E_m vs pH to pH = 0 (section 7.6).

stabilise the oxidised form makes the redox couple a better electron donor and results in a more negative redox potential. Conversely, any factor which acts to stabilise the reduced form makes the couple a better electron acceptor and gives rise to a more positive redox potential.

These effects can be regarded as alterations in the free energy levels of the oxidised and reduced forms (G_{ox} and G_{red}). Although information is not usually available on the absolute values of G_{ox} and G_{red}, it is possible to measure a change in free energy (ΔG_{stab} - a free energy of stabilisation) as a change in redox potential (ΔE).

$$\Delta G_{stab} = -nF\Delta E \tag{7.3}$$

Notice that because we know only the energy change, and not the values of G_{ox} and G_{red}, we cannot say which of these terms is altered. For example, a lowering of the redox potential may be due either to a stabilisation of the oxidised state or to a destabilisation of the reduced state. Nevertheless from other considerations, we may be able to deduce which of the two states is affected.

The importance of these stabilising factors in setting the potential of a redox protein is illustrated by the case of cytochrome *c*. The midpoint potential (E_m) of free protoheme IX in aqueous solution (pH 7) is -115 mV (Kassner, 1972) while substituted protoheme is found in some cytochromes *c* with E_m as low as -400 mV and in others with E_m high as +400 mV (Pettigrew & Moore, 1987: Chapter 1). The former value indicates a relative stabilisation of the oxidised form of 285 mV (27.5 kJ mol^{-1}) while the latter reflects a relative stabilisation of the reduced form of 515 mV (49.7 kJ mol^{-1}).

The free energy of stabilisation (ΔG_{stab}) is a composite term (Moore *et al.*, 1986a):

$$\Delta G_{stab} = \Delta G^{lig} + \Delta G^{coord} + \Delta G^{el} + \Delta G^{conf} \tag{7.4}$$

ΔG^{lig} is a stabilisation conferred by ligand binding to the protein; ΔG^{coord} is due to coordination at the redox centre; ΔG^{el} incorporates electrostatic interactions between the charge on the redox centre and polar species within both the protein and the solvent; ΔG^{conf} is a conformational stabilisation. An assessment of an individual component of ΔG_{stab} can be made as long as the others are held constant. For example, in the class I cytochromes *c* with histidyl-methionyl axial coordination of the iron and in which conformational differences between the two redox states are very small, it can be assumed that differences in ΔG_{stab} are due predominantly to ΔG^{el}.

However, ΔG^{el} itself is a complex factor:

$$\Delta G^{el} = \Delta G^{ion} + \Delta G^{dip} + \Delta G^{int} + \Delta G^{surf} \tag{7.5}$$

ΔG^{ion} incorporates effects of ions in solution; ΔG^{dip} is the energy of dipolar interactions both within the protein and with the aqueous phase; ΔG^{int} is the energy of interaction of buried charges (which may include other redox centres in the same protein) with the charge of the redox centre; ΔG^{surf} is the energy of interaction of solvent-exposed charges with the charge of the redox centre. This range of influences on the redox potential is represented diagrammatically in Fig. 7.1. Some workers have chosen to investigate individual free energy components of ΔG^{el} while others have tried to produce a model which integrates the interplay between these individual components. Both types of analyses are described in the present chapter.

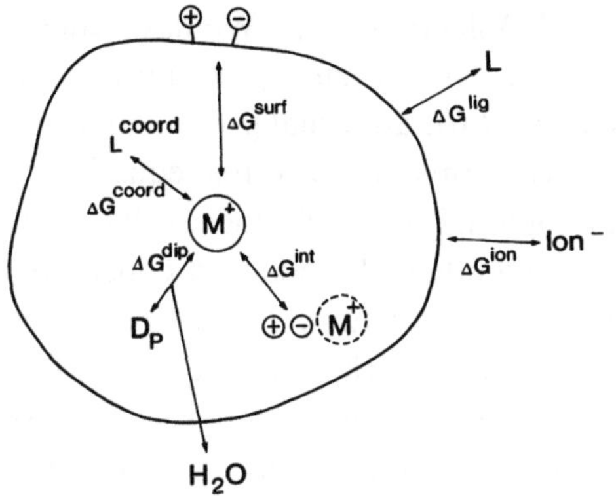

Fig. 7.1 Factors affecting the stability of a metal ion (M^+) buried in a redox protein. Free energy changes are shown which result from interaction of the metal ion with coordination ligands (L^{coord}, ΔG^{coord}), dipoles within and surrounding the protein (D_P, H_2O, ΔG^{dip}), buried charges (M^+,⊕,⊖ , ΔG^{int}), surface charges (⊕,⊖ , ΔG^{surf}), ions in solution (Ion^-, ΔG^{ion}) and ligands which bind to the protein (L, ΔG^{lig})

We have not addressed the question of whether investigations of the enthalpy (ΔH_{stab}) and entropy (ΔS_{stab}) contributions to ΔG_{stab} help define how proteins influence the level of the redox potential. Although ΔH_{stab} and ΔS_{stab} values may be determined from the temperature dependence of the redox potential or from calorimetric experiments there are not many reported studies and their interpretation is not clear. Various studies of horse, tuna and yeast cytochromes *c* at pH 7 and ionic strengths in the range 0.01 to 0.1 M yield ΔG_{stab}, ΔH_{stab} and ΔS_{stab} values respectively of -6.0 kJ mol^{-1}, -14 to -17 kJ mol^{-1} and -113 to -121 J K^{-1} mol^{-1} (Margalit and Schejter, 1973; Watt and Sturtevant, 1969; Taniguchi *et al.*, 1980). At higher ionic strengths, different values of ΔH_{stab} and ΔS_{stab} are obtained, probably because of complications due to ion binding. However, Kreishman *et al.* (1978) suggested that the biphasic behaviour that they observed for the temperature dependence of the redox potential of horse cytochrome *c* in 0.1 M phosphate and 0.1 M NaCl is a result of a change in the structure of the bulk water. Further thermodynamic investigations are needed to ascertain whether this approach is useful in characterising protein redox potentials.

7.2 Ligand Binding

The effect of ligand binding on redox potential is analysed in Scheme 7.2, based on the treatment of Dutton and Wilson (1974). We have chosen to use dissociation rather than association constants because proton effects, to which these equations can be applied (section 7.6), are usually discussed in terms of pK values. It is useful to generate two forms of the Nernst equation (7.12 and 7.17). In the former, the midpoint potential in the absence of ligand is altered by a factor containing redox-state dependent dissociation constants and the ligand concentration. This form simplifies to Eqn. 7.13 at saturating ligand which shows that a ten fold difference in the affinity of the ligand for the two redox states gives rise to a 60 mV change in redox potential. The second form of the Nernst equation (7.17) is expressed in terms of the midpoint potential in the presence of bound ligand ($E_{m,L}$) and is useful in the treatment of proton binding effects (section 7.6) where $E_{m,L}$ is the midpoint potential at low pH.

Binding of Small Ions. In the presence of a binding buffer the ionic strength dependence of redox potential for horse cytochrome *c* has a positive slope (Fig. 7.2) indicating that the factor $(Z^2_{red} - Z^2_{ox})$ must be positive (Margalit and Schejter, 1973: see section 7.4.1). These authors calculate that, to account for the change in $(Z^2_{red} - Z^2_{ox})$, the oxidised form must bind the equivalent of two negative charges while the reduced form must bind one positive charge. These specific binding effects are to be distinguished from the general effects of the ionic environment on redox potential to be discussed in section 7.4.1.

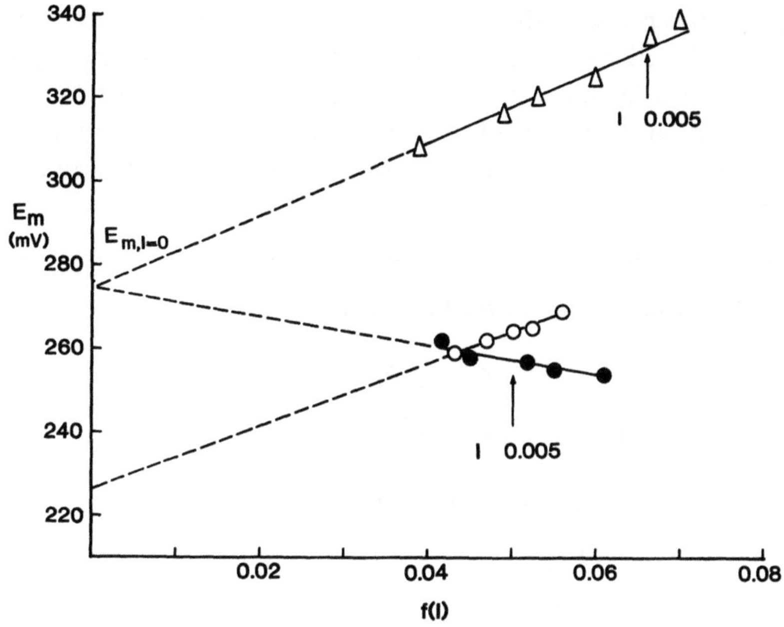

Fig. 7.2 Ionic strength dependence of redox potential of horse heart cytochrome *c* and *Euglena gracilis* cytochrome *c*-552. $f(I) = I/(1 + 6\sqrt{I})$ for horse heart cytochrome *c* (O, ●). $f(I) = I/(1 + \sqrt{I})$ for *Euglena gracilis* cytochrome *c*-552 (Δ). Redox potentials for Δ & ● were measured in the 'non-binding buffer', tris-cacodylate, pH 7 at 25°C using ferro-ferricyanide mixtures. Redox potentials for O were measured in 2 mM tris-cacodylate, pH 7 but with added NaCl as the predominant ionic strength component (Margalit and Schejter, 1973; Goldkorn and Schejter, 1976). $E_{m,I=0}$ values are obtained by extrapolation to f(I) = 0

Table 7.1 The effect of ligand binding on redox potential

Cytochrome *c*	Ligand	ΔE_m(mV)	Refs.
Horse	liposomes	(-40 to -50)	Dutton *et al.*, 1970; Kimelberg & Lee, 1970; Vanderkooi *et al.*, 1973
Horse	phosvitin	-35	Yoshimura *et al.*, 1980
Horse	mitochondria	(-47 to -50)	Dutton *et al.*,1970;Vanderkooi *et al.*, 1973
S. cerevisiae iso-1	mitochondria	-20	Vanderkooi *et al.*, 1973
Horse	cytochrome aa_3	-40	Vanderkooi & Erecinska, 1974
Horse	cytochrome bc_1	-40	Vanderkooi & Erecinska, 1974
Horse	YCCP [a]	0	Vanderkooi & Erecinska, 1974
Horse (x-linked)	YCCP	-20	Pettigrew & Seilman, 1982
Horse (x-linked)	plastocyanin	-25	Geren *et al.*,1983

[a] *S. cerevisiae* cytochrome *c* peroxidase

Binding of Macromolecules. Eqn. 7.13 shows that a perturbation in the redox potential will be observed if there is differential binding by the two redox states to a ligand. Such perturbations are observed for cytochrome *c* binding to lipid vesicles, mitochondria and purified respiratory complexes (Table 7.1) and are of a magnitude consistent with a five to seven times tighter binding of the oxidised state.

Scheme 7.2 General ligand binding equation

$$\text{ox} + \text{L} \rightleftharpoons \text{ox-L} \qquad K_{ox} = \frac{[\text{ox}][\text{L}]}{[\text{ox-L}]} \tag{7.6}$$

$$\text{red} + \text{L} \rightleftharpoons \text{red-L} \qquad K_{red} = \frac{[\text{red}][\text{L}]}{[\text{red-L}]} \tag{7.7}$$

since $ox_t = ox + ox\text{-}L$, $red_t = red + red\text{-}L$, and

$$E_h = E_m + 0.06 \log \frac{ox_t}{red_t} \tag{7.8}$$

(1) Form A

$$\text{then: } E_h = E_m + 0.06 \log \frac{\left([\text{ox}] + \frac{[\text{ox}][\text{L}]}{K_{ox}}\right)}{\left([\text{red}] + \frac{[\text{red}][\text{L}]}{K_{red}}\right)} \tag{7.9}$$

$$E_h = E_m + 0.06 \log \frac{[\text{ox}]\left(1 + \frac{[\text{L}]}{K_{ox}}\right)}{[\text{red}]\left(1 + \frac{[\text{L}]}{K_{red}}\right)} \tag{7.10}$$

When $[\text{ox}] = [\text{red}]$; $E_h = E_{m,NL}*$

$$\therefore \quad E_{m,NL} = E_m + 0.06 \log \frac{\left(1 + \frac{[\text{L}]}{K_{ox}}\right)}{\left(1 + \frac{[\text{L}]}{K_{red}}\right)} \tag{7.11}$$

$$\text{or} \quad \mathbf{E_m = E_{m,NL} - 0.06 \log \frac{K_{red}(K_{ox} + L)}{K_{ox}(K_{red} + L)}} \tag{7.12}$$

In the limit, at high [L]

$$\mathbf{E_{m,L} = E_{m,NL} - 0.06 \log \frac{K_{red}}{K_{ox}}} \tag{7.13}$$

if $K_{ox} < K_{red}$ (i.e. higher affinity for ox state), $E_m < E_{m,NL}$

if $K_{ox} > K_{red}$ (i.e. higher affinity for red state), $E_m > E_{m,NL}$

Rees (1985) suggests that this type of effect is of importance in matching the potentials of the redox centres in an electron transfer complex composed of a lower potential acidic protein and a higher potential basic protein. This seems unlikely as a general mechanism for two reasons. Firstly, Nicholls (1974) notes that the observed perturbations tend to be larger for the ligands high in lipid and considers that the lowering of potential is due to non-specific lipid binding rather than specific active site binding. Secondly, the proposed pattern of increasing basicity accompanying more positive redox potential is not supported by actual electron transport sequences. The basic mitochondrial cytochromes *c* for example, must interact with acidic surfaces, not only on its reductase, but also on cytochrome oxidase.

As noted in Pettigrew & Moore (1987: p42), a redox state dependent binding of cytochrome *c* to its redox partners could contribute to the

Scheme 7.2 continued

(2) Form B

then: $$E_h = E_m + 0.06 \log \frac{\left([\text{ox-L}] + K_{ox}\frac{[\text{ox-L}]}{[L]}\right)}{\left([\text{red-L}] + K_{red}\frac{[\text{red-L}]}{[L]}\right)} \qquad (7.14)$$

$$E_h = E_m + 0.06 \log \frac{[\text{ox-L}]\left(1 + \frac{K_{ox}}{[L]}\right)}{[\text{red-L}]\left(1 + \frac{K_{red}}{[L]}\right)} \qquad (7.15)$$

when $[\text{ox-L}] = [\text{red-L}]$, $E_h = E_{m,L}$

$$\therefore \quad E_{m,L} = E_m + 0.06 \log \frac{\left(1 + \frac{K_{ox}}{[L]}\right)}{\left(1 + \frac{K_{red}}{[L]}\right)} \qquad (7.16)$$

or $$\mathbf{E_m = E_{m,L} - 0.06 \log \frac{(L + K_{ox})}{(L + K_{red})}} \qquad \mathbf{(7.17)}$$

In the limit, at low [L]

$$\mathbf{E_m = E_{m,L} - 0.06 \log K_{ox}/K_{red}} \qquad \mathbf{(7.18)}$$

if $K_{ox} < K_{red}$ (i.e. higher affinity for ox form), $E_{m,NL} > E_{m,L}$

if $K_{ox} > K_{red}$ (i.e. higher affinity for red form), $E_{m,NL} < E_{m,L}$

* $E_{m,L}$ is the midpoint potential for: $\text{ox-L} + e^- \rightarrow \text{red-L}$

$E_{m,NL}$ is the midpoint potential for: $\text{ox} + e^- \rightarrow \text{red}$

efficiency of electron transfer if the ferricytochrome *c* has a high affinity for the reductase and a low affinity for the oxidase while the ferrocytochrome *c* shows the converse pattern. However, the experimental results suggest that both reductase and oxidase bind the oxidised protein more strongly than the reduced.

7.3 The Coordination Structure of the Metal Ion

The two redox states of a metal centre may have different preferences for the type of coordination ligand and for the geometry of the coordination. For example, the oxidised ion favours negative ligands or those with greater σ-donor power, while the reduced ion favours ligands with greater π-acceptor power (Williams, 1959; Perrin, 1959; Moore and Williams, 1976). Changes in spin state and steric factors can alter the relative importance of these influences, as can redox dependent changes in the structure of the metal complex. However, ligand replacement during the oxidation reduction process is rare in simple electron transfer proteins and conformational changes are often small or absent (section 7.7; Chapter 4.5).

Building on the work with small compounds, Vallee and Williams (1968) proposed that metal ions in proteins were held in a constrained site that is a compromise between what would normally be preferred by the two individual oxidation states. The precise nature of the compromise will define the relative stabilisation of the two oxidation states and influence the level of the redox potential. This proposal, known as the entatic state hypothesis, is considered further in section 7.3.4.

7.3.1 Porphyrin as an Fe Ligand

A porphyrin is a weak dibasic acid which loses two pyrrole-N-protons on coordinating iron and becomes a divalent anion (Chapter 1.4). Thus the net charge of the Fe(II) porphyrin redox centre is zero and that of the Fe(III) porphyrin is +1. Coordination of the iron by this negatively charged ligand results in a relative stabilisation of the oxidised state and lowers the midpoint redox potential from the +771 mV of $Fe(III)_{aq}/Fe(II)_{aq}$ to -115 mV for protoheme IX (Table 7.2). This is an effect shared by all cytochromes.

Table 7.2 Redox potentials of hemes - influence of axial ligands

	Redox potential (mV)	Coordination structure	L (6th ligand)
	$E_{m,7}$		
Protoheme IX	-115 [b]		
Cytochrome *c* heme octapeptide	-207 [c]	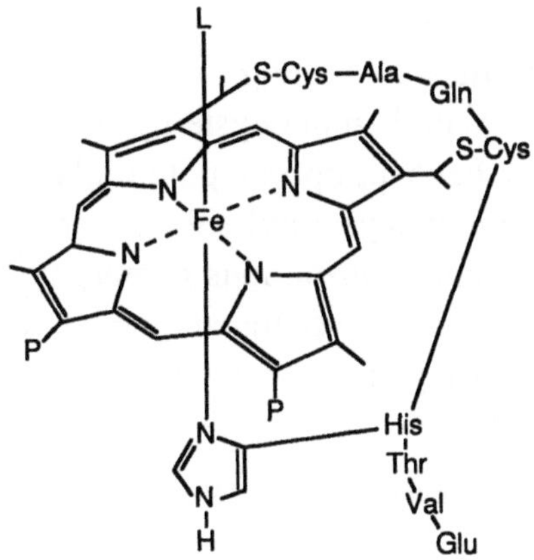	Vacant
Imidazole cytochrome *c* heme octapeptide	-210 [c]		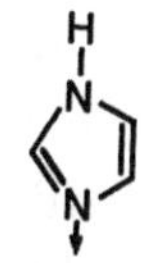
N-acetyl methionine cytochrome *c* heme octapeptide	-50 [d]		(Ac)Met(S) ↓
	E_{SCE}		
(N-MeIm)(C_5Im)TPP [a]	+18 [e]	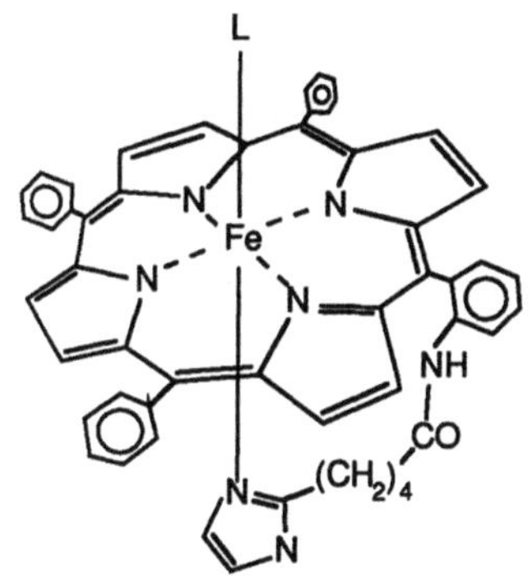	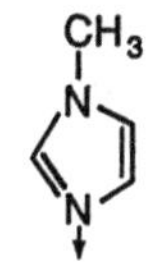
(PMS)(C_5Im)TPP [a]	+186 [e]		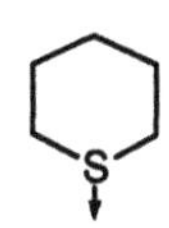

[a] (C_5Im)TPP is a tetraphenylporphyrin with a 'tail' of C_5-imidazole which provides the 5th iron ligand. The 6th ligand is provided by N-methyl imidazole (N-MeIm) or pentamethylene sulfide (PMS). These hydrophobic systems were studied in tetrahydrofuran and potentials are expressed relative to the standard calomel electrode (SCE) in this solvent.
[b]Shack & Clark (1947); [c]Harbury & Loach (1960); [d]Harbury *et al.* (1965); [e]Marchon *et al.* (1982)

7.3.2 Effects of Porphyrin Substituents

Substituents at the periphery of the porphyrin ring which are electron-withdrawing lead to increasing acidity (lower pK) of the pyrrole nitrogens (Table 7.3). The consequent decrease in the donor power of these nitrogens results in poorer stabilisation of the oxidised state and more positive redox potentials (Williams, 1959; Falk, 1964). However, the effect is not a large one for mesoheme and protoheme and it would be expected to contribute a stabilisation energy of only 20 - 40 mV to the difference in redox potentials between cytochromes containing protoheme IX and those containing heme *c*. In fact, the monovinyl, monocysteinyl-substituted cytochromes *c* found in certain protozoa have midpoint potentials slightly less, rather than more, positive than their biscysteinyl-substituted counterparts (Pettigrew, 1979). Thus, although small contributions of porphyrin substituents to ΔG_{stab} can be detected in simple model systems, they may be obscured or outweighed within the complex stabilising environment of a protein.

Table 7.3 Redox potentials of hemes - influence of porphyrin substituents [a]

	Side Chains at position: 2	4	pK (porphyrin methyl ester)	$E_{m,9.6}$ (bis CN^-)
mesoheme	C_2H_5	C_2H_5	5.8	-229
hematoheme	$CH(OH)CH_3$	$CH(OH)CH_3$		-200
protoheme IX	$CH{=}CH_2$	$CH{=}CH_2$	4.8	-183
chlorocruoroheme	CHO	$CH{=}CH_2$	3.7	-113

[a] Data from Falk (1964)

7.3.3 Effects of Axial Ligands

As with the planar pyrrole nitrogens, differences in the electron donor-acceptor power of ligands in the axial positions also influence redox potential. Thus histidine, which is a good electron donor, will tend to stabilise the oxidised state. In fact histidine in one of the two axial positions is a constant structural *motif* in *c*-type cytochromes. Where there are differences in different cytochromes *c* is in the second axial ligand position which can be a second histidine, a methionine, a lysine, or can be vacant. The methionine sulfur is a good electron acceptor and therefore favours the relatively electron-rich reduced state (Chapter 1.4) resulting in a more positive redox potential. This is shown by the two model systems in Table 7.2. The heme octapeptide is a proteolytic fragment of cytochrome *c* which retains the histidine axial ligand but in which the second axial position can be filled by a variety of ligands. Coordination by methionine sulfur gives rise to a midpoint potential more positive by 160 mV compared to that observed for imidazole coordination (Harbury *et al.*, 1965). Similar results are obtained for the synthetic tetraphenylporphyrin to which a covalently attached imidazole provides the first axial ligand (Marchon *et al.*, 1982). Coordination of this compound by the sulfur of pentamethylene sulfide shifts the redox potential 168 mV more positive compared to the value for bisimidazole coordination. These are large shifts and represent a major influence on redox potential. The class I cytochromes, with histidyl-methionyl coordination, have E_m values in the range 0 to +400 mV while the class III cytochromes *c*, with bishistidyl coordination, occupy the range -400 to -100 mV. Using the medians of these ranges, replacement of histidine by methionine as axial ligand accounts for about a third of the difference in redox potentials between the two classes of cytochromes *c*.

The histidinate anion (His^-) would be expected to stabilise the ferric state even more strongly but is not found in cytochromes *c* at physiological pH values (Chapter 1.4). However, hydrogen-bonding to the histidine ligand, which is a common feature of the class I *c*-type cytochromes, confers partial histidinate character (Chapter 1.4) and Valentine *et al.* (1979) have proposed that this hydrogen-bonding may modulate the redox potential. However, differences in hydrogen-bonding should be reflected in differences in the NMR resonances of the histidine ligand and this was found not to be the case for a range of class I cytochromes *c* (G.R.Moore, unpublished observations). Similarly, the proposal of Korszun *et al.* (1982), that the orientation of the histidine relative to the heme plane is a major cause of the variation in redox potentials of class I cytochromes *c*, is not supported by NMR data.

The contribution of ΔG^{coord} to ΔG_{stab} may be determined spectroscopically by study of the electronic charge transfer transitions (Chapter 2.3.1.1). Makinen and Churg (1983) and Schejter and Eaton (1984) argue that the energy of a charge transfer transition and the reduction potential are related because they share the feature of electron promotion from a porphyrin π orbital into the iron d_{yz} orbital. Schejter and Eaton assume that the CT band of ferricytochrome *c* at ~1700 nm and of its imidazole derivative at ~1500 nm arises from the transition $a_{2u}(\pi) \rightarrow d_{yz}$ (transition 4 in Fig. 2.2.1), although this has not been unambiguously demonstrated. As Table 2.4 shows, this wavelength shift is almost solely dependent on the axial ligands. The increase in CT frequency between native cytochrome *c* and imidazole cytochrome *c* is proposed by Schejter and Eaton to correspond to a decrease in E_m of 100 mV, in reasonable agreement with the figures already discussed for histidyl-methionyl and bishistidyl coordination of porphyrins. A comparison of *c*-type cytochromes with a common tertiary structure but different axial ligation is awaited with interest. Studies of proteins with different tertiary structures and common axial ligation show that there is no relationship between E_m and the CT band.

Ligand displacement is not thought to be a feature of the electron transfer reactions of most cytochromes *c*. They operate by outer sphere processes, with retention of coordination around the iron (Chapter 8.2). At alkaline or acid pH values or at elevated temperatures, ligand displacement does occur (Chapter 4.4) but these effects are not thought to be of physiological significance. The displacement at alkaline pH of the methionine ligand (probably by a lysine, Chapter 4.4) leads to a marked relative stabilisation of the oxidised state and a more negative redox potential. In cytochrome *c* oxidases, oxygen (and its competitive mimic, carbon monoxide) do bind to the heme group. Because both bind much more strongly to the reduced state, they cause a marked elevation in midpoint potential (this effect is only experimentally accessible with carbon monoxide because oxygen itself is immediately reduced). The shift to more positive E_m values has been used to identify potential *b*-type oxidases in bacterial membranes where multiple *b*-type cytochromes exist (Reid and Ingledew, 1979). The magnitude of the observed shift in E_m is related quantitatively to the relative affinity of the ligand for the two redox states by Eqn 7.13, Scheme 7.2.

7.3.4 Geometry of Coordination

We have seen that the nature of the coordinating ligand influences the relative stability of the reduced and oxidised states of the metal ion. In addition, these two redox states may have different preferences in coordination geometry. For example, Cu(I) and Cu(II) favour tetrahedral and planar tetragonal geometries respectively. Fe(II) and Fe(III) share a preference for octahedral coordination but the Fe(III) state favours a shorter bond length (Moore and Williams, 1976).

As noted in the introduction to this section, the coordination geometry of a redox protein may be a compromise (an entatic state). A good example of this principle is the copper protein, plastocyanin, in which the copper is held in a distorted tetrahedron. The consequent relative stabilisation of Cu(I) leads to a midpoint potential 190 mV more positive than the potential of the $Cu(II)_{aq}/Cu(I)_{aq}$ couple (Colman *et al.*, 1978).

In the cytochromes, entatic coordination geometries are more difficult to detect because the differences in preferred bond lengths are very small, below the resolution achieved by most protein x-ray crystallographic analyses. Moore and Williams (1977) proposed that the NMR chemical shift of the methionine methyl resonance is an indicator of the Fe-S bond length in class I ferrocytochromes *c*. As the bond length shortens, the methyl protons should move further into the electron cloud of the heme group experiencing greater shielding and an upfield shift. These workers found a correlation between this chemical shift and the redox potential of the class I cytochromes *c*. The greater the shift (and by implication, the shorter the Fe-S bond length), the more negative the redox potential (consistent with the predicted stabilising effect on the oxidised state as the methionine sulfur becomes a better donor).

However this proposal has not been supported by further NMR work (Moore *et al.*, 1980c) nor by determination of the Fe-S bond length using EXAFS, a method which is sensitive to small differences in atomic spacing around a fluorescing metal centre (Chapter 2.2.7). Korszun *et al.* (1982) found that the Fe-S bond was virtually constant in length in the different class I cytochromes *c*. Thus differences in Fe-S bond length are probably not important influences on redox potential in the cytochromes *c*.

7.4 Electrostatic Influences on Redox Potential

In this section we consider electrostatic influences beyond the coordination shell of the metal ion that contribute to the term ΔG^{el}. We note that this distinction is somewhat artificial since the effects considered in the previous section are partly electrostatic in nature. However, these cannot be treated by simple electrostatic equations since the distribution and bonding of electrons are affected.

Simple models in which the protein is treated as a homogeneous sphere are quite successful in describing some of the electrostatic properties of cytochromes. These models, which are described as macroscopic, are considered in the following three sections. However, in some respects they prove to be inadequate and more recent 'microscopic' and 'semi-macroscopic' models are discussed in the final section.

7.4.1 The Influence of Ions in Solution

Both acidic cytochromes *c* (for example *Euglena* cytochrome *c*-552) and basic cytochromes *c* (for example horse heart cytochrome *c*) show an ionic strength dependence of midpoint potential (Margalit and Schejter, 1973; Goldkorn and Schejter, 1976; Fig. 7.2). This dependence is due to changes in the activity coefficients, f(ox) and f(red), of the two redox states:

$$E_m = E_{m,I=0} + \frac{RT}{nF} \ln \frac{f(ox)}{f(red)} \qquad (7.19)$$

where $E_{m,I=0}$ is the extrapolated value of E_m for I=0. Using Debye-Huckel theory, the ratio of activity coefficients can be expressed in terms of net charge (Z_{red} and Z_{ox}) and ionic strength (I).

$$E_m = E_{m,I=0} + 2.3\frac{RT}{nF}(Z^2_{red} - Z^2_{ox})\frac{0.5\sqrt{I}}{1 + 0.33A\sqrt{I}} \qquad (7.20)$$

where A is the 'effective ionic diameter', usually taken as the simple sum of the ionic radii (Tanford, 1961; Dutton and Wilson, 1974; Margalit and Schejter, 1973).

Qualitatively, the redox form carrying the larger net charge will experience greater shielding of its charge by the ionic atmosphere, and therefore a relative stabilisation. For a basic protein, the oxidised form will be stabilised, giving rise to more negative E_m values with rising ionic strength; for an acidic protein, the reduced form will be stabilised, giving rise to more positive E_m values with rising ionic strength. These features are apparent in Fig. 7.2 for 'non-binding' conditions ('binding conditions' have been considered in section 7.2).

The mathematical model represented by Eqn. 7.20 is naive in that it is based on a representation of the protein as a low dielectric cavity with charges smeared on the surface. However, it has been surprisingly effective in describing the behaviour of horse cytochrome *c* (Fig. 7.2) and its modified derivatives (Fig. 7.3). For A = 18.5, f(I) in Fig. 7.2 is $I/(1 + 6\sqrt{I})$ and, from the slope, a value of -11.2 for $(Z_{red}^2 - Z_{ox}^2)$ is obtained which is consistent with $Z_{red} = +5$ and $Z_{ox} = +6$ compared to the theoretical figures of (+6 to +8) and (+7 to +9) respectively depending on how the heme propionates are taken into account (Chapter 4.4.2). When four lysines have been maleylated (a net charge change of -8), the midpoint potential is almost independent of ionic strength (Aviram *et al.*, 1981; Fig. 7.3) indicating that the average net charge is zero.

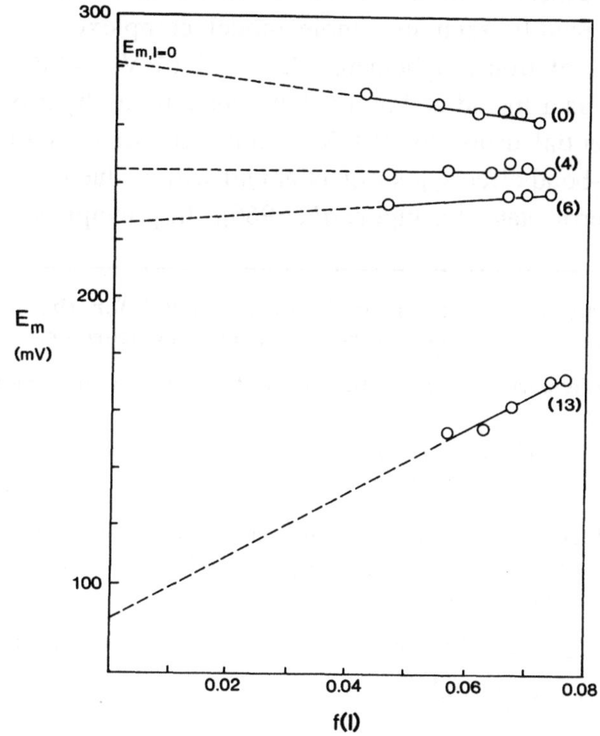

Fig. 7.3 The effect of maleylation on the redox potential of horse heart cytochrome *c*. $f(I) = I/(1 + 6\sqrt{I})$. Figures in parentheses are the numbers of maleylated lysines. Each maleylation results in a net charge change of -2. $E_{m,I=0}$ values are obtained by extrapolation to f(I) = 0 (Schejter *et al.*, 1982)

Unfortunately, although *Euglena* cytochrome *c*-552 qualitatively obeys the model, sensible values of Z_{ox} (-8) and Z_{red} (-9) could only be obtained if f(I) was altered to $I/(1 + \sqrt{I})$ (Goldkorn and Schejter, 1976). This is a purely pragmatic adjustment with no theoretical justification and detracts from the general application of the equation. The effect of ion-shielding on redox potential thus varies with the individual protein and can produce substantial alterations in the value of $E_{m,I=0}$. For *Euglena* cytochrome *c*-552 $E_{m,I=0}$ is 59 mV more negative than E_m at I = 0.005 while the corresponding figure for horse heart cytochrome *c* is +16 mV (Fig. 7.2).

7.4.2 The Influence of Surface Charge

The influence of intramolecular electrostatic interactions on redox potential can be readily seen in simple model complexes of iron. The midpoint potential of iron-trisphenanthroline is 1.14 V. With 4,7-dihydroxyphenanthroline as a ligand and at a pH at which all six hydroxyl groups are ionised, the potential drops to -0.1 V. Although some of the fall will be due to through-bond effects, a substantial part will be due to electrostatic stabilisation of the ferric state (George *et al.*, 1966). In principle such effects are expected

Scheme 7.3 The smeared charge model for the influence of surface charge on the oxidation-reduction of a cytochrome.

Difference in electrostatic energy of the reduced and oxidised forms of a redox couple:

$$\Delta G = \frac{Ne^2(Q_{red}^2 - Q_{ox}^2)}{8\pi e_o DR} \tag{7.21}$$

N, Avogadro's number; e, the charge on the electron (1.6 x 10^{-19} C); e_o, the permittivity of free space (8.8542 x 10^{-12} $C^2 N^{-1} m^{-2}$); D, the dielectric constant; R, the cavity radius.

Since $Q_{red} = Q_{ox} - 1$ (assuming no oxidation-linked ligand binding):

$$\Delta G = \frac{Ne^2(1 - 2Q_{ox})}{8\pi e_o DR} \tag{7.22}$$

For a change in net charge (ΔQ), the change in ΔG is:

$$\Delta\Delta G^{surf} = \frac{Ne^2}{8\pi e_o DR}[1 - 2Q_{ox} - \{1 - 2(Q_{ox} - \Delta Q)\}] \tag{7.23}$$

$$\Delta\Delta G^{surf} = -\frac{Ne^2\Delta Q}{4\pi e_o DR} \tag{7.24}$$

$$\therefore \Delta E^{surf} = \frac{Ne^2\Delta Q}{4\pi e_o FDR} \text{ (volts)} \tag{7.25}$$

which simplifies to:

$$\Delta E^{surf} = 12\Delta Q \text{ (millivolts)} \tag{7.26}$$

for a sphere of radius, 15 Å in water (D = 78.5 at 20°C)

for redox proteins carrying a net charge; in practice they can be difficult to evaluate separately from other factors.

According to the smeared charge model (Tanford, 1961; Rees, 1985), the difference in electrostatic energy of the reduced and oxidised forms of a redox couple can be expressed by Eqn. 7.21 (Scheme 7.3). This scheme shows that the loss of one positive charge will theoretically result in a 12 mV fall in redox potential for a sphere of radius 15Å in water, consistent with a relative stabilisation of the oxidised state.

One experimental approach for investigating the influence of surface charge is to examine the midpoint potentials of a variety of redox proteins carrying different net charge. Rees (1985) found a correlation between net charge and E_m which could be expressed by $dE_m/dQ = 23$ mV. Although in moderate agreement with the theoretically predicted value, this is almost entirely fortuitous because the other components of ΔG_{stab} cannot be constant in the varied group of proteins included in the analysis. The danger of this approach is illustrated by the cases of *Euglena* cytochrome *c*-552 and horse heart cytochrome *c* in Fig. 7.2. These are respectively very acidic and very basic proteins but their midpoint potentials at zero ionic strength are the same. The original interpretation of this finding by Goldkorn and Schejter (1976) was that surface charge had no influence on redox potential. However, an alternative view, later endorsed by Schejter *et al.*(1982), is that electrostatic and non-electrostatic components of ΔG_{stab} happen to balance in the case of the two cytochromes.

A more reasonable approach to the influence of surface charge on redox potential is to study the effect of altering the net charge of an individual protein. This may be done either by chemical modification or by pH titration. The results of chemical modification studies are ambivalent. Rees (1980) chose to emphasise the fall of 15 mV in E_m caused by trifluoromethylphenylcarbamoylation of lysines 13, 27, 72 or 79 in horse cytochrome *c* (Smith *et al.*, 1977), but ignored the fact that trifluoroacetylation (Staudenmeyer *et al.*, 1977), carboxydinitrophenylation (Ferguson-Miller *et al.*, 1978) or trinitrophenylation (Osheroff *et al.*, 1983) of some of the same lysines had no effect on E_m. This is especially noteworthy for the CDNP group, which results in a net charge change of -2. Maleylation of horse heart cytochrome *c* results in a lowering of $E_{m,I=0}$ to a degree dependent on the extent of modification (Aviram *et al.*, 1981; Schejter *et al.*, 1982; Fig. 7.3). An average value of $dE/dQ = 6$ mV can be calculated from the data which, although lower than the theoretically predicted value of 12 mV, is certainly an indicator that surface charge influences redox potential. This is also indicated by the ionic strength dependence of E_m for *Euglena* cytochrome c-552 studied at different pH values (Schejter *et al.*, 1982). For a calculated charge change of +9 between pH 7 and pH 4.8, the midpoint potential at I=0 became 83 mV more positive, giving dE/dQ = 9 mV. However, in some of these studies it must be borne in mind that the chemical modifications may lead to a change in the conformation that affects the potential.

To summarise, there is evidence that a single charge change can alter the midpoint potential of cytochromes *c* at zero ionic strength by about 5-10 mV, rather lower than the theoretical prediction of the smeared charge

Scheme 7.4 The energetics of heme burial in a hydrophobic shell

The free energy difference in going from water to the dielectric of the protein is given by:

$$\Delta G_{W\rightarrow P} = (G^o_{Fe(II),W} - G^o_{Fe(III),W}) - (G^o_{Fe(II),P} - G^o_{Fe(III),P}) \quad (7.27)$$

$$= \Delta G^o_{Fe(II),W\rightarrow P} + \Delta G^o_{Fe(III),W\rightarrow P} \quad (7.28)$$

These terms will have electrostatic and non-electrostatic components:

$$\Delta G_{W\rightarrow P} = \Delta G^{o,el}_{Fe(II),W\rightarrow P} + G^{o,non\text{-}el}_{Fe(II),W\rightarrow P} - G^{o,el}_{Fe(III),W\rightarrow P} - G^{o,non\text{-}el}_{Fe(III),W\rightarrow P} \quad (7.29)$$

By far the dominant factor is considered to be $G^{o,el}_{Fe(III),W\rightarrow P}$ since the ferrous heme is electrically neutral and the non-electrostatic terms are predicted to be very similar.

$$\Delta G_{W\rightarrow P} \approx G^{o,el}_{Fe(III),W\rightarrow P} \quad (7.30)$$

This is the difference in the electrostatic energy of the ferric heme in the two environments. This can be calculated in a two step process.

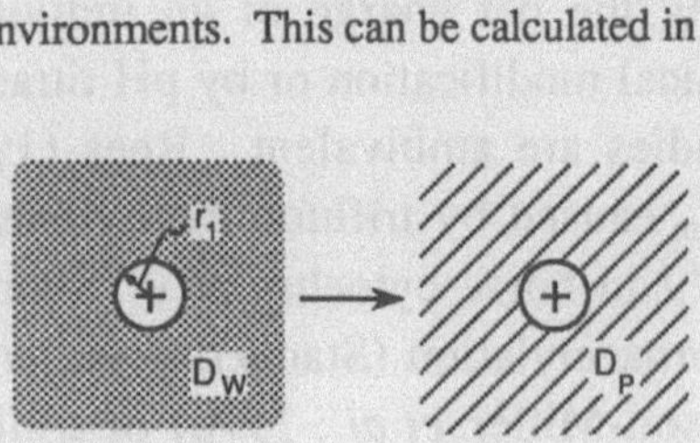

$$\Delta G = \frac{Ne^2Q^2}{8\pi e_o r_1}\left(\frac{1}{D_P} - \frac{1}{D_W}\right) \quad (7.31)$$

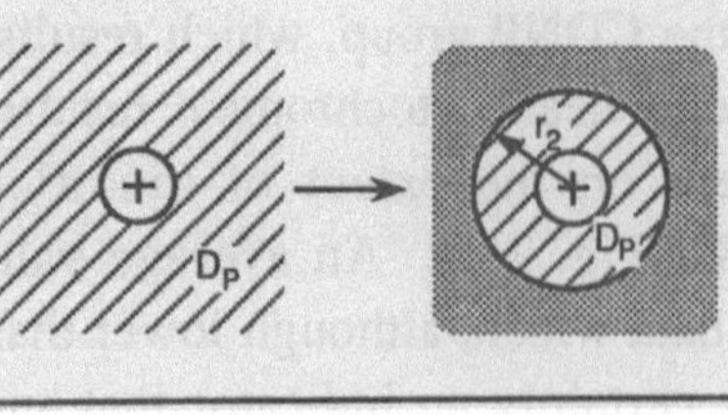

$$\Delta G = \frac{Ne^2Q^2}{8\pi e_o r_2}\left(\frac{1}{D_W} - \frac{1}{D_P}\right) \quad (7.32)$$

model. A problem with the use of this simple model is that, in reality, the heme charge is buried within the protein, which is a local medium with a complex dielectric response. This problem is considered in the following section. In experimental investigations, care must be taken to ensure that other components of ΔG_{stab} are constant and that ionic strength effects are accounted for.

7.4.3 The Influence of Redox Centre Burial within the Protein

The class I cytochromes *c* contain heme buried within the protein with little solvent exposure (Chapter 4.2). Their redox potentials are much more positive than that for the methionyl heme octapeptide model system, in spite of a common coordination of iron, and Kassner (1973) proposed that this reflects a relative destabilisation of the singly charged ferric heme in the hydrophobic environment of the protein.

Kassner's calculation of this destabilisation energy involves two steps (given in detail in Scheme 7.4). The first is the transfer of the ferric heme (radius r_1) to a hydrophobic solvent with dielectric constant D_P. The second is the transfer of a shell of hydrophobic solvent (radius r_2) containing the heme back into water (D_W).

Scheme 7.4 continued

$$\therefore \quad \Delta G_{Fe(III),W\rightarrow P} = \frac{Ne^2Q^2}{8\pi e_o r_1}\left(\frac{1}{D_P} - \frac{1}{D_W}\right) + \frac{Ne^2Q^2}{8\pi e_o r_2}\left(\frac{1}{D_W} - \frac{1}{D_P}\right) \tag{7.33}$$

$$\Delta G_{Fe(III),W\rightarrow P} = \frac{Ne^2Q^2}{8\pi e_o}\left(\frac{1}{r_1} - \frac{1}{r_2}\right)\left(\frac{1}{D_P} - \frac{1}{D_W}\right) \tag{7.34}$$

$$\Delta E_{W\rightarrow P} = -\frac{Ne^2Q^2}{nF8\pi e_o}\left(\frac{1}{r_1} - \frac{1}{r_2}\right)\left(\frac{1}{D_P} - \frac{1}{D_W}\right) \tag{7.35}$$

If $r_1 = 0.5$ nm and $r_2 = 1.5$ nm and $D_P = 3.5$* and $D_W = 78.54$ then:

$\Delta G^o = 25.1 \text{ kJ mol}^{-1}$

$\Delta E = 260 \text{ mV}$

* Note that Kassner (1972) chose a lower value for D_P (2.27) and slightly different values for r_1 and r_2 which gave rise to a ΔE of 468 mV. D_P is often assumed to be between 2 and 5 and we have chosen 3.5 as a reasonable value for use in several of the calculations that appear in this chapter and for comparison with the Warwicker and Watson model.

With $D_P = 3.5$, $D_W = 78.54$ and values for r_1 (5Å) and r_2 (15Å), a destabilisation energy of 260 mV is obtained from Eqn. 7.34. This is in good agreement with the 300 mV difference between the redox potentials of the methionine heme octapeptide and mitochondrial cytochrome *c*. A similar shift to more positive redox potential was observed for bispyridyl-Fe-mesoporphyrin in benzene compared to the value in water (Kassner 1972).

Table 7.4 Redox potential and heme exposure

Crystallography		$E_{m,7}$ (mV)	% heme [a] exposure	Refs.
Rhodospirillum rubrum	cytochrome c_2	320	6	d
Mitochondrial	cytochrome *c*	260	4	d
Paracoccus denitrificans	cytochrome *c*-550	250	5	d
Hemoglobin	α-chain	113	14	d
	ß-chain	53	20	d
Myoglobin		47	18	d
Cytochrome b_5		20	23	d
Solvent perturbation		$E_{m,7}$ (mV)	% heme [b] exposure	Refs.
R. rubrum	ferricytochrome c_2	323	21	e
	ferrocytochrome c_2		17	e
Horse	ferricytochrome *c*	260	26	e
	ferrocytochrome *c*		20	e
Prosthecochloris aestuarii	ferricytochrome *c*-555	103	17	f
Desulfuromonas acetoxidans	ferricytochrome *c*-551.5 [c]	-102(1)	36	g
	ferrocytochrome *c*-551.5	-177(2)	33	g

[a] Calculated from XRD data by examination of 6 nearest neighbour atoms in a cubic grid.
[b] Calculated from the increase in absorbance at the Soret band due to the presence of 20% ethylene glycol. This is expressed relative to the increase observed for the heme octapeptide in imidazole.
[c] Cytochrome *c*-551.5 is a class III cytochrome with 3 hemes (see Chapter 5.3).
[d]Stellwagen (1978); [e]Schlauder & Kassner (1979); [f]Fiechtner & Kassner (1978); [g]Fiechtner & Kassner (1979)

The degree of hydrophobicity of the heme environment, and therefore the redox potential, might be expected to be related to the degree of exposure of the heme. By representing a hemeprotein as a grid of atoms, and identifying those that were not bounded on all sides by other heme or protein atoms, Stellwagen (1978) found the high potential cytochromes *c* contained a less exposed heme than the globins and cytochrome b_5, consistent with their more positive redox potentials (Table 7.4). However, it seems of doubtful validity to compare such structurally and functionally diverse proteins. Perhaps a more reasonable approach is comparison within the cytochrome *c* family. Stellwagen obtained very similar and low heme exposures for the three cytochromes *c* that he studied, but investigation by solvent perturbation yielded more discriminating results (Table 7.4). This method is based on the small increase in absorbance of the Soret band in the presence of perturbants such as ethylene glycol. This perturbation is linear with concentration of ethylene glycol so that an unfolding effect is not involved. Heme octapeptide with added imidazole is taken as the model for fully exposed heme and gives the maximal increase in absorbance against which less exposed hemes are measured.

Comparing cytochrome c_2 and cytochrome *c*, which are close structural relatives, Schlauder and Kassner (1979) concluded that the less positive redox potential of cytochrome *c* is associated with a greater heme exposure. The exposures are much greater than those obtained by Stellwagen, perhaps because the reference heme of the heme octapeptide is in fact considerably shielded by the peptide chain.

The small class III cytochrome *c*-551.5 of *Desulfuromonas acetoxidans* contains only 68 amino acids to envelope three heme groups and, not surprisingly, gives a high figure (33 - 36%) for heme exposure (Fiechtner and Kassner, 1979; Table 7.4). However, the cytochrome *c*-555 from *Prosthecochloris aestuarii* is anomalous in having a lower degree of heme exposure than horse heart cytochrome *c*, but a less positive redox potential (Fiechtner and Kassner, 1978).

Thus within a structurally homologous series there may be a detectable correlation between heme exposure and redox potential but, because the latter is influenced by multiple factors, anomalies like cytochrome *c*-555 will appear and will become more frequent, the less homologous the grouping. Use of heme exposures to assign E_m values to redox centres in multiheme proteins should be treated cautiously. Certainly, the correlation equation of Stellwagen (1978) should not be used for this purpose.

7.4.4 The Influence of Buried Charge

We have seen in section 7.4.2 that changes in surface charge perturb the redox potential by only a few millivolts, implying a high dielectric constant between the surface groups and the iron. Yet in the previous section we found that the properties of the heme in a cytochrome are consistent with burial in a medium of low dielectric constant. Thus ionisable groups that are also buried in the protein would be expected to influence strongly the relative stabilities of the reduced and oxidised states and therefore the redox potential.

Such groups are rare in proteins, but examples in cytochromes are the heme propionic acids. The rear heme propionate (HP-7) is situated at the bottom of the heme crevice in an environment including several hydrogen-bonding or charged groups (Chapter 4). For example, in most of the cytochromes c_2 and mitochondrial cytochromes, HP-7 lies close to the conserved basic residue, arginine 38, which would be expected to stabilise the propionate anion and lead to a low pK_a (Moore *et al.*, 1984a). In just two members of this family - the cytochromes c_2 of *Rhodomicrobium vannielii* and *Rhodospirillum rubrum* - arginine 38 is replaced (by a glutamine and an asparagine respectively) and the pK_a of the propionate is raised to near 7 (Moore *et al.*, 1984a) (see Chapter 4.4.2 for further discussion of this point).

A different pattern of residues forms the bottom of the heme crevice in the small class I cytochromes *c*, which include the cytochrome *c*-551 of the Pseudomonads. Again however, HP-7 lies close to a basic residue, which is

Scheme 7.5 Coulomb's Law

Coulomb's Law states that the force (F) acting between two charges is proportional to the amount of each charge and inversely proportional to the square of the distance (R) that separates them:

$$F \propto \frac{Ne^2 Q_1 Q_2}{R^2} \qquad (7.36)$$

Introducing a constant of proportionality, gives (for charges in a vacuum):

$$F = \frac{1}{4\pi e_o} \frac{Ne^2 Q_1 Q_2}{R^2} \qquad (7.37)$$

where e_o is the permittivity of free space.

When the charges are placed in a medium, the force is reduced due to the insulating properties of the medium. The dielectric constant (D) is a reflection of those insulating properties and is defined as:

$$D = \frac{F_{vacuum}}{F_{medium}} \qquad (7.38)$$

Introducing this into equation 7.37 and recasting in terms of work (W) gives:

$$W = \frac{1}{4\pi e_o D} \frac{Ne^2 Q_1 Q_2}{R} \qquad (7.39)$$

Since the work done for a thermodynamically reversible system is a Gibbs free energy term, Eqn.7.39, with insertion of constants, becomes:

$$\Delta G = \frac{1389 Q_1 Q_2}{DR} \text{ kJ mol}^{-1} \qquad (7.40)$$

arginine 47 in the case of *Ps.aeruginosa* cytochrome *c*-551; this residue is a histidine in other Pseudomonads. In this cytochrome group the propionic acid ionises near pH 7 (Moore *et al.*, 1980c; Leitch *et al.*, 1984). The identification of the propionic acid ionisations and their effect on redox potential will be discussed in the final section. We concentrate here on the interaction between the ionisable group and the ferric heme.

Table 7.5 The energetics of propionate ionisation in cytochromes

Cytochrome	EXPERIMENTAL [a b]		THEORETICAL MODELS [b c]						
			Uniform dielectric			*Cavity*		*Microscopic*	
	ΔE_m (obs) mV	D_{eff} (calc)	r Å	D	ΔE_m (calc) mV	ΔE_m (calc) mV	D_{eff} (calc)	ΔE_m (calc) mV	D_{eff} (calc)
Rm.vannielii c_2	65	21.9	(10.1)	3.5	406				
R.rubrum c_2	47	30.2	10.1	3.5	406	148	9.6		
						(119)	(12.0)		
Ps.aeruginosa *c*-551	65	27.0	8.2	3.5	501	90	19.5		
						(70)	(25.0)		
Horse *c*								259	11.1

[a] Values of ΔE_m (observed) are derived from the pH dependence of redox potential described in section 7.6.
[b] Values of D_{eff} (calculated) are obtained using the values for ΔE_m obtained either experimentally or by calculation of the interaction energy. Rearrangement of Coulomb's equation gives:

$$D_{eff} = \frac{NQ_1Q_2e^2}{nF4\pi e_o r_{12}\Delta E_m} \tag{7.41}$$

[c] For the uniform dielectric model, values of ΔE_m (calculated) are obtained using Coulomb's Law with the appropriate value of r and a dielectric constant of 3.5. For the cavity model (Rogers and Moore, 1988) and the microscopic model (Churg and Warshel, 1986), calculation of ΔE_m is described in the text. Figures in parentheses for the cavity model are obtained assuming a redistribution of charge on the propionate by association with an arginine in *Ps. aeruginosa* cytochrome *c*-551 and with a histidine in *R. rubrum* cytochrome c_2 (Rogers and Moore 1988). In the case of the microscopic model we have assumed that each propionate contributes equally to the ΔE_m of 259 mV in order to calculate D_{eff} of 11.1.

The energy of this interaction can be determined by the effect of ionisation on redox potential. The ionisation of HP-7 is associated with a fall in redox potential of 47 - 65 mV in both the large and small class I cytochromes *c* (Table 7.5). This is consistent with a relative electrostatic stabilisation of the oxidised state. Coulomb's Law (Scheme 7.5) describes the interaction energy between two charges in a medium of dielectric constant D. Using a low dielectric constant for the protein of 3.5 leads to a calculated value for the fall in redox potential almost an order of magnitude greater than the observed figure (Table 7.5). It is disagreements of this type between experimental observations and calculations based on a uniform dielectric that have led some authors to develop structurally more realistic models with which to calculate electrostatic interactions in proteins.

7.4.5 Electrostatic Models for Redox Proteins - Recent Developments

We have seen that a macroscopic model incorporating a uniform and low dielectric constant for the protein interior gives rise to erroneous estimates of the energy of interaction of buried charges. The problem can be further illustrated by considering an ionisable group with radius 2Å (for example a carboxylate) placed in a shell of 10Å radius and dielectric constant of 3.5. The Kassner analysis (Scheme 7.4) for calculation of the electrostatic energy of such a system yields the very large value of 75 kJ mol^{-1} which is equivalent to a pK shift of 13 pH units (Warshel *et al.*, 1984). Carboxylates, like the heme propionates we are considering here, are occasionally found in the interior of proteins but their pK values are not shifted by anything like this degree.

The problem lies in the application of the macroscopic concept of a dielectric constant, D, to the microscopic scale of a protein molecule in aqueous solution. D is a measure of the polarizability of a medium and is influenced by the presence and mobility of permanent and induced dipoles and charges. The value of 3.5, which we have adopted in the preceding calculations is a median figure for the range 2 - 5, measured with a capacitor, for hydrocarbons and polyamides. However, protein interiors contain many polar groups apart from the polypeptide backbone itself. Also, the dielectric function governing the interaction of two charged groups in a protein will be influenced not only by the polarity of intervening and surrounding protein groups but also by the solvent itself, which will set up a reaction field opposing the electrostatic field between the two charges.

One way to emphasise the importance of these factors is to turn the calculation on its head and use Coulomb's Law and the experimentally determined interaction energies to calculate an 'effective dielectric constant' (D_{eff}, Table 7.5). This concept may be quite a useful one if used with care. It is important to realise that D_{eff} incorporates effects of the solvent and the whole protein and therefore does not give a measure of the polarity of the region between the two charges. Also, it applies only to the particular interaction under consideration; other interactions in the same protein may be governed by different effective dielectric constants.

Two distinct approaches have been developed to deal with this complexity at the microscopic level.

The first is the cavity dielectric model employed by Rogers *et al.* (1985) that is based on the method of Warwicker and Watson (1982). This considers the protein-solvent system as a three dimensional cubic grid, each box of which is assigned a dielectric constant of either 3.5 (protein) or 80 (solvent), with charges located at grid points. This model can be used to calculate the extent to which the electrostatic field between two buried charges is modified by the reaction field generated in the solvent. Since this reaction field opposes that of the internal charges, estimates of interaction energy are obtained which are lower than those for the simple uniform dielectric model. For example, the method gives an interaction energy of 90 mV between HP-7 and the ferric heme of cytochrome *c*-551 which is in good agreement with the experimental value of 65 mV and corresponds to an 'effective dielectric constant' of 19.5 (Table 7.5). Indeed if the proximity of the arginine charge to the propionate is taken into account, a figure of 70 mV for the theoretical interaction energy is obtained (Rogers and Moore, 1988).

However, less satisfactory agreement with experimental results is obtained for *Rhodospirillum rubrum* cytochrome c_2 (Table 7.5). Rogers and Moore consider that this discrepancy may reflect a conformational change occurring in this protein on ionisation of the propionic acid. Even though the NMR data indicate a conformational change, this must be regarded as a speculative suggestion given that the model is so dependent on the assumption of $D_P = 3.5$. In spite of these problems, there is no doubt that the cavity dielectric model illustrates the important influence of the high dielectric response of the solvent.

The second approach is the microscopic polarizability model of Warshel and coworkers in which no dielectric constants are assumed, although a number of other assumptions concerning the dielectric response of the system are made (Warshel and Russell 1984; Rogers 1986). In the model of Churg and Warshel (1986) the change in solvation energy accompanying oxidation:

$$\Delta\Delta G_{sol} = \Delta G_{sol,ox} - \Delta G_{sol,red} \tag{7.42}$$

is given by the sum of several components evaluated at the detailed microscopic level:

$$\Delta\Delta G_{sol} = \Delta\Delta G_{PD} + \Delta\Delta G_{ID} + \Delta\Delta G_{conf} + \Delta\Delta G_{LD} + \Delta\Delta G_{bulk\ H_2O} \tag{7.43}$$

$\Delta\Delta G_{PD}$ and $\Delta\Delta G_{ID}$ are due to interaction of the ferric heme with permanent and induced dipoles in the protein. $\Delta\Delta G_{conf}$ is due to conformational differences between the redox states and is considered small and negligible. $\Delta\Delta G_{LD}$ and $\Delta\Delta G_{bulk\ H_2O}$ are due to protein solvent interactions, the former involving an inner layer of polarisable water molecules treated as Langevin dipoles and the latter involving the outer bulk water phase.

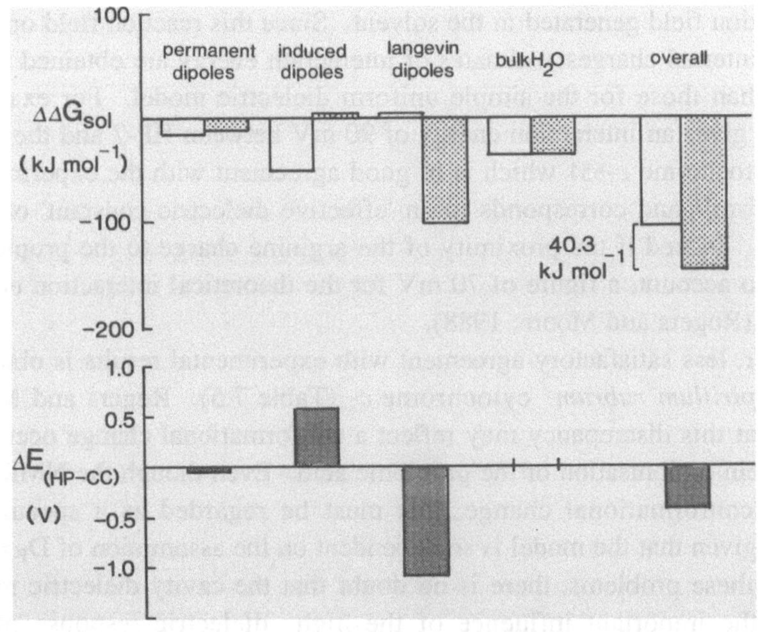

Fig. 7.4 Components of the solvation energy ($\Delta\Delta G_{sol}$) for the oxidation of horse heart cytochrome *c* and the heme octapeptide. The difference in solvation energy between the oxidised and reduced forms ($\Delta\Delta G_{sol} = \Delta G_{sol,ox} - \Delta G_{sol,red}$), shown in the upper part of the diagram, contains a number of contributing factors which were individually analysed as described by Churg and Warshel (1986) and summed to give the overall energy difference for the oxidation of horse heart cytochrome *c* (open boxes) and the heme octapeptide (stippled boxes). Theoretical differences in redox potential ($\Delta E_{HP\text{-}CC}$) between the two systems, heme peptide (**HP**) and cytochrome *c* (**CC**), can be calculated from the equation:

$$\Delta E_{HP\text{-}CC} = -\frac{\Delta\Delta G_{sol,HP} - \Delta\Delta G_{sol,CC}}{nF} \tag{7.44}$$

and are shown in the lower part of the diagram

The contributions of these components to the overall energy changes for oxidation of the heme octapeptide and cytochrome *c* are shown in Fig. 7.4. Note that the largest contribution to the stabilisation of the ferric state is that of induced protein dipoles in the case of the cytochrome *c* but the shell of Langevin dipoles in the case of the heme octapeptide. The latter are considerably more effective stabilising forces than the former and this gives rise to the difference in overall stabilisation of the ferric states of 40.3 kJ mol^{-1} which converts to a redox potential difference of 418 mV. We recall that Kassner's model for solvation of the heme in a protein shell with a dielectric constant of 3.5 gave rise to a redox potential difference of 260 mV. The value of 418 mV obtained using Churg and Warshel's model would correspond to a dielectric constant of 2.22 in Kassner's calculation. Thus we can conclude that the protein microenvironment must destabilise the ferric heme charge relative to its energy level in water and that the protein is therefore a medium of effectively low dielectric as far as heme solvation is concerned.

The microscopic approach of Churg and Warshel can also be used to compare the solvation energies for the oxidation of horse cytochrome *c* with the propionic acids in the ionised and unionised states (Fig. 7.5). This analysis shows that the strong intrinsic stabilisation of the propionates on the ferric heme is more than half compensated by unfavourable changes in the induced polarisation and orientation of water molecules and protein groups. The difference in overall energies of stabilisation indicates an interaction energy between the propionates and the ferric heme of 25 kJ mol, which corresponds to an effective dielectric constant of 11.1 (Table 7.5). Thus the compensation of charge-charge interaction energy by solvation energy leads to a higher effective dielectric for the ferriheme-propionate interaction (Churg and Warshel, 1986).

We can conclude that protein interiors are not of uniformly low dielectric but provide polar regions in which ionised groups can be stabilised by permanent dipoles in spite of low solvent access. The ferric heme group appears unusual in this respect in that its positive charge does seem to be accommodated in a region of low dielectric constant. This may be because the charge is delocalised over its whole structure and therefore does not exert individually large forces on the protein environment.

In summary, the interaction of the heme propionates with the ferric heme of cytochrome *c* is weakened by the polarity of the environment of the carboxylates and by the reaction field produced in the surrounding solvent, which together result in an effective dielectric constant of between 10 and 30.

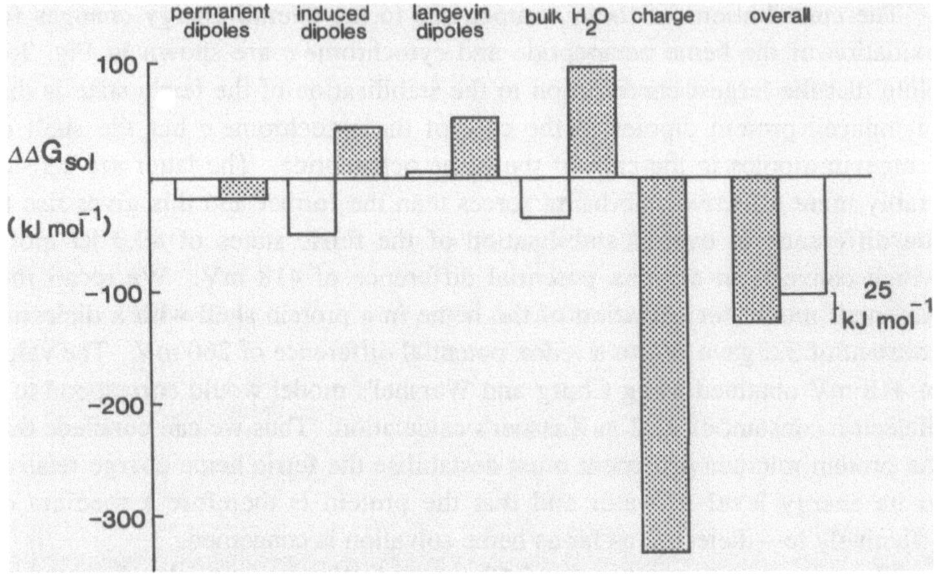

Fig. 7.5 Components of the solvation energy ($\Delta\Delta G_{sol}$) for the oxidation of horse heart cytochrome *c* assuming unionised propionic acids (open boxes) and ionised propionates (stippled boxes). The analysis of the components of $\Delta\Delta G_{sol}$ for the oxidation of horse heart cytochrome *c*, which assumes heme propionic acids are unionised is from Fig. 7.4. The stippled bars show the corresponding analysis for horse heart cytochrome *c* with both propionates ionised. Their stabilisation of the ferric state is given by the bar for 'charge' and was calculated using:

$$\Delta\Delta G_{sol} = 138.7 \frac{Q_1 Q_2}{r_{12}} \tag{7.45}$$

The difference in the overall stabilisation energies corresponds to a $\Delta\Delta G_{sol}$ of 25 kJ mol^{-1}

The cavity model allows calculation of this interaction energy in fair agreement with that obtained experimentally from redox potential measurements. Thus the influence of the buried heme propionates on redox potential is less than expected from a uniform dielectric model, but nevertheless is considerably greater than that associated with charge change on the protein surface. Different cytochromes may differ in the degree to which the ferric heme experiences the charge on the propionate. Some of the differences in redox potential between different cytochromes *c* may be due to the precise placement and nature of polar or charged groups relative to propionate-7. This will be discussed further in section 7.6 (see Fig. 7.15).

7.5 Multiple Redox Centres

One special case of charge interaction within a redox protein is when multiple redox centres are present. These may act independently, or they may interact, thus introducing complexity into their redox behaviour. These two possibilities are difficult to distinguish if the centres are spectroscopically identical (Malmström, 1973; Scheme 7.6).

Scheme 7.6 Independent and interacting models for a diheme cytochrome

$E_{m,1+}$, $E_{m,2+}$, $E_{m,2O}$, $E_{m,1O}$ are microscopic midpoint potentials for the reduction of a particular heme (1 or 2). The redox state of the second heme is defined by +, oxidised and O, reduced.

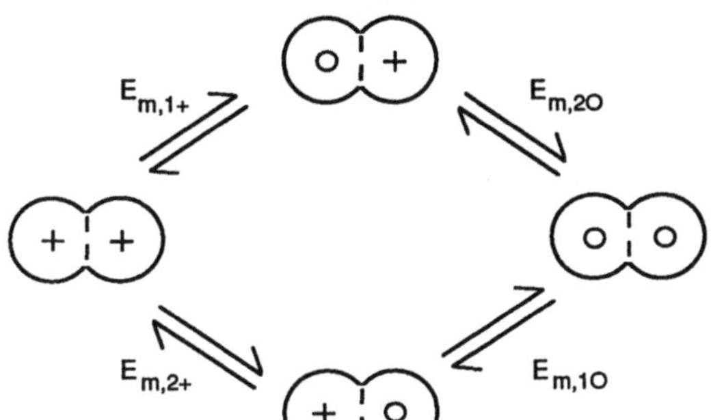

$$K_{1+}K_{2O} = K_{2+}K_{1O} \qquad (7.46)$$

$$E_{m,1+} + E_{m,2O} = E_{m,2+} + E_{m,1O} \qquad (7.47)$$

$$E_{m,1O} - E_{m,1+} = E_{m,2O} - E_{m,2+} = \partial \qquad (7.48)$$

∂ is the measure of interaction and is the same for the two sites. $\Delta E_{m,Av}$ is the difference in average potentials for the two sites.

$$\Delta E_{m,Av} = \frac{(E_{m,1+} + E_{m,1O})}{2} - \frac{(E_{m,2+} + E_{m,2O})}{2} \qquad (7.49)$$

When $\Delta E_{m,Av}/2$ is set at 0 V:

		Condition 1 (independent) $\Delta E_{m,Av} = 100$, $\partial = 0$	Condition 2 (interacting) $\partial = -100$, $\Delta E_{m,Av} = 0$
$E_{m,1+}$	$= +(\Delta E_{m,Av}/2) - \partial/2$	+50	+50
$E_{m,1O}$	$= +(\Delta E_{m,Av}/2) + \partial/2$	+50	-50
$E_{m,2+}$	$= -(\Delta E_{m,Av}/2) - \partial/2$	-50	+50
$E_{m,2O}$	$= -(\Delta E_{m,Av}/2) + \partial/2$	-50	-50

For condition 1 the top part of the diagram is the preferred route and heme 1 is high potential. For condition 2, neither route is preferred and there is a negative cooperativity of -100 mV for addition of the second electron.

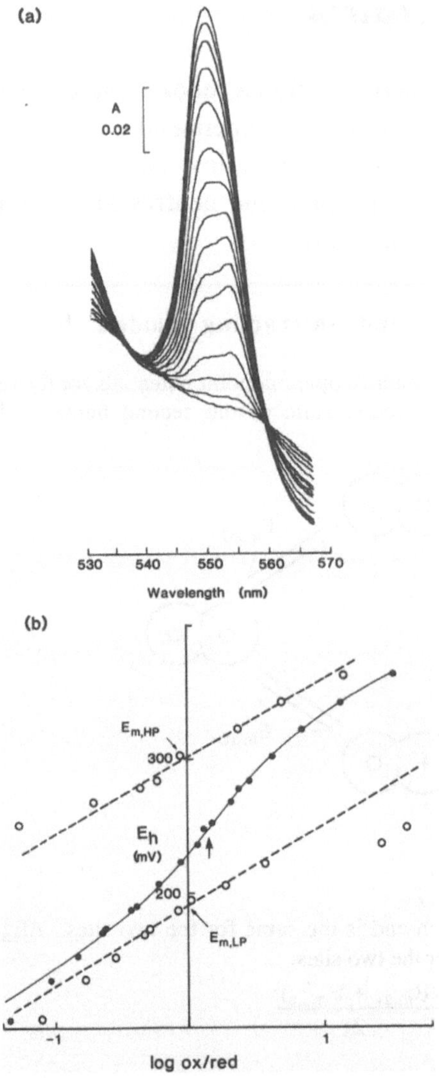

Fig. 7.6 Potentiometric and spectrophotometric titration of *Pseudomonas stutzeri* cytochrome c_4. (a) absorbance changes in the α-band. **(b)** analysis of the Nernst plot. Filled circles - experimental data; the larger arrow indicates the point equidistant from the two component midpoint potentials and, on the abscissa, represents the logarithm of the ratio of the contribution of the higher potential component (HP) to that of the lower potential component (LP). The open circles are the experimental data replotted using the new values for the fully reduced higher potential component and the fully oxidised lower potential component. The broken lines are slopes of 60 mV passing through the replotted data from which midpoint potentials of 300 mV (41%) and 190 mV (59%) can be derived. Using these figures a theoretical curve can be constructed (solid line) which provides a good fit to the original experimental data.

In the non-interacting case, heme 1 is high potential and receives the first electron; in the interacting case the hemes are potentiometrically identical in the oxidised state and the entry of the first electron to either heme imposes a negative interaction potential on its oxidised neighbour, making further reduction less favourable. The problem in distinguishing these two possibilities is illustrated by the case of the diheme cytochrome c_4 of *Pseudomonas stutzeri*. Fig. 7.6(b) shows the analysis, in terms of two independent heme groups, of the spectrophotometric redox titration in Fig. 7.6(a). According to this interpretation, at the centre of the sigmoidal Nernst plot is the point at which the reduction of the higher potential heme (HP) is complete and the reduction of the lower potential heme (LP) begins. From this it is possible to calculate and replot log ox_{HP}/red_{HP} and log ox_{LP}/red_{LP} versus E_h to give two 60 mV slopes with intercepts $E_{m,HP}$ and $E_{m,LP}$ (broken lines). A composite theoretical curve (solid line) can then be constructed to fit the experimental data.

However, the same curve is obtained for the case of identical hemes experiencing a negative interaction potential equal to the potential difference between the two hemes in the non-interacting case. This situation of negative or anti-cooperativity is illustrated in Fig. 7.7, where the Nernst plot, rotated through 90 degrees, is seen to be the formal equivalent of the Hill plot traditionally used for the detection of cooperativity in ligand interactions (Malmström, 1973). The cytochrome c_4 data approximate to a Hill coefficient of 0.5.

Several methods may be used to determine whether an individual case is due to interacting or independent redox centres. One is to use a spectroscopic method which does distinguish the redox centres. The visible spectroscopy of Fig. 7.6 almost satisfies this requirement because the high potential ($n = 1$) phase of the titration is associated with appearance of an asymmetric α-peak, whereas, with further reduction, the asymmetry disappears. This is consistent with independent hemes, one of high potential and asymmetric $\alpha = 554$nm and the second, low potential and symmetric $\alpha = 551$ nm. With an interacting system of potentiometrically identical but spectroscopically distinct hemes, the titration of both hemes would occur in a non-Nernstian fashion throughout a broad potential span and the shape of the composite spectrum would not depend on the degree of reduction. Other spectroscopic methods such as EPR and MCD spectroscopy may offer better discrimination between redox centres.

Even greater resolution of the system can be achieved if the individual redox transitions can be identified spectroscopically. In Scheme 7.6 for example, this would require discrimination between reduction of centre 1 when centre 2 is oxidised (microscopic potential $E_{m,1+}$) and reduction of centre 1 when centre 2 is reduced (microscopic potential $E_{m,1O}$).

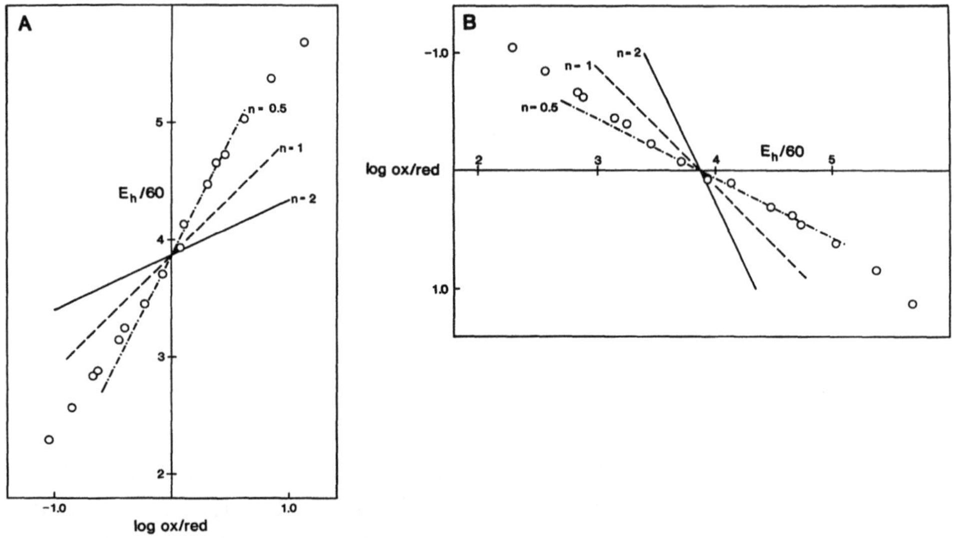

Fig. 7.7 Complex Nernst plots and cooperativity. The Nernst equation is plotted in the form:

$$\frac{E_h}{60} = \frac{E_m}{60} + \frac{1}{n}\log\frac{ox}{red} \tag{7.50}$$

using the data of *Ps. stutzeri* cytochrome c_4 from Fig. 7.6 (a). In this form the data fit best to a value of n = 0.5 (-·--·-). n = 1 (- - - -) and n = 2 (——) lines are shown for comparison. Cooperativity in enzyme systems is often analysed by the Hill plot and the Nernst equivalent is shown in **B** which is obtained by simple rotation of **A** through 90°

Cytochrome c_3 contains four chemically identical hemes in close proximity but in unique protein environments (Chapter 5.3). This system might be expected to show heme potentials which are both intrinsically different and interacting. In the analysis of Santos *et al.* (1984) of the cytochrome c_3 from *Desulfovibrio gigas*, a matrix of intermediate redox states was set up (Fig. 7.8). The matrix involved 32 redox pairs and therefore 32 Nernst equations, with each Nernst equation being governed by the relative populations of two redox states and a microscopic midpoint potential. If the electron affinity for a heme group was affected by the redox state of the other hemes, the microscopic midpoint potential was adjusted by an interaction energy term.

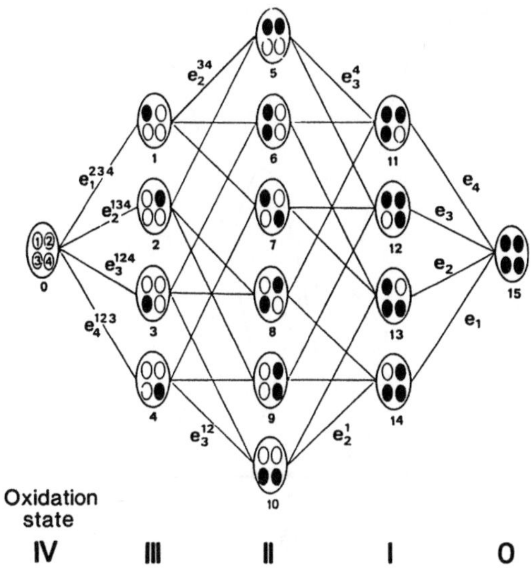

Fig. 7.8 Electron distribution scheme for a tetraheme cytochrome. Forms **0** and **15** refer respectively to the fully oxidised and fully reduced protein. e_i^{jkl} is the microscopic midpoint potential of heme i when hemes j, k and l remain oxidised. (O - reduced heme; ● - oxidised heme). (Santos *et al.*, 1984)

This theoretical description of the population distribution for the different redox states as the ambient potential was varied was compared with the actual distribution obtained from NMR analysis of the downfield heme methyl resonances. A best fit (Fig. 7.9) contained both positive and negative interaction energies (Fig. 7.10(a)).

Related analyses of *D. vulgaris* (Miyazaki and Hildenborough) and *D. desulfuricans* (Norway) cytochromes c_3 have been carried out with electrochemical methods (Sokol *et al.*, 1980; Niki *et al.*, 1984), and by EPR (Gayda *et al.*, 1988; Benosman *et al.*, 1989). The macroscopic and microscopic potentials of these proteins were found to be different indicating some degree of cooperativity, as was also found for *D. gigas* cytochrome c_3 (Fig. 7.10(a)). The values of the interaction potentials were not fully evaluated from these measurements but they clearly indicated substantial differences between the *D. vulgaris* and *D. gigas* cytochromes. However, the EPR measurements (Benosman *et al.*, 1989) and electrochemical studies (Niki *et al.*, 1984) of *D. vulgaris* (Miyazaki) cytochrome c_3 yielded conflicting results.

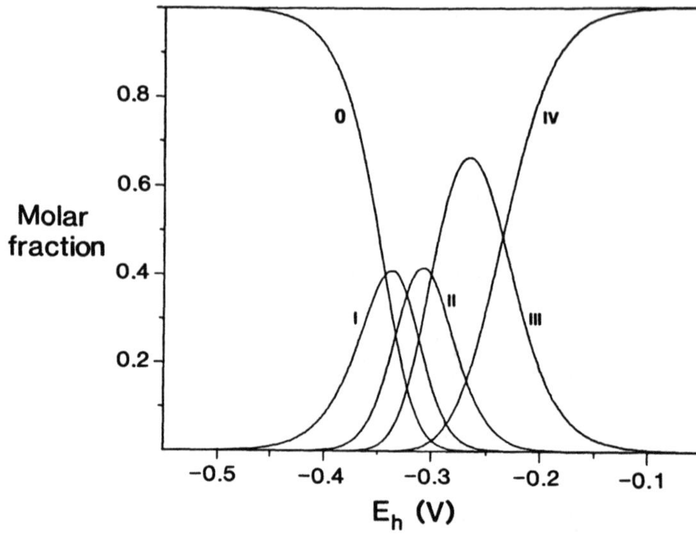

Fig. 7.9 Molar fractions for oxidation steps 0, I, II, III, IV (Fig. 7.8) as a function of ambient redox potential for cytochrome c_3 from *Desulfovibrio gigas*. Redox potentials: e_1, -345 mV; e_2, -315 mV; e_3, -314 mV; e_4, -290 mV. Interaction potentials: I_{12}, 19 mV; I_{13}, -26 mV; I_{14}, 6 mV; I_{23}, 42 mV; I_{24}, -24 mV; I_{34}, -18 mV (Santos *et al.*, 1984)

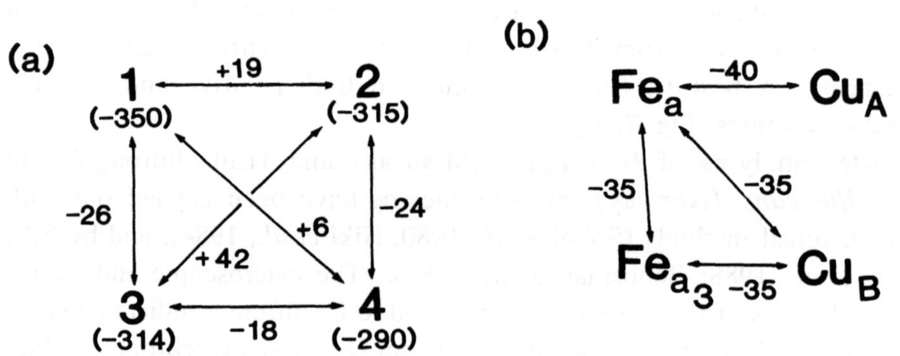

Fig 7.10 Interaction between redox centres in cytochromes. (a) The interaction energies in millivolts for *Desulfovibrio gigas* cytochrome c_3. Intrinsic midpoint potentials are given in parentheses (Santos *et al.*, 1984). Note that **1** - **4** merely indicate four different hemes. Correlation between the structure of cytochrome c_3 (Chapter 5.3) and the redox potentials remains to be done. **(b)** The interaction energies in millivolts for the redox centres of mitochondrial cytochrome *c* oxidase (Blair *et al.* 1986)

Cytochrome *c* oxidase of mitochondria also contains four redox centres, but of two different kinds - heme *a* and copper. Potentiometric titration monitored at 605 nm yielded a sigmoidal Nernst plot with $n_{apparent}<1$ (Wilson *et al.*, 1972). The interpretation of this behaviour depends on whether the 605 nm absorbance is considered to be due predominantly to heme *a* or equally to heme *a* and a_3. Wilson *et al.* assumed the latter and analysed the titration in terms of independent contributions of heme *a* (n = 1, E_m 220 mV) and heme a_3 (n = 1, E_m 380 mV). The alternative explanation is that heme *a* is the major contributor to the 605 nm band and that the broad redox titration is therefore due to negative cooperativity between heme *a* and other redox centres in the protein (Malmström, 1973; Nicholls and Petersen, 1974; Wikström *et al.*, 1976).

Most evidence favours the second of these alternative explanations. Thus MCD allows spectroscopic resolution of the two hemes and shows broad titrations and comparable reduction levels for both. This precludes high and low potential Nernstian components and the results are best fitted by two roughly equipotential hemes (E_m 350 mV) with a reciprocal negative interaction of 90 mV (Carithers and Palmer, 1981). Studies with mixed valence enzymes, in which one or more of the redox centres is trapped in a redox inactive state by ligand binding, indicate that negative interactions must also occur between heme *a* and the copper centres. For the CO enzyme, in which the ligand binding pole is trapped in the a_3^{2+} Cu_B^{1+} state (Wilson and Nelson, 1982), non-Nernstian behaviour persists, consistent with a negative interaction between heme *a* and Cu_A (Ellis *et al.*, 1986). The apparent midpoint potential of 255 mV for heme *a* is due to the strong negative interaction potential of the reduced ligand binding pole.

Using the ligand binding results and those of spectral titration of the native enzyme at 443 and 605 nm, where heme a_3 and heme *a* are respectively the main contributors, Blair *et al.* (1986) propose the pattern of interactions shown in Fig. 7.10(b).

The structural basis for redox interactions between centres in the same protein is largely unknown. One possible origin for the negative interactions is electrostatic. For a diheme cytochrome for example, reduction of the first heme, when the second is reduced, will be less favourable (lower potential) than when the second is oxidised. The magnitude of such a simple electrostatic effect could be estimated using Eqn. 7.36 if the effective dielectric constant between the redox centres is known. We have seen that an effective dielectric constant of approximately 20 governs the energy of interaction of the ferric iron with the buried heme propionate in cytochrome *c* (section 7.4.4). If a similar value is applicable to interactions between redox centres in cytochrome c_3, then interaction potentials of 36 - 72 mV would be obtained for centres 20 - 10Å apart. Although this is certainly of sufficient magnitude to account for the observed patterns, they cannot fully explain the cytochrome c_3 data, where both positive and negative interactions are present. It seems that conformational changes must be invoked to account for positive interactions. Also, in the case of cytochrome oxidase, most emphasis has been on conformational rather than electrostatic origins for the interactions (Carithers & Palmer, 1981; Blair *et al.*, 1986; Copeland *et al.*, 1987). However, from electrostatic considerations alone Moore and Rogers (1985) calculated an interaction potential of -80 mV for two Fe(III) ions 25Å apart, and Krishtalik (1985) calculated an interaction potential of -105 mV to -178 mV for two Fe(III) ions 12 - 16Å apart. Both sets of calculations attempted to take account of the membrane environment of cytochrome oxidase.

7.6 Proton Binding

The binding of protons to redox proteins is one particular case of ligand binding and we will be using here the general ligand binding equations derived in section 7.2. However, proton binding deserves special consideration because of the established association of bioenergetic processes with changes in proton concentration. The kinetics and thermodynamics of electron transport may be influenced by changes in proton concentration and certain specialised redox centres may be involved in translocating protons across the bioenergetic membrane.

7.6.1 Theoretical Considerations

For pH to influence the redox potential, protons must appear in the redox equation:

$$AH \rightleftharpoons A + H^+ + e^-$$

which gives the Nernst equations:

$$E_h = E_{m,0} + 0.06 \log \frac{[A][H]}{[A]} \qquad (7.51)$$

$$E_h = E_{m,0} + 0.06 \log \frac{[A]}{[AH]} + 0.06 \log [H^+] \qquad (7.52)$$

when [A] = [AH], $E_h = E_m$:

$$\therefore\ E_m = E_{m,0} - 0.06\ pH \qquad (7.53)$$

This equation describes the pH dependence of a redox couple which loses both a proton and an electron on oxidation. In principle, a related equation could be written for proton loss on reduction, but electrostatic considerations lead us to expect that the more positively charged oxidised state will have a lower affinity for the proton. For cytochromes, the iron itself undergoes simple electron gain and loss without associated proton uptake and release. Yet pH dependence of redox potential is of quite common occurrence in cytochromes. In these cases the proton must associate with a group (R) on the protein rather than with the redox centre itself:

$$\text{(Fe}^{\circ}\text{)—RH}^+ \rightarrow \text{(Fe}^{+}\text{)—R} + H^+ + e^- \qquad (7.54)$$

However, it is unlikely that Eqn 7.53 could be a complete description of the pH dependence of the system because it implies that R is completely unprotonated in the oxidised state. More likely is a model which incorporates a lower, but finite, affinity of the proton for the oxidised state and a higher affinity for the reduced state. Such a model would then be defined by three redox equations and would exhibit three regions of pH influence (Fig. 7.11). At low pH, R is protonated in both the oxidised and reduced states and at high pH, R is unprotonated in both states. In both these pH regions the redox potential is independent of pH. In the intermediate pH range (between the two pK values), R is protonated in the reduced form but unprotonated in the oxidised form and the redox potential is pH dependent, approaching a slope of -60 mV/pH unit (Eqn. 7.52).

A quantitative description of the composite curve can be obtained using the treatment developed by Clark (1960) and Dutton (1978). For protons, the general ligand binding equation 7.17 (Scheme 7.2) becomes:

$$E_m = \tilde{E} - 0.06 \log \frac{K_{ox} + [H^+]}{K_{red} + [H^+]} \qquad (7.55)$$

where $\tilde{E}$ replaces $E_{m,L}$ as the midpoint potential of the fully protonated form.

This equation defines the curve of Fig 7.11 and in certain regions simplifies to the three limiting equations as shown in Fig. 7.12. The intersections of these limiting equations can be used to estimate pK_{ox} and pK_{red}.

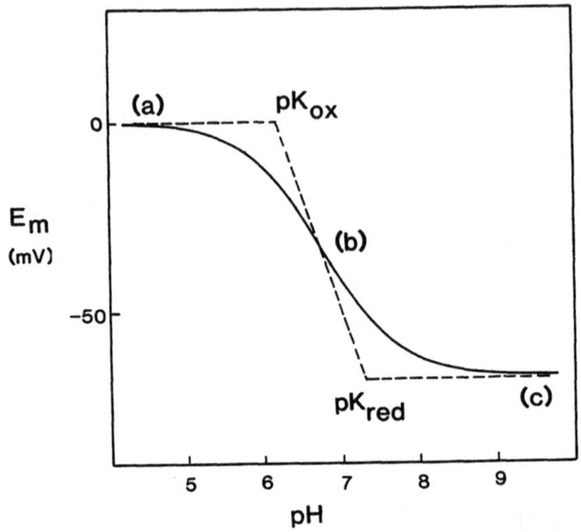

Fig. 7.11 pH dependence of redox potential and contributing half cell equations.

(a) $(\text{Fe}^{\circ})\text{—RH}^+ \rightarrow (\text{Fe}^{+})\text{—RH}^+ + e^-$ (7.56)

(b) $(\text{Fe}^{\circ})\text{—RH}^+ \rightarrow (\text{Fe}^{+})\text{—R} + e^- + H^+$ (7.57)

(c) $(\text{Fe}^{\circ})\text{—R} \rightarrow (\text{Fe}^{+})\text{—R} + e^-$ (7.58)

The group R which gives rise to pK_{ox} and pK_{red} can be described as heme-linked. Proton loss from R gives rise to a relative stabilisation of the positively charged, oxidised heme and this is reflected in the fall in redox potential ($\Delta E_m = \tilde{E} - E_{m,NL}$). On the other hand, proton loss from R is itself influenced by the redox state of the heme group. Thus the positively charged ferric heme results in a lower affinity (lower pK) for the proton. These constitute reciprocal and equal effects so that:

$$-nF\Delta E_m = -RT \ln K_{ox}/K_{red} \qquad (7.59)$$

More complex curves can be analysed by extensions of this treatment. In general, a change to greater slope as the pH is raised indicates a pK in the oxidised state and a change to lesser slope, a pK in the reduced state (Clark,

1960). Thus by visual inspection, pK values can be assigned and a theoretical equation can be derived to fit the experimental data. These equations have the general form:

$$E_m = \tilde{E} - 0.06 \log \frac{\left([H^+]^x+[H^+]^{x-1}K_{O1}+[H^+]^{x-2}K_{O1}K_{O2}...K_{O1}K_{O2}...K_n\right)}{\left([H^+]^y+[H^+]^{y-1}K_{R1}+[H^+]^{y-2}K_{R1}K_{R2}...K_{R1}K_{R2}...K_m\right)} \quad (7.60)$$

for n ionisations in the oxidised state and m ionisations in the reduced state.

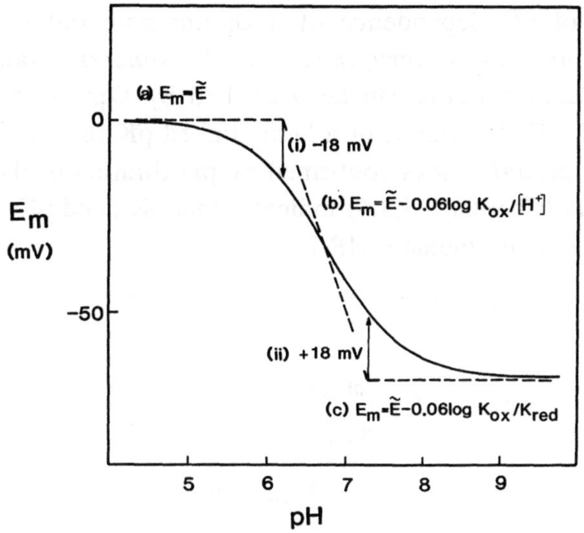

Fig. 7.12 pH dependence of redox potential and contributing Nernst equations. General ligand binding equation:

$$E_m = \tilde{E} - 0.06 \log \frac{K_{ox} + [H^+]}{K_{red} + [H^+]} \quad (7.55)$$

Limiting equations: **(a)** when $[H^+] > K_{ox}$ and K_{red}; **(b)** when $K_{ox} > [H^+] > K_{red}$; and **(c)** when K_{ox} and $K_{red} > [H^+]$

Intersections: **(i)** $\tilde{E} = \tilde{E} - 0.06 \log \frac{K_{ox}}{[H^+]}$, $\therefore \frac{K_{ox}}{H^+} = 1$, $\therefore$ pH = pK_a;

(ii) $\tilde{E} - 0.06 \log \frac{K_{ox}}{[H^+]} = \tilde{E} - 0.06 \log \frac{K_{ox}}{K_{red}}$, $\therefore \frac{K_{ox}}{[H^+]} = \frac{K_{ox}}{K_{red}}$,

$\therefore$ pH = pK_{red}.

Theoretical curve at pK values: **(i)** $E_m = \tilde{E} - 0.06 \log \frac{2[H^+]}{[H^+]} = \tilde{E} - 0.018$;

(ii) $E_m = \tilde{E} - 0.06 \log \frac{K_{ox}}{2K_{red}} = \tilde{E} - 0.06 \log \frac{K_{ox}}{K_{red}} + 0.018$

7.6.2 Structural Basis for the pH Dependence of Redox Potential

In the previous section we have derived equations which satisfactorily describe the pH dependence of redox potential controlled by a single heme-linked group with different pK values in the oxidised and reduced states. Here we discuss particular instances of pH dependence in cytochromes and attempts to identify the responsible heme-linked groups.

The Cytochromes c-551 of the Pseudomonads. Fig. 7.13 shows the patterns of pH dependence of midpoint potential for cytochromes *c*-551 from *Pseudomonas aeruginosa* and *Ps. stutzeri* (strain 221) (Leitch *et al.*, 1984). Both patterns can be described by Eqn. 7.55 using the pK values indicated. The existence of a heme-linked pK for the ferrocytochrome *c*-551 from *Ps. aeruginosa* is confirmed by pH titration of the α-band (Fig. 7.14). NMR data (Chapter 2.3.3.4) indicates that pK_O and pK_R are the ionisations of the rear heme propionate (HP-7).

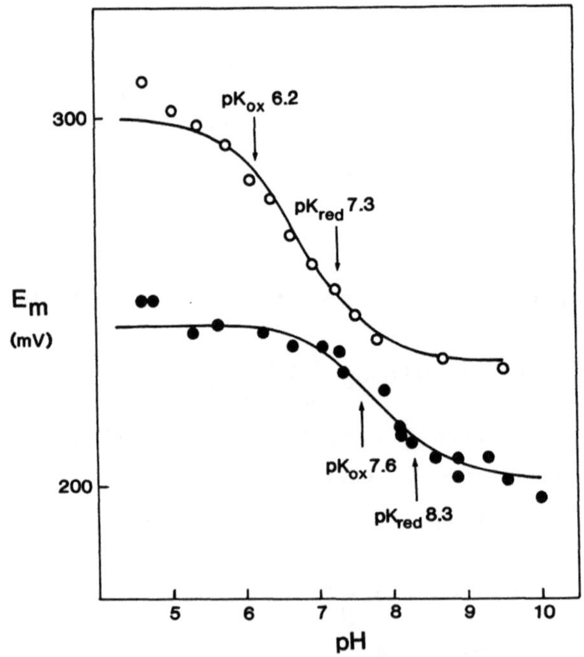

Fig. 7.13 The pH dependence of redox potential of cytochromes *c*-551 from *Pseudomonas stutzeri* (221) and *Pseudomonas aeruginosa*. *Ps. aeruginosa* cytochrome *c*-551 (○); *Ps. stutzeri* cytochrome *c*-551 (●). Solid lines are theoretical curves defined by Eqn. 7.55 with pK values as shown (Leitch *et al.*, 1984)

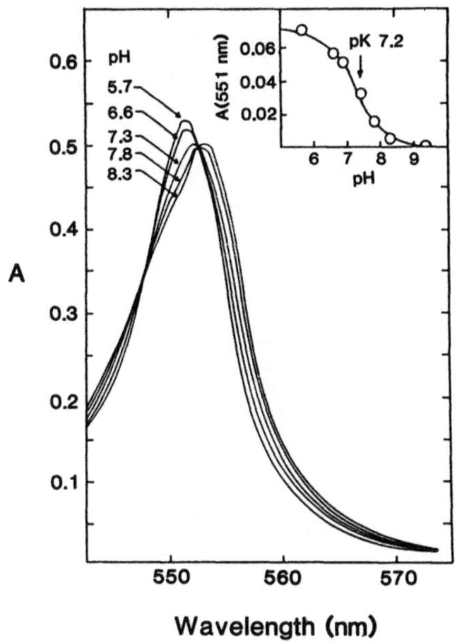

Fig. 7.14 Spectrophotometric titration of the α-band of *Pseudomonas aeruginosa* ferrocytochrome *c*-551. The solid line in the inset is a theoretical curve for pK 7.2 (Moore *et al.*, 1980c)

However the two cytochromes *c*-551 differ in the degree of separation of the pK_O and pK_R values (Fig. 7.13) and in the magnitude of the shift of the propionate NMR resonances with pH. These differences are consistent with the substitution of arginine 47 by histidine in *Ps. stutzeri* cytochrome *c*-551. Like the heme propionate, this histidine also titrates with pK_O and pK_R yet the overall pattern of pH dependence is compatible with the loss of just one proton. It is proposed that this single proton is held between the histidine and the propionate (Fig. 7.15(d)) and its loss perturbs the resonances of both groups. Since the propionate group is undergoing a fractional charge change, the effect on both the β-methylene proton NMR peaks and the redox potential is less than that observed for *Ps. aeruginosa* cytochrome *c*-551, for which an integral charge change occurs.

The Cytochromes c_2 of the Rhodospirillaceae and Mitochondrial Cytochromes c. We can identify three different patterns of pH dependence of redox potential in the cytochromes c_2 and the mitochondrial cytochromes *c*, illustrated by the three curves of Fig. 7.16 (Moore *et al.*,

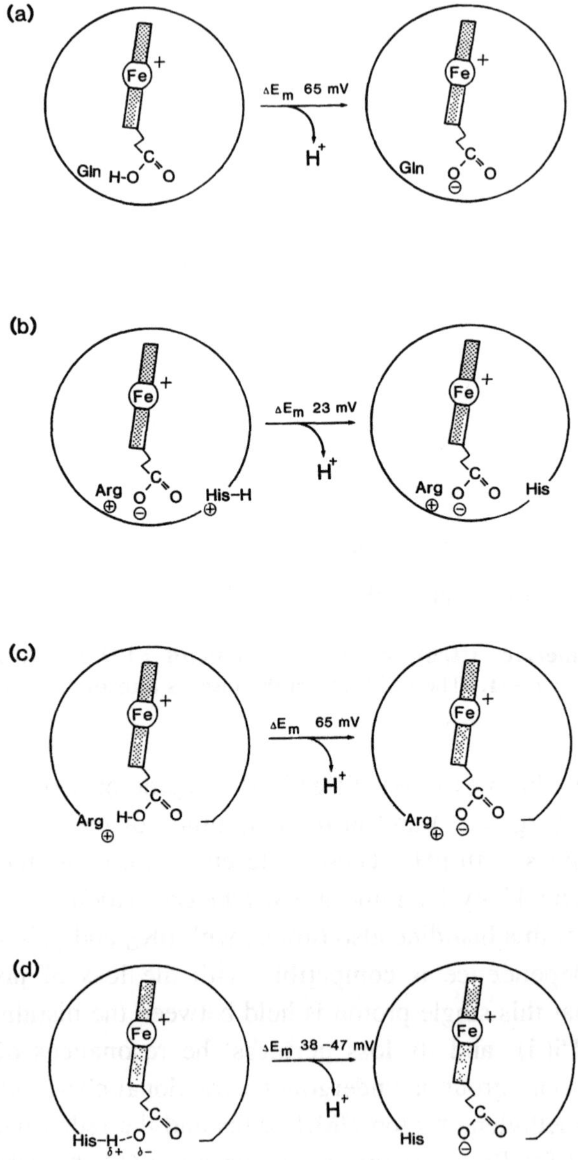

Fig. 7.15 Ionisations influencing the redox potential of cytochromes *c*. The heme group is shown edge-on in the ferric state (**Fe^+**) with only the rear heme propionate (HP-7) included. **(a)** ***Rhodomicrobium vannielii*** cytochrome c_2 with glutamine 38; **(b)** ***Rhodopseudomonas viridis*** cytochrome c_2 with arginine 38 and histidine 39; **(c)** ***Pseudomonas aeruginosa*** cytochrome *c*-551 with arginine 47; **(d)** other Pseudomonad cytochromes *c*-551 with histidine 47

1984a). Cytochromes of group A, which includes only *Rhodomicrobium vannielii* and *Rhodospirillum rubrum* cytochromes c_2, have a pH dependence of redox potential defined by three pK values, in order: pK_{O1}; pK_R; pK_{O2}. pK_{O1} and pK_R are separated by approximately one pH unit. The relevant Nernst equation, derived using the methods of Clark (1960) is:

$$E_m = \tilde{E} - 0.06 \log \frac{[H^+]^2 + K_{O1}[H^+] + K_{O1}K_{O2}}{[H^+]^2 + K_R[H^+]} \qquad (7.61)$$

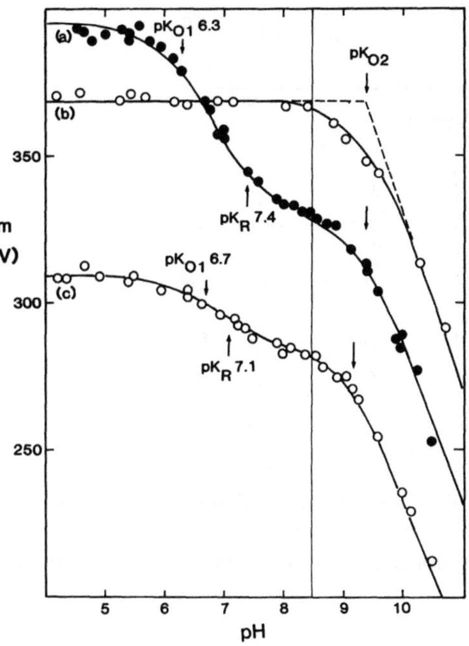

Fig. 7.16 The pH dependence of midpoint potential of the cytochromes c_2. **(a)** *Rm. vannielii* cytochrome c_2; **(b)** *Rhodobacter (f.Rhodopseudomonas) capsulatus* cytochrome c_2; **(c)** *Rps. viridis* cytochrome c_2. Solid lines are theoretical curves defined by Eqn. 7.61 in the case of **(a)** and **(c)**, and by Eqn. 7.62 in the case of **(b)**

Cytochromes of group B, which includes *Rhodopseudomonas viridis* cytochrome c_2 and *S. cerevisiae* iso-1 cytochrome *c*, have the same pattern of pH dependence as those of group A (and therefore the same Nernst equation) but pK_O and pK_R are separated by only 0.4 pH unit. Group C is the largest group and contains cytochromes *c* with a pH dependence of redox potential defined by only one pK in the oxidised state (Eqn. 7.62).

$$E_m = \tilde{E} - 0.06 \log \frac{[H^+] + K_O}{[H^+]} \qquad (7.62)$$

This pK, near 9 in the oxidised state, is common to all members of all three groups and is associated with loss of methionine iron coordination. The nature of this ligand replacement has been discussed fully in Chapter 4.4.4 and will be considered no further here.

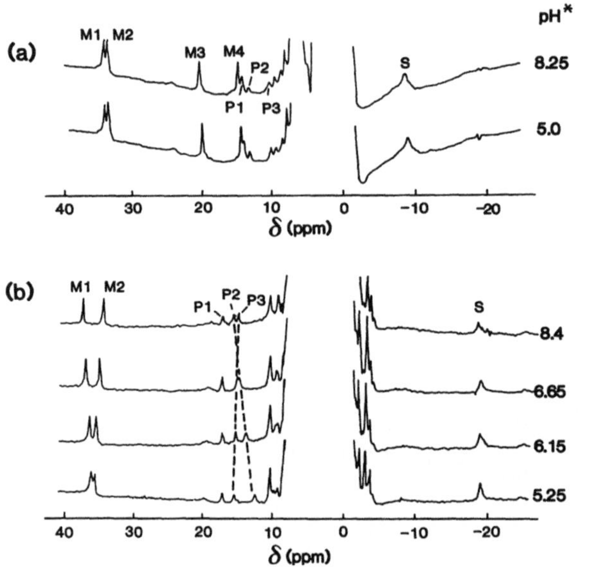

Fig. 7.17 pD titration of the ^{1}H NMR spectra of ferricytochromes c_2. (a) *Rb. capsulatus* and **(b)** *Rm. vannielii*. pH* is the uncorrected pH meter reading in D_2O (pD). Heme methyl protons (M); heme propionic acid β-methylene protons (P); methionine methyl protons (S)

Downfield single proton resonances are found in the NMR spectra of the ferricytochromes *c* of all three groups (Fig. 7.17) some of which are due to the β-methylene protons of the heme propionates. Two of these resonances (P2 and P3) in *Rm. vannielii* cytochrome c_2 shift with a pK which matches pK_{O1} (Figures 7.17 and 7.18). Although these resonances are not resolved in the spectrum of the ferrocytochrome, resonances assigned to tryptophan 59 are resolved and shift with a pK which matches pK_R. Thus, as we found for the cytochromes *c*-551, the pH dependent features of the NMR spectra of *Rm. vannielii* cytochrome c_2 are consistent with the ionisation of the rear heme propionate with pK_{O1} in the oxidised state and pK_R in the reduced state (Fig.

7.15). On the other hand, none of the heme substituent resonances of cytochromes in groups B and C show such pH dependence.

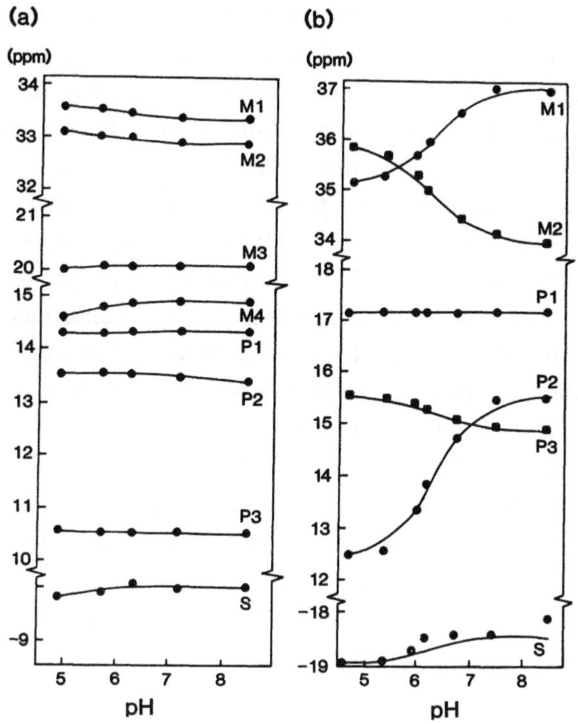

Fig. 7.18 pH titration of the ^{1}H NMR spectra of ferricytochromes c_2. (a) *Rb. capsulatus* and **(b)** *Rm. vannielii*. The solid lines in **(b)** for **M1, M2, P2, P3** and **S** are theoretical curves for a pK_a of 6.3. (pH* is defined in Fig. 7.17)

Arginine 38 is a conserved residue in the cytochrome c_2-mitochondrial cytochrome *c* family which lies close enough to the rear heme propionate to form a salt link. It is present in all members of the family, except the two cytochromes c_2 of group A which have asparagine or glutamine. The effect of arginine 38 appears to be to lower the pK of the rear heme propionic acid to a pH outside the range 4 to 9. The absence of arginine 38 in the cytochromes c_2 of group A is then proposed to have the effect of raising the pK to the values observed for pK_{O1} and pK_R. However, site-directed mutagenesis of arginine 38 in yeast cytochrome *c* indicates that substitution of this residue alone does not shift the propionic acid pK to within the pH range 4 to 9 (Cutler *et al.*, 1989). Thus it may be that changes in two or more residues which provide the environment of the propionate, are required before a large pK shift can be observed.

Although the heme substituent resonances of the cytochromes of group B show no pH dependent shift, the C-4 protons of a histidine titrate with pK values corresponding to pK_{O1} in the ferricytochrome and pK_R in the ferrocytochrome. Histidine 39 is the sole unliganded histidine in *Rps. viridis* cytochrome c_2 and a histidine in this position is also a feature of the mitochondrial cytochromes *c* from *Saccharomyces cereviseae* and *Crithidia oncopelti,* which have a similar pattern of pH dependence of redox potential (Moore *et al.*, 1984a). Modification of histidine 39 abolishes this pH dependence. The proximity of histidine 39 to the rear heme propionate and to arginine 38 suggests that its influence on redox potential is a consequence of its specific location in a region that undergoes a redox state conformation change (Chapter 4.4.3) rather than due to a general electrostatic effect (Fig. 7.15b).

In summary it is the rear heme propionate and its surrounding protein environment that is the structural locus of pH dependence of redox potential. In almost all the large class I cytochromes *c*, an arginine stabilises the propionate anion and lowers the pK so that the midpoint potential is either pH independent in the range pH 4 - 9 or only slightly perturbed by proton loss from a neighbouring histidine. Rare substitution of the arginine in cytochrome c_2 leads to ionisation of the propionic acid group in the physiological pH range. In the small class I cytochromes *c*, an arginine close to the rear heme propionate does not have such a dramatic lowering effect on its pK, perhaps because the bottom of the heme crevice is more open to solvent. Replacement of the arginine by histidine gives rise to apparently elevated pK values for the propionate, but the proton lost is probably one shared between the imidazole ring and the carboxyl group.

Class II Cytochromes c. Although there is only limited data concerning the pH dependence of the midpoint redox potential of class II cytochromes *c*, what there is available shows there is a strong pH dependence with at least one ionisation in the oxidised form of the protein (Fig. 7.19) (Barakat and Strekas, 1982). This was originally ascribed to the ionisation of water bound to the iron, by analogy with the globins, but it has now been shown by spectroscopic studies to be the ionisation of the histidine ligand to a histidinate (Chapter 2.3.2.5). The six coordinate class II protein, *Agrobacterium tumefaciens* cytochrome *c*-556, experiences a similar fall in redox potential with increasing pH (J.van Beeumen and G.W.Pettigrew, unpublished data) and its histidine ligand also ionises (Chapter 5.2.4).

Proteins where an iron ligand ionises in the oxidised form have a strong pH dependence to their redox potential because the corresponding pK_R is usually shifted to much higher pH as a result of the decreased Lewis acidity of Fe^{2+} compared to Fe^{3+}. Thus the slope of the graph should approach -60 mV/pH unit (Eqn. 7.53).

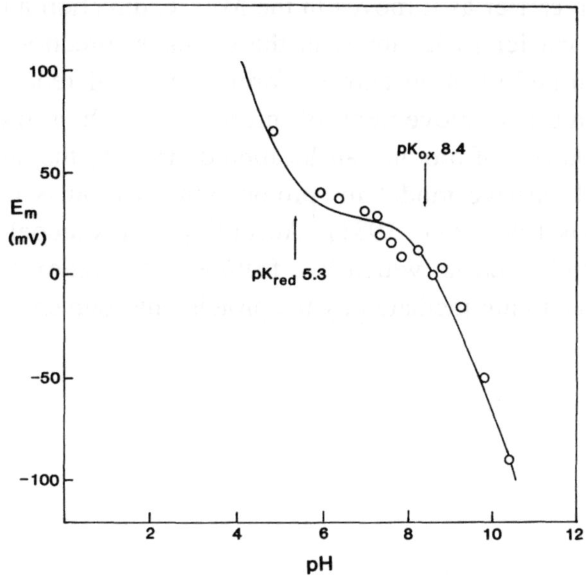

Fig. 7.19 The pH dependence of midpoint potential for *Chromatium* cytochrome *c'*. The solid line is a theoretical curve using:

$$E_m = \tilde{E} - 0.06 \log \frac{[H^+] + K_O}{[H^+]^2 + [H^+]K_R} \qquad (7.63)$$

with values of E = 340 mV; $pK_R = 5.3$ and $pK_O = 8.4$. (Barakat and Strekas, 1982)

7.6.3 pH Dependence of Redox Potential and Proton Translocation

The oxidation-linked proton release described in the previous section is a scalar effect occurring in solution. Much interest at present centres around vectorial proton effects coupled to electron transport and occurring across a bioenergetic membrane.

There is general acceptance (see the following reviews and references therein: Papa, 1976; Wikström *et al.*, 1981; Slater, 1987; Malmström, 1989) that energy conservation during oxidative phosphorylation results from the transmembrane movement of protons from the matrix side (M) to the cytoplasmic side (C) as a consequence of redox reactions (Mitchell, 1961, 1980: Williams, 1961b) Where there is less agreement, and less information, is in the mechanism of this proton translocation.

Using the principle of ligand conduction (Mitchell, 1980), Mitchell has proposed that a single reducing equivalent travels as a hydrogen atom on a hydrogen carrier as it moves in the M → C direction and as an electron on an electron carrier as it moves in the opposite direction. Alternating pairs of hydrogen and electron carriers form a series of redox loops which catalyse transmembrane movement of protons. Such a model implies a fixed stoicheiometry of proton translocation defined by the number of redox loops.

An alternative model for proton translocation is proton pumping (Papa, 1976; Wikström *et al.*, 1981). According to this model, the direct coupling of proton and electron, which is a feature of the redox loop, is replaced by an indirect coupling mediated by the protein environment of the redox centre.

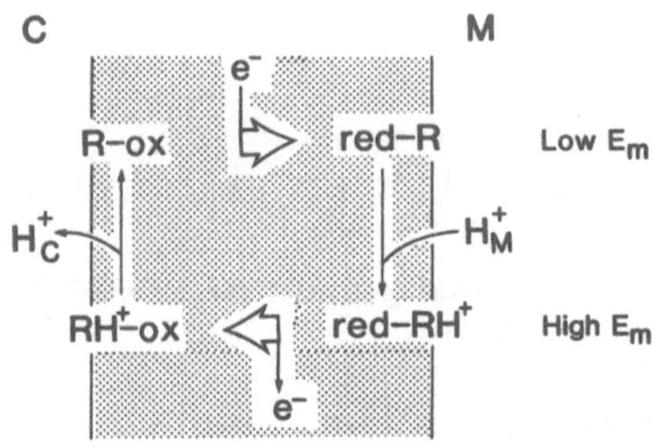

Fig. 7.20 A model for electron transfer coupled to proton translocation. Reduction involves a switch in group **R** from cytoplasmic **C**-facing to matrix **M**-facing; oxidation involves a switch in group **RH⁺** from **M**-facing to **C**-facing. Redox couple **R-ox/red-R** is in equilibrium with the low potential electron input side; redox couple **RH⁺-ox/red-RH⁺** is in equilibrium with high potential electron output side. Differential uptake of protons at M side and release of protons at C side is due to the redox state-linked pK values of group R which has a low pK in the oxidised state and a high pK in the reduced state

A simple model for such a process is shown in Fig. 7.20, in which a redox centre with an associated group (**R**) can take up a proton on reduction and lose the proton on re-oxidation. **R** therefore resembles the heme propionate in some soluble cytochromes in having a heme-linked pK. Where the model differs from the scalar process described in the previous section is in the switch of orientation of group **R** from one side of the membrane to the other. If proton channels are present, such a switch need not involve a large movement.

Such a model will exhibit pH dependence of redox potential. The redox couple **R-ox/red-R** will operate at low potentials in near equilibrium with the electron donor, while the couple **RH^+-ox/red-RH^+** will operate at higher potentials in near equilibrium with the electron acceptor. By this means, a free energy span is available to drive proton movement, while at the same time high concentrations of the reduced and oxidised states can be maintained to avoid kinetic insufficiency (Wikström, 1982a).

pH dependence of redox potential has therefore been seen as a necessary, if not a sufficient, property of a redox centre involved in proton translocation and, on this basis, attention has focused on the *b*-type cytochromes and cytochrome *a* of the mitochondrial electron transport chain, both of which show such pH dependence. We concentrate here on cytochrome *a*.

Cytochrome *c* oxidase contains two coppers and two heme *a* groups, all of which are electron carriers. Thus in Mitchell's original formulation of mitochondrial proton translocation, cytochrome *c* oxidase was the electron carrying arm of a redox loop which originated in Complex III (Mitchell 1980). Since then, evidence has accumulated that the enzyme catalyses not only the transmembrane oxidation of cytochrome *c* by oxygen, but also the translocation of protons from the matrix to the intermembrane space (Wikström, 1977; Krab and Wikström, 1978; Casey *et al.*, 1979; Sigel and Carafoli, 1978; 1980) and there is now agreement that the enzyme is a proton translocator (Mitchell *et al.*, 1985).

However, the structural features within the enzyme concerned with proton translocation remain poorly characterised. These features include the nature of the redox centre 'trigger' and the proton carrying group(s), and the process of coupling between them. The recent identification by Wikström (1989) that only two of the four electron transfers internal to cytochrome oxidase are linked to proton translocation, together with the observation that the heme a_3 - Cu_B site is readily accessible to M protons but not to C protons (Wikström, 1988) should help define the proton-pumping mechanism. Indeed, Wikström (1989) states "...that it seems that the bimetallic centre is itself critically involved in the mechanism." However, he goes on to consider possible scenarios in which heme *a* and Cu_A might also be involved. We too consider the involvement of heme *a* and Cu_A since detailed chemical proposals have been made regarding them. However, we emphasise that the experimental support for these proposals is limited.

The first two models for proton translocation implicate heme *a* as the redox trigger. This is supported by several lines of evidence. Firstly, in the antimycin-blocked electron transport chain, three oxidising equivalents (1.5c, $0.5c_1$, 0.5R, $0.5QH_2/QH$) limit further H^+ translocation which suggests that the proton pump must be the electron acceptor from cytochrome *c* (Wikström and Casey 1985). Secondly, although interpretation is complicated by the cooperative effects within the enzyme and the disagreement as to the contribution of heme *a* and heme a_3 to the visible absorption spectrum (section 7.5), most analyses show that the midpoint potential of heme *a* is pH dependent (Wilson *et al.*, 1972; Arzatbanov *et al.*, 1978; Carithers and Palmer, 1981; Wikström, 1982b). For example, the cyano-enzyme with heme a_3 in the ferric state shows a pH dependence of redox potential of -30 mV/pH unit (Arzatbanov *et al.*, 1978). Thirdly, energisation of the membrane produces a lowering of the midpoint potential of heme *a* (Wikström *et al.*, 1976). This would occur if heme *a* lay proximal to an M-facing well in which energisation produces a locally raised pH (Wikström and Krab, 1979). It is consistent with the observation of Arzatbanov *et al.*(1978) that the redox state of heme *a* in mitochondria poised at 330 mV and blocked with CN^- is rapidly perturbed by ferricyanide but only slowly, and in an uncoupler-sensitive fashion, by pH. Thus when heme a_3 is oxidised (CN^- bound), heme *a* is in communication with matrix protons but cytoplasmic electrons. This feature is incorporated in the general model of Fig 7.21a in which the orientation of heme *a* and an associated proton-carrying group R alternates between a C-facing and an M-facing proton well.

The first model of proton pumping implicating heme a involves a heme propionate as the proton carrying group (Wikström, 1982a; Fig. 7.21b). This group switches between a position proximal to the ferric iron and unprotonated to a position distal to the ferrous iron and protonated. Consistent with a central role for a heme propionate is the Ca^{2+} perturbation of the spectrum of both heme *a* in the oxidase and free bisimidazole heme *a* (Wikström and Saari, 1975). In the latter case the Ca^{2+} effect was abolished by esterification of the propionates (Saari *et al.*, 1980). This model bears a close resemblance to the heme-linked ionisation of propionic acid observed in the soluble cytochromes *c*, with the additional feature of movement of the propionate group due to electrostatic attraction to the oxidised redox centre. However, the propionates of heme *a* of bovine monomeric subunit III-depleted cytochrome oxidase do not ionise over the pH range 6.6 to 8.6, according to NMR studies (Rigby *et al.*, 1989).

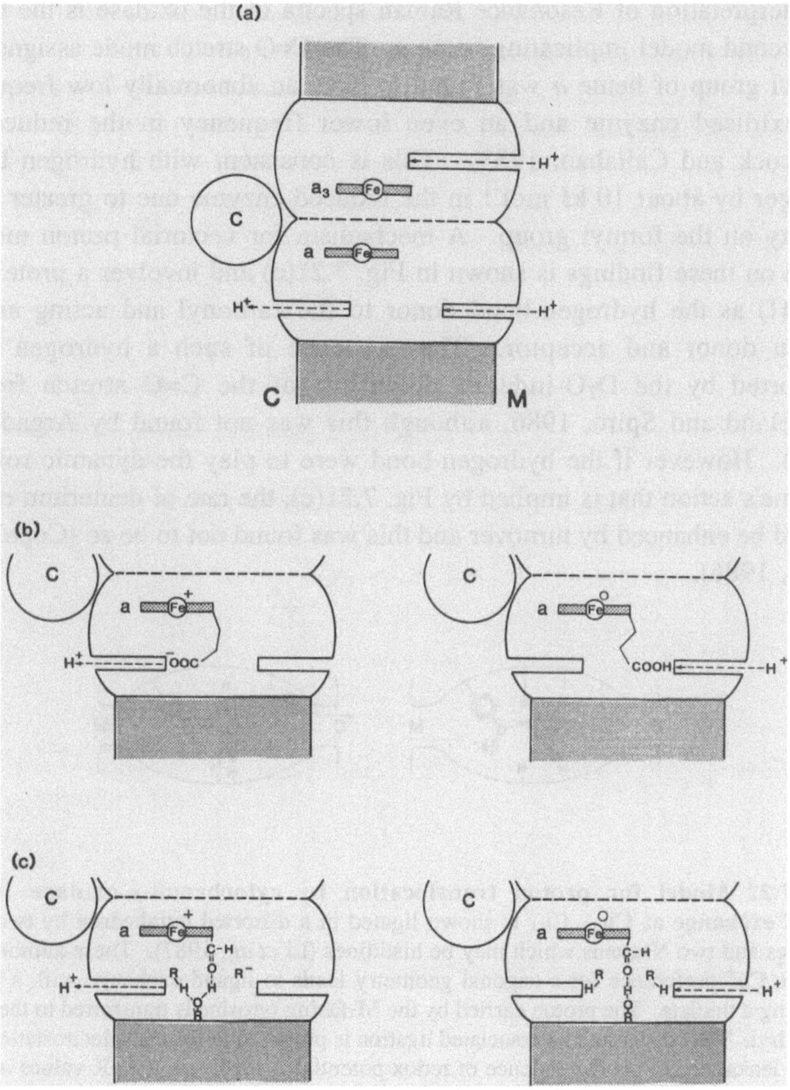

Fig. 7.21 Models for proton translocation by cytochrome *c* oxidase. **(a)** General diagram showing proton wells allowing proton entry to heme a_3 and heme *a* and proton exit from heme *a*. **(b)** Heme propionic acid as a mobile proton carrier. In the oxidised state the propionate is attracted to the positively charged iron and the proton is lost at the C-surface; in the reduced state the propionate swings to the **M**-facing well and picks up a proton due to the raised pK of this state (Wikström, 1982a). **(c)** Hydrogen-bonding network involving heme formyl as proton carrier (Babcock and Callahan, 1983)

Interpretation of Resonance Raman spectra of the oxidase is the basis for the second model implicating heme *a*. The C=O stretch mode assigned to the formyl group of heme *a* was found to have an abnormally low frequency in the oxidised enzyme and an even lower frequency in the reduced state (Babcock and Callahan, 1983). This is consistent with hydrogen bonding, stronger by about 10 kJ mol^{-1} in the reduced enzyme due to greater electron density on the formyl group. A mechanism for vectorial proton movement based on these findings is shown in Fig. 7.21(c) and involves a protein group (R-OH) as the hydrogen-bond donor to the carbonyl and acting as both a proton donor and acceptor. The existence of such a hydrogen bond is supported by the D_2O-induced downshift of the C=O stretch frequency (Copeland and Spiro, 1986, although this was not found by Argade *et al.*, 1986). However if the hydrogen-bond were to play the dynamic role in the enzyme's action that is implied by Fig. 7.21(c), the rate of deuterium exchange should be enhanced by turnover and this was found not to be so (Copeland and Spiro, 1986).

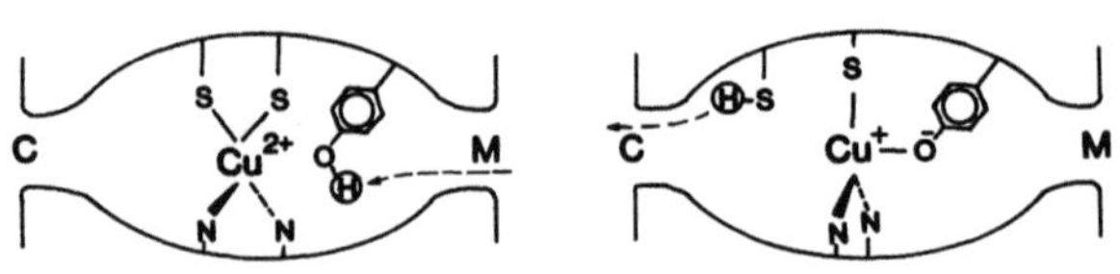

Fig. 7.22 Model for proton translocation by cytochrome *c* oxidase involving ligand exchange at Cu_A. Cu_A is shown ligated in a distorted tetrahedron by two cysteine thiolates and two N-atoms which may be histidines (Li *et al.*, 1987). These authors propose that the Cu^+ preference for a trigonal geometry leads to ligand exchange with a phenolate replacing a thiolate. The proton carried by the M-facing tyrosine is transferred to the C-facing sulfhydryl. The copper and its associated ligation is proposed to form an electrostatic barrier to proton leakage. No pH dependence of redox potential is involved: the pK values of tyrosine and sulfhydryl are determined by their ligation to copper

The third model for proton translocation by cytochrome oxidase implicates Cu_A rather than heme *a* as the redox trigger. This is consistent with the location of the proton translocation site at the electron accepting pole of the enzyme (Wikström and Casey, 1985), but runs counter to the view that the redox centre trigger should have a pH dependent redox potential. However Gelles *et al.*(1987) argue that the latter is not an absolute requirement and propose that proton translocation is initiated by the redox state-dependent ligand exchange at Cu_A, shown in Fig 7.22. According to this model, Cu_A switches from a low potential tetrahedral geometry, in equilibrium with cytochrome *c*, to a high potential trigonal geometry, in equilibrium with the ligand binding pole. This switch involves replacement of a C-facing cysteine by an M-facing tyrosine as Cu_A ligand and the associated transfer of a proton from the latter to the former. This proton transfer may be a weakness of the model since no mechanism for its vectorial movement is proposed and it must pass the electrostatic barrier of the central copper ion, which the authors view as a gate to prevent proton leakage.

In summary, three models have been proposed for proton pumping by cytochrome oxidase, but without strong enough experimental evidence to favour one choice. Two place heme *a* as the redox trigger and involve pH dependence of its redox potential; the third involves a pH independent ligand exchange process at Cu_A. The proton channel, which converts these proposed proton movements into a transmembrane proton gradient, may be subunit III. The reagent dicyclohexylcarbodiimide (DCCD), which inhibits proton translocation, binds to the enzyme at a glutamate of subunit III which occurs in a sequence having some resemblance to the proposed proton channel subunit of the ATP synthase (Casey *et al.*, 1980; Prochaska *et al.*, 1981). Subunit III can be removed from the oxidase with retention of electron transfer activity but either total or partial loss of H^+ translocation (Brunori *et al.*, 1987).

Whatever the proton channel is, a common feature of all the proposed mechanisms is that conformational changes take place within cytochrome oxidase to assist proton translocation (e.g. Wikström *et al.*, 1981; Malmström, 1989). Williams (1987) has drawn attention to a possible structural similarity between cytochrome oxidase and other metalloproteins - namely, the occurrence of α-helices in key functional units - to propose a model in which proton pumping is linked to the movement of α-helices.

7.7 Conformational Stabilisation

The preceding analyses of coordination and electrostatic influences on redox potential have, for the most part, considered the protein conformation as fixed during the redox process. However, there are cases where there are redox linked changes in the nature of the heme ligation (Chapter 2.3.1.2) and, even with mitochondrial cytochromes *c*, small differences in the conformations of the two redox states are evident (Chapter 4.5).

Although mitochondrial cytochrome *c* is structurally well-characterised, it is poorly characterised energetically. Nevertheless, we offer the following comments concerning the relationship of the redox-linked conformational change with the redox potential. The oxidation-linked changes in tuna cytochrome *c* observed by Takano and Dickerson (1981b; 1982) have been outlined in Chapter 4.5 (Fig. 4.11) and involve a 0.15Å shift of the heme group into the solvent and a 1Å approach of a water molecule towards the iron. Both these changes would tend to increase the polarity of the heme environment and stabilise the oxidised state.

No quantitative assessment of the effect on redox potential of such small conformational changes has been made, principally because such assessments involve calculations of electrostatic energies and, as we have discussed in section 7.4, such calculations are still relatively crude. The only relevant work we are aware of that attempts to estimate the energies of redox-linked conformational changes in proteins is by Rogers and Moore (1988). These authors dealt with conformational changes accompanying pH dependent oxidation and reduction of *R. rubrum* cytochrome c_2 and *Ps. aeruginosa* azurin. For cytochrome c_2 the pH dependent drop in redox potential associated with ionisation of a heme propionate was calculated to be 119 mV (Table 7.5), considerably higher than the observed value of 47 mV. Rogers and Moore (1988) suggested that part of the difference was due to protein conformational changes but further work on these systems is needed to verify the suggestion that these changes required an energy of 7 to 11 kJ mol^{-1}.

Chapter 8 Electron-Transfer Mechanisms

8.1 Introduction

The investigation of electron-transfer mechanisms is an interdisciplinary area, more so perhaps than any other area of cytochrome research, in which physicists, chemists and biochemists have combined to produce an extensive literature. This work has eliminated a number of proposed mechanisms but it has not replaced them with a generally accepted mechanism for any protein. In fact many of the fundamental questions about electron transfer within a protein matrix have not yet been answered. However, the interplay between different disciplines has led to the development of a theoretical framework for electron transfer and indicated some of the factors important in controlling electron-transfer rates. In the present chapter we describe these mechanistic influences and give a brief guide to the biochemical relevance of particular physical concepts of electron transfer.

8.2 General Aspects of Electron Transfer

8.2.1 Classification of Electron-Transfer Reactions in Metal Complexes

There are two general classes of redox reactions involving metal complexes: inner-sphere and outer-sphere reactions (Taube, 1984). In the former class the metal ions share a bridging ligand that generally plays a role in transmitting the electron during reaction. In contrast to this, in outer-sphere reactions, the individual coordination shells of the metal ions remain intact. The reactions of simple electron-transfer proteins tend to be outer-sphere.

Electron-transfer reactions between reactants in solution proceed in three main stages: formation of a bimolecular precursor complex; electron transfer; and dissociation of the bimolecular product complex. The electron-transfer stage involves the formation and relaxation of an activated complex. These steps are required to allow the electron transfer to occur isoenergetically. Thus the simplest reaction scheme is that given in Fig. 8.1.

Association:	$A_o + B_r \rightarrow (A_o \text{---} B_r)$	(8.1)
Equalization of energy levels:	$(A_o \text{---} B_r) \rightarrow (A_o \text{---} B_r)^*$	(8.2)
Electron transfer:	$(A_o \text{---} B_r)^* \rightarrow (A_r \text{---} B_o)^*$	(8.3)
Relaxation:	$(A_r \text{---} B_o)^* \rightarrow (A_r \text{---} B_o)$	(8.4)
Dissociation:	$(A_r \text{---} B_o) \rightarrow A_r + B_o$	(8.5)

Fig. 8.1 Individual steps of a simple outer-sphere electron-transfer reaction. o and r indicate oxidised and reduced species respectively. When **A** and **B** are the same the reaction is self-exchange; where they are different it is a cross-reaction. This distinction is important because, as will become clear, the reaction rates of self-exchange reactions are simpler to analyse than those of cross-reactions

The overall rate for the reaction scheme of Fig. 8.1 is given by:

$$\text{rate} = k\,[A_o][B_r] \tag{8.6}$$

$$\text{where: } k = K_a.k_{et} \tag{8.7}$$

K_a is the association constant for the formation of the precursor complex (Eqn. 8.1), and k_{et} is the first order electron-transfer rate constant for the reaction $(A_o \text{---} B_r) \rightarrow (A_o \text{---} B_r)^*$.

The association-dissociation steps, in which unactivated collision complexes are formed, are of central importance to many protein electron-transfer reactions. We have considered these elsewhere (Pettigrew & Moore, 1987: Chapter 2). The main concerns of the present chapter are the role of the activated complex and the electron-transfer step itself. Before moving onto these subjects we consider a topic important to the understanding of electron transfer.

8.2.2 What is an Electron?

Many chemical reactions can be described satisfactorily without needing to delve too deeply into the physical nature of the reacting species. Thus with enzyme catalysed reactions, for example, it is generally sufficient to know whether the reaction proceeds by, say, acid-base catalysis, and whether the transition-state is stabilised by hydrogen-bonds and salt-bridges. However, to describe electron transfer adequately, we need to take account of the fact that an electron can be considered to act both as a particle and as a wave.

One way of describing the behaviour of electrons is to use the quantum-mechanical concept of the electron as a probability wave (see, for example: Gribbin, 1984; Moore, 1972). This is the basis of Schrödingers wave equation (Eqn. 8.8) which relates the location and energy of an electron to its wave function (ψ) or orbital.

$$\left(\frac{\delta^2\psi}{\delta x^2} + \frac{\delta^2\psi}{\delta y^2} + \frac{\delta^2\psi}{\delta z^2}\right) + \frac{8\pi^2 m}{h^2}(E - U)\psi = 0 \qquad (8.8)$$

Where x, y and z are the Cartesian coordinates of the electron (Fig. 8.2a), m is its mass, h is Plancks constant, and E and U are the total and potential energies of the electron respectively.

The equation can be rewritten into spherical polar coordinates (Fig. 8.2a) and separated into two parts. One part, the angular part, is a function of θ and ϕ only, and the other, the radial part, is a function of r only. The angular part determines the shape of the orbital (Fig. 1.1 shows the shape of the d-orbitals) and the radial part determines the distance between an electron and its nucleus.

The radial part (R) decays exponentially, with the exact form of the decay depending on what orbital the electron is in. For a 1s orbital the decay is:

$$R_{(1s)} = \frac{1}{\sqrt{\pi}}\left(\frac{Z}{0.53}\right)^{3/2}.\exp\left(\frac{-Zr}{0.53}\right) \qquad (8.9)$$

where Z is the atomic number of the nucleus. This function is shown graphically in Fig. 8.2b for hydrogen, together with the corresponding graph for $R_{(2p)}$.

The probability of finding an electron in a given location is proportional to ψ^2. Thus the radial distribution function, $|4\pi r^2 R^2|$, gives the probability as a function of the distance r. Fig. 8.2c shows this function for $R_{(1s)}$ and $R_{(2p)}$ plotted against r. These plots indicate that there is a certain probability of finding an electron anywhere from within the nucleus to a considerable distance away. For example, Moore (1972: p639) has calculated that the relative probability of the hydrogen 1s electron being 5.29Å from its nucleus, instead of 0.529Å, is 1.52×10^{-6} and although this is a low probability, it may be significant in the context of electron transfer. The maximum probability corresponds to the radius of the orbit in the solar-system model of an atom, which considers electrons as revolving like planets around a sun, the nucleus.

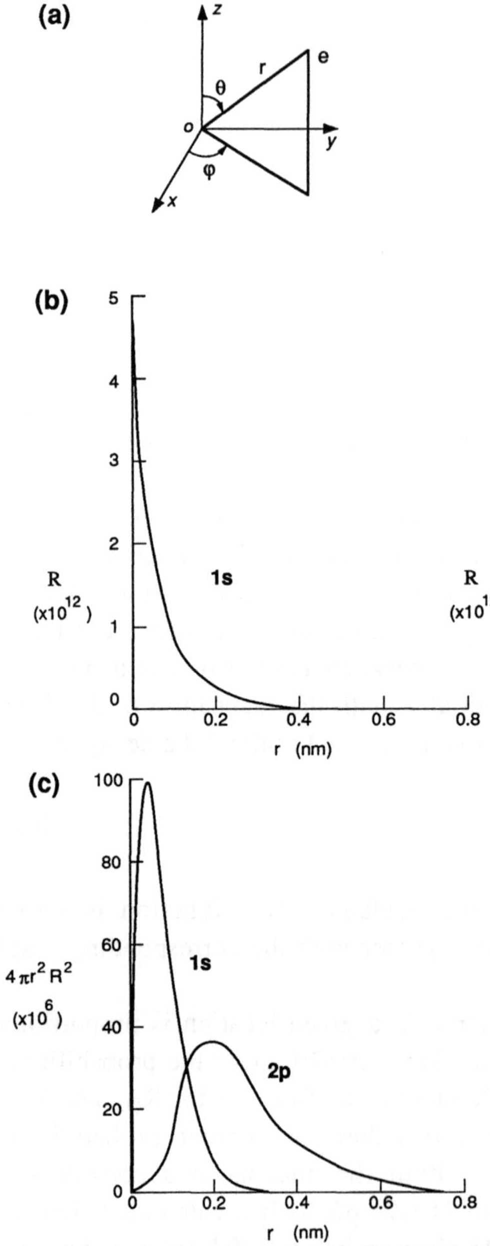

Fig. 8.2 Representations of the radial part of a wavefunction. a Position of electron (**e**) from nucleus (***o***) in spherical polar coordinates. **b** Radial part of wave function for **1s** and **2p** orbitals of a hydrogen atom. **c** Radial distribution functions for the **1s** and **2p** orbitals. The distribution function gives the probability of finding the electron in a spherical shell of thickness **dr** at a distance **r** from the nucleus. (Adapted from Moore, 1972)

In order to transfer an electron from one atom to another the orbital donating the electron and the orbital accepting it must interact; i.e. the probability waves must overlap to some extent (Fig. 8.3). The greater the overlap, the greater the probability of electron transfer. Thus many investigations of electron-transfer reactions are directed towards determining the shape and extent of the donor and acceptor orbitals.

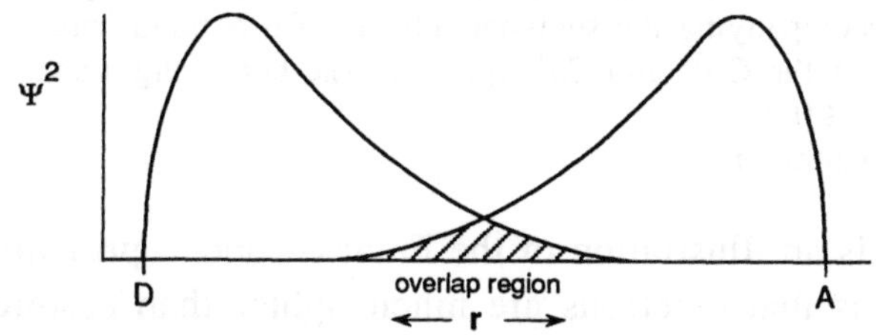

Fig. 8.3 Overlap of donor (D) and acceptor (A) orbitals. **D** and **A** are the nuclear centres of the two orbitals. **r** is the distance between **D** and **A**

8.2.3 Electron-Transfer Reactions of Inorganic Compounds

Before considering electron-transfer reactions involving proteins we discuss briefly some outer-sphere electron-transfer reactions of simple inorganic compounds (Table 8.1), because these illustrate many of the features we wish to stress. All of the reactant pairs are similarly charged so we assume that the work terms for formation of the transient reactive complexes will be about the same, and therefore differences in the observed bimolecular rate constant, k, reflect differences in the unimolecular rate constant k_{et} (Eqn. 8.7). Furthermore, since all the reactions are self-exchange reactions, the redox driving energies are zero. Therefore, differences in k reflect differences in the ease of activating the precursor complexes and/or differences in the probability of electron-transfer within the activated complexes (Fig. 8.1).

We can usefully make three comparisons:

1. $[Co(NH_3)_6]^{3+/2+}$ with $[Ru(NH_3)_6]^{3+/2+}$

The rate for the Ru compound is at least 10^6 times faster than that for the Co compound. The main reason for this is that the Co compound undergoes a structural change on reaction, with the Co-N bond length changing by 0.2Å (Table 8.1), whereas the Ru compound does not. Thus less energy is required to activate the Ru precursor compound.

Table 8.1 Electron transfer self exchange reactions of inorganic complexes

Reactants [a]	Approximate ionic radii (Å)	k ($M^{-1}\ s^{-1}$)
$[Ru(NH_3)_6]^{3+/2+}$	~3.6	8.2×10^2
$[Co(NH_3)_6]^{3+/2+}$	~3.6	$\leq 10^{-7}$
$[Ru(phen)_3]^{3+/2+}$ [b]	~5.2	$\geq 10^7$
$[Co(phen)_3]^{3+/2+}$	~5.2	1.1

[a] All the reactants are low-spin except $[Co(NH_3)_6]^{2+}$ which is high spin. There is a small structural change accompanying the spin-state change: Co-N bond lengths are 1.96Å and 2.16Å respectively for the Co^{3+} and Co^{2+} species. The bond length changes in the other reactions are probably ≤ 0.05Å.

[b] phen is 1,10-phenanthroline.

This example is an illustration of the Franck-Condon principle. The basis of this principle is that electrons are much lighter than atomic nuclei, and hence electron transfer occurs in a time much shorter than that required for a change in nuclear positions (Libby, 1952). Thus electron transfer takes place in a virtually static nuclear framework and occurs between reactants whose nuclear configurations cannot respond immediately to the changed oxidation state. As a result, activated complexes are essential intermediates to allow the reactants to match their energies prior to the transfer (Fig. 8.4).

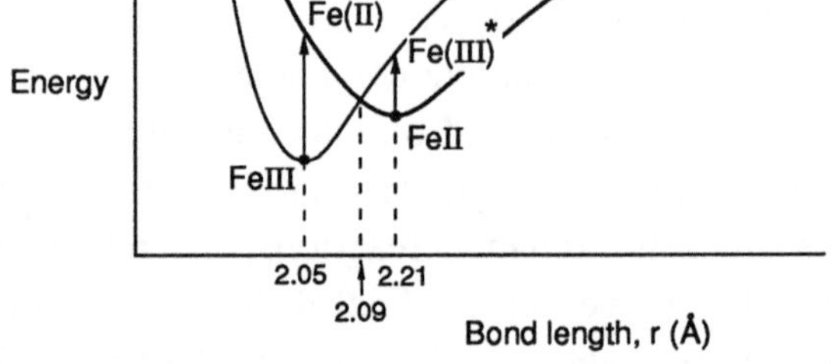

Fig. 8.4 Activated complex in the self-exchange reaction between $[Fe(H_2O)_6]^{3+}$ and $[Fe(H_2O)_6]^{2+}$. The diagram is a two-dimensional potential energy illustration of the energy dependence of the Fe-O bond length. The ground-state Fe-O bond length in $[Fe(H_2O)_6]^{2+}$ (2.21Å) is longer than that in $[Fe(H_2O)_6]^{3+}$ (2.05Å) and so if electron transfer took place between the ground-state configurations both products (indicated by *) would be energetically excited; the product $[Fe^*(H_2O)_6]^{2+}$ would contain a compressed **Fe(II)*** ion and the product $[Fe^*(H_2O)_6]^{3+}$ would contain a stretched **Fe(III)*** ion. This is thermodynamically impossible. Therefore, the reactants must match their energies by bond compression, stretching and bending before electron transfer can occur. This requires an activation energy and results in each Fe-O bond length in the activated complex becoming 2.09Å, the energetically equivalent distance. (Adapted from Tobe, 1972)

Another restriction of the Franck-Condon principle is that spin-angular momentum should be conserved. This suggests that if spin-state changes do occur, they will lead to slower rates, even when there is not a significant structural change. However, this effect is difficult to quantify experimentally, since structural changes nearly always accompany spin-state changes. The $[Co(NH_3)_6]^{3+/2+}$ example is often cited as an illustration of a reaction slowed by a change in spin-state (e.g. see Tobe, 1972), but Geselowitz & Taube (1982) question whether this is really a significant factor in reducing the rate.

2. $[Ru(NH_3)_6]^{3+/2+}$ with $[Ru(phen)_3]^{3+/2+}$

The trisphenanthroline compound is larger than the hexammine (Table 8.1) so one might have supposed that this would lead to poorer overlap of the metal donor-acceptor orbitals with a consequent reduction in the rate. However, the rate for $[Ru(phen)_3]^{3+/2+}$ is much faster than that for $[Ru(NH_3)_6]^{3+/2+}$. The explanation for this is that phen is an aromatic group whose π-orbitals can mix with the metal donor-acceptor orbitals, thereby increasing their radial extent. Thus phen is a conducting ligand (called by Sutin a 'π-way') and NH_3 is a non-conducting ligand.

3. $[Co(phen)_3]^{3+/2+}$ with $[Ru(phen)_3]^{3+/2+}$

Now with neither compound having a redox-state linked spin-state or conformational change, and both having a conducting ligand, the difference in rate is still at least 10^7 (Table 8.1) The reason for this is that the transferring electron of the Ru compound enters or leaves a t_{2g} orbital whereas that of the Co compound enters or leaves an e_g orbital (Fig. 8.5). Overlap of the metal orbitals in the activated complex is significantly greater for the t_{2g} than the e_g orbitals thereby leading to a higher rate.

These examples serve to illustrate that subtle electronic effects can have significant influences on the rate. They also act as indicators of the kind of features we should look for in electron-transfer proteins, and this we do in section 8.2.4.

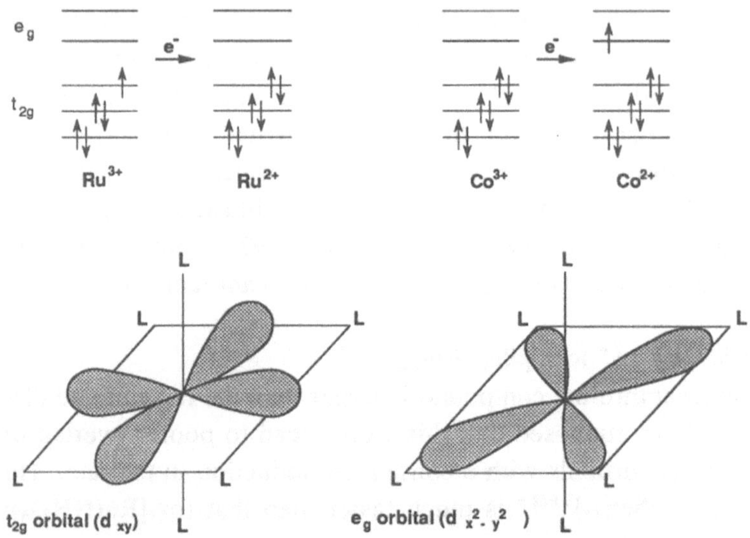

Fig. 8.5 Comparison of the electronic structures of low-spin Ru and Co. The donor-acceptor orbital for Ru is a $\mathbf{t_{2g}}$ orbital, which has a greater radial extension than an $\mathbf{e_g}$ orbital

8.2.4 General Features of Electron-Transfer Proteins

The preceding discussion of relatively simple inorganic complexes emphasised two features that are important for rapid intramolecular or intracomplex electron-transfer. First, structural changes should be minimised; second, good overlap between the donor and acceptor orbitals is required. In this section we briefly review various electron-transfer proteins to see how these requirements are met. Note that for rapid bimolecular electron-transfer, non-productive collisions need to be minimised. Possible mechanisms for achieving this are described by Pettigrew & Moore, (1987: Chapter 2) and we shall not consider this subject further here.

8.2.4.1 Spin-state and Conformational Changes in Electron-Transfer Proteins

In view of the discussion of inorganic compounds in section 8.2.3, one might expect that many reactants in biological electron-transfer systems have evolved so as to minimise the extent of electronic and nuclear rearrangement necessary to attain the activated state. This is the underlying principle of the entatic state hypothesis proposed by Vallee & Williams (1968). These authors suggested that electron-transfer proteins have redox centres with strained geometries, so that the ground-state structure of the protein resembles the transition-state of small molecule reactants. An illustration of this is the distorted redox centre of the blue copper proteins azurin and plastocyanin (Chapter 7.3) (Williams, 1971).

It is not clear whether the entatic state hypothesis is useful when applied to cytochromes, however, because there is only a small difference in the geometric requirements of six-coordinated Fe(II) and Fe(III). The geometric demands of sulfur ligation to heme iron are especially low (Scheidt & Reed, 1981). Comparison of the x-ray structures of cytochromes and small porphyrin complexes (Chapter 1.3) shows that the bond lengths and angles of the iron-ligand bonds in the proteins are in the same range as those expected from small molecule studies.

Coupled conformational changes away from the redox centre could also lower the electron-transfer rate and it may be because of this that the class I cytochromes *c* have, at most, only a small redox state conformation change (Chapter 4.5).

It is not clear whether the spin-state of the iron is an important rate determining factor, provided there is not a high-spin to low-spin conversion accompanying the electron transfer. However, it is notable that virtually all the simple electron-transfer hemeproteins are low-spin. Electron transfer between the Fe(II) and Fe(III) states of a low-spin hemeprotein requires less electronic rearrangement than the corresponding high-spin reaction, although the magnitude of the difference is difficult to gauge. Mayo *et al.* (1986) have approached the problem by comparing electron-transfer rates of cytochrome *c* with those of deoxymyoglobin and they found that for comparable reactions cytochrome *c* is $\sim 10^2$ times faster. However, this comparison is not ideal because myoglobin has not evolved to be an electron-transfer protein and the reduction of metmyoglobin to deoxymyoglobin requires the displacememt of H_2O coordinated to the Fe(III).

Unfortunately there is no simple high-spin heme protein that could be used to assess the effect of a high-spin state. Studies of cytochrome *c'* would be interesting, although there is a redox-linked spin-state change from the ferric $S = 3/2, 5/2$ admixed state to the ferrous $S = 2$ state (Table 1.1).

8.2.4.2 Electron Transmission in Proteins

The question of how electrons are transmitted in proteins is bound up with the extent of donor-acceptor orbital overlap and the nature of the electron transmission pathway. We consider both topics in this section.

Redox Centres and Donor-Acceptor Orbitals in Proteins. To achieve optimum orbital overlap, metals are commonly found in proteins in structures where their orbitals can be greatly extended, as in the cases involving unsaturated ligands, like porphyrin and histidine, and multimetal centres like the ferredoxins.

Also, it appears that redox centres involved in intermolecular reactions are not generally found at the centre of proteins a long way from their surface but are displaced to one side, as is seen in the class I cytochromes *c* (Chapter 4), and the blue copper proteins (Adman, 1979), so that in interprotein complexes the centres may be brought relatively close together. However, it is unlikely that they will be in Van der Waals contact. In the proposed cytochrome c/b_5 complex, the shortest interheme distance is 8Å (Salemme, 1976; Pettigrew & Moore, 1987: Fig. 2.20), while in the cytochrome *c* peroxidase:cytochrome *c* complex an electron has to travel over distances up to 20Å between heme edges (Poulos & Kraut, 1980; Vanderkooi *et al.*, 1980; Pettigrew & Moore, 1987). Interestingly, when the hemes are in Van der Waals contact, as in the cytochromes c_3 (Fig. 5.3), the electron-transfer rates appear to be little faster than those for the longer range electron-transfer. This may be due to a variety of reasons but clearly short distances do not automatically mean rapid electron transfer.

In addition to the influence of distance between redox centres, their relative orientations may also be important. The orientation of the hemes in the proposed cytochrome *c* complexes is approximately coplanar, a feature that has been suggested to be necessary for the correct alignment of the electron-transfer orbitals (Salemme, 1976; Poulous & Finzel, 1984). However, coplanar hemes are not essential for electron transfer as demonstrated by the structures of cytochrome c_3 (Fig. 5.3) and the photosynthetic reaction centre cytochrome *c* (Fig. 5.5). Also, the π-orbital extension is greatest above and below the plane of the heme and thus orbital overlap will be greater for two parallel hemes compared to two coplanar hemes with the same separation. One reason parallel hemes have not been implicated in interprotein complexes may be that there are major steric constraints to having two parallel low-spin hemes with a relatively small interheme separation.

Hop Centres and Electron Transfer. Even though simple electron-transfer proteins are constructed to assist donor-acceptor orbital overlap there has been a reluctance to accept that long-range electron-transfer in proteins is possible without a transient intermediary electron trap between the redox centres. This type of mechanism can be represented as:

Redox Centre 1---→ Electron Trap 2---→ Electron Trap 3---→ Redox Centre

where the broken lines indicate electron transfer involving the transient traps and the solid line indicates the direct electron transfer. The distinction between the direct electron transfer and the transfer indicated by broken lines is that in the latter the electron resides for a period of time in the electron traps. Hop centres should therefore be detectable by chemical trapping methods or spectroscopically. Such electron transfer is an example of a hop mechanism (Winfield, 1965; Williams, 1969; Dickerson *et al.*, 1971). Possible non-metal electron traps in proteins are: aromatic groups and methionines (as free radicals); or disulfides (-S-S- + $2e^-$ + $2H^+$ $\rightarrow$ 2-SH). Partly because of the historical role of proposed hop mechanisms for mitochondrial cytochrome *c* based on aromatic free radicals we discuss this aspect below. However, we emphasise that there is no evidence as yet that electron transfer involving cytochromes lacking an enzymic function operates via intermediary electron traps.

Free Radicals. The first detailed mechanism for mitochondrial cytochrome *c* was the free radical proposal of Dickerson *et al.* (1971), later modified by Takano *et al.* (1973). In this mechanism electron entry into the protein was suggested to occur at Tyr 74 with Tyr 67, Tyr 74 and Trp 59 becoming free radicals at some stage of the process. This proposal led many groups to explore experimentally the role of the aromatic residues. It is now clear that this mechanism does not operate because Tyr 67 and Tyr 74 can be replaced by leucines without seriously affecting the reactivity of cytochrome *c* (Chapter 4.6).

Even before these chemical modification studies the mechanism had been criticised on the grounds that to generate free radicals from aromatic amino acids requires redox potentials around 1 volt (Burns *et al.*, 1976; Moore &

Williams, 1976; Salemme, 1977; Butler *et al.*, 1982; De Felippis *et al.*, 1989). Such potentials are not readily attainable with the Fe^{2+}/Fe^{3+} couple though they are with higher oxidation states of iron. This problem was recognised by Winfield (1965) in his original discussion of free radicals, which was restricted to high potential states of oxidases and peroxidases. Winfield's specific suggestion for cytochrome *c* was that it might have a surface aromatic group readily oxidizable by virtue of its proximity to the oxidase redox centre in the protein complex.

Are Aromatic Amino Acids Essential for Electron Transfer? As the comparison of $Ru(NH_3)_6]^{3+/2+}$ with $[Ru(phen)_3]^{3+/2+}$ demonstrated (section 8.2.3), aromatic groups can act as 'π-ways' to assist electron transmission. The question we address in this section is: are there any data to support such a role for non-ligand aromatic amino acids in proteins?

The clearest set of data come from the elegant work of Lode *et al.* (1974a;b). These workers semisynthetically replaced the sole aromatic residue of *Clostridium* ferredoxin by leucine and found unchanged activities in *in vitro* enzymic assays. Thus in this system at least aromatic groups are not essential.

The only comparable work to that of Lode *et al.* (1974a;b) in the cytochrome *c* field involves chemical modification and site-directed mutagenesis of mitochondrial cytochrome *c* (Chapter 4.6). Recent work in this area has concentrated on the role of the unvaried Phe 82 of cytochrome *c* in the yeast cytochrome *c* peroxidase (YCCP) oxidation of ferrocytochrome *c* (Pettigrew & Moore, 1987: Chapter 2). A major feature of this reaction is the formation of a bimolecular cytochrome *c*:YCCP complex. Poulos & Kraut (1980) have proposed a structure for this complex; the interheme region is shown in Fig. 8.6.

Poulos & Kraut (1980) recognised that the residues between the hemes may play a role in electron transmission and they pointed out that Phe 82 of cytochrome *c* and His 181P were only 7Å apart in the complex, and approximately parallel to each other. Together with Arg 48P and Trp 51P, which are not in the interaction region, the hemes, Phe 82 and His 181P were proposed by Poulos & Kraut to form a conduction pathway.

This proposal has been investigated with mutants of *S. cerevisiae* iso-1 cytochrome *c* in which Phe 82 was replaced by a variety of residues (Table 8.2). The relative second order rate constants for the oxidation of the ferrocytochromes by native YCCP (Activity parameter 1) suggest there is not a unique electron transfer role for an aromatic group at position 82, although the rate constants for the $ZnYCCP^+$ oxidation of the reduced cytochromes *c* (Activity parameter 3) indicate the opposite. Interpretation of both sets of data are complicated by the possibility that the mutations produce different cytochrome *c*:YCCP complexes and thus further work is required to establish whether Phe 82 really forms an important 'π-way' (see also section 8.5.3).

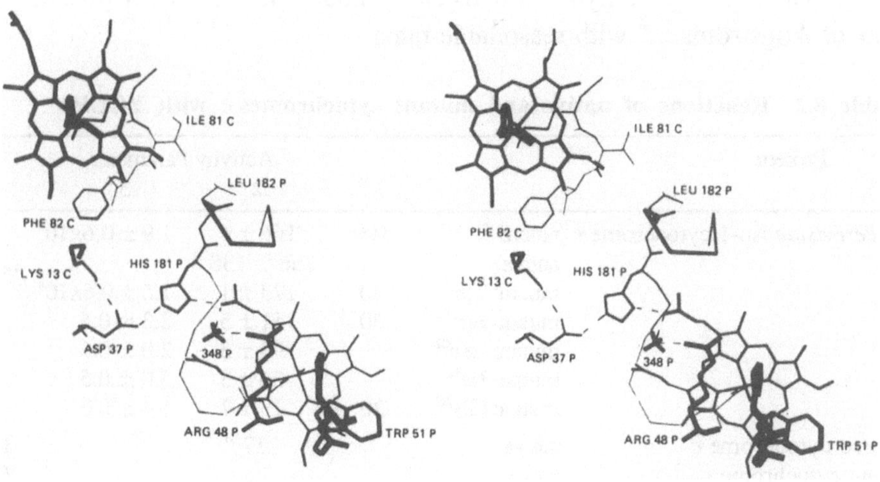

Fig. 8.6 Stereoscopic view of the interheme region of the proposed cytochrome *c*:cytochrome *c* peroxidase (YCCP) complex. Cytochrome *c* residues are labelled with a C, and YCCP residues are labelled with a P. The broken lines represent hydrogen bonds and ionic interactions. As with the cytochrome c/b_5 complex (Salemme, 1976) the area between the proteins does not contain solvent. Space-filling models show that the two protein surfaces are touching. (Reproduced from Poulos & Finzel, 1984)

The suggestion that electrons can travel over relatively large distances without the involvement of 'π-ways' is not new. It is well-established that rapid electron transfer over distances of 5 - 10Å can occur in non-polar media such as inert organic solvents at room temperature (Minday *et al.*, 1972; Noda & Keran, 1974). The work of Miller *et al.* (1984) is a more recent example (Fig. 8.7).

Through-space or Through-bond Electron Transmission? Closs and Miller (1988) addressed directly the question of how an electron gets from a donor to an acceptor when the intervening medium has no apparent intrinsic electron affinity. These authors stressed that there are two ways to get electronic coupling, and hence electron transmission, between the donor and acceptor. First, the direct through-space coupling provided by overlap of the donor and acceptor wave-function (Fig. 8.3); and second, the through-bond mechanism which requires an electronic interaction between the donor and acceptor wave-functions with those of intervening groups. Closs and Miller stated that for compounds of the type shown in Fig. 8.7, "... the through-bond mechanism has won acceptance by most investigators ...", and they go on to claim that "... saturated hydrocarbons can conduct electrons over [distances of] tens of Ångströms ..." with reasonable rates.

Table 8.2 Reactions of native and mutant cytochromes *c* with YCCP

Protein		Activity Parameter 1 [a]	2 [b]	3 [c]	4 [d]
S. cerevisiae iso-1 cytochrome *c*	native [e]	100	166 ± 4	1.9 ± 0.6x10^4	
	native		266 [i]; 138 [ii]		3.4
	mutant Tyr^{82}	30	173 ± 1	1.5 ± 0.6x10^4	
	mutant Ser^{82}	70	151 ± 5	2.3 ± 0.5	
	mutant Leu^{82}		93 ± 5	2.0 ± 0.5	
	mutant Ile^{82}		56 ± 3	3.0 ± 0.5	
	mutant Gly^{82}	20	13 ± 2	1.4 ± 0.3	
Horse cytochrome *c*	native		17 [ii]		0.3
Tuna cytochrome *c*	native		25 [i]		0.2

[a] Relative activity for YCCP oxidation of ferrocytochrome *c* assay (0.02 M Tris-HCl, pH 7) (Pielak *et al.*, 1985).
[b] Rate (s^{-1}) of oxidation of ZnYCCP by C_o indicated by k_1 in Eqn. 8.29 of Scheme 8.2 [1.0 mM phosphate, pH 7, 0°C (Liang *et al.*, 1988); except i, which were 1.0 mM phosphate, pH 7 (Liang *et al.*, 1987);, 25°C and ii, which were 10 mM phosphate, pH 7, 20°C (Ho *et al.*, 1985)].
[c] Rate (s^{-1}) of reduction of $(ZnYCCP)^{\ddagger}_{\bullet}$ by C_r, indicated by k_2 in Eqn. 8.29 of Scheme 8.2 [1.0 mM phosphate, pH 7, 0°C (Liang *et al.*, 1988)].
[d] Rate (s^{-1}) of reaction indicated by k_1 in Eqn. 8.27 of Scheme 8.2 [1.0 mM phosphate, pH 7, 24°C (Cheung *et al.*, 1986)].
[e] Native cytochrome *c* has Phe 82

Although these chemical systems indicate how facile electron transfer can be they do not address directly the topic of the route of electron transmission in a protein. This is being studied using proteins containing fixed donor and acceptor sites (section 8.5). Gray and his colleagues have adopted this approach and in their study of derivatives of myoglobin, each containing $[Ru(NH_3)_5]^{3+\text{ or }2+}$ covalently attached to a histidine, they conclude the ease of electron transfer governed by the medium is (Mayo *et al.*, 1986): *aromatic* amino acids>*non-polar aliphatic* amino acids>*polar* amino acids>H_2O.

However, theoretical studies of this reaction system suggest aromatic groups are not rate enhancing (Kuki & Wolynes, 1987). This is also indicated by experimental studies of ruthenated yeast cytochrome *c* (Bowler *et al.*, 1989). Future work in this area involving the use of site-directed mutagenesis to change the nature of the medium may allow the question of how electrons are transmitted to be fully explored.

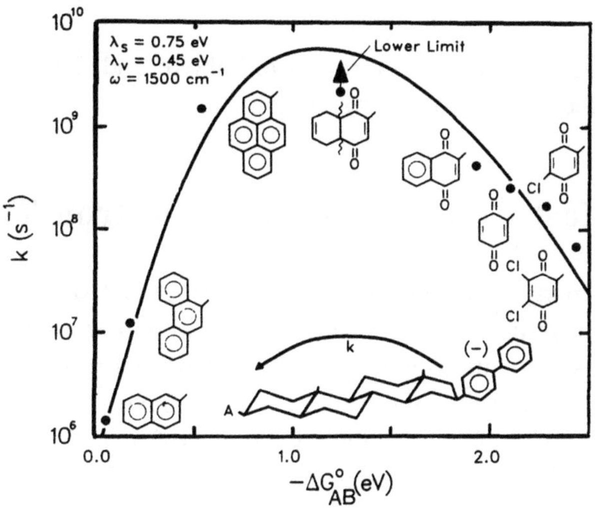

Fig. 8.7 Intramolecular electron-transfer rate constants as a function of free energy change in 2-methyl tetrahydrofuran solution at 297 K. Electrons transferred from biphenyl ions to the eight different acceptor groups, **A** (shown labelling the points) in eight bifunctional molecules having the general structure shown in the center of the figure. The solid line is a theoretical line calculated assuming particular values for the solvent and internal reorganisation energies. (Reproduced from Miller *et al.*, 1984)

8.2.5 Summary of Reactivity Determinants

Although we have emphasised the role of the activated complex it is important to note that there are a number of factors that may be rate determining for a bimolecular reaction of the type illustrated in Fig. 8.1. Indeed, much of the analysis of a given reaction is aimed at discovering which factor is rate limiting. Possible rate limiting factors are:

1. The protein association-dissociation steps. These are governed by the interaction work terms.

2. The reactant rearrangement energy. This may be electronic or conformational in origin and is not restricted to events at the redox centres.
3. The redox driving energy. This is usually taken to be the difference in redox potentials between the acceptor and donor.
4. The electron transmission coefficient within the activated complex. This depends upon the nature of the medium between the redox centres, and their separation.

The relationships between the terms listed above and the rate of electron transfer are described in the following section.

8.3 Theoretical Considerations of Electron Transfer

There have been many contributions to the development of electron-transfer theory, and these have been extensively described in numerous review articles. Therefore we make no attempt to provide either a complete or historical account of this area. Articles the present authors have found useful include those by Marcus (1964, 1979), Marcus & Sutin (1975, 1985, 1986), Cannon (1980), De Vault (1980), Sutin (1982), McLendon (1988) and the conference proceedings edited by Chance *et al.* (1979).

Two general classes of electron-transfer mechanism have been widely discussed with regard to biological systems: classical mechanisms, in which the reactant system is activated to surmount the energy barrier resulting from the Franck-Condon principle; and electron tunnelling, in which the reactant system tunnels through the barrier. These mechanisms are not mutually exclusive, nor are they always distinguishable, but for clarity we shall consider them separately.

8.3.1 Classical Description of Electron Transfer

The classical description starts by considering the role of nuclear motion in electron transfer, and a convenient representation of this is an energy diagram such as that for the $[Fe(H_2O)_6]^{2+}$ - $[Fe(H_2O)_6]^{3+}$ self-exchange reaction given in Fig. 8.4. However, this diagram is much simplified because it only considers one bond length, namely the Fe-O distance. For a full analysis it is necessary to consider the potential energy of all the nuclear coordinates in the system, including the solvent. Since even for a reaction between small molecules this will lead to a complex, multi-dimensional energy surface, we shall follow normal practice and use, as illustrations, profiles of such a surface. Fig. 8.8 shows the free energy profiles for a variety of reaction types. Most theoretical discussions of electron transfer consider potential energy surfaces but, provided a rigorous treatment is not required, the main features of the potential energy analysis apply to a free energy analysis.

Fig. 8.8 illustrates three general classes of mechanism which are discussed in the following sections:

1. *The thermal mechanism* in which electron transfer is achieved by thermally activated transition of **R** to **P** by passage through point (C).
2. *The tunnel mechanism* in which electron transfer is achieved by the transition of **R** to **P** below the level of point (C); indicated by the broken lines in Fig. 8.8(a).
3. *The photoinduced mechanism* in which electron transfer is achieved by the vertical transition marked O_p in Fig. 8.8(a). This corresponds to the absorption of a photon.

The Thermal Mechanism. Fig. 8.8(a) shows two parabolic free energy curves for the reactant system (curve **R**) and for the product system (curve **P**) when there is a negligible probability of electron transfer. Fig. 8.8(b) illustrates the case where there is a high probability of electron transfer. Electron transfer occurs by the system jumping from **R** to **P**. To satisfy the Franck-Condon principle the electron transfer must be a vertical transition between R and P and, for radiationless transfer, conservation of energy requires it also be a horizontal transition. The only position in Fig 8.8b where these two conditions are met substantially is at the intersection point with nuclear configuration (C). (There is a small probability of the system being located below (C) and this allows tunnelling to occur). Thus the path of reaction is activation from configuration (A) to (C). Then, depending on the electron transfer probability, the system may jump from **R** to **P**, followed by relaxation to configuration (B), or remain on curve **R** with a subsequent relaxation back to configuration (A). The former, adiabatic pathway, illustrated by Fig. 8.8(b), occurs in a system where there is an appreciable electronic interaction between **R** and **P**. The interaction results in the energy levels being split so that there is a smooth transition from **R** to **P** on the lower energy surface. The latter, non-adiabatic, pathway (Fig. 8.8(a)) occurs where there is a only a small, or no, electronic coupling between **R** and **P**.

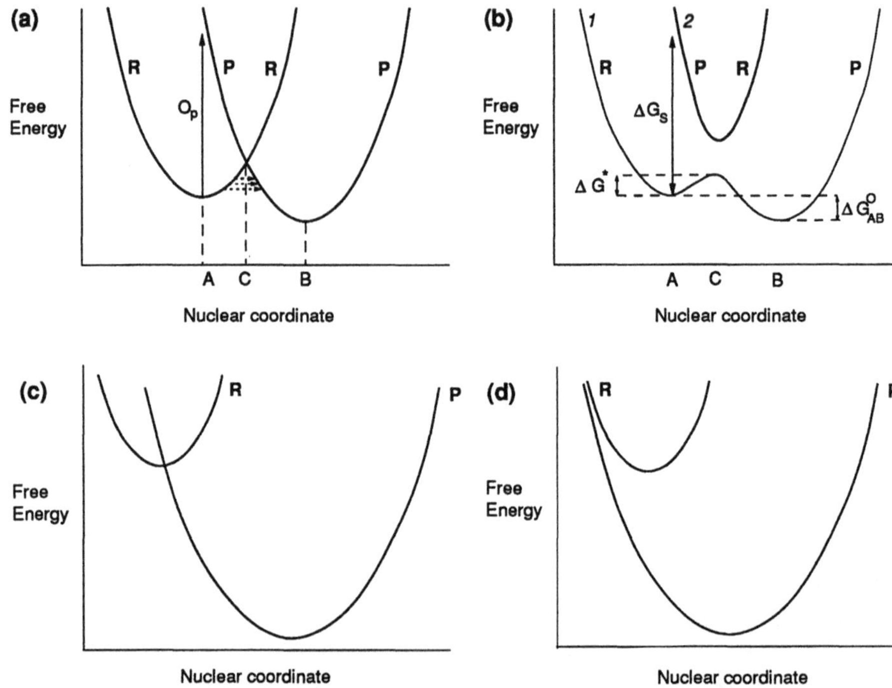

Fig. 8.8 Free energy profiles for electron transfer reactions. The ordinates represent the free energy of the nuclei of the reactants, products and solvent. The abscissa represents a combination of the positions of all the nuclei affected by the electron transfer and runs from the ground-state configuration of the reactants A to the ground-state configuration of the products B via the nuclear configuration representing the activated complex C. Configuration C is at the crossover or intersection point. Curve **R** represents the energy of the nuclei when the electron donor is reduced and the acceptor oxidized, and curve **P** represents the nuclear energy when the electron donor is oxidized and the acceptor reduced. The broken lines linking **R** and **P** in **(a)** indicate tunnelling pathways. $\mathbf{O_p}$ is the energy for photoinduced electron transfer, this can take place at A, as shown, or at B. $\mathbf{\Delta G^*}$ is the free energy of activation; $\mathbf{\Delta G^o_{AB}}$ is the free energy change for the reaction within the complex (this is the redox driving energy; $\Delta G^o_{AB} = \Delta G^o - \Delta G^o_A + \Delta G^o_B$ where ΔG^o_A and ΔG^o_B are the free energies of formation of the precursor and successor complexes respectively, and $\Delta G^o = -nF\Delta E^o$); and ΔG_s is the free energy required to transform the geometry of the reactant complex into that of the product complex. **(a)** The non-adiabatic case. The interaction between **R** and **P** is so weak that there is litle electronic coupling between them. Passage through C does not usually cause transition from **R** to **P**. **(b)** The adiabatic case for a slightly exergonic reaction. The interaction between **R** and **P** is so strong that appreciable splitting into curves **1** and **2** occurs at the intersection point. Passage through C usually causes transition from **R** to **P**. **(c)** Profile for a strongly exergonic reaction with a significantly large change in ground-state nuclear configurations of reactants and products (for simplicity the diagram does not illustrate the splitting that accompanies adiabaticity - see **(b)**). **(d)** Profile for a strongly exergonic reaction with a smaller difference in ground-state nuclear configuration of reactants and products. The intersection point occurs at high energy

In adiabatic reactions where the electronic coupling is not too large (as in most bimolecular reactions between small reagents) the classical description of electron transfer is sufficient to describe the reaction. However, in non-adiabatic reactions other formal descriptions, such as electron tunnelling, may be more appropriate.

The classical adiabatic formalism satisfactorily relates the rate of electron transfer to the free energy required by the reactants to reach the activated complex at the intersection region (ΔG^*). This is governed by two free energy terms (Eqn. 8.12): the free energy required for the change in nuclear configuration on going from **R** to **P** (ΔG_S); and the redox driving energy (ΔG^o_{AB}). The details of this analysis are considered in the following section. Here we wish to point out that there are other types of free energy profile that are important for some biological systems.

Fig. 8.8(c) illustrates a similar reaction to that occurring in Fig. 8.8(b), but one which is more exergonic. This usually leads to an increased reaction rate because the barrier, ΔG^*, is reduced. However, increasing the difference in nuclear coordinates with a fixed exergonicity, or increasing exergonicity with a fixed set of nuclear coordinates, will eventually lead to a situation where the rate of reaction is reduced because the intersection point occurs at higher free energy (compare Fig. 8.8(c) with Fig. 8.8(d)).

The two general types of reaction exemplified by Fig. 8.8(c) and 8.8(d) can be distinguished by the relative magnitudes of ΔG^o_{AB} and ΔG_S. When $-\Delta G_S \leq \Delta G^o_{AB} < \Delta G_S$, the reaction is in the normal free energy region and the rate increases with increasing exergonicity. The maximum theoretical rate occurs when $\Delta G^o_{AB} = -\Delta G_S$, because then $\Delta G^* = 0$. The rate at this point will be diffusion limited. When $\Delta G^o_{AB} < -\Delta G_S$, the reaction is in the abnormal or inverted region (Fig. 8.8(d)) and increasing exergonicity reduces the rate. An example of such behaviour is shown in Fig. 8.7 for electron transfer involving biphenyl radical anions. For bimolecular reactions the maximum observed rate may be less than the predicted rate because the reaction is diffusion limited.

Reaction-Rate Theory. The rate constant for electron transfer is given by an equation of the type:

$$k_{et} = \text{(electronic factors)} \times \text{(nuclear factors)} \qquad (8.10)$$

Reaction-rate theory indicates the full equation is:

$$k_{et} = (pz).(e^{-\Delta G^*/RT}) \qquad (8.11)$$

where p is a transmission coefficient that describes the probability of electron transfer once the appropriate nuclear configuration has been obtained. If the reaction is a homogeneous, solution-phase bimolecular reaction, z is the collision frequency. Between two uncharged molecules at 25°C it is $\sim 10^{11}$ $M^{-1}\ s^{-1}$. If the reaction is intramolecular, z is the vibrational frequency $\mathbf{KT}/h$ (where **K** is the Boltzmann constant, **T** the absolute temperature, and h Plancks constant) which is $6 \times 10^{12}\ s^{-1}$ at 25°C.

The two general classes of mechanism we are mainly concerned with - thermally activated electron-transfer assisted by nuclear motions, and electron tunnelling - are both represented in Eqn. 8.11.

The Electron Transmission Coefficient. When p<1 (the non-adiabatic case) electron transmission may become an important rate limiting factor and the formalism of electron tunnelling be more appropriate. However, when p = 1 (the adiabatic case) electron transmission is not rate limiting and the classical theory with its emphasis on nuclear motions affecting ΔG^* is sufficient to account for the reaction rate.

Marcus' Theory. Marcus has derived the following important relationship:

$$\Delta G^* = \frac{(\Delta G_S + \Delta G^o_{AB})^2}{4\Delta G_S} \qquad (8.12)$$

This simplifies for a self-exchange reaction ($\Delta G^o_{AB} = 0$) to:

$$\Delta G^* = \frac{\Delta G_S}{4} \qquad (8.13)$$

Equation 8.12 is a quantitative description of the relationships, illustrated earlier (Fig. 8.8), responsible for the occurence of reduced rates in the inverted free energy region.

From the analysis of cytochrome self-exchange and cross-reactions given in sections 8.4 and 8.5.3 it appears that the ΔG_S term may be important for some protein:protein reactions. Marcus has considerably refined the analysis of this term for some small reactants by separating it into two components: λ_i and λ_o. λ_i is the reorganisational energy of the inner shell of atoms and λ_o that for the surrounding solvent molecules. (Note that for hemeproteins the iron solvation cage is comprised largely of the protein itself). Equations have been derived to calculate these terms but a consideration of these is beyond the scope of the present review.

For our purposes the most useful form of the Marcus theory is the correlation equation that relates an electron cross-reaction (referred to by the subscript 12) to its component self-exchange reactions (referred to by the subscripts 11 and 22). This equation is (assuming $\Delta G^{o}_{AB} = \Delta G^{o}_{12}$):

$$\Delta G^{*}_{12} = \frac{\Delta G^{*}_{11}}{2} + \frac{\Delta G^{*}_{22}}{2} + \frac{\Delta G^{*}_{22}}{2} + (1 + \alpha) \qquad (8.14)$$

$$\text{where } \alpha = \frac{\Delta G^{o}_{12}}{4(\Delta G^{*}_{11} + \Delta G^{*}_{22})}$$

In its rate form, Eqn. 8.14 becomes:

$$k_{12} = (k_{11}k_{22}K_{12}f_{12})^{1/2} \qquad (8.15)$$

where K_{12} is the equilibrium constant, and:

$$\log f_{12} = \frac{(\log K_{12})^2}{\left(4 \log \left[\frac{k_{11}k_{22}}{z^2}\right]\right)}$$

z is the bimolecular collision frequency. Unless the overall cross-reaction free energy change, ΔG^{o}_{12}, is large, $\alpha \ll 1$ and $f_{12} \approx 1$. They can then be disregarded.

The strength of this correlation approach is that the parameters used are relatively easy to determine experimentally. The ΔG_S terms, which are difficult to measure, do not appear in the final equations. There is, however, an experimental complication. The theoretical treatment assumes that the experimental data are corrected for the work involved in bringing the reactants together and separating the products. Various proposals have been made in attempts to overcome this problem and the most satisfactory of these are referred to in section 8.5, which deals with the application of Marcus' theory to cross-reactions of cytochrome *c*.

8.3.2 Tunnelling in Electron Transfer

Tunnelling is the process whereby a particle disappears from one side of a barrier and simultaneously reappears on the other side, even though it has insufficient energy to surmount the barrier. This process is classically forbidden and it requires quantum mechanics to describe it.

The transmission coefficient (p) for a tunnelling process depends upon the nature of the barrier (Fig. 8.9) and the properties of the tunnelling particle. It takes the general form:

$$p \propto e^{-[b^2 m (V-B)]^{1/2}} \tag{8.16}$$

when V - B >0. In this relationship, b is the width of the barrier at the height of penetration, V is the height of the barrier, B is the kinetic energy of the particle, and m its mass. Thus Eqn. 8.16 describes the case when the particle has insufficient energy to overcome the barrier. When $V \leq B$, the transfer can occur classically with a transmission probability of one. We need to consider two tunnelling processes: nuclear tunnelling and electron tunnelling.

The nuclear motions required for classical electron-transfer may include a contribution from nuclear tunnelling, although the classical description itself cannot take this into account. In Fig. 8.8 nuclear tunnelling means a horizontal movement from **R** to **P** below the intersection point. This does not violate the Franck-Condon principle because, as quantum mechanics shows, there is a small probability that the system can be found at nuclear configuration C below the level of the barrier. Nuclear tunnelling may make a significant contribution to some reactions; for example, to reactions occurring in the inverted free energy region (Fig. 8.8(d)). In some of these reactions, nuclear tunnelling is so great that the classical prediction of increasing exergonicity and decreasing rate is not always observed.

The free energy diagrams of Fig. 8.8 do not give a good indication of the barrier to electron tunnelling because they represent the electronic energy of the system as determined by nuclear coordinates. The potential energy diagram for electron coordinates is more difficult to define (De Vault, 1980; discussion in Chance *et al.*, 1979: pp 281 - 295). Fig. 8.9 presents a simplified representation of the barrier to electron transmission and illustrates electron tunnelling.

The diagram represents the potential well of the transferring electron at different stages of the reaction. The classical thermal mechanism involves progressive changes in the nuclear coordinates that change the shape of the potential energy surface and equalises donor and acceptor energy levels. Electron tunnelling between the isoenergetic levels can then occur. The electronic energy levels are coupled to vibronic motions of the system so that there will be a range of isoenergetic levels. Tunnelling can take place between any pair of levels with a probability dependent upon the occupancy of the level. This introduces a thermally activated step - it is equivalent to stating that the term $[b^2m(V - B)]^{1/2}$ in Eqn. 8.16 is temperature dependent - and such a process is termed vibronically coupled electron tunnelling (Hopfield, 1974; 1977). The magnitude of the vibronic coupling, expressed as the vibronic coupling parameter (Δ), can be obtained from photoinduced charge transfer spectra (see section 8.3.3).

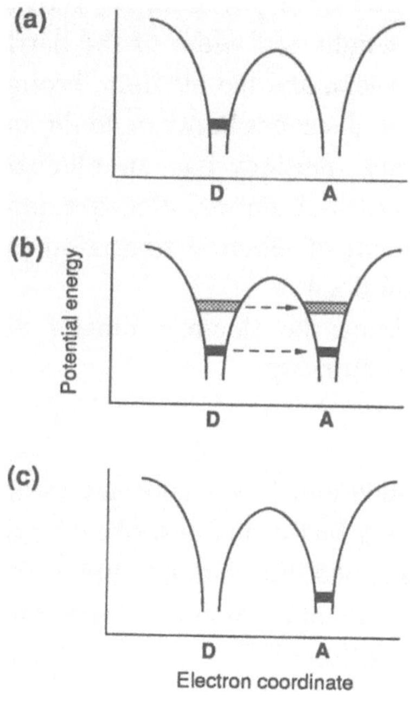

Fig. 8.9 Electron tunnelling. The diagram represents the potential energy well of the transferring electron at different stages of the reaction. **(a)** represents the precursor reactant system with the electron localised on the donor in a particular energy level (indicated by the solid line). **(b)** represents the activated complex in which the donor energy level indicated in **(a)** is at the same energy as the corresponding level in the acceptor. The electron may then tunnel through the barrier (---→). There are also equivalent higher energy levels (indicated by the stippled lines). These are coupled to the vibronic motions of the system and are separated from the lower energy levels by the activation energy ΔE. Tunnelling may occur between these levels. **(c)** represents the product complex

Tunnelling appears to be important in many biological systems (section 8.5). This is indicated by two characteristics that are often readily determined experimentally: the temperature dependence of the rate; and the distance between the donor and acceptor centres. A third characteristic - whether the reaction is adiabatic - is not easily determined.

The temperature dependence of the reaction rate is an important guide because whereas the classical mechanism is an activated process, simple tunnelling is not and therefore it is not temperature dependent. However, temperature dependence is not an infallible guide to mechanism because there are certain conditions when the classical mechanism is temperature independent (e.g. when $\Delta H^* = 0$) and, as we have seen, other conditions when electron tunnelling is temperature dependent (Fig. 8.9).

The general form of the tunnelling probability (Eqn. 8.16) shows there is a strong dependence of the rate on the height and width of the barrier. These depend on the distance the electron tunnels, and the medium through which it tunnels. DeVault (1980) suggests that at distances between donor and acceptor sites >5Å, and when the intervening medium has an electron affinity significantly less than that of the original donor, electron tunnelling is important. This touches on the problem of electron migration in a protein matrix, which was considered in section 8.2.4.

An empirical approach to determining the electron-transfer distance for intramolecular reactions exploits the relationship:

$$k_{et} \propto e^{-\beta R} \tag{8.17}$$

where R is the electron-transfer distance and β is a constant for a particular type of system. $-\beta R$ can be regarded as a factor that describes the exponential decay of the donor and acceptor orbitals, and thus is a measure of the extent of orbital overlap (section 8.2.2). For systems such as those illustrated in Fig. 8.8, β is 11 - 12 nm^{-1} (Marcus & Sutin, 1985).

Gray has adopted this approach for derivatives of metalloproteins covalently labelled with redox active groups (section 8.5.2) and has found that for cytochrome *c* and myoglobin β is 7 - 9 nm^{-1} (Mayo *et al.*, 1986). However, if data for blue copper proteins and bacterial cytochromes are included in this analysis the variation of k_{et} with R appears to be random (Sykes, 1988; Osvath *et al.*, 1988). Further work is needed in this area.

Finally, although we have discussed electron tunnelling and nuclear tunnelling separately from each other and separately from the classical thermal mechanism, we should emphasise that these processes often occur simultaneously. Which one dominates in any given case depends on a variety of factors. The excellent discussion in Chance *et al.* (1979: pp 281-295) describes the interrelationship between nuclear tunnelling and electron tunnelling in particular, and it emphasises the point that temperature independent electron-transfer at low temperature is most probably due to nuclear tunnelling whilst with increasing temperature the classical thermal mechanism generally becomes more important. Electron tunnelling may be significant over the whole temperature range and an important criterion for determining this is the adiabaticity of the reaction. Only for non-adiabatic reactions is electron tunnelling a useful concept.

8.3.3 The Photoinduced Mechanism

Electron transfer resulting from the absorption of a photon, indicated by the vertical transition marked O_p in Fig. 8.8(a), is photoinduced electron-transfer. This transition obeys the Franck-Condon principle because the nuclear configuration is not changed during the electron transfer. Although this is an important biological mechanism (for example in the light capture reactions of photosynthesis) it is not the way that cytochromes *c* normally transfer electrons. However, it is important to consider this process because it provides an experimental test for the occurrence of electron tunnelling.

A charge-transfer (CT) absorption band results from photoactivated electron-transfer and, as Hopfield (1977) and others noted, this assists characterisation of the electron-transfer mechanism. In Hopfields' experimental procedure the donor (D)-acceptor (A) complex is in its ground-state (D^+A^-) at nuclear configuration (B) on curve **P** in Fig. 8.8(a); and it is excited to curve **R**, the excited state (DA), by the absorption of a photon at a wavelength corresponding to the CT band. The electron-transfer rate is for the thermodynamically favourable reaction (i.e. for $DA \rightarrow D^+A^-$) rather than for the photoinduced electron-transfer ($D^+A^- \rightarrow DA$). This is because the intensity of the CT-band depends on the steady-state concentration of complexes in the excited state and this in turn is dependent upon the rate of the forward reaction. The strength, shape and energy of the CT band gives information about the electron transfer. The energy of the band (E_{peak}) is a function of the redox energies of the reactants (E_D and E_A), which Hopfield assumes to be given by their redox potentials, and of the strength of the vibronic coupling between them. This is expressed as the vibronic coupling parameter (Δ) that influences electron tunnelling. Thus:

$$E_{peak} = E_D - E_A + \Delta \qquad (8.18)$$

Hopfield also derives equations relating the intensity and width of the absorption band to specific electronic features of the two reactants and an electron-transfer rate.

The main experimental problem is that the molar extinction coefficient of the CT band is very low in proteins for donor-acceptor separations ≥8Å; it is of the order of 1 M^{-1} cm^{-1}. This makes detection by standard static spectroscopy difficult and has led to the development of sensitive, dynamic detection systems (Potasek & Hopfield, 1977; Potasek, 1978; Austin & Hopfield, 1982; Goldstein & Bearden, 1984). Partly because of this problem there are only a few reports identifying photoinduced CT bands. These are described in section 8.5. In this context it is relevant to note that the CT band observed by Potasek & Hopfield (1977) in their study of the reaction between cytochrome *c* and $[Fe(CN)_6]^{3-/4-}$ has been shown by Austin & Hopfield (1982) to come from a binuclear cyanobridged species formed between $[Fe(CN)_6]^{3-}$ and $[Fe(CN)_6]^{4-}$ and not to be associated with electron transfer involving cytochrome *c*, as was originally claimed.

8.3.4 Summary and Comparison of Theory with Experimental Data

The theoretical descriptions in the previous sections lay a foundation for the experimental determination of mechanism. Three reactivity characteristics have been stressed: the redox driving energy (ΔG^{o}_{12}); the rearrangement energy (ΔG_S); and the electron transmission coefficient (p). A fourth important characteristic, which we have not yet considered, is the magnitude of the work terms involved in the protein association steps.

It is generally experimentally straightforward to measure ΔG^{o}_{12}, overall bimolecular reaction rates (although uncorrected for work terms), intra-molecular electron-transfer rates, activation parameters and certain structural features such as the transfer distance. Occasionally CT bands can be observed but p is rarely possible to obtain. Because of this, evidence for the type of mechanism operating in a bimolecular reaction is usually fragmentary at best.

When a set of data are satisfactorily described by the Marcus equations (Eqns. 8.12 - 8.15) it is probable, but not certain, that the cross-reaction is adiabatic. The uncertainty arises because non-adiabatic reactions in which there is a fortuitous matching of electronic interaction energies may satisfy the equations. Thus it is common for a series of chemically similar, but energetically different, reactions to be analysed in order to assess this possibility.

Even when the type of mechanism has been identified, understanding of its structural basis is usually lacking. The main uncertainties arise around the electronic properties of proteins, which control how electrons travel in a protein matrix, and the nature of the activation process (section 8.2.4). As with small molecules (Marcus, 1964; DeVault, 1980), activation of proteins includes rearrangement of the solvent as well as a change in the atomic positions of the redox reagent itself. We have discussed the dynamic properties of cytochrome *c* (Chapter 4), some of which may contribute to the activation process, but a fuller account of this aspect is given by Williams *et al.* (1985c).

8.4 Self-Exchange Reactions of Cytochromes *c*

Self-exchange reactions not only provide essential data for the Marcus correlation equation (Eqn. 8.15), they are also usually simpler to analyse than cross-reaction rates. On modifying Eqn. 8.13 to take account of the work involved in forming the reactive complex (w), the free energy of activation for self-exchange is given by:

$$\Delta G^* = \frac{\Delta G_s}{4} + w \qquad (8.19)$$

Table 8.3 lists those cytochromes for which self-exchange rates (given the symbol k_{11}) have been experimentally determined. The most thoroughly studied protein is horse cytochrome *c* (Gupta *et al.*, 1972; Gupta, 1973; Concar *et al.*, 1986) whose rate is strongly ionic strength dependent at pH 7, but is pH independent from pH 5 to high pH values (Fig 8.10). Above pH 8 the rate increases with increasing pH until close to the isoelectric point (~pH 10), when it begins to fall. At the isoelectric point the rate is independent of ionic strength. Gupta *et al.* (1972) concluded from these data that at pH 7 and low ionic strength, electrostatic repulsion was limiting the electron exchange: but close to the isoelectric point, where the electrostatic work terms are negligible, the rate was not limited by this mechanism. However, at this point the rate is still much lower than that of other cytochromes *c*, such as the cytochromes *c*-551 (Table 8.3). The reason for this is that at high pH the electron exchange reaction is coupled to a redox state change in the iron coordination (Chapter 4.4) that makes a substantial contribution to ΔG_S.

Table 8.3 Self-exchange rates (k_{11}) of class I cytochromes *c*.

Cytochrome		k_{11} (M^{-1} s^{-1}) [a]	Conditions			Refs.
Horse	*c*	10^3	25°C	pD 7	I = 0.1 M (NaCl)	Gupta *et al.* (1972)
	c	10^4	25°C	pD 7	I = 1.0 M (NaCl)	Gupta *et al.* (1972)
	c	6 x 10^2	27°C	pD 7	I = 0.24 M (cac) [b]	Concar *et al.* (1986)
	CDNP-K13 [b]		27°C	pD 7	I = 0.24 M (cac) [b]	Concar *et al.* (1986)
	CDNP-K72 [b]		27°C	pD 7	I = 0.24 M (cac) [b]	Concar *et al.* (1986)
C. krusei	*c*	10^2	25°C	pD 7	I = 0.1 M (NaCl)	Gupta (1973)
	c	10^3	25°C	pD 7	I = 1.0 M (NaCl)	Gupta (1973)
Ps. aeruginosa	*c*-551	1.2 x 10^7	42°C	pD 7	0.05 phos [b]	Keller & Wüthrich (1976)
Ps.stutzeri	*c*-551	6 x 10^6	27°C	pD 5.5	no added salt	Leitch *et al.* (1984)
E. gracilis	*c*-552	5 x 10^6	29°C	pD 7	0.05 phos [b]	Keller & Wüthrich (1977)

[a] These measurements were made by NMR with D_2O solutions and since the pH values are uncorrected for any isotope effect they are denoted by pD.
[b] cac = sodium cacodylate; phos = sodium phosphate; CDNP-K13 and CDNP-K72 are the horse cytochrome *c* 4-carboxy-2,6-dinitrophenyl derivatives of Lys 13 and Lys 72 (Pettigrew & Moore, 1987; Chapter 2.4)

Yoshimura *et al.* (1980) confirmed that the self-exchange rate of mitochondrial cytochrome *c* at pH 7 is limited by repulsive work terms, from their study of the exchange between horse and *C. krusei* cytochromes *c* bound to phosvitin. The maximim rate they observed was 2 x 10^6 M^{-1} s^{-1} compared to the rate for the free cytochromes of 2 - 3 x 10^4 M^{-1} s^{-1} (both at 20°C in 20 mM Tris-HCl at pH 7.4). The rate enhancement was due to an increased frequency of productive collisions.

Concar *et al.* (1986) showed that the charge distribution and not just the overall charge is important for determining the repulsive work terms, by measuring the self-exchange rates of the CDNP-Lys 13 and CDNP-Lys 72 derivatives of horse cytochrome *c* (Table 8.3). Both derivatives had higher rates than native cytochrome *c* but the Lys 13 derivative had a slower rate than the Lys 72 derivative. However, in the presence of polyanions, such as ATP^{4-} and $[Co(CN)_6]^{3-}$, the electron self-exchange was enhanced (Concar *et al.*, 1986). At saturation a limiting value of 1.2 - 1.6 x 10^5 M^{-1} s^{-1} was observed for all three proteins.

The activation energies measured by Gupta (1973) and Yoshimura *et al.* (1980) of 30 - 54 KJ mol^{-1}, depending upon ionic strength, do not relate directly to the electron-transfer step because the formation of the transient complex is an activated process. Therefore the experimental activation energy cannot be used to distinguish between particular mechanisms of electron transfer.

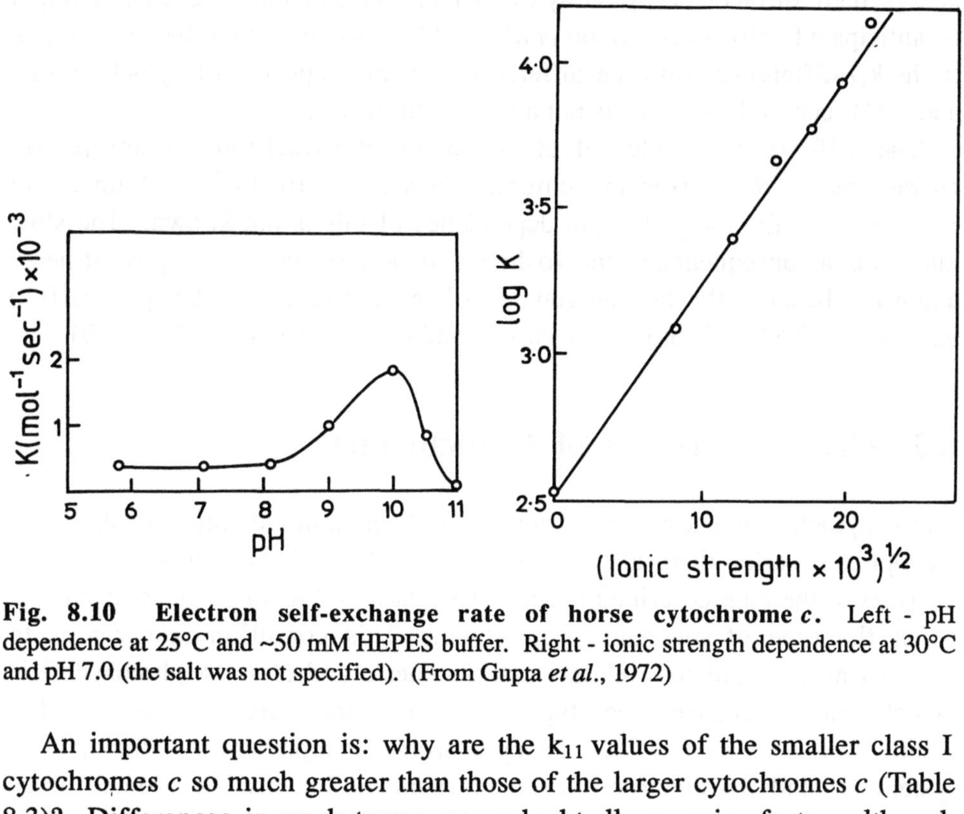

Fig. 8.10 Electron self-exchange rate of horse cytochrome *c*. Left - pH dependence at 25°C and ~50 mM HEPES buffer. Right - ionic strength dependence at 30°C and pH 7.0 (the salt was not specified). (From Gupta *et al.*, 1972)

An important question is: why are the k_{11} values of the smaller class I cytochromes *c* so much greater than those of the larger cytochromes *c* (Table 8.3)? Differences in work terms are undoubtedly a major factor, although even at high ionic strength (Gupta *et al.*, 1972), or in the presence of polyanions (Yoshimura *et al.*, 1980; Concar *et al.*, 1986) horse cytochrome *c* only attained a rate of 10^5 - 10^6 M^{-1} s^{-1}. Thus differences in work terms are unlikely to be the whole explanation of the differences in rate.

Part of the difference may be related to differences in ΔG_S because the two classes of protein respond differently to the redox state change (Chapter 4.5); although the indications from small molecule cross-reactions are that this is not a major factor (section 8.5.1). Another factor that may be important is the relative size of the electron-transfer sites on the proteins. From calculations of the diffusion limited k_{11} of horse cytochrome *c*, Gupta *et al.* (1972) calculated that the site occupied ~7Å^2 of the protein surface, in reasonable agreement with the amount of the surface taken up by the exposed heme edge (Stellwagen, 1978). Calculations by Marcus & Sutin (1985) also give good agreement on this point. Since the smaller cytochromes have a greater relative area of their surfaces occupied by the heme, a greater rate of exchange might be anticipated. However, Dixon *et al.* (1989), who quantified the contribution to the k_{11} difference from the difference in heme exposure of cytochromes *c* and *c*-551, conclude that this is not a significant factor.

The self-exchange rates of other classes of cytochromes *c* are poorly characterised. *R. rubrum* cytochrome *c'* has a k_{11} >10^4 M^{-1} s^{-1} (Emptage *et al.*, 1981) but the ionic strength dependence of this is not known. The slow rate is not a consequence of the 4α-helical bundle structure or degree of heme exposure, because the homologous *E coli* cytochrome b_{562} (Chapter 6.10.2) has a k_{11} >10^6 M^{-1} s^{-1} (at pH 7, I <0.1 M and 42°C) (Moore *et al.*, 1985b).

8.5 Cross-Reactions of Cytochromes *c*

Many cytochrome *c* cross-reactions have been studied and we shall not attempt to describe them all. We have selected reactions in which energetic analyses of the type described in section 8.3 have led to some clear statements about the type of mechanism involved. This condition restricts us to considering in detail only class I cytochromes *c* and class IV photosynthetic cytochromes *c* because other types of cytochrome *c* are generally insufficiently well-described. Class III cytochromes *c* represent an almost ideal system for studying electron transfer mechanisms because they contain three or four hemes constrained in close proximity by one polypeptide chain. This circumvents the problem of interaction work terms because once one or two electrons have entered the fully oxidised protein the intramolecular rates and their activation parameters should define the mechanism. However, these studies have not been carried through to the necessary level of detail and therefore we do not consider them here.

The problem of the initial reduction of the protein has generally been overcome by using a small molecule redox reagent. Two relatively recent developments are: the use of electrochemical methods (Eddowes & Hill, 1981; Armstrong *et al.*, 1988), which are also used for redox potential determinations; and photoxidation of an electron donor within a tightly formed complex (e.g. Ho *et al.*, 1985; McLendon *et al.*, 1985). The electrochemical techniques are an important practical development but they are outside the scope of this review since the electrode surface introduces additional factors to consider. We note however that conditions have been established for redox equilibrium between electrodes and cytochrome *c* and also between electrodes and membrane systems such as those found in mitochondria and bacterial protoplasts (Coleman *et al.*, 1983). Examples of the photooxidation procedure, using *S. cerevisiae* cytochrome *c* peroxidase and cytochrome *c*, are given in section 8.5.3.

The reaction systems we have chosen for further discussion illustrate that:

1. Some bimolecular reactions of class I cytochromes *c* with small molecule reagents satisfy the Marcus correlation equations and are probably best described by the classical thermal outer-sphere mechanism. These reactions also illustrate the complexities introduced by interaction work terms.

2. Some protein:protein reactions involving class I cytochromes *c* also satisfy the Marcus correlation equation but the best characterised physiological reaction of eukaryotic cytochrome *c*, that with *S. cerevisiae* cytochrome *c* peroxidase, appears to have an electron tunnelling mechanism.

3. Some of the reactions of bacterial photosynthetic reaction centres have mainly tunnelling mechanisms. These reactions also illustrate one of the ways in which control over electron flow can be achieved.

8.5.1 Bimolecular Reactions with Non-Physiological Small Molecules

The problem of work term corrections has been tackled by, among others, Sutin (1977) and Wherland & Gray (1977). The simplest approach is to assume that the work required to bring together reactants in the three constituent reactions of the correlation equation, Eqn. 8.15, cancel each other. This is the case for a cross-reaction involving similarly charged species where the work terms are unfavourable for both the formation of the cross-reaction precursor complex, and for the corresponding self-exchange reactions (Scheme 8.1). Sutin (1977), however, described two types of reaction relevant to those of cytochrome *c* where this simplifying assumption is not valid.

Scheme 8.1 Cancellation of work terms in the Marcus correlation approach

Definition of Symbols

The subscripts $_{11}$ and $_{22}$ refer to self-reactions; $_{12}$ and $_{21}$ refer to cross-reactions. The superscripts * and ^{o} refer to free energies, not including work terms; ** and ^{oo} refer to free energies including work terms. The work terms used to describe the formation and dissociation of reactant complexes (reactions 8.1 & 8.5 of Fig. 8.1) are denoted by w. The activation energy ΔG^{*}_{12} is the energy required to equalise the donor and acceptor energy levels within the reactant complex (reaction 8.2 of Fig. 8.1).

Marcus Correlation

Eqn. 8.14 can be written:

$$\Delta G^{*}_{12} = 1/2(\Delta G^{*}_{11} + \Delta G^{*}_{22} + \Delta G^{o}_{12}) \quad (8.20)$$

assuming α<<1, and that each activation free energy refers to the complex as defined above. However, the experimentally observed activation energies:

$$\Delta G^{**}_{11}, \Delta G^{**}_{22}, \Delta G^{**}_{12}$$

include the work terms. Thus:

$$\Delta G^{**}_{11} = \Delta G^{*}_{11} + w_{11} \quad (8.21)$$

$$\Delta G^{**}_{22} = \Delta G^{*}_{22} + w_{22} \quad (8.22)$$

Also the redox driving energy in Eqn. 8.20, ΔG^{o}_{12}, refers to the complex. The experimental measurement of this term is usually taken from the redox potentials of the two reactants. These are not corrected for the interaction work. Thus:

$$\Delta G^{oo}_{12} = \Delta G^{o}_{12} - w_{12} + w_{21} \quad (8.23)$$

(where w_{12} and w_{21} are the association and dissociation work terms respectively). Substituting Eqns. 8.21 - 8.23 into 8.20 gives:

$$\Delta G^{*}_{12} = 1/2(\Delta G^{**}_{11} + \Delta G^{**}_{22} + \Delta G^{oo}_{12} - w_{11} - w_{22} - w_{12} + w_{21}) \quad (8.24)$$

If all the reagents have charges of the same sign so that there are no attractive work terms:

$$w_{21} - w_{12} \approx w_{11} + w_{22}$$

and thus, Eqn. 8.24 becomes:

$$\Delta G^{*}_{12} = 1/2(\Delta G^{**}_{11} + \Delta G^{**}_{22} + \Delta G^{oo}_{12}) \quad (8.25)$$

One is the case of reactants of opposite charge. Here, precursor complex formation in the cross-reaction is aided by attractive electrostatic interaction, while the self-exchange reactions remain affected by repulsive electrostatic interactions. Hence the work terms for the three components of Eqn. 8.15 do not cancel and the reaction will proceed faster than predicted by this equation.

Second, when the cross-reaction is between a hydrophilic and a hydrophobic reactant, there are unfavourable non-electrostatic interactions that are not present in their constituent self-exchange reactions. Thus the cross-reaction proceeds more slowly than predicted by Eqn. 8.15.

The assumption that work terms for similarly charged reactants cancel was tested by Sutin (1977) for a series of reactions betwen transition metal complexes and cytochrome *c*. The relevant data are given in Table 8.4 which shows that the Marcus approach has been remarkably successful in predicting the cross-reaction rates and activation parameters for some reactions.

Table 8.4 Calculated rate constants and activation parameters for reactions of horse cytochrome *c* with small positively charged reactants. [a]

Reaction [b]	k_{12}^{calc} ($M^{-1}s^{-1}$)	$\frac{k_{12}^{calc}}{k_{12}^{obs}}$	ΔH^{*}_{calc} ($kJ\ mol^{-1}$)	$\frac{\Delta H^{*}_{calc}}{\Delta H^{*}_{obs}}$	ΔS^{*}_{calc} ($J\ K^{-1}\ mol^{-1}$)	$\frac{\Delta S^{*}_{calc}}{\Delta S^{*}_{obs}}$
$[Ru(NH_3)_6]^{2+} + C_o$	1.2×10^5	3.1	29.3	2.4	-58.5	2.0
$[Ru(NH_3)_5(bm)]^{2+} + C_o$	6.3×10^4	1.1	n.d.	n.d.	n.d.	n.d.
$Co(phen)_3]^{3+} + C_r$	2×10^3	1.3	46.4	1.0	-25.1	1.0

[a] From Sutin (1977).

[b] C_r and C_o are horse ferro and ferricytochrome *c* respectively; bm and phen are benzimidazole and 1,10-phenanthroline respectively. The redox potentials of the reactants are: $[Ru(NH_3)_6]^{3+/2+}$, 51 mV; $[Ru(NH_3)_5(bm)]^{3+/2+}$, 150 mV; $[Co(phen)_3]^{3+/2+}$, 370 mV.

The work term problems for reactions between anionic transition metal complexes and mitochondrial cytochrome *c* have not been satisfactorily resolved (Sutin 1977; Moore *et al.*, 1984b). Wherland & Gray (1977) advocate a procedure whereby the ionic strength dependence of a reaction rate is used to correct the observed rate but, aside from the fact that this procedure does not correct for non-electrostatic work terms, there are many inherent problems in the interpretation of ionic strength data (Moore *et al.*, 1984b; and references therein).

Another approach is to compare a range of similar reactions in which only one term is varied, usually ΔG^{o}_{12}. This relatively simple approach is not of general application to inorganic reagents because only rarely can their redox potentials be changed without the attendant alteration of some other reactivity characteristic. This is not the case with quinols however and Rich (1982) has exploited this to show that there is a simple correlation between k_{12} and ΔG^{o}_{12}

for a series of reactions between cytochrome *c* and substituted p-benzoquinols.

At 25°C the logarithmic form of Eqn. 8.15 is:

$$\log k_{12} = \frac{\Delta E_m}{118} + \frac{1}{2}(\log k_{11} + \log k_{22}) \qquad (8.26)$$

when f = 1. Provided that both ($\log k_{11} + \log k_{22}$) and the interaction work terms are constant over the range of reactions, a plot of $\log k_{12}$ against ΔE_m should give a straight line of slope $^1/_{118}$. Fig. 8.11, covering a ΔE_m range of 600 mV and k_{12} range of 10^4 shows that the reactions of two cytochromes *c* with 13 substituted p-benzoquinols fit the theoretical expectations well.

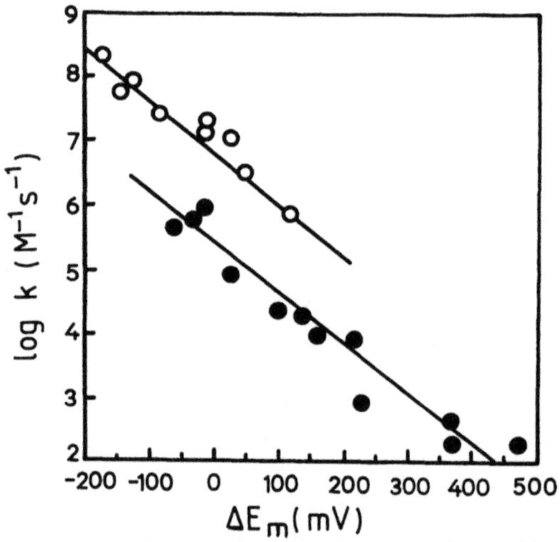

Fig. 8.11 Marcus plot of rate constant against redox potential difference for the quinol reduction of cytochromes. The reactions are:

$$QH^- + heme^{3+} \xrightarrow{k_{12}} QH^{\bullet} + heme^{2+}$$

where QH^- and $QH^{\bullet}$ are the anionic quinol reductant and its corresponding semiquinone respectively. The reactions of various substituted p-benzoquinols with horse cytochrome *c* (●) and plant cytochrome *f* (○) were studied. Both sets of data have slopes close to $^1/_{118}$. The displacement of the two lines occurs because k_{11} and/or work terms are different for the two proteins. (Adapted from Rich, 1982)

Cusanovich has followed the analogous approach of investigating the reactions between one small reagent and a group of different cytochromes with a range of redox potentials. In his initial work Cusanovich used anionic reagents, such as $[Fe(CN)_6]^{3-/4-}$ (Ohno & Cusanovich, 1981), for which large interaction work terms make the analysis difficult (Sutin, 1977; Moore *et al.*,

1984b); but recently he and his coworkers have explored the reactions of photoreduced flavins with a range of proteins including class I, II and III cytochromes *c*, Fe/S proteins, blue Cu proteins and cytochrome b_5 (Meyer *et al.*, 1983;1984). Using the ionic strength dependence of k_{12} to obtain adjusted rate constants (k_{12}*), Meyer *et al.* correlate ΔG^{o}_{12} with log k_{12}*. Although the adjustment of rate constant and the empirical fitting procedure used can be criticised (Pettigrew & Moore, 1987: Chapter 2.7), it is remarkable that such a wide range of redox proteins obey the predicted free energy relationships.

There are no uncertainties attached to the analyses of Table 8.4 and Fig. 8.11 which, by indicating that cytochrome *c* can follow a classical pathway, strongly suggest that there are no large intrinsic barriers to the reaction; i.e. the redox state conformation change is not so large that it prevents adiabatic electron-transfer.

The available evidence indicates that the small reagents react at the exposed heme edge of cytochrome *c* (Sutin, 1977; Butler *et al.*, 1983; Moore *et al.*, 1984b) where direct contact between the reagent and the heme allows a strong electronic interaction. However, it seems probable that there are multiple interaction sites on most cytochromes *c* with some being far from the heme (Moore *et al.*, 1984b; Cheddar *et al.*, 1989). Reactions of cytochrome *c* with larger reagents, such as homologues of $[Co(phen)_3]^{3+}$ made with bulky substituted 1,10-phenanthrolines (Wherland & Gray, 1977), appear to be non-adiabatic; an indication that the electronic interaction between the donor and acceptor sites is relatively weak, largely because they are far apart.

An implicit assumption in all these studies is that the electron-transfer sites on the protein are the same for the small molecule reactions and the protein self-exchange reaction. This is difficult to confirm. However, there are unlikely to be unique reaction sites for the small molecule reagents.

NMR studies of horse cytochrome *c* indicated that $[Fe(CN)_6]^{3-/4-}$ and $[Fe(edta)(H_2O)]^{1-/2-}$ interact differently with the protein (Williams *et al.*, 1982). Both bind close to the exposed heme edge, but with different preferences for particular lysines. This was subsequently confirmed by kinetic studies of CDNP derivatives of cytochrome *c* (Pettigrew & Moore, 1987: Chapter 2.4; Armstrong *et al.*, 1986) and native cytochrome *c* (Cho *et al.*, 1988). Similarly, NMR studies of the $[Fe(CN)_6]^{4-}$ reduction of native and CDNP-lysine horse ferricytochrome *c* indicate that there is not a single preferred interaction site (Moore *et al.*, 1984b; D.W.Concar, G.R.Moore, R.J.P.Williams, unpublished data). The effect of modifying a lysine close to the heme is to cause the reactive hexacyanide to move to another region, which is also close to the heme. Thus not only is the association constant altered, but also the electron-transfer constant within the complex (Table 8.5). Fortunately the effect on k_{et} is rather small.

Table 8.5 **$[Fe(CN)_6]^{4-}$ reduction of horse ferricytochrome *c*.**

Protein	Reaction Conditions		K_a (M^{-1})	k_{et} (s^{-1})
Native [a]	I = 0.12 M (NaCl)	25°C	90 ± 20	785 ± 80
	I = 0.04 M (cac) [b]	32°C	910 ± 100	485 ± 50
	I = 0.04 M (cac)	37°C	850 ± 100	570 ± 60
CDNP-Lys 13 [b c]	I = 0.04 M (cac)	32°C	330 ± 40	415 ± 40
	I = 0.04 M (cac)	37°C	370 ± 40	490 ± 50
CDNP-Lys 72 [c]	I = 0.04 M (cac)	32°C	410 ± 50	165 ± 20
	I = 0.04 M (cac)	37°C	600 ± 70	210 ± 20

[a] From Eley *et al.* (1984).
[b] cac = sodium cacodylate; CDNP = carboxydinitrophenyl.
[c] From D.W.Concar, G.R.Moore & R.J.P.Williams, unpublished data.

8.5.2 Reactions of Cytochromes *c* Covalently Labelled with Non-Physiological Redox Reagents

Reactions of cytochrome *c* covalently labelled with redox active reagents are currently being explored in an attempt to define further its electron-transfer properties without having to deal with the complications resulting from the association step. The first such system studied was the Cr^{2+} reduction of horse ferricytochrome *c*. Cr^{2+} binds tightly to cytochrome *c* in an inner sphere complex, and since Cr^{3+} is inert to hydrolysis, the ferrocytochrome *c* product contains covalently bound Cr^{3+} (Kowalsky, 1969; Dawson *et al.*, 1972; Yandell *et al.*, 1973). Attempts to identify the location of the bound Cr^{3+}, and hence the electron-transfer site, have produced conflicting results (Grimes *et al.*, 1974; Peterson & Gupta, 1979), and additional work is needed to characterise further this sytem.

An alternative approach makes use of derivatives of cytochromes *c* with ruthenium compounds attached to histidines. These derivatives are structurally and electrochemically well characterised (Table 8.6). The rates of intramolecular electron-transfer have been measured (Table 8.6) and some of these are relatively rapid considering the small driving force and large distances the electrons have to traverse. However, the data of Bechtold *et al.*

Table 8.6 Electron-transfer rate constants for Ru labelled cytochromes *c*.

Reactants [a]		ΔE_m (mV) [b]	d (Å)	k_1 (s^{-1})	k_2 (s^{-1})	Refs.
$[Ru(NH_3)_5]$-His 33	cytochrome *c*	110	12	20		e
		110	12	55		f
$[Ru(NH_3)_4(isn)]$ -His 33	cytochrome *c* [c]	180	12		$<10^{-4}$	g
$[Ru(NH_3)_5]$-His 33	cytochrome *c*-551 [d]	200	8	13		h

[a] The reactions are: $Ru^{2+}\text{-}C_o \underset{k_2}{\overset{k_1}{\rightleftharpoons}} Ru^{3+}\text{-}C_r$

[b] The redox potentials of $[Ru(NH_3)_5]$-His and $[Ru(NH_3)_4(isn)]$-His attached to horse cytochrome *c* are 150 mV and 440 mV respectively.

[c] isn is isonicotinamide

[d] From *Ps. stutzeri*.

[e] Winkler *et al.* (1982); [f] Bechtold *et al.* (1986a); [g] Bechtold *et al.* (1986b); [h] Osvath *et al.* (1988).

(1986a), showing the $Fe^{2+} \rightarrow Ru^{3+}$ rate with a driving force of 0.11 V is negligible compared to the $Ru^{2+} \rightarrow Fe^{3+}$ rate with a similar driving force, is anomalous. Bechtold *et al.* interpret this result to indicate that the protein matrix of cytochrome *c* exerts a directional control over electron transfer, possibly due to conformational changes accompanying reaction. However, it is surprising that virtually no electron transfer for the reaction $Fe^{2} \rightarrow Ru^{3+}$ was recorded. A larger number of examples of different systems of this kind are needed since the experiments of Bechtold *et al.* (1986a) raise fundamental problems. This need may be satisfied by the many systems now under study. These include: horse cytochrome *c* with its lysines converted to Ru polypyridine derivatives (Pan *et al.*, 1988) or modified with a cobalt-cage complex (Conrad & Scott, 1989); zinc substituted horse cytochrome *c* with His 33 modified by Ru amines (Elias *et al.*, 1988; Meade *et al.*, 1989); ruthenated yeast cytochrome *c* (Bowler *et al.*, 1989); and *Ps. stutzeri* cytochrome *c*-551 with His 47 modified by Ru pentammine (Osvath *et al.*, 1988). A clearer example of conformational control over the rate of electron transfer is given in the following section, for the reaction of cytochrome *c* with cytochrome b_2 (Fig. 8.12).

Electron transfer in these covalently labelled cytochromes probably involves tunnelling. This underpins the assumption that the measured rate gives information on the value of β in Eqn. 8.17 (section 8.3.2) and is supported by the observation that the electron-transfer rate for $Ru(NH_3)_5$-His 33 cytochrome *c* is independent of temperature over the range 0 - 37°C (Winkler *et al.*, 1982).

8.5.3 Protein:Protein Reactions in Aqueous Solution

Class I Cytochromes c and Blue Copper Proteins. The reactions we consider here are those between various class I cytochrome *c* and the blue copper proteins, azurin and plastocyanin. These may be important in some organisms (Pettigrew & Moore, 1987: Chapter 3) but most of the studied reactions are not physiologically relevant. The aspect of these reactions that we are mainly concerned with is: are their rates described by the Marcus equation (Eqn. 8.15)? The data in Table 8.7 show that in many cases they are (Wherland & Pecht, 1978; Wilson *et al.*, 1979).

An interesting feature of the Wherland and Pecht study is that there was generally good agreement even though the cross-reaction rates were not adjusted to compensate for the interaction work terms. This was probably because, with one exception (horse cytochrome *c*), all the proteins had acidic pI values which led to the cancellation of work term contributions.

Table 8.7 Calculated rate constants for protein:protein cross-reactions involving class I cytochromes *c* and blue copper proteins.

Oxidant		Reductant		k_{12}^{calc} (M^{-1} s^{-1})	$k_{12}^{calc}/k_{12}^{obs}$ [c]
Alcaligenes [a]	azurin	*B. filiformis* [b]	cytochrome *c*-553	2.11×10^6	1.1
Parsley	plastocyanin	*B. filiformis*	cytochrome *c*-553	4.00×10^5	1.3
Horse	cytochrome *c*	*B. filiformis*	cytochrome *c*-553	3.96×10^4	1.4
Alcaligenes	azurin	*Ps. aeruginosa*	cytochrome *c*-551	2.96×10^6	0.8
Horse	cytochrome *c*	*Ps. aeruginosa*	azurin	1.60×10^3	1.0
Parsley	plastocyanin	Horse	cytochrome *c*	1.00×10^6	2×10^{-3}

[a] *Alcaligenes* sp(Iwasaki-II).

[b] *Bumilleriopsis filiformis*.

[c] These ratios were either taken directly from Wherland & Pecht (1978) or calculated using data presented by them. The k_{12}^{obs} values were measured at pH 7 (0.01 - 0.05 M phosphate), 25°C and I = 0.1 M (NaCl) by Wood (1974) and Wherland & Pecht (1978). The self exchange rates (k_{11}) of the proteins listed in the table were treated by Wherland & Pecht as adjustable fitting parameters. There is good agreement between their calculated values and experimentally determined values, where available, which lends support to their procedure. k_{11}^{calc}(k_{11}^{obs}): cytochrome *c*-553, 2.8×10^8 M^{-1} s^{-1} [not available]; cytochrome *c*-551, 4.6×10^7 M^{-1} s^{-1} [Table 8.3]; horse cytochrome *c*, 1.5×10^2 M^{-1} s^{-1} [Table 8.3]; *Ps. aeruginosa* azurin, 9.9×10^5 M^{-1} s^{-1} [2×10^6 M^{-1} s^{-1} at pD 6 (20 mM phosphate + 100 mM NaCl) and 50°C; Canters *et al.*, 1984]; *Alcaligenes* azurin, 2.6×10^5 M^{-1} s^{-1} [not available]; plastocyanin, 6.6×10^2 M^{-1} s^{-1} [$<<2 \times 10^4$ M^{-1} s^{-1} at pD 7.4 (10 mM phosphate) and 50°C; Beattie *et al.*, 1975].

The reaction between the azurin and cytochrome *c*-551 from *Ps. aeruginosa* has been studied in detail because of its possible physiological role (Pettigrew & Moore, 1987: Chapter 3). Both of these proteins have redox dependent ionisable groups (Chapter 7.6) and at pH 7 electron transfer is coupled to proton transfer. Additionally, there appears to be a number of conformational states of azurin (Silvestrini *et al.*, 1981; Adman *et al.*, 1982; Corin *et al.*, 1983). This plethora of different protein forms might be expected to make the energetics of this reaction more complex than those of the small molecule reactions discussed previously. However, they do not, and although this may be partly fortuitous, it is also an indication of the strength of the correlation approach.

One feature that makes the energetics of most of these reactions open to simple analysis is that stable collision complexes are usually not formed. The *Ps. aeruginosa* azurin:cytochrome *c*-551 system is an example: fluorescence spectroscopy (Corin *et al.*, 1983) and NMR (G.R.Moore, unpublished observations) have failed to detect complex formation. However, the poor agreement for the reactions of horse cytochrome *c* with azurin and plastocyanin (Table 8.7) may be due to non-cancellation of work terms, especially in the latter case. Here, the rate enhancement probably results from the highly dissimilar charges of the two reactants, a view consistent with the preparation of a cross-linked complex of the two proteins by Geren *et al.* (1983) and the NMR identification of a collision complex by King *et al.* (1985).

There is no direct evidence bearing on the identity of the interaction sites for the azurin:cytochrome *c*-551 system but kinetic studies of CDNP modified horse cytochrome *c* show that *Ps. aeruginosa* azurin and parsley plastocyanin react at the front face of the cytochrome (Augustin *et al.*, 1983).

The main conclusion from these studies is that protein:protein reactions can be analysed within the framework of the Marcus theory. However, whether or not they are truly classical outer-sphere reactions remains uncertain. The redox centres of the copper proteins are not exposed at the molecular surface (Adman, 1979) and so even if there were close contact between the heme of the cytochromes and the copper proteins in the activated complex, it is difficult to see how a strong electronic coupling between the redox centres could result. Another important conclusion is that formation of a stable collision complex is not always needed to obtain a fast rate of reaction.

Cytochrome c and Cytochrome c Peroxidase. The cytochrome *c*: cytochrome *c* peroxidase (YCCP) system is a well characterised physiological system; kinetic and binding studies (Pettigrew & Moore, 1987: Chapter 2) show that there is a stable bimolecular YCCP:cytochrome *c* complex in which electron transfer takes place. The hypothetical structure of the complex based on computer modelling (Fig. 8.6; Poulos & Kraut, 1980) shows that the Fe-Fe distance is 25 - 26Å and the heme edge-to-edge distance is 17 - 18Å. These distances are too great for classical outer-sphere electron-transfer, and even if it is assumed that the structures of the proteins in the activated complex are slightly distorted it is unlikely that the electron-transfer distance can be significantly shortened. Therefore it seems that tunnelling must make a major contribution to electron transfer in this system.

Numerous recent studies of the YCCP:cytochrome *c* complex have been reported and a variety of elegant reaction systems developed to investigate electron transfer in the complex. Some of these are summarised in Scheme 8.2.

Potasek (1978) investigated the reaction in Eqn. 8.27 (Scheme 8.2) in an attempt to determine if a CT band was formed by the ferriYCCP:C_r complex. Potasek reported such a band with an extinction coefficient of 0.35 M^{-1} cm^{-1} and from its properties calculated an electron tunnelling distance of 7 - 10Å. This is significantly less than the heme edge-to-edge distance in the proposed complex. The conflict between the two values may arise in the assumptions used to derive a distance from the CT band data. However, it may also be because the reported band is not associated with the YCCP:C_r complex; observation of such a weak band is difficult and a number of artefacts can arise (Potasek & Hopfield, 1977; Austin & Hopfield, 1982). For this reason it is desirable for the presence of the CT band to be confirmed.

The rate of reduction of ferricytochrome *c* by ferroYCCP (k_1 in Eqn. 8.27, Scheme 8.2) has been measured in two laboratories. Potasek (1978) reports a value of 11.7 x 10^4 s^{-1} and Cheung *et al.* (1986) a value of 0.23 s^{-1}. It appears as though the lower figure is the correct one since k_{CO} in the reaction of Eqn. 8.28 (Scheme 8.2), which has a value of 5 s^{-1}, is greater than k_1 (Cheung *et al.*, 1986).

Ho *et al.* (1985) have reported photoinduced electron-transfer in the ZnYCCP:native cytochrome *c* complex (Eqn. 8.29, Scheme 8.2). The Zn-protoporphyrin triplet state generated by the photoexcitation can decay slowly to the ground state or it can react with cytochrome *c* to generate a cation radical and ferrocytochrome *c*. Subsequent reduction of the cation radical by ferrocytochrome *c* is analogous to the physiological reaction.

Scheme 8.2 Electron-transfer reactions involving metal-free, zinc-substituted and native cytochromes

1. *Generation of reactive metal-free and zinc-substituted species*

$$\text{a)}\quad \text{Zn - cyt} \xrightarrow{h\nu} {}^{3}\text{Zn - cyt} \rightarrow (\text{Zn - cyt})^{\ddagger} + e^{-}$$

Photoexcitation of the zinc-substituted cytochrome (Zn - cyt) generates the triplet state zinc protein which can be oxidised to produce a cation radical. Reduction of the radical generates the original cytochrome. Metal-free cytochromes (porphyrin-cytochromes; denoted by p - C) undergo analogous reactions.

$$\text{b)}\quad \text{p - cyt} \overset{e^{-}_{aq}}{\rightleftharpoons} (\text{p - cyt})^{\bar{\bullet}}$$

The porphyrin cytochrome anion radical is generated by pulse radiolysis.

2. *Redox potentials*

$(\text{ZnYCCP})^{\ddagger}$ / ^{3}ZnYCCP	~- 0.64 V	Ho *et al.* (1985)
$(\text{ZnYCCP})^{\ddagger}$ / ZnYCCP	~ 1.20 V	Ho *et al.* (1985)
$(\text{Zn - C})^{\ddagger}$ / ^{3}ZnC	~- 0.75 V	McLendon & Miller (1985)
$(\text{p - C})^{\ddagger}$ / ^{3}p - C	~- 0.35 V	McLendon & Miller (1985)
p - C / $(\text{p - C})^{\bar{\bullet}}$	~- 1.10 V	McLendon & Miller (1985)
C_o / C_r	0.26 V	Chapter 7

3. *Electron-transfer reactions*

Examples of reactions between cytochrome *c* and YCCP are given below. Analogous reactions between cytochromes *c* and b_5 have also been reported (see text). For example, in Eqn. 8.29 Zn-C replaces ZnYCCP and ferricytochrome b_5 replaces C_o.

$$\text{Fe(II)YCCP} + C_o \rightleftharpoons \text{Fe(II)YCCP:}C_o \underset{k_2}{\overset{k_1}{\rightleftharpoons}} \text{Fe(III)YCCP:}C_r \rightleftharpoons \text{Fe(III)YCCP} + C_r \quad (8.27)$$

$$\text{Fe(II)YCCP(CO):}C_o \underset{k_{CO}}{\overset{h\upsilon}{\rightleftharpoons}} \text{Fe(II)YCCP:}C_o + \text{CO} \xrightarrow{k_1} \text{Fe(III)YCCP:}C_r + \text{CO} \quad (8.28)$$

$$\text{ZnYCCP:}C_o \xrightarrow{h\upsilon} {}^{3}\text{ZnYCCP:}C_o \xrightarrow{k_1} (\text{ZnYCCP})^{\ddagger}\text{:}C_r \quad (8.29)$$

(with k_2 returning $(\text{ZnYCCP})^{\ddagger}\text{:}C_r$ to ZnYCCP:C_o)

$$\text{Fe(III)YCCP:p-C} \xrightarrow{e^{-}_{aq}} \text{Fe(III)YCCP:(p-C)}^{-} \xrightarrow{k_1} \text{Fe(II)YCCP:p-C} \quad (8.30)$$

Hoffman and his colleagues have studied a variety of cytochromes *c* with this system and have produced some intriguing results (Table 8.2). They draw attention to the very different values of k_1 and k_2 for native *S. cerevisiae* cytochrome *c* (166 s^{-1} and 1.9 x 10^4 s^{-1} respectively) and point out that since these reactions have the same driving force of 0.90 - 0.95 V their reorganisation energies are very different. The differences between the different native cytochromes *c* (Table 8.2) imply that the structures of the YCCP:cytochrome complexes are different since the appropriate redox driving energies are the same. Interestingly, the predicted complex structure uses the x-ray coordinates of tuna cytochrome *c* (Poulous & Kraut, 1980), which is only 10% as active in the reaction of Eqn. 8.29 as the yeast protein. Whether it is a structural difference that is reponsible for the variation in k_2 with the type of residue at position 82 of cytochrome *c* (Table 8.2), or whether the residue plays a key mechanistic role, is yet to be established (Liang *et al.*, 1988). Further data on this system is keenly awaited.

McLendon and his coworkers have adopted a similar approach to Hoffman in their studies of native YCCP reacting with zinc substituted cytochrome *c* and porphyrin cytochrome *c* (Cheung *et al.*, 1986; Conklin & Miller, 1986). The latter reaction is Eqn. 8.30 (Scheme 8.2) and has an electron-transfer rate k_1 of 150 s^{-1}. Yet another reaction system is the oxidation of the Fe(IV) form of YCCP by native horse cytochrome *c* and its zinc-substituted and porphyrin derivatives (Hazzard & Tollin, 1987: Conklin & McLendon, 1988). An analysis of the rates and driving forces of these reactions with the Marcus theory (section 8.3.1) yielded a reactant reorganisation energy of ~140 kJ mol^{-1} (Conklin & McLendon, 1988). However this, and the value of ~75 kJ mol^{-1} for the comparable series of reactions of cytochrome b_5 with native and modified horse cytochrome *c* (Fig. 8.12) appear rather high.

Cytochrome c and Cytochromes b_5 and b_2. McLendon and his colleagues have applied the same methods as described in the previous section for reactions of YCCP to those of cytochrome *c* with cytochrome b_5 and b_2. Their results are summarised in Fig. 8.12.

Assuming the cytochrome c/b_5 complexes have the same structure, the variation in rate with redox driving energy may be an indication of electron transfer in the inverted region (Fig. 8.7 & 8.8d), as McLendon & Miller (1985) assume; or it may reflect different reorganisation energies for the different reactions. Given the variation in reorganisation energies observed in similar reactions involving YCCP (see previous subsection) the latter explanation seems possible. Further data are needed to clarify this.

The lack of variation in k_{et} with redox driving energy in the cytochrome c/b_2 reactions is surprising. McLendon *et al.* (1987) conclude that the observed rate constants are not simple electron-transfer rates but are affected by conformational changes within the reactive complex. Presumably there is a slow conformational change of cytochrome b_2 coupled to the electron transfer. Such a result is consistent with the x-ray structure of cytochrome b_2 which shows that two of the heme domains are ordered and two are disordered (Xia *et al.*, 1987).

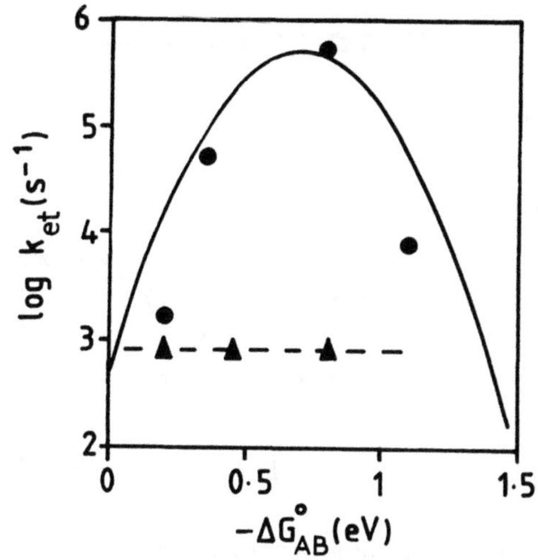

Fig. 8.12 Electron transfer in the cytochromes *c/b*$_5$ and *c/b*$_2$ complexes. Plot of log k_{et} vs ΔG^{o}_{AB} for cytochrome *c/b*$_5$ (●) and cytochrome *c/b*$_2$ (▲) reactions. The reactions studied were those of native cytochrome b_5 and cytochrome b_2 with cytochrome *c*, zinc cytochrome *c* and porphyrin cytochrome *c*. The solid line is the theoretical fit to the cytochrome *c/b*$_5$ data using Eqn. 8.12 and taking ΔG_s to be 75 kJ M^{-1}. (Data from McLendon & Miller, 1985; McLendon *et al.*, 1987)

8.5.4 Bacterial Photosynthetic Reactions

Bacterial photosynthetic reaction centres (RC) constitute a well characterised group of intact biochemical redox complexes (Deisenhofer *et al.*, 1984: Feher *et al.*, 1989; Pettigrew & Moore, 1987: Chapter 3), and since the classic paper by DeVault & Chance (1966), many features of their electron-transfer mechanisms have been defined (DeVault, 1980; Marcus & Sutin, 1985; Dracheva *et al.*, 1988). There are two important experimental features of these systems which have assisted the mechanistic characterisation. First, they are organised within a membrane complex, so that molecular collisions of the type important for class I cytochromes *c* are not relevant to the intracomplex electron-transfer events being studied. Second, an exogenous source of reducing electrons is not required for the initial charge separation because electron transfer is initiated by irradiation of the complex with light of the appropriate wavelength. RC electron-transfer reactions are described in Fig. 8.13 using the nomenclature of DeVault (1980) and the temperature dependence of two of the reactions is given in Fig. 8.14.

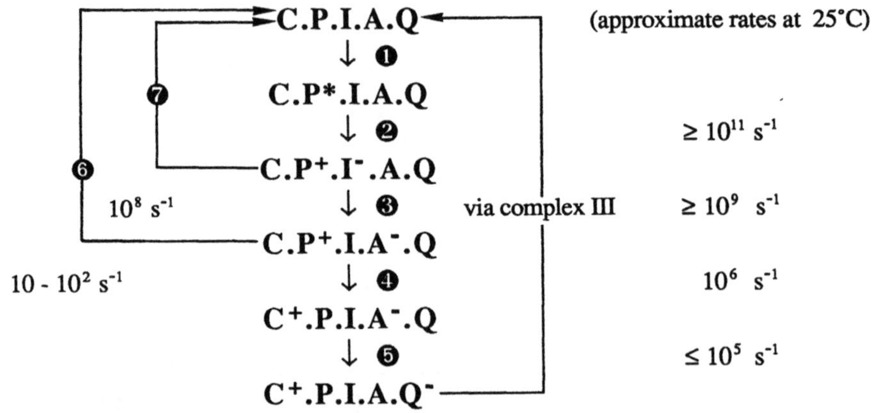

Fig. 8.13 Bacterial photosynthetic reaction centre electron transfer. These equations represent some of the main electron transfer events leading to charge separation in a bacterial photosynthetic reaction centre. The reaction centre contains the primary electron donor, a bacteriochlorophyll dimer (**P**); the intermediate primary electron acceptor, a bacteriopheophytin (**I**); the primary electron acceptor, probably a non-heme iron quinone complex (**A**); secondary electron acceptors, quinones (**Q**); secondary electron donors, cytochromes (**C**). Photon absorption by the antenna pigments and subsequent energy transfer results in excitation of the bacteriochlorophyll dimer; reaction ❶. Electron transfer then proceeds in the sequential manner illustrated by reactions ❷ - ❺. When cytochrome oxidation cannot occur, the reversed primary reaction, step ❻, takes place. Charge recombination by direct electron transfer, step ❼, can also occur. The approximate rates shown do not refer to a single organism or type of reaction centre preparation. They were obtained from a variety of systems (see the compilations of DeVault, 1980, and Gunner & Dutton, 1989). The temperature dependence of the rate of cytochrome oxidation, reaction ❹, and the rate of the reversed primary reaction, reaction ❻, are given in Fig. 8.14

The key features of Fig. 8.14 are the biphasic nature of the plots and the temperature independent rate at low temperature. The latter feature strongly suggests a tunnelling process. The thermally activated process at higher temperature, with an activation energy of 14 KJ mol^{-1}, was originally ascribed to some movement of the donor and acceptor sites relative to each other (DeVault & Chance, 1966). However, more recently a single electron tunnelling model containing a component of vibronic coupling (Fig. 8.9) has been shown to satisfactorily describe the data over the whole temperature range (DeVault, 1980, and references therein). The recent detection of a photoinduced CT band for the RC oxidation of *Chromatium* cytochrome *c* at 10 K and 85 K supports the proposed tunnelling mechanism (Goldstein & Bearden, 1984).

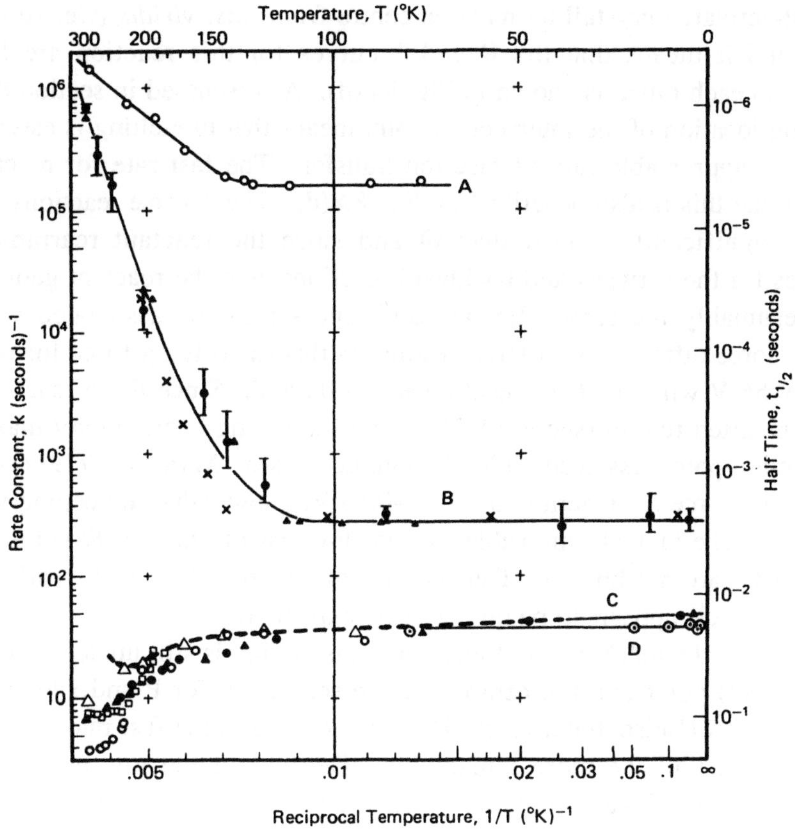

Fig. 8.14 Temperature dependence of bacterial photosynthetic electron transfer reactions. Curves **A** and **B** are photoinduced cytochrome oxidation (reaction ❹ in Fig. 8.13) for: **A** *Rhodopseudomonas* sp N.W., and **B** *Chromatium vinosum* (D). Curves **C** and **D** are the reversed primary reaction (reaction ❻ in Fig. 8.13) for: **C** *R. rubrum* (---, ●, ▲), and **D** *Rb. sphaeroides* (○, □, Δ and ○). The different symbols represent different studies reported by a number of groups on a variety of types of preparation including whole cells, subchromatophore preparations and reaction centre preparations. Note that the horizontal scale to the left of 1/T = 0.01 is different to that on the right. We are indebted to DeVault for permission to reproduce this diagram. His 1980 review should be consulted for the sources and details of the experimental data

Another feature of the electron-transfer rate data that has been considered is the relative rates of some of the reactions and their relationship to their redox driving energies (Marcus, 1979; DeVault, 1980; Feher *et al.*, 1989; Gunner & Dutton, 1989 and references therein). Reaction ❷ is only a factor of 10^2 slower than the theoretical maximum, indicating that the reactant rearrangement energy is relatively low. This is consistent with FT-IR studies of light-activated crystalline reaction centre from *Rps. viridis* (Gerwert *et al.*, 1988) and it means that the **R** and **P** curves for this reaction are located relative to each other as shown in Fig. 8.8(d). As discussed in section 8.3, the resulting location of the intersection point means that tunnelling is essential to obtain an appreciable rate of electron transfer. The fast rate for reaction ❸ implies that this is also described by Fig. 8.8(d). The reverse reactions, ❻ and ❼, are significantly slower than ❷ and since the reactant rearrangement energies for the forward and backward reactions, and the reactant geometries, are presumably the same, the rate differences must be associated with the driving force differences. Marcus estimates that the driving force for reaction ❷ is ~0.56 V while that for reaction ❼ is ~0.85 V. Since these reactions are in the inverted region (section 8.3), where increasing exergonicity makes the intersection point less accessible, the rate decreases. Marcus (1979) estimates the driving force for reaction ❸ to be ~0.35 V, a low value in keeping with its high rate. The factors reponsible for the slowness of reaction ❸ compared to reaction ❷, are not known. Thus control of electron flow at the early stages of the RC cycle appears to be qualitatively explicable.

However, the mechanism of electron transmission is still unclear. As Feher *et al.* (1989) point out, the centre - centre separation for P and I tetrapyrrole rings is ~17Å (Deisenhofer *et al.*, 1984) which makes the fast rate for reaction ❷ (Fig. 8.13) all the more remarkable. Because of the low probability of significant P and I wave-function overlap, Feher *et al.* (1989) consider groups such as tyrosine play a role in electron transmission by bridging the gap between the P and I centres.

References

Adar F (1978) In : Dolphin D (ed) The Porphyrins vol. 3A. Academic Press, London New York pp 167 - 209

Adman ET (1979) A comparison of the structures of electron transfer proteins. Biochim Biophys Acta 549 : 107 - 144

Adman ET, Canters GW, Hill HAO, Kitchen NA (1982) The effect of pH and temperature on the structure of the active site of azurin from *Pseudomonas aeruginosa*. FEBS Lett 143 : 287 - 292

Aitken A (1976) Protein evolution in cyanobacteria. Nature (London) 26 : 792 - 796

Aitken A (1977) Purification and primary structure of cytochrome *f* from the cyanobacterium, *Plectonema boryanum*. Eur J Biochem 78 : 273 - 279

Aitken A (1978) Plastocyanin and the evolutionary relationship between chloroplasts and cyanobacteria. In : Matsubara H, Yamanaka T (eds) Evolution of Protein Molecules. Japan Scientific Societies Press pp 251 - 263

Aitken A (1979) Purification and primary structure of cytochrome *c*-552 from the cyanobacterium, *Synechococcus* PCC 6312. Eur J Biochem 101 : 297 - 308

Akutsu H, Kyogoku Y, Horio T (1983) Conversion of spin state and protein structure in *R. rubrum* cytochrome *c′* coupled with dissociation of dimer. Biochemistry 22 : 2055 - 2061

Alleyne TA, Wilson MT (1987) Zinc cytochrome *c* fluorescence as a probe for conformational changes in cytochrome oxidase. Biochem J 247 : 475 - 484

Almassy RJ, Dickerson RE (1978) *Pseudomonas* cytochrome *c*-551 at 2.0Å resolution. Proc Natl Acad Sci USA 75 : 2674 - 2678

Alt J, Hermann RG (1984) Nucleotide sequence of the gene for pre-apocytochrome *f* in the spinach plastid chromosome. Curr Genet 8 : 551 - 557

Amati BB, Goldschmidt MC, Wallace CJA, Rochaix JD (1988) c-DNA and deduced amino acid sequence of cytochrome *c* from *Chlamydomonas reinhardtii* : Unexpected functional and phylogenetic implications. J Mol Evol 28 : 151 - 160

Ambler RP (1963) The amino acid sequence of *Pseudomonas* cytochrome *c*-551. Biochem J 89 : 349 - 378

Ambler RP (1968) The amino acid sequence of cytochrome c_3 from *Desulfovibrio vulgaris*. Biochem J 109 : 47P - 48P

Ambler RP (1971) The amino acid sequence of cytochrome c-551.5 (cytochrome c_7) from the green photosynthetic bacterium, *Chloropseudomonas ethylica*. FEBS Lett 18 : 351 - 353

Ambler RP (1973a) Bacterial cytochromes *c* and molecular evolution. Syst Zool 22 : 554 - 565

Ambler RP (1973b) The amino acid sequence of cytochrome *c′* from *Alcaligenes* sp NCIB 11015. Biochem J 135 : 751 - 758

Ambler RP (1974) The evolutionary stability of cytochrome *c*-551 in *Pseudomonas aeruginosa* and *Pseudomonas fluorescens* Biotype C. Biochem J 137 : 3 - 14

Ambler RP (1978) Amino acid sequences and bacterial phylogeny. In : Matsubara H, Yamanaka T (eds) Evolution of Protein Molecules. Japan Scientific Societies Press pp 311 - 322
Ambler RP (1980) The structure and classification of cytochromes *c*. In Robinson AB, Kaplan NO (eds) From cyclotrons to cytochromes. Academic Press, London New York pp 263 - 279
Ambler RP, Bartsch RG (1975) Amino acid sequence similarity between cytochrome *f* from a blue green bacterium and from algal chloroplasts. Nature (London) 253 : 285 - 288
Ambler RP, Murray S (1973) N-terminal amino acid sequences of cytochromes c_4 from *Azotobacter vinelandii* and denitrifying Pseudomonads. Biochem Soc Trans 1 : 107 - 109
Ambler RP, Taylor E (1973) Amino acid seuqence of cytochrome c_5 from *Pseudomonas mendocina*. Biochem Soc Trans 1 : 111 - 113
Ambler RP, Wynn M (1973) The amino acid sequences of cytochromes *c*-551 from three species of *Pseudomonas*. Biochem J 131 : 485 - 498
Ambler RP, Bruschi M, Le Gall J (1969) The structure of cytochrome c_3 from *Desulfovibrio gigas* (NCIB 9332). FEBS Lett 5 : 115 - 117
Ambler RP, Bruschi M, Le Gall J (1971) The amino acid sequence of cytochrome c_3 from *Desulfovibrio desulfuricans* (strain El Agheila Z, NCIB 8380). FEBS Lett 18 : 347 - 350
Ambler RP, Meyer TE, Kamen MD (1976) Primary structure determination of two cytochromes c_2: close similarity to functionally unrelated mitochondrial cytochrome *c*. Proc Natl Acad Sci USA 73 : 472 - 475
Ambler RP, Daniel M, Hermoso J, Meyer TE, Bartsch RG, Kamen MD (1979a) Cytochrome c_2 sequence variation among recognised species of purple non-sulfur photosynthetic bacteria. Nature (London) 278 : 659 - 660
Ambler RP, Meyer TE, Kamen MD (1979b) Anomalies in the amino acid sequences of small cytochromes *c* and cytochromes *c'* from two species of purple photosynthetic bacteria. Nature (London) 278 : 661 - 662
Ambler RP, Daniel M, Meyer TE, Bartsch RG, Kamen MD (1979c) The amino acid sequence of cytochrome *c'* from the purple photosynthetic bacterium, *Chromatium vinosum*. Biochem J 177 : 819 - 823
Ambler RP, Bartsch RG, Daniel M, Kamen MD, McLellan L, Meyer TE, Van Beeumen J (1981a) Amino acid sequences of bacterial cytochromes *c'* and *c*-556. Proc Natl Acad Sci USA 78 : 6854-6857
Ambler RP, Meyer TE, Kamen MD, Schichman SA, Sawyer L (1981b) A reassessment of the structure of *Paracoccus* cytochrome *c*-550. J Mol Biol 147 : 351 - 356
Ambler RP, Daniel M, Melis K, Stout CD (1984) The amino acid sequence of the dihaem cytochrome c_4 from the bacterium *Azotobacter vinelandii*. Biochem J 222 : 217 - 227
Ambler RP, Meyer TE, Trudinger PA, Kamen MD (1985) The amino acid sequence of the cytochrome *c*-554(547) from the chemolithotrophic bacterium, *Thiobacillus neapolitanus*. Biochem J 227 : 1009 - 1013
Ambler RP, Dalton H, Meyer TE, Bartsch RG, Kamen MD (1986) The amino acid sequence of cytochrome *c*-555 from the methane oxidising bacterium *Methylococcus capsulatus*. Biochem J 233 : 333 - 37
Ambler RP, Meyer TE, Cusanovich MA, Kamen MD (1987a) The amino acid sequence of the cytochrome c_2 from the phototrophic bacterium, *Rhodopseudomonas globiformis*. Biochem J 246 : 115 - 120

Ambler RP, Daniel M, McLellan L, Meyer TE, Cusanovich MA, Kamen MD (1987b) Amino acid sequences of cytochrome *c*-554(548) and cytochrome *c'* from a halophilic denitrifying bacterium of the genus *Paracoccus*. Biochem J 248 : 365 - 371

Andersson KK, Babcock GT, Hooper AB, (1984) Diheme cytochrome *c*-554 from *Nitrosomonas*: Soret resonance Raman indication of an unusual ferric 5-coordinate structure. FEBS Lett 170 : 331 - 334

Andersson KK, Lipscomb JD, Valentine M, Münck E, Hooper AB (1986) Tetraheme cytochrome *c*-554 from *Nitrosomonas europaea*: heme - heme interactions and ligand binding. J Biol Chem 261 : 1126 - 1138

Ando K, Matsubara H, Okunuki K (1966) Carboxymethylation of beef and human cytochromes *c* in the oxidized and reduced forms. Biochim Biophys Acta 118 : 256 - 267

Ångström J, Moore GR, Williams RJP (1982) The magnetic susceptibility of ferricytochrome *c*. Biochim Biophys Acta 703 : 87 - 94

Antalis TM, Palmer G (1982) Kinetic characterisation of the interaction between cytochrome *c* oxidase and cytochrome *c*. J Biol Chem 257 : 6194 - 6206

Antonini E, Brunori M (1971) Hemoglobin and myoglobin in their reactions with ligands. North-Holland, Amsterdam, London

Arai H, Sanbongi Y, Igarashi Y, Kodama T (1990) Cloning and sequencing of the gene encoding cytochrome *c*-551 from *Pseudomonas aeruginosa*. FEBS Lett 261 : 196 - 198

Argade PV, Ching YC, Sassaroli M, Rousseau DL (1986) Accessibilty of the cytochrome *a* heme in cytochrome *c* oxidase to exchangeable protons. J Biol Chem 261 : 5969 - 5973

Armstrong FA, Hill HAO, Walton NJ (1988) Direct electrochemistry of redox proteins. Acc Chem Res 21 : 407 - 413

Armstrong GD, Chambers JA, Sykes AG (1986) Preferred sites for electron exchange between cytochrome *c* and $[Fe(edta)]^{2-}$ and $[Co(sep)]^{2+}$ complexes. J Chem Soc Dalton Trans 755 - 758

Aron J, Baldwin DA, Marques HM, Pratt JM, Adams PA (1986) Preparation and analysis of the heme containing octapeptide (microperoxidase-8) and identification of the monomeric form in aqueous solution. J Inorg Biochem 27 : 227 - 244

Arzatbanov VY, Konstantinov AA, Skulachev VP (1978) Involvement of intramitochondrial protons on redox reactions of cytochrome *a*. FEBS Lett 87 : 180 - 185

Augusteyn RC, (1973) Primary structure of cytochrome *c* from the emu, *Dromaeus novaehollandiae*. Biochim Biophys Acta 303 : 1 - 7

Augusteyn RC, McDowall MA, Webb EC, Zerner B (1972) Primary structure of cytochrome *c* from the elephant seal, *Mirounga leonina*. Biochim Biophys Acta 257 : 264 - 272

Augustin MA, Chapman SK, Davies DM, Sykes AG, Speck SH, Margoliash E (1983) Interaction of cytochrome *c* with the blue copper proteins, plastocyanin and azurin. J Biol Chem 258 : 6405 - 6409

Austin RH, Hopfield JJ (1982) Attempted time-resolved measurement of photo-assisted charge transfer of cytochrome *c* - iron hexacyanide. In : Ho C (ed) Electron transport and oxygen utilization. Elsevier, New York London 73 - 80.

Aviram I, Schejter A (1969) Physicochemical properties of yeast iso-1-cytochrome *c*. J Biol Chem 244 : 3773 - 3778

Aviram I, Schejter A (1971) Modification of the tryptophanyl residue of horse heart cytochrome *c*. Biochim Biophys Acta 229 : 113 - 118

Aviram I, Myer YP, Schejter A (1981) Stepwise modification of the electrostatic charge of cytochrome *c*. J Biol Chem 256 : 5540 - 5544

Ayala FJ (1986) On the virtues and pitfalls of the molecular evolutionary clock. J Hered 77 : 226 - 235

Baba ML, Darga LL, Goodman M, Czelusniak J (1981) Evolution of cytochrome *c* investigated by the maximum parsimony method. J Molec Evol 17 : 197 - 213

Babcock GT, Callahan PM (1983) Redox linked hydrogen bond strength changes in cytochrome *a*. Implications for a cytochrome oxidase proton pump. Biochemistry 22 : 2314 - 2318

Babul J, Stellwagen E (1972) Participation of the protein ligands in the folding of cytochrome *c*. Biochemistry 11 : 1195 - 1200

Bahl OP, Smith EL (1965) Amino acid sequence of rattlesnake heart cytochrome *c*. J Biol Chem 240 : 3585 - 3593

Baldwin DA, Marques HM, Pratt JM (1986) The pH-dependent equilibria of microperoxidase-8 and characterization of the coordination sphere of Fe(III). J Inorg Biochem 27 : 245 - 254

Barakat R, Strekas TC (1982) pH variation of midpoint potentials for three photosynthetic bacterial cytochromes *c'*. Biochim Biophys Acta 679 : 393 - 401

Barlow CM, Maxwell JC, Wallace WJ, Caughey WS (1973) Elucidation of the mode of binding of oxygen to iron in oxyhemoglobin by infrared spectroscopy. Biochem Biophys Res Commun 55 : 91 - 95

Barrett CP, Peterson J, Greenwood C, Thomson AJ (1986) Optical detection of paramagnetic resonance by magnetic circular dichroism. Applications to aqueous solutions of metalloproteins. J Am Chem Soc 108 : 3170 - 3177

Barstow LE, Young RS, Yakali E, Sharp JJ, O'Brien JC, Berman PW, Harbury HA (1977) Semisynthetic cytochrome *c*. Proc Natl Acad Sci USA 74 : 4248 -4 250

Bartsch RG (1978) Cytochromes. In : Clayton RK, Sistrom WR (eds) The Photosynthetic Bacteria. Plenum, New York pp 249 - 279

Bartsch RG, Kamen MD (1960) Isolation and properties of two soluble heme proteins in extracts of the photoanaerobe *Chromatium*. J Biol Chem. 235 : 825 - 831

Baumgartner CP, Sellars Sr. M, Nassif R, May L (1974) Mössbauer studies of some complexes of the undecapeptide of cytochrome *c*. Eur J Biochem 46 : 625 - 629

Bearden AJ, Dunham WR (1970) Iron electronic configurations in proteins: studies by Mossbauer spectroscopy. Structure & Bonding 8 : 1 - 52

Beattie JK, Fensom DJ, Freeman HC, Woodcock E, Hill HAO, Stokes AM (1975) An nmr investigation of electron transfer in the copper protein, plastocyanin. Biochim Biophys Acta 405 : 109 -114

Becam AM, Lederer F (1981) Amino acid sequence of the cytochrome *c* from the yeast, *Hansenula anomala*. Eur J Biochem 118 : 295 - 302

Bechtold R, Kuehn C, Lepre C, Isied SS (1986a) Directional electron transfer in ruthenium-modified horse heart cytochrome *c*. Nature (London) 322 : 286 - 288

Bechtold R, Gardineer MB, Kazmi A, van Hemelryck B, Isied SS (1986b) Ruthenium-modified horse heart cytochrome *c*: effect of pH and ligation on the rate of intramolecular electron transfer between Ru(II) and Heme(III). J Phys Chem 90 : 3800 - 3804

Becker ED (1980) High Resolution NMR : Theory and Chemical Applications. Academic Press, New York

Benosman H, Asso M, Bertrand P, Yagi T, Gayda JP (1989) EPR study of the redox interactions in cytochrome c_3 from *Desulfovibrio vulgaris* Miyazaki. Eur J Biochem 182 : 51 - 55

Bertini I, Luchinat C (1986) NMR of paramagnetic molecules in biological systems. Benjamin/Cummings, Menlo Park California

Bhatia GE, Finzel BC, Kraut J (1984) 2Å resolution X-ray coordinates of *R. rubrum* cytochrome c_2. Deposited with the protein data bank.

Billeter M, Kline AD, Braun W, Huber R, Wüthrich K (1989) Comparison of the high-resolution structures of the α-amylase inhibitor tendamisat determined by nuclear magnetic resonance in solution and by x-ray diffraction in single crystals. J Mol Biol 206 : 677 - 687

Bitar KG, Vinogradov SN, Nolan C, Weiss LJ, Margoliash E (1972) The primary structure of cytochrome *c* from the rust fungus *Ustilago sphaerogena*. Biochem J 129 : 561 - 569

Blackmore RS, Brittain T, Gadsby PMA, Greenwood C, Thomson AJ (1987) EPR and MCD studies of a hexa-heme nitrite reductase from *Wolinella succinogenes*. FEBS Lett 219 : 244 - 248

Blair DF, Ellis WR, Wang H, Gray HB, Chan HB (1986) Spectroelectrochemical study of cytochrome *c* oxidase: pH and temperature dependence of the cytochrome potentials. J Biol Chem 261 : 11524 - 11534

Blasie JK, Ereciǹska M, Samuels S, Leigh JS (1978) The structure of a cytochrome oxidase-lipid model membrane. Biochim Biophys Acta 501 : 33 - 52

Blumberg WE, Peisach J (1971) A unified theory for low-spin forms of all ferric heme proteins as studied by EPR. In : Chance B, Yonetani T, Mildvan AS (eds) Probes of Structure and Function of Macromolecules and Membranes. Vol II 215 - 229. Academic Press, New York

Boon PJ (1981) Semisynthesis of cytochrome *c* analogues. Doctoral thesis : Catholic University of Nijmegen

Boon PJ, Tesser GI, Nivard RJF (1979a) Semisynthetic horse heart [65-homoserine] cytochrome *c* from three fragments. Proc Natl Acad Sci USA 76 : 61 - 65

Boon PJ, van Raay AJM, Tesser GI, Nivard RJF (1979b) Semi-synthesis, conformation and cytochrome *c* oxidase activity of eight cytochrome *c* analogues. FEBS Lett 108 : 131 - 135

Boots HA (1986) Covalent and non-covalent semisynthesis of cytochrome *c* analogues. Doctoral thesis : Catholic University of Nijmegen

Bosshard HR (1981) Alkaline isomerisation of ferricytochrome *c*. J Mol Biol 153 : 1125 - 1149

Bosshard HR, Zürrer M (1980) The conformation of cytochrome *c* in solution. J Biol Chem 255 : 6694 - 6699

Boswell AP, McClune GJ, Moore GR, Williams RJP, Pettigrew GW, Inubishi T, Yonetani T, Harris DE (1980) NMR study of the interaction of cytochrome *c* with cytochrome *c* peroxidase. Biochem. Soc. Trans. 8 : 637 - 638

Boswell AP, Moore GR, Williams RJP, Harris DE, Wallace CJA, Bocieck S, Welti D (1983) Ionization of tyrosine and lysine residues in native and modified horse cytochrome *c*. Biochem J 213 : 679 - 686

Boswell AP, Moore GR, Williams RJP, Wallace CJA, Boon PJ, Nivard RJF, Tesser GI (1981) Structural studies of eukaryotic cytochrome *c* modified at Met 65. Biochem J 193 : 493 - 512

Bowler BE, Meade TJ, Mayo SL, Richards JH, Gray HB (1989) Long-range electron transfer in structurally engineered pentaammineruthenium (histidine-62) cytochrome *c*. J Am Chem Soc 111 : 8757 - 8759

Brandt KG, Parks PC, Czerlinski GH, Hess GP (1966) On the elucidation of the pH dependence of the oxidation -reduction potential of cytochrome *c* at alkaline pH. J Biol Chem 241 : 4180 - 4185

Braterman PS, Davies RC, Williams RJP (1964) The properties of metal-porphyrin and similar complexes. Adv Chem Phys 7 : 359 - 407

Brautigan DL, Feinberg BA, Hoffman BM, Margoliash E, Peisach J, Blumberg WE (1977) Multiple low-spin forms of the cytochrome *c* ferrihemochrome. J Biol Chem 252 : 574 - 582

Brems DN, Stellwagen E (1981) The effect of methylation on cytochrome *c* fragment complementation. J Biol Chem 256 : 11688 - 11690

Brems DN, Cass R, Stellwagen E (1982) Conformational transitions of frog heart ferricytochrome *c*. Biochemistry 21 : 1488 - 1493

Broger C, Salardi S, Azzi A (1983) Interaction between isolated cytochrome c_1 and cytochrome *c*. Eur J Biochem 131 : 349 - 352

Brown LR, Wüthrich K (1977) NMR and ESR studies of the interactions of cytochrome *c* with mixed cardiolipin-phosphatidylcholine vesicles. Biochim Biophys Acta 468 : 389 - 410

Brown R, Boulter D (1973a) The amino acid sequence of cytochrome *c* from *Allium porrum* L. (leek). Biochem J 131 : 247 - 251

Brown R, Boulter D (1973b) The amino acid sequence of cytochrome *c* from *Nigella damascena* L. (love-in-a-mist). Biochem J 133 : 251 - 254

Brown R, Boulter D (1974) The amino acid sequences of cytochrome *c* from four plant sources. Biochem J 137 : 93 - 100

Brown R, Boulter D (1975) A re-examination of the amino acid sequence data of cytochromes *c* from potato and tomato. FEBS Lett 51 : 66 - 67

Brown R, Richardson M, Boulter D, Ramshaw JAM, Jefferies RPS (1972) The amino acid sequence of cytochrome *c* from *Helix aspersa* Müller (garden snail). Biochem J 128 : 971 - 974

Brown R, Richardson M, Scogin R, Boulter D (1973) The amino acid sequence of cytochrome *c* from *Spinacea oleracea* L. (spinach). Biochem J 131 : 253 - 256

Brunori M, Antonini G, Malatesta F, Sarti P, Wilson MT (1987) Cytochrome *c* oxidase subunit structure and proton pumping. Eur J Biochem 169 : 1 - 8

Bruschi M (1981) The primary structure of the tetraheme cytochrome c_3 from *Desulfovibrio desulfuricans* (strain Norway 4). Biochim Biophys Acta 671 : 219 - 224

Bruschi M, Le Gall J (1972) C-type cytochromes of *Desulfovibrio vulgaris* - the primary structure of cytochrome c-553. Biochim Biophys Acta 271 : 48 - 60

Bruschi M, Loutfi M, Bianco P, Haladjian J (1984) Correlation studies between structural and redox properties of cytochromes c_3. Biochem Biophys Res Commun 120 : 384 - 389

Bull HN, Breese K (1966) Ionization of cytochrome *c*. Biochem Biophys Res Commun 24 : 74 - 78

Burch AM, Rigby SEJ, Funk W, MacGillivray RTA, Mauk AG, Moore GR (1988) A ^{13}C NMR investigation of the interaction between cytochrome *c* and cytochrome b_5. Biochem Soc Trans 16 : 844 - 845

Burch AM, Rigby SEJ, Funk W, MacGillivray RTA, Mauk MR, Mauk AG, Moore GR (1990) NMR characterization of surface interactions in the cytochrome b_5-cytochrome *c* complex. Science 247 : 831 - 833

Burns PD, La Mar GN (1979) Detection of localized conformational flexibility in horse heart cytochrome *c* by proton NMR. J Amer Chem Soc 101 : 5844 - 5846

Burns PD, La Mar GN (1981) Characterization of conformational heterogeneity in the heme pocket of ferricytochrome *c* using high field proton nuclear magnetic resonance spectroscopy. J Biol Chem 256 : 4934 - 4939

Burns PS, Harrod JF, Williams RJP, Wright PE (1976) A study of redox reactions of biological importance between Fe(III) complexes and aromatic moieties. Biochim Biophys Acta 428 : 261 - 268

Buse G, Hensel S, Fee JA (1989) Evidence for a cytochrome oxidase subunit and a cytochrome *c*-subunit II fused protein in the cytochrome 'c_1aa_3' of *Thermus thermophilus*. Eur J Biochem 181 : 261 - 268

Butler J, Koppenol WH, Margoliash E (1982) Kinetics and mechanism of the reduction of ferricytochrome *c* by the superoxide anion. J Biol Chem 257 : 10747 - 10750

Butler J, Chapman SK, Davies DM, Sykes AG, Speck SH, Osheroff N, Margoliash E (1983) Preferred sites for electron transfer between cytochrome *c* and iron and cobalt complexes. J Biol Chem 258 : 6400 - 6404

Butt WD, Keilin D (1962) Absorption spectra and some other properties of cytochrome *c* and of its compounds with ligands. Proc Royal Soc Lond B152 : 429 - 458

Byrn MP, Katz BA, Keder NL, Levan KR, Magurany CJ, Miller KM, Pritt JW, Strouse CE (1983) Low-spin ferric porphyrin complexes: Analysis of the electronic structure based on single-crystal electron spin resonance measurements. J Am Chem Soc 105 : 4916 - 4922

Campbell ID, Dobson CM, Moore GR, Perkins SJ, Williams RJP (1976) Temperature dependent molecular motion of a tyrosine residue of ferrocytochrome *c*. FEBS Lett 70 : 96 - 100

Campbell ID, Dwek RA (1984) Biological Spectroscopy , Benjamin-Cummings, California

Cannon RD (1980) Electron Transfer Reactions , Butterworths, London Boston

Canters GW, Hill HAO, Kitchen NA, Adman ET (1984) A proton NMR study of the electron exchange between reduced and oxidized azurin from *Pseudomonas aeruginosa*. J Mag Res 57 : 1 - 23

Cantor CR, Schimmel PR (1980) Biophysical Chemistry part II : Techniques for the study of biological structure and function. Freeman, San Francisco

Carithers RP, Palmer G (1981) Characterisation of the potentiometric behaviour of soluble cytochrome *c* oxidase by magnetic circular dichroism. J Biol Chem 256 : 7967 - 7976

Carlson SS, Moss GA, Wilson AC, Mead RT, Wolin LD, Bowers SF, Foley NT, Muijsers AO, Margoliash E (1977) Primary structure of mouse, rat, and guinea pig cytochrome *c*. Biochemistry 16 : 1437 - 1442

Carter KR, T'sai A, Palmer G (1981) The coordination environment of mitochondrial cytochromes *b*. FEBS Letters 132 : 243 - 246

Carter DC, Melis KA, O'Donnell SE, Burgess BK, Furey Jr WF, Wang BC, Stout CD (1985) Crystal structure of *Azotobacter* cytochrome c_5 at 2.5Å resolution. J Mol Biol 184 : 279 - 295

Casey RP, Chappell JB, Azzi A (1979) Limited turnover studies on proton translocation in reconstituted cytochrome *c* oxidase containing vesicles. Biochem J 182 : 149 - 156

Casey RP, Thelen M, Azzi A (1980) Dicyclohexylcarbodiimide binds specifically and covalently to cytochrome *c* oxidase while inhibiting its H^+ translocating activity. J Biol Chem 255 : 3994 - 4000

Chakrabarti P (1989) Geometry of interaction of metal ions with sulfur-containing ligands in protein structures. Biochemistry 28 : 6081 - 6085

Champion PM, Lipscomb JD, Münck E, Debrunner P, Gunsalas IC (1975) Mössbauer investigations of high-spin ferrous heme proteins.1. Cytochrome P-450. Biochemistry 14 : 4151 - 4158

Chan SK (1970) Biochemical studies in the developing thoracic muscles of the tobacco horn worm. IV. Primary structure of cytochrome *c*. Biochim Biophys Acta 221 : 497 - 501

Chan SK, Margoliash E (1966a) Properties and primary structure of the cytochrome *c* from the flight muscles of the moth, *Samia cynthia*. J Biol Chem 241 : 335 - 348

Chan SK, Margoliash E (1966b) Amino acid sequence of chicken heart cytochrome *c*. J Biol Chem 241 : 507 - 515

Chan SK, Tulloss I, Margoliash E (1966) Primary structure of the cytochrome *c* from the snapping turtle, *Chelydra serpentina*. Biochemistry 5 : 2586 - 2597

Chan SK, Walasek OF, Barlow GH, Margoliash E (1967) Structure of bullfrog cytochrome *c*. Fed Proc 26 : 723 abstract 2603

Chance B, Schoener B, DeVault D (1964) Reaction velocity constants for electron transfer and transport reactions. In: King TE, Mason HS, Morrison M (eds) Oxidases and Related Redox Systems Wiley, New York London vol 2. pp 907 - 942

Chance B, Lee CP, Mela L, Wilson DF (1968) Some properties of the 695 nm band of cytochrome *c*. In : Okunuki K, Kamen MD, Sekuzu I (eds) Structure and function of cytochromes. Univ Park Press, Baltimore pp353 - 356

Chance B, DeVault D, Fraunfelder H, Marcus RA, Schrieffer JR, Sutin N (eds) (1979) Tunneling in Biological Systems. Academic Press, London New York

Chao YYH, Bersohn R, Aisen P (1979) ^{1}H-NMR and ESR studies of oxidized cytochrome *c*-551 from *Pseudomonas aeruginosa*. Biochemistry 18 : 774 - 779

Cheddar G, Meyer TE, Cusanovich MA, Stout CD, Tollin G (1989) Redox protein electron-transfer mechanisms : Electrostatic interactions as a determinant of reaction site in *c*-type cytochromes. Biochemistry 28 : 6318 - 6322

Cheung E, Taylor K, Kornblatt JA, English AM, McLendon G, Miller JR (1986) Direct measurements of intramolecular electron transfer rates between cytochrome *c* and cytochrome *c* peroxidase : effects of exothermicity and primary sequence on rate. Proc Natl Acad Sci USA 83 : 1330 - 1333

Cho KC, Chu WF, Choy CL, Che CM (1988) Kinetics of electron transfer between cytochrome *c* and iron hexacyanides. Evidence for two electron-transfer sites. Biochim Biophys Acta 934 : 161 - 168

Chothia C, Lesk A (1985) Helix movements and the reconstruction of the haem pocket during the evolution of the cytochrome *c* family. J Mol Biol 182 : 151 - 158

Churg AK, Warshel A (1986) Control of the redox potential of cytochrome *c* and microscopic dielectric effects in proteins. Biochemistry 25 : 1675 - 1681

CIBA Foundation Symposium No. 93 (1983) Mobility and function in proteins and nucleic acids.

Clark WM (1960) Oxidation reduction potential of organic systems. Williams and Wilkins, Baltimore, MD

Clayden NJ, Moore GR, Williams G (1987) Metal-porphyrin induced NMR dipolar shifts and their use in conformational analysis. Metal Ions in Biological Systems 21 : 187 - 227

Closs GL, Miller JR (1988) Intramolecular long-distance electron transfer in organic molecules. Science 240 : 440 - 447

Coates M, Stone S (1981) Simulation of protein evolution by random fixation of allowed codons. J Molec Evol 17 : 311 - 328

Cohen H, Strampp A (1976) Bacterial synthesis of a substance similar to human chorionic gonadotrophin. Proc Soc Exptl Biol Med 152 : 408 - 410

Cohen JS, Hayes MB (1974) NMR titration curves of histidine ring protons. Comparative study of cytochrome *c* from three species. J Biol Chem 249 : 5472 - 5477

Cohen JS, Fisher WR, Schechter AN (1974) Spectroscopic studies of the conformation of cytochrome *c* and apocytochrome *c*. J Biol Chem 249 : 1113 - 1118

Cohn CL, Hermondson MA, Krogmann DW (1989a) The amino acid sequence of cytochrome *c*-553 from *Microcystis aeruginosa*. Arch Biochem Biophys 270 : 219 - 226

Cohn CL, Sprinkle JR, Alain J, Hermondson M, Meyer T, Krogmann DW (1989b) The amino acid sequence of low potential cytochrome *c*-550 from the cyanobacterium *Microcystis aeruginosa*. Arch Biochem Biophys 270 : 227 - 235

Coleman JOD, Hill HAO, Walton NJ, Whatley FR (1983) Electrochemically driven respiration in mitochondria and *Paracoccus denitrificans*. FEBS letters 154 : 319 - 322

Collawn JF, Wallace CJA, Proudfoot AEI, Paterson Y (1988) Monoclonal antibodies as probes of conformational changes in protein-engineered cytochrome *c*. J Biol Chem 263 : 8625 - 8634

Collins DM, Countryman R, Hoard JL (1972) Stereochemistry of low-spin iron porphyrins. Bis(imidazole)-tetraphenylporphinato iron(III) chloride. J Am Chem Soc 94 : 2066 - 2071

Collman JP, Hoard JL, Kim N, Lang G, Reed CA (1975) Synthesis, stereochemistry and structure-related properties of tetraphenylporphinato iron(III). J Am Chem Soc 97 : 2676 - 2685

Colman PM, Freeman HC, Guss JM, Murata M, Norris VA, Ramshaw JAM, Ventakappa M P (1978) X-ray structure analysis of plastocyanin at 2.7Å resolution. Nature (London) 272 : 319 - 324

Colosimo A, Brunori M, Andreasi F, Mobilio S (1981) The K edges of the heme iron in the X-ray absorption spectra of native and carboxymethylated cytochrome *c*. J Inorg Biochem 15 : 179 - 184

Concar DW, Hill HAO, Moore GR, Whitford D, Williams RJP (1986) The modulation of cytochrome *c* electron self-exchange by site-specific chemical modification and anion binding. FEBS Lett 206 : 15 - 19

Conklin KT, McLendon G (1986) Photoinduced long-range electron transfer in the zinc substituted cytochrome *c* - cytochrome *c* peroxidase complex. Inorg Chem 25 : 4804 - 4806

Conklin KT, McLendon G (1988) Free energy effects on biological electron transfer: reactions of iron(IV) cytochrome *c* peroxidase (ES) with metallocytochromes *c*. J Am Chem Soc 110 : 3345 - 3350

Conrad DW, Scott RA (1989) Long-range electron transfer in a cytochrome *c* derivative containing a covalently attached cobalt-cage complex. J Am Chem Soc 111 : 3461 - 3463

Cookson DJ, Moore GR, Pitt RC, Williams RJP, Campbell ID, Ambler RP, Bruschi M, LeGall J (1978) Structural homology of cytochromes *c*. Eur J Biochem 83 : 261 - 275

Cooper HM, Jemmerson R, Hunt DF, Griffin PR, Yates JR, Shabanowitz J, Zhu N Z, Paterson Y (1987) Site-directed chemical modification of horse cytochrome *c* results in changes in antigenicity due to local and long-range conformational perturbations. J Biol Chem 262 : 11591 - 11597

Copeland RA, Spiro TG (1986) Resonance Raman evidence for an exchangeable protein hydrogen associated with the heme *a* group of cytochrome *c* oxidase. FEBS Lett 197 : 239 - 243

Copeland RA, Smith PA, Chan SI (1987) Cytochrome *c* oxidase exhibits a rapid conformational change on reduction of Cu_A -a tryptophan fluorescence study. Biochemistry 26 : 7311 - 7316

Corin AF, Bersohn R, Cole PE (1983) pH dependence of the reduction-oxidation reaction of azurin with cytochrome *c*-551 : Role of histidine-35 of azurin in electron transfer. Biochemistry 22 : 2032 - 2038

Corker GA, Sharpe SA (1975) Influence of light on the epr detectable electron transport components of *Rhodospirillum rubrum.* Photochem Photobiol 21 : 49 - 61

Corradin G, Harbury HA (1971) Reconstitution of horse heart cytochrome *c*. Interaction of the components obtained upon cleavage of the peptide bond following Met residue 65. Proc Natl Acad Sci USA 68 : 3036 - 3039

Corradin G, Harbury HA (1974) Reconstitution of horse heart cytochrome *c*. Reformation of the peptide bond linking residues 65 & 68. Biochem Biophys Res Commun 61 : 1400 - 1406

Cotton FA, Wilkinson G (1980) Advanced Inorganic Chemistry 4th ed. Wiley - Interscience New York pp 625 - 628

Cramer SP, Hodgson KO (1979) X-ray absorption spectroscopy : a new structural method and its applications to bioinorganic chemistry. Prog Inorg Chem 25 : 1 - 39

Crick FHC (1953) The packing of α-helices: simple coiled coils. Acta Cryst 6 : 689 - 697

Cronin JR, Farringer BA, Nieman RA, Gust D (1985) Ionization of tyrosine residues in horse heart cytochrome *c* and its guanidated and acetylated-guanidated derivatives. Biochim Biophys Acta 828 : 325 - 335

Cronquist A (1976) The taxonomic significance of the structure of plant proteins: a classical taxonomists view. Brittonia 28 : 1 - 27

Crowson RA (1972) A systematist looks at cytochrome *c*. J Molec Evol 2 : 28 - 37

Cusanovich MA (1971) Molecular weights of some cytochromes *cc'*. Biochim Biophys Acta 236 : 238-241

Cusanovich MA, Gibson QH (1973) Anomalous ligand binding by a class of high-spin *c*-type cytochromes. J Biol Chem 248 : 822 - 834

Cusanovich MA, Tedro SM, Kamen MD (1970) *Pseudomonas denitrificans* cytochrome *cc'*. Arch Biochem Biophys 141 : 557 - 570

Cutler RL, Davies AM, Creighton S, Warshel A, Moore GR, Smith M, Mauk AG (1989) Role of arginine 38 in regulation of the cytochrome *c* oxidation reduction equilibrium. Biochemistry 28 : 3188 - 3197

Czerlinski G, Bracokova V (1973) Kinetics and equilibria among the mutiple forms of ferricytochrome *c*. Biochim Biophys Acta 295 : 480 - 489

Daldal F, Cheng S, Applebaum J, Davidson E, Prince RC (1986) Cytochrome c_2 is not essential for photosynthetic growth. Proc Natl Acad Sci USA 83 : 2012 - 2016

Das G, Hickey DR, Principio L, Conklin KT, Short J, Miller JR, McLendon G, Sherman F (1988) Replacements of Lys 32 in yeast cytochrome *c*. J Biol Chem 263 : 18290 - 18297

Das G, Hicket DR, McLendon D, McLendon G, Sherman F (1989) Dramatic thermo-stabilization of yeast iso-1 cytochrome *c* by an asparagine to isoleucine replacement at position 57. Proc Natl Acad Sci USA 86 : 496 - 499

Davidson E, Daldal F (1987) Primary structure of the bc_1 complex of *Rhodopseudomonas capsulata*. Nucleotide sequence of the pet operon encoding the Rieske cytochrome *b* and cytochrome c_1 apoproteins. J Mol Biol 195 : 13 - 24

Davis DJ, Frame MK, Johnson DA (1988) Resonance Raman spectroscopy indicates a lysine as the sixth iron ligand in cytochrome *f*. Biochim Biophys Acta 936 : 61 - 66

Davis LA, Schejter A, Hess GP (1974) Alkaline isomerization of oxidized cytochrome *c* J Biol Chem 249 : 2624 - 2632

Dawson JW, Gray HB, Holwerda RA, Westhead EW (1972) Kinetics of the reduction of metalloproteins by chromous ion. Proc Natl Acad Sci USA 69 : 30 - 33

Dayhoff MO (1972) Atlas of Protein Sequence and Structure. Vol 5 : National Biomedical Research Foundation Silver Spring, Maryland

Dayhoff MO (1979) Atlas of Protein Sequence and Structure. Vol 5 Supplement 3 : National Biomedical Research Foundation Silver Spring, Maryland

Dayhoff MO, Schwartz RM (1981) Evidence on the origin of eukaryotic mitochondria from protein and nucleic acid sequences. Annals NY Acad Sci 361 : 92 - 102

DeFelippis MR, Murthy CP, Faraggi M, Klapper MH (1989) Pulse radiolytic measurement of redox potentials: the tyrosine and tryptophan radicals. Biochemistry 28 : 4847 - 4853

Deisenhofer J, Epp O, Miki K, Huber R, Michel H (1984) X-ray structure analysis of a membrane protein complex : electron density map at 3Å resolution and a model of the chromophores of the photosynthetic reaction center from *Rhodopseudomonas viridis*. J Mol Biol 180 : 385 - 398

Deisenhofer J, Epp O, Miki K, Huber R, Michel H (1985) Structure of the protein sub-units in the photosynthetic reaction centre of *Rhodopseudomonas viridis* at 3Å resolution. Nature (London) 318 : 618 - 624

DeKlerk H, Bartsch RG, Kamen MD (1965) Atypical soluble haem proteins from a strain of *Rhodopseudomonas palustris* sp. Biochim Biophys Acta 97 : 275 - 280

Delange RJ, Glazer AN, Smith EL (1969) Presence and location of an unusual amino acid, ε-N-trimethyllysine, in cytochrome *c* of wheat germ and *Neurospora*. J Biol Chem 244 : 1385 - 1388

Demoulin V (1979) Protein and nucleic acid sequence data and phylogeny. Science 205 : 1036 - 1038

DerVartanian DV, Le Gall J (1974) A monomolecular electron transfer chain : structure and function of cytochrome c_3. Biochim Biophys Acta 346 : 79 - 99

DerVartanian DV, Xavier AV, Le Gall J (1978) EPR determination of the oxidation-reduction potentials of the hemes in cytochrome c_3 from *Desulfovibrio vulgaris*. Biochimie 60 : 321 - 325

DeVault D (1980) Quantum mechanical tunnelling in biological systems. Quart Rev Biophys 13 : 387 - 564

DeVault D, Chance B (1966) Studies of photosynthesis using a pulsed laser. I. Temperature dependence of cytochrome oxidation rate in *Chromatium*. Evidence for tunneling. Biophys J 6 : 825 - 847

Dickerson RE (1971) The structure of cytochrome *c* and the rates of molecular evolution. J Molec Evol 1 : 26 - 45

Dickerson RE (1978) Structural conservatism in proteins over three billion years - cytochrome with a touch of collagen. In: Srinivasan R (ed) International Symposium on Biomolecular Structure Conformation, Function and Evolution. University of Madras, Pergamon Press, New York
Dickerson RE (1980a) Evolution and gene transfer in purple photosynthetic bacteria. Nature (London) 283 : 210 - 214
Dickerson RE (1980b) Cytochrome *c* and the evolution of energy metabolism. Sci Amer 242 : 136 - 153
Dickerson RE (1980c) The cytochromes *c*. An exercise in scientific serendipity. In : Sigman DS, Brazier MA (eds) The evolution of protein structure and function. Academic Press, London New York pp 174 - 202
Dickerson RE, Timkovich R (1975) Cytochromes *c*. In : Boyer PD (ed) The Enzymes, 3rd edn, Vol 11. Academic Press, London New York pp 397 - 547
Dickerson RE, Takano T, Eisenberg D, Kallai OB, Samson L, Cooper A, Margoliash E (1971) General features of the horse and bonito proteins at 2.8Å resolution. J Biol Chem 246 : 1511 - 1535
Dickerson RE, Takano T, Kallai OB, Samson L (1972) Ferricytochrome *c*: chain flexibility and a possible reduction mechanism. In Åkeson Å, Ehrenberg A (eds) The Structure and Function of Oxidation-Reduction Enzymes. Pergamon Press, Oxford pp 69 - 83
Dickinson LC, Chien JCW (1974) Cobalt cytochrome *c*: preparation and characterization. Biochem Biophys Res Commun 58 : 236 - 241
Dickinson JC, Chien JCW (1975a) Cobalt-cytochrome *c*: Magnetic resonance spectra and conformational transitions. Biochemistry 14 : 3534 - 3542
Dickinson JC, Chien JCW (1975b) Cobalt-cytochrome *c*: Preparation, properties and enzymic activity. Biochemistry 14 : 3526 - 3534
Dixit BPSN, Waring AJ, Wells III KO, Wong PS, Woodrow III GV, Vanderkooi JM (1982) Rotational motion of cytochrome *c* derivatives bound to membranes measured by fluorescence and phosphorescence anisotropy. Eur J Biochem 126 : 1 - 9
Dixon DW, Hong X, Woehler SE (1989) Electrostatic and steric control of electron self-exchange in cytochromes *c*, *c*-551 and b_5. Biophys J 56 : 339 - 351
Dobson CM, Hoyle NJ, Geraldes CF, Bruschi M, Le Gall J, Wright PE, Williams RJP (1974) Outline structure of cytochrome c_3 and consideration of its properties. Nature 249 : 425 - 429
Dobzhansky T, Ayala FJ, Stebbins GL, Valentine JW (1977) Evolution. Freeman, San Francisco
Dohler K, Huss VAR, Zumft WG (1987) Transfer of *Pseudomonas perfectomarinus* (Baumann, Bowditch, Baumann and Beaman 1983) to *Pseudomonas stutzeri* (Lehman and Neumann 1896) Sijderus 1946. Int J Syst Bacteriol 37 : 1 - 3
Dolphin D (1973 - 1979) (ed) The Porphyrins. Academic Press, New York London
Donohue TJ, McEwan AG, Kaplan S (1986) Cloning, DNA sequence and expression of the *Rhodobacter sphaeroides* cytochrome c_2 gene. J Bact 168 : 962 - 972
Doolittle WF, Bonen L (1981) Molecular sequence data indicating an endosymbiotic origin for plastids. Annals NY Acad Sci 361 : 248 - 256
Doyle ML, Gill SJ (1985) Stoichiometry determination for carbon monoxide binding to *R. molischianum* cytochrome *c'*. J Biol Chem 260 : 9534 - 9536
Doyle ML, Gill SJ, Cusanovich MA (1986) Ligand-controlled dissociation of *Chr. vinosum* cytochrome *c'*. Biochemistry 25 : 2509 - 2516

Dracheva SM, Drachev LA, Konstantinov AA, Semenov AY, Skulachev VP, Arutjunjan AM, Shuvalov VA, Zaberezhnaya SM (1988) Electrogenic steps in the redox reactions catalyzed by photosynthetic reaction-centre complex from *Rhodopseudomonas viridis*. Eur J Biochem 171 : 253 - 264

Drew HR, Dickerson RE (1978) The unfolding of the cytochromes *c* in methanol and acid. J Biol Chem 253 : 8420 - 8427

Drott HR, Lee CP, Yonetani T (1970) Spin label studies of hemoproteins: cytochrome *c*. J Biol Chem 245 : 5875 -5879

Drucker H, Campbell LL, Woody RW (1970) Optical rotatory properties of the cytochrome c_3 from three species of *Desulfovibrio*. Biochemistry 9 : 1519 - 1527

Drummond M (1979) Crown gall disease. Nature (London) 281 : 343 - 347

Dus K, Sletten K, Kamen MD (1968) Cytochrome c_2 of *Rhodospirillum rubrum*, complete amino acid sequence and phylogenetic relationships. J Biol Chem 243 : 5507 - 5518

Dus K, DeKlerk H, Bartsch RG, Horio T, Kamen MD (1967) On the monoheme nature of cytochrome *c'* (*Rhodopseudomonas palustris*) Proc Natl Acad Sci USA 57 : 367 - 370

Dutton PL (1978) Redox potentiometry determination of midpoint potentials of oxidation reduction components of biological electron transfer systems. Meth Enzymol 54 : 411 - 434

Dutton PL, Wilson DF (1974) Redox potentiometry in mitochondrial and photosynthetic bioenergetics. Biochim Biophys Acta 346 : 165 - 212

Dutton PL, Wilson DF, Lee CP (1970) Oxidation reduction potentials of cytochromes in mitochondria. Biochemistry 9 : 5077 - 5082

Dwivedi A, Toscano Jr WA, Debrunner PG (1979) Mössbauer studies of cytochrome *c*-551: intrinsic heterogeneity related to g-strain. Biochim Biophys Acta 576 : 502 - 508

Dyson HJ, Beattie JK (1982) Spin-state and unfolding equilibria of ferricytochrome *c* in acidic solutions. J Biol Chem 257 : 2267 - 2273

Eaton WA, Charney E (1969) Near-infrared absorption and circular dichroism spectra of ferrocytochrome *c* : dd transitions. J Chem Phys 51 : 4502 - 4505

Eaton WA, Hochstrasser RM (1967) Electronic spectrum of single crystals of ferricytochrome *c*. J Chem Phys 46 : 2533 - 2539

Eddowes MJ, Hill HAO (1981) Investigations of electron transfer reactions of proteins by electrochemical methods. Biosci Rep 1 : 521 - 532

Eden D, Matthew JB, Rosa JJ, Richards FM (1982) Increase in apparent compressibility of cytochrome *c* upon oxidation. Proc Natl Acad Sci USA 79 : 815 - 819

Ehrenberg A (1962) Some oxidative enzymes and related compounds with unpaired electrons. Svensk Kemisk tidskrift 74 : 52 - 68

Ehrenberg A, Theorell H (1955) On the stereochemical structure of cytochrome *c*. Acta Chem Scand 9 : 1193 - 1205

Ehrenberg A, Kamen MD (1965) Magnetic and optical properties of some bacterial haem proteins. Biochim Biophys Acta 102 : 333 - 340

Eisenberg D, Hill CP (1989) Protein crystallography: more surprises ahead. Trends Biochem Sci 14 : 260 - 264

Eley CGS, Moore GR (1983) ^{1}H-NMR investigation of the interaction between cytochrome *c* and cytochrome b_5. Biochem J 215 : 11 - 21

Eley CGS, Moore GR, Williams RJP, Neupert W, Boon PJ, Brinkoff HHK, Nivard RJF, Tesser GI (1982a) Structural role of the tyrosine residues of cytochrome *c*. Biochem J 205 : 153 - 165

Eley CGS, Moore GR, Williams G, Williams RJP (1982b) NMR studies of the electron exchange between cytochrome *c* and iron hexacyanides. Eur J Biochem 124 : 295 - 303

Eley CGS, Ragg E, Moore GR (1984) Kinetics of electron transfer between mitochondrial cytochrome *c* and iron hexacyanides. J Inorg Biochem 21 : 295 - 310

Elias H, Chou MH, Winkler JR (1988) Electron-transfer kinetics of Zn-substituted cytochrome *c* and its $Ru(NH_3)_5$ (His-33) derivative. J Am Chem Soc 110 : 429 - 434

Ellis WR, Wang H, Blair DF, Gray HB, Chan SI (1986) Spectroelectrochemical study of the cytochrome *a* site in carbon monoxide inhibited cytochrome *c* oxidase. Biochemistry 25 : 161 - 167

Emptage MH (1978) Physical studies of *c*-type cytochromes. Ph.D. Thesis, University of Illinois

Emptage MH, Zimmerman R, Que L, Münck E, Hamilton WD, Orme-Johnson WH (1977) Mössbauer studies of cytochrome *c'* from *R.rubrum*. Biochim Biophys Acta 495 : 12 - 23

Emptage MH, Xavier AV, Wood JM, Alsaadi BM, Moore GR, Pitt RC, Williams RJP, Ambler RP, Bartsch RG (1981) NMR studies of *R.rubrum* cytochrome *c'*. Biochemistry 20 : 58 - 64

Endo S, Nagayama K, Wada A, (1985) Comparison of protease susceptibility and thermal stability of cytochromes *c*. J Biomol Structure & Dynamics 3 : 409 - 421

Englander SW, Wand AJ (1987) Main chain directed strategy for the assignment of ^{1}H NMR spectra of proteins. Biochemistry 26 : 5953 - 5958

English DR, Hendrickson DN, Suslick KS, Eigenbrot Jr CW, Scheidt WR Low-spin five-coordinate ferric porphyrin complex : [5,10,15,20-Tetrakis (4-methoxyphenyl) porphyrinato]-(hydrosulfido)iron(III). J Am Chem Soc 106 : 7258 - 7259

Erecinska M (1975) Cytochrome *c* interaction with membranes: formylated cytochrome *c*. Arch Biochem Biophys 169 : 199 - 208

Erecinska M, Vanderkooi JM (1978) Modification of cytochrome *c*. Meth Enzymol LIII.D : 165 - 181

Erecinska M, Wilson DF, Blasie JK (1978) Studies on the orientations of the mitochondrial redox carriers. I. Orientation of the hemes of cytochrome *c* oxidase with respect to the plane of a cytochrome oxidase-lipid model membrane. Biochim Biophys Acta 501 : 53 - 62

Erman JE, Vitello LB (1980) The binding of cytochrome *c* peroxidase and ferricytochrome *c*. J Biol Chem 255 : 6224 - 6227

Ernst JF, Stewart JW, Sherman F (1981) The cyc1-11 mutation in yeast reverts by recombination with a non-allelic gene - composite genes determining the isocytochromes *c*. Proc Natl Acad Sci USA 78 : 6334 - 6338

Ernst JF, Hampsey DM, Stewart JW, Rackovosky S, Goldstein D, Sherman F (1985) Substitutions of proline 76 in yeast iso-1 cytochrome *c*. J Biol Chem 260 : 13225 - 13236

Fabry ME, Eisenstadt M (1971) The mechanism of water proton nuclear magnetic resonance relaxation in the presence of mammalian and *Aplysia* metmyoglobin fluoride. J Biol Chem 249 : 2915 - 2919

Falk JE (1964) Porphyrins and Metalloporphyrins. Elsevier

Falk KE, Jovall PA, Ångström J (1981) NMR and EPR characterisation of 4-carboxy-2,6-dinitrophenyl lysine cytochromes *c*. Biochem J 193 : 1021 - 1024
Fanger MW, Harbury HA (1965) Trifluoroacetylated cytochrome *c*. Biochemistry 4 : 2541 - 2545
Feher G, Allen JP, Okamura MY, Rees DC (1989) Structure and function of bacterial photosynthetic reaction centres. Nature (London) 339 : 111 - 116
Feng Y, Roder H, Englander SW, Wand AJ, Stefano DLD (1989) Proton resonance assignments of horse ferricytochrome *c* . Biochemistry 28 : 195 - 203
Ferguson-Miller S, Brautigan DL, Margoliash E (1978) Definition of cytochrome *c* binding domains by chemical modification. J Biol Chem 253 : 149 - 159
Ferguson-Miller S, Brautigan DL, Margoliash E (1979) The electron transfer function of cytochrome *c*. In: Dolphin D (ed) The Porphyrins vol 7. Academic Press, London New York pp 149 - 240
Fermi G, Perutz MF (1981) Atlas of Molecular Structures in Biology. 2: Haemoglobin and Myoglobin. Clarendon Press, Oxford
Fernandez-Sousa JM, Gavilanes JG, Municio AM, Paredes JA, Perez-Aranda A, Rodriguez R (1975) Primary structure of cytochrome *c* from the insect *Ceratitis capitata*. Biochim Biophys Acta 393 : 358 - 367
Fiechtner MD, Kassner RJ (1978) Axial ligation and heme environment in cytochrome *c*-555 from *Prosthecochloris aestuarii*. Investigation by absorption and solvent perturbation difference spectroscopy. Biochemistry 17 : 1028 - 1031
Fiechtner MD, Kassner RJ (1979) The redox properties and heme environment of cytochrome *c*-551.5 from *Desulfuromonas acetoxidans*. Biochim Biophys Acta 545 : 424 - 428
Finzel BC, Poulos TL, Kraut J (1984) Crystal structure of yeast cytochrome *c* peroxidase refined at 1.7Å resolution. J Biol Chem 259 : 13027 - 13036
Finzel BC, Weber PC, Hardman KD, Salemme FR (1985) Structure of ferricytochrome *c'* from *R. molischianum* at 1.67Å resolution. J Mol Biol. 186 : 627 - 643
Fisher MT, Sligar SG (1985) Control of heme protein redox potential and reduction rate : linear free energy relation between potential and ferric spin state equilibrium. J Am Chem Soc 107 ; 5018 - 5019
Fisher WR, Taniuchi H, Anfinsen CB (1973) On the role of heme in the formation of the structure of cytochrome *c*. J Biol Chem 248 : 3188 - 3195
Fita I, Rossmann MG (1985) The active center of catalase. J Mol Biol 186 : 21 - 37
Fitch WM (1971) Towards defining the course of evolution: minimum change for a specific tree topology. Syst Zool 20 : 406 - 416
Fitch WM (1976) The molecular evolution of cytochrome *c* in eukaryotes. J Molec Evol 8 : 13 - 40
Fitch WM, Langley CH (1976) Protein evolution and the molecular clock. Fed Proc 35 : 2092 - 2097
Fitch WM, Margoliash E (1967) Construction of phylogenetic trees. Science 155 : 279 - 284
Fitch WM, Margoliash E (1969) The construction of phylogenetic trees. How well do they reflect past history? In: Structure, Function and Evolution of Proteins. Brookhaven Symposia in Biology. Brookhaven National Laboratory, Upton NY Vol 21 : pp 217 - 242
Fitch WM, Markowitz E (1970) An improved method for determining codon variability in a gene and its application to the rate of fixation of mutations in evolution. Biochem Genet 4 : 579 - 593

Flatmark T, Robinson AB (1968) Circular dichroic absorption spectra of cytochrome *c* from bovine heart and cytochrome c_2 from *Rhodospirillum rubrum*. In : Okunuki K, Kamen MD, Sekuzu I (eds) Structure and function of cytochromes. Univ Park Press, Baltimore pp 318 - 327

Foote N, Peterson J, Gadsby PMA, Greenwood C, Thomson AJ (1984) A study of the oxidized form of *Pseudomonas aeruginosa* cytochrome *c* peroxidase with the use of magnetic circular dichroism. Biochem J 223 : 369 - 378

Foote N, Gadsby PMA, Greenwood C, Thomson AJ (1989) pH-dependent forms of the ferryl haem in myoglobin peroxide analysed by variable-temperature magnetic circular dichroism. Biochem J 261 : 515 - 522

Fox GE, Stackebrándt E, Hespell BB, Gibson J, Manikoff J, Dyer TA, Wolfe RS, Balch WE, Tanner RS, Magrum LJ, Zablen LB, Blakemore R, Gupta R, Bonen L, Lewis BJ, Stahl DA, Luehrsen KR, Chen KN, Woese CR (1980) The phylogeny of prokaryotes. Science 209 : 457 - 463

Fritzsch G, Buchanan S, Michel H (1989) Assignment of cytochrome hemes in crystallized reaction centers from *Rhodopseudomonas viridis*. Biochim Biophys Acta 977 : 157 - 162

Gabellini N, Sebald W (1986) Nucleotide sequence and transcription of the fbc operon from *Rhodopseudomonas sphaeroides*. Eur J Biochem 154 : 569 - 579

Gadsby PMA, Thomson AJ (1982) Identification of the imidazolate anion as a ligand in met-myoglobin by near-infrared magnetic circular dichroism spectroscopy. FEBS Lett 150 : 59 - 63

Gadsby PMA, Thomson AJ (1986) Low-temperature EPR and near-infrared MCD studies of highly anisotropic low-spin ferrihaem complexes. FEBS Lett 197 : 253 - 257

Gadsby PMA, Thomson AJ (1990) Assignment of the axial ligands of ferric iron in low-spin haemoproteins by near-infrared MCD and EPR spectroscopy. J Am Chem Soc *in press*

Gadsby PMA, Peterson J, Foote N, Greenwood C, Thomson AJ (1987) Identification of the ligand exchange process in the alkaline transition of horse heart cytochrome *c*. Biochem J 246 : 43 - 54

Gadsby PMA, Hartshorn RT, Moura JJG, Sinclair-Day JD, Sykes AG, Thomson AJ (1989) Redox properties of the diheme cytochrome c_4 and characterisation of the two hemes by NMR, MCD and EPR spectroscopy. Biochim Biophys Acta 994 : 37 - 46

Gao Y, Lee ADJ, Williams RJP, Williams G (1989) The effects of multiple amino acid substitutions on the polypeptide backbone of tuna and horse cytochromes *c* . Eur J Biochem 182 : 57 - 65

Gast P, Wasielewski MR, Schiffer M, Norris JR (1983) Orientation of the primary donor in single crystals of *Rhodopseudomonas viridis* reaction centres. Nature (London) 305 : 451 - 452

Gayda JP, Bertrand P, More C, Guerlesquin F, Bruschi M (1985) EPR potentiometric titration of c_3-type cytochromes . Biochim Biophys Acta 829 : 262 - 267

Gayda JP, Benosman H, Bertrand P, More C, Asso M (1988) EPR determination of interaction redox potentials in a multiheme cytochrome: cytochrome c_3 from *Desulfovibrio desulfuricans* Norway. Eur J Biochem 177 : 199 - 206

Gelles J, Blair DF, Chan SI (1987) The proton pumping site of the cytochrome *c* oxidase: a model of its structure and mechanism. Biochim Biophys Acta 853 : 205 - 236

George P, Beetlestone J, Griffith JS (1961) Ferrihaemoprotein hydroxides : a correlation between magnetic and spectroscopic properties. In: Falk JE, Lemberg R, Morton RK (eds) Haematin enzymes, IUB symposium series 19 : 105 - 141

George P, Hanania GIH, Irvine DH, Abu-Issa I (1964) The effect of coordination on ionization. Part IV : Imidazole and its ferrimyoglobin complex. J Chem Soc 5689 - 5694

George P, Hanania GIH, Eaton WA (1966) Effect of electrostatic environment on redox potentials. In: Chance B, Estabrook R, Yonetani T (eds) Hemes and Hemoproteins. Academic Press, London New York pp 267 - 270

Geren LM, Stonehuerner J, Davies DJ, Millet F (1983) The use of a water soluble carbodiimide to cross-link cytochrome *c* to plastocyanin. Biochim Biophys Acta 724 : 62 - 68

Gerwert K, Hess B, Michel H, Buchanan S (1988) FT-IR studies on crystals of photosynthetic reaction centers. FEBS Lett 232 : 303 - 307

Geselowitz D, Taube H (1982) Reexamination of the $Co(NH_3)_6^{3+/2+}$ self-exchange reaction. Adv Inorg Bioinorg Mech 1 : 391 - 407

Gest H (1980) The evolution of biological energy transducing systems. FEMS Lett 7 : 73 - 77

Gibson JF, Ingram DJE (1957) Binding in haemoglobin azide as determined by electron resonance. Nature (London) 180 : 29 - 30

Gibson QH, Ainsworth S (1957) Photosensitivity of haem compounds. Nature (London) 180 : 1416 - 1417

Gibson QH, Kamen MD (1966) Kinetic analysis of the reaction of cytochrome *cc'* with carbon monoxide. J Biol Chem 241 :1969 - 1976

Gilson MK, Honig B (1989) Destabilization of an α-helix-bundle protein by helix dipoles. Proc Natl Acad Sci USA 86 : 1524 - 1528

Goldkorn T, Schejter A (1976) The redox potential of cytochrome *c*-552 from *Euglena gracilis* - a thermodynamic study. Arch Biochem Biophys 177 : 39 - 45

Goldstein RF, Bearden A (1984) Tunneling in *Chromatium* chromatophores: detection of a Hopfield charge-transfer band. Proc Natl Acad Sci USA 81 : 135 - 139

Goldstone A, Smith EL, (1966) Amino acid sequence of whale heart cytochrome *c*. J Biol Chem 241 : 4480 - 4486

Goldstone A, Smith EL, (1967) Amino acid sequence of the cytochrome *c* from the dogfish, *Squalus sucklii*. J Biol Chem 242 : 4702 - 4710

Gouterman M (1978) In : Dolphin D (ed) The Porphyrins vol 3A Academic Press, London New York pp 1 - 165

Gray JC (1978) Purification and properties of monomeric cytochrome *f* from charlock, *Sinapis arvensis* L. Eur J Biochem 82 : 133 - 141

Gray MW, Doolittle WF (1982) Has the endosymbiont hypothesis been proven? Microbiol Reviews 46 : 1 - 142

Greenwood C, Palmer G (1965) Evidence for the existence of two functionally distinct forms of cytochrome *c* monomer at alkaline pH. J Biol Chem 240 : 3660 - 3663

Greenwood C, Wilson MT (1971) Effect of pH, ionic strength and protein denaturants on the spectra of ferricytochrome *c*. Eur J Biochem 22 : 5 - 10

Greenwood C, Foote N, Peterson J, Thomson AJ (1984) The nature of species prepared by photolysis of half-reduced, fully reduced and fully reduced carbonmonoxy-cytochrome *c*-551 peroxidase from *Pseudomonas aeruginosa*. Biochem J 223 : 379 - 391

Greenwood C, Foote N, Gadsby PMA, Thomson AJ (1988) A di-haem cytochrome *c* peroxidase (*Pseudomonas aeruginosa*) : its activation and catalytic cycle. Chemica Scripta 28A : 79 - 84

Gribbin J (1984) In search of Schrödinger's cat. Corgi, London

Griffith JS (1957) Theory of electron resonance in ferrihaemoglobin azide. Nature (London) 180 : 30 - 31

Griffith JS (1961) The Theory of Transition Metal Ions. Cambridge University Press, Cambridge

Grimes CJ, Piszkiewicz D, Fleischer EB (1974) Electron transfer reactions in biological systems : The reduction of ferricytochrome *c* by chromous ions. Proc Natl Acad Sci USA 71 : 1408 - 1412

Guerlesquin F, Bovier-Lapierre G, Bruschi M (1982) Purification and characterisation of cytochrome c_3 (M_r 26000) isolated from *D. desulfuricans* (Norway). Biochem Biophys Res Commun. 105 : 530 - 538

Guerlesquin F, Noailly M, Bruschi M (1985) Preliminary ^{1}H-NMR studies of the interaction between cytochrome c_3 and ferredoxin I from *Desulfovibrio desulfuricans* Norway. Biochem Biophys Res Commun 130 : 1102 - 1108

Guiard B, Lederer F (1979) The cytochrome b_5 fold - structure of a novel protein superfamily. J Mol Biol 135 : 639 - 650

Gunner MR, Dutton PL (1989) Temperature and -$\Delta G°$ dependence of the electron transfer from $BPh^{\cdot -}$ to Qa_A in reaction centre protein from *Rhodobacter sphaeroides* with different quinones as Qa_A. J Am Chem Soc 111 : 3400 - 3412

Gupta RK (1973) Electron transfer in cytochrome *c*. Role of the polypeptide chain. Biochim Biophys Acta 292 : 291 - 295

Gupta RK, Koenig SH (1971) Some aspects of pH and temperature dependence of the NMR spectra of cytochrome *c*. Biochem Biophys Res Commun 45 : 1134 - 1143

Gupta RK, Yonetani T (1973) Nuclear magnetic resonance studies of the interaction of cytochrome *c* with cytochrome *c* peroxidase. Biochim Biophys Acta 292 : 502 - 508

Gupta RK, Koenig SH, Redfield AG (1972) On the electron transfer between cytochrome *c* molecules as observed by nuclear magnetic resonance. J Mag Res 7 : 66 - 73

Gürtler L, Horstmann HJ (1970) The aminosäure-sequenz vom cytochrom *c* des karpfens, *Cyprinus carpio*. Eur J Biochem 12 : 48 - 57

Gürtler L, Horstmann HJ (1971) Zur primärstruktur des cytochromes *c* des steppenzebras (*Equus quagga boehmi*). FEBS Lett 18 : 106 - 108

Hampsey DM, Das G, Sherman F (1986) Amino acid replacements in yeast iso-1 cytochrome *c*. J Biol Chem 261 : 3259 - 3271

Hampsey DM, Das G, Sherman F (1988) Yeast iso-1 cytochrome *c* genetic analysis of structural requirements. FEBS Lett 231 : 275 - 283

Hantgan RR, Taniuchi H (1977) Formation of a biologically active ordered complex from two overlapping fragments of cytochrome *c*. J Biol Chem 252 : 1367 - 1374

Hantgan RR, Taniuchi H (1978) Conformational dynamics in cytochrome *c*. J Biol Chem 253 : 5373 - 5380

Harami T, Maeda Y, Morita Y, Trautwein A, Gonser U (1977) Mössbauer spectroscopic determination of the electronic structure of highly oxidized iron in hemoproteins. J Chem Phys 67 : 1164 - 1169

Harbury HA (1978) Cytochrome *c* reconstitution, formation of hybrids and semisynthesis. In: Offord RE, DiBello C (eds) Semisynthetic Peptides and Proteins. Academic Press, London New York pp 74 - 88

Harbury HA, Loach PA (1960) Oxidation linked proton functions in heme octa- and undecapeptides from mammalian cytochrome *c*. J Biol Chem 235 : 3640 - 3645

Harbury HA, Marks RHL (1973) Cytochromes *b* and *c*. In: Eichorn GL (ed) Inorganic Biochemistry vol 2. Elsevier, Amsterdam pp 902 - 954

Harbury HA, Cronin JR, Fanger MW, Hettinger TP, Murphy AJ, Myer YP, Vinogradov SN (1965) Complex formation between methionine and a heme peptide from cytochrome *c*. Proc Natl Acad Sci USA 54 : 1658 - 1664

Harris DE (1979) Semisynthetic analogues of cytochrome *c* prepared by the noncovalent association of modified tryptic fragments. In: Gross E, Meienhofer J (eds) Peptides Structure and Biological Function Proc 6th Amer Peptide Symp Pierce Chemical Co pp 613 - 616

Harris DE, Offord RE (1977) A functioning complex between tryptic fragments of cytochrome *c*. Biochem J 161 : 21 - 25

Hartshorn RT, Moore GR (1989) A denaturation-induced proton-uptake study of horse ferricytochrome *c*. Biochem J 258 : 595 - 598

Hartshorn RT, Mauk AG, Mauk MR, Moore GR (1987) NMR study of the interaction between cytochrome b_5 and cytochrome *c*. FEBS Lett 213 : 391 - 395

Haser R (1981) Structural approach of the electron pathways in the multiheme cytochrome c_3. Biochimie 63 : 945 - 949

Haser R, Pierrot M, Frey M, Payan F, Astier JP, Bruschi M, Le Gall J (1979a) Structure and sequence of the multihaem cytochrome c_3. Nature (London) 282 : 806 - 810

Haser R, Payan F, Bache R, Bruschi M, Le Gall J (1979b) Crystallisation and preliminary crystallographic data for cytochrome *c*-551.5 from *Desulfuromonas acetoxidans*. J Mol Biol 130 : 97 - 98

Hasumi H (1980) Kinetic studies on isomerization of ferricytochromes *c* in alkaline and acid pH ranges by the circular dichroism stopped-flow method. Biochim Biophys Acta 626 : 265 - 276

Hazzard JT, Tollin G (1985) Proton NMR study of the cytochrome *c* flavodoxin electron transfer complex. Biochem Biophys Res Commun 130 : 1281 - 1286

Hazzard JT, Poulos TL, Tollin G (1987) Kinetics of reduction by free flavin semiquinones of the components of the cytochrome *c* - cytochrome *c* peroxidase complex and intra-complex electron transfer. Biochemistry 26 : 2836 - 2848

Heller J, Smith EL (1966) *Neurospora crassa* cytochrome *c*. J Biol Chem 241 : 3165 - 3180

Henderson RW, Nankiville DD (1966) Electrophoretic and other studies on haem pigments from *Rhodopseudomonas palustris*. Biochem J 98 : 587 - 593

Hendrich MP, Debrunner PG (1988) EPR spectra of quintet ferrous myoglobin and a model heme compound. J Mag Res 78 : 133 - 141

Hennessey JP, Johnson WC (1981) Information content in the circular dichroism of proteins. Biochemistry 20 : 1085 - 1094

Hennig B (1975) Change of cytochrome *c* structure during development of the mouse. Eur J Biochem 55 : 167 - 183

Hettinger TP, Harbury HA (1964) Guanidated cytochrome *c*. Proc Natl Acad Sci USA 52 : 1469 - 1476

Higuchi Y, Bando S, Kusunoki M, Matsuura Y, Yasuoka N, Kakudo M, Yamanaka T, Yagi T, Inokuchi H (1981a) The structure of cytochrome c_3 from *D. vulgaris* Miyazaki at 2.5Å resolution. J Biochem 89 : 1659 - 1662

Higuchi Y, Kusunoki M, Yasuoka N, Kakudo M, Yagi T (1981b) On cytochrome c_3 folding. J Biochem 90 : 1715 - 1723

Higuchi Y, Kusunoki M, Matsuura Y, Yasuoka N, Kakudo M (1984) Refined structure of cytochrome c_3 at 1.8Å resolution. J Mol Biol 172 : 109 - 139

Hill GC, Pettigrew GW (1975) Evidence for the amino acid sequence of *Crithidia fasciculata* cytochrome *c*-555. Eur J Biochem 57 : 265 - 271

Ho C, Russu IM (1987) How much do we know about the Bohr effect of hemoglobin? Biochemistry 26 : 6299 - 6305

Ho PS, Sutoris C, Liang N, Margoliash E, Hoffman BM (1985) Species specificity of long-range electron transfer within the complex between zinc-substituted cytochrome *c* peroxidase and cytochrome *c*. J Am Chem Soc 107 : 1070 - 1071

Hoard JL (1971) Stereochemistry of hemes and other metalloporphyrins. Science 174 : 1295 - 1302

Hochman JH, Schindler M, Lee JG, Ferguson-Miller S (1982) Lateral mobility of cytochrome *c* on intact mitochondrial membranes as determined by fluorescence redistribution after photobleaching. Proc Natl Acad Sci USA 79 : 6866 - 6870

Holloway PW, Mantsch HH (1988) Infrared spectroscopic analysis of salt bridge formation between cytochrome b_5 and cytochrome *c*. Biochemistry 27 : 7991 - 7993

Holmquist B (1978) The magnetic optical activity of hemoproteins. In: Dolphin D (ed) The Porphyrins vol 3A Academic Press, London New York pp 249 - 270

Holmquist R (1972) Empirical support for a stochastic model of evolution J Molec Evol 1 : 211 - 222

Holmquist R, Jukes TH (1975) Species specific effects and the evolutionary clock. J Molec Evol 4 : 377 - 381

Holmquist R, Cantor C, Jukes T (1972) Improved procedures for comparing homologous sequences in molecules of proteins and nucleic acids. J Mol Biol 64 : 145 - 161

Holzschu D, Principio L, Concklin KT, Hickey DR, Short J, Rao R, McLendon G, Sherman F (1987) Replacement of the invariant lysine 77 by arginine in yeast iso-1-cytochrome *c* results in enhanced and normal activities *in vitro* and *in vivo*. J Biol Chem 262 : 7125 - 7131

Homer RB, Johnson CD (1970) Acid-base and complexing properties of amides. In : Zabicky J (ed) The chemistry of amides. Interscience Publishers, London New York pp 187 - 243

Hong X, Dixon DW (1989) NMR study of the alkaline isomerization of ferricytochrome *c*. FEBS Lett 246 : 105 - 108

Hopfield JJ (1974) Electron transfer between biological molecules by thermally activated tunneling. Proc Natl Acad Sci USA 71 : 3640 - 3644

Hopfield JJ (1977) Photo-induced charge transfer: A critical test of the mechanism and range of biological electron transfer processes. Biophys J 18 : 311 - 321

Hori H, Morimoto H (1970) Haem-plane orientation in single crystal of bonito cytochrome *c*. Biochim Biophys Acta 200 : 581 - 583

Horio T, Higashi T, Sasagawa M, Kusai K, Nakai M, Okunuki K (1960) Crystalline *Pseudomonas* cytochrome *c*-551. Biochem J 77 : 194 - 201

Horio T, Kamen MD (1961) Preparation and properties of three pure crystalline bacterial haem proteins. Biochim Biophys Acta 48 : 266 - 286

Howard NL, Joubert FJ, Strydom DJ (1974) The amino acid sequence of ostrich (*Struthlo camelus*) cytochrome *c*. Comp Biochem Physiol 48B : 75 - 85

Hubbard JAM, Evans MCW (1989) Electron donation by the high-potential haems in *Rhodopseudomonas viridis* reaction centres at low temperatures. FEBS Lett 244 : 71 - 75

Hvidt A, Nielsen SO (1966) Hydrogen exchange in proteins. Adv Prot Chem 21 : 287 - 386

Imai Y, Imai K, Sato R, Horio T (1969a) Three spectrally different states of cytochromes *cc'* and *c'* and their interconversion. J Biochem 65 : 225 - 237

Imai Y, Imai K, Ikeda K, Hamaguchi K, Horio T (1969b) Circular Dichroism of cytochrome *cc'* and cytochrome *c'*. J Biochem 65 : 629 - 637

Imhoff JF, Truper HG, Pfennig N (1984) Rearrangement of the species and genera of the photosynthetic purple nonsulfur bacteria. Int J Syst Bact 34 : 340 - 343

Imoto T, Fukuda K, Yagashita K (1974) A study of the native $\rightleftharpoons$ denatured transition in lysozyme. Biochim Biophys Acta 336 : 264 - 269

Inoue S, Matsubara H, Yamanaka T (1985) Complete amino acid sequence of cytochrome *c* from the Honeybee *Apis mellifera* and evolutionary relationship of the honeybee to other insects on the basis of the amino acid sequence. J Biochem 97 : 947 - 954

Jackson JT, La Mar GN, Bartsch RG (1983) Proton NMR studies of the ligation states of the monomeric ferricytochrome *c'* from *Rps. palustris*. J Biol Chem 258 : 1799 - 1805

Jacobs EE, Sanadi DA (1960) The reversible removal of cytochrome *c* from mitochondria. J Biol Chem 235 : 531 - 534

Jemmerson R, Margoliash E (1979) Topographic antigenic determinants on cytochrome *c*. J Biol Chem 254 : 12706 - 12716

Jemmerson R, Morrow PR, Klinman NR, Paterson Y (1985) Analysis of an evolutionarily conserved antigenic site on mammalian cytochrome *c* using synthetic peptides. Proc Natl Acad Sci USA 82 : 1508 - 1512

John P, Whatley FR (1975) *Paracoccus denitrificans* and the evolutionary origin of the mitochondrion. Nature (London) 254 : 495 - 498

Juillerat M, Parr GR, Taniuchi H (1980) A biologically active, three fragment complex of horse heart cytochrome *c*. J Biol Chem 255 : 845 - 853

Kägi JHR, Ulmer DD (1968) Hydrogen-deuterium exchange of cytochrome *c*: Effect of pH. Biochemistry 7 : 2718 - 2724

Kakuno T, Bartsch RG, Nishikawa K, Horio T (1971) Redox components associated with chromatophores of *Rhodospirillum rubrum*. J Biochem 70 : 79 - 94

Kalkinnen N, Ellfolk N (1978) 11th IUPAC Int Symp Chem Nat Prod pp 79 - 82

Kaminsky LS, Henderson JJ, Ivanetich KM (1973a) Fluorescence studies of a cytochrome *c* mixed phospholipid complex. Biochem Biophys Res Commun 51 : 40 - 45

Kaminsky LS, Miller VJ, Davison AJ (1973b) Thermodynamic studies of the opening of the heme crevice of ferricytochrome *c*. Biochemistry 12 : 2215 - 2221

Kaminsky LS, Chiang YL, King TE (1975) Some properties of mammalian cardiac cytochrome c_1. J Biol Chem 250 : 7280 - 7287

Kassner RJ (1972) Effects of nonpolar environments on the redox potentials of heme complexes. Proc Natl Acad Sci USA 69 : 2263 - 2267

Kassner RJ (1973) A theoretical model for the effects of local nonpolar heme environments on the redox potentials in cytochromes. J Am Chem Soc 95 : 2674 - 2677

Kassner RJ, Kykta MG (1985) Anion binding to a cytochrome *c'*. Rev Port Quim 27 : 206 - 207Keller RM, Wüthrich K (1976) Structural studies of cytochrome *c*-551 by ^{1}H-NMR spectroscopy at 360MHz. FEBS Lett 70 : 180 - 184

Keller RM, Wüthrich K (1976) Structural studies of cytochrome *c*-551 by ^{1}H-NMR spectroscopy at 360MHz. FEBS Lett 70 : 180 - 184

Keller RM, Wüthrich K (1977) ^{1}H-NMR studies at 360 MHz of the aromatic amino acid residues in ferrocytochrome *c*-552 from *Euglena gracilis*. Biochim Biophys Acta 491 : 416 - 422

Keller RM, Wüthrich K (1978) Evolutionary change of the heme-c electronic structure: ferricytochrome *c*-551 from *Pseudomonas aeruginosa* and horse heart cytochrome *c*. Biochem Biophys Res Commun 83 : 1132 - 1139

Keller RM, Wüthrich K (1981) ^{1}H-NMR studies of structural homologies between the heme environments in horse cytochrome *c* and in cytochrome *c*-552 from *Euglena gracilis*. Biochim Biophys Acta 668 : 307 - 320

Kihara H, Saigo S, Nakatani H, Hiromi K, Ikeda-Saito M, Iizuka T (1976) Kinetic study of isomerisation of ferricytochrome *c* at alkaline pH. Biochim Biophys Acta 430 : 225 - 243

Kim C-S, Kueppers F, Dimaria P, Farooqui J, Kim S, Paik WK (1980) Enzymatic trimethylation of residue-72 lysine in cytochrome *c*. Biochim Biophys Acta 622 : 144 - 150

Kimelberg HK, Lee CP (1970) Interactions of cytochrome *c* with phospholipid membranes. II - reactivity of cytochrome *c* bound to phospholipid liquid crystals. J Memb Biol 2 : 252 - 262

Kimura M (1969) The rate of molecular evolution considered from the standpoint of population genetics. Proc Natl Acad Sci USA 63 : 1181 - 1188

Kimura M (1987) Molecular evolutionary clock and neutral theory. J Molec Evol 26 : 24 - 33

King GC, Binstead RA, Wright PE (1985) NMR and kinetic characterization of the interaction between French bean plastocyanin and horse cytochrome *c*. Biochim Biophys Acta 806 : 262 - 271

King JL, Jukes TH (1969) Non-Darwinian evolution. Science 164 : 788 - 798

Kitigawa T, Ozaki Y, Kyogoku Y, Horio T (1977a) Resonance Raman study of the pH-dependent and detergent-induced alterations in the heme moiety of *R. rubrum* cytochrome *c'*. Biochim Biophys Acta 495 : 1 - 11

Kitigawa T, Ozaki Y, Teraoka J, Kyogoku Y, Yamanaka T (1977b) The pH dependence of the resonance Raman spectra and structural alterations at heme moieties of various *c*-type cytochromes. Biochim Biophys Acta 494 : 100 - 114

Knapp JA, Pace CN (1974) Guanidine HCl and acid denaturation of horse, cow and *Candida krusei* cytochromes *c*. Biochemistry 13 : 1289 - 1294

Ko K, Strauss NA (1987) Sequence of the apocytochrome *f* gene encoded by the *Vicia fabia* chloroplast genome. Nucl Acids Res 15 : 2391 - 2396

Kobayashi N, Nozawa T, Hatano M (1981) Near-infrared magnetic circular dichroism studies on iron(III) horse heart cytochrome *c*. Bull Chem Soc Jpn 54 : 919 - 920

Koeppe RE, Stroud RM (1976) Mechanism of hydrolysis by serine proteases: Direct determination of the pK's of asp-102 and asp-194 in bovine trypsin using difference infrared spectroscopy. Biochemistry 15 : 3450 - 3458

Koloczek H, Horie T, Yonetani Y, Anni H, Maniara G, Vanderkooi JM (1987) Interaction between cytochrome *c* and cytochrome *c* peroxidase: excited-state reactions of zinc- and tin-substituted derivatives. Biochemistry 26 : 3142 - 3148

Kon H (1969) Electron paramagnetic resonance of nitric oxide cytochrome *c*. Biochem Biophys Res Commun 35 : 423 - 427

Kon H (1975) An interpretation of the three line EPR spectrum of nitric oxide hemeproteins and related model systems : The effect of the heme environment. Biochim Biophys Acta 379 : 103 - 113

Koppenol WH, Margoliash E (1982) The asymmetric distribution of charges on the surface of horse cytochrome *c*. J Biol Chem 257 : 4426 - 4437

Korszun ZR, Salemme FR (1977) Structure of cytochrome *c*-555 of *Chlorobium thiosulfatophilum*: Primitive low-potential cytochrome *c*. Proc Natl Acad Sci USA 74 ; 5244 - 5247

Korszun ZR, Moffat K, Frank K, Cusanovich MA (1982) EXAFS on cytochrome *c*: structural aspects of oxidation reduction. Biochemistry 21 : 2253 - 2258

Koul AK, Nasserman GF, Warme PK (1979) Semi-synthetic analogues of cytochrome *c* at positions 67 and 74. Biochem Biophys Res Commun 89 : 1253 - 1259

Kowalsky A (1965) Nuclear magnetic resonance studies of cytochrome *c*. Possible electron delocalization. Biochemistry 4 : 2382 - 2388

Kowalsky A (1969) A study of the mechanism of electron transfer in cytochrome *c*. J Biol Chem 244 : 6619 - 6625

Krab K, Wikström M (1978) Proton translocating cytochrome *c* oxidase in phospholipid vesicles. Biochim Biophys Acta 504 : 200 - 214

Kreil G (1963) Über die Artspezifität con cytochrome *c*. Vergleich der Aminosäuresequenz des Thunfisch-Cytochroms *c* mit der des Pferde-cytochroms *c*. Z Physiol Chemie 334 : 154 - 166

Kreil G (1965) Die *c*-terminale Aminosäuresequenz des Thunfisch-Cytochroms *c*. Z Physiol Chemie 340 : 86 - 87

Kreishman GP, Anderson CW, Su CH, Halsall HB, Heineman WR (1978) The effect of bulk solvent structure on the temperature dependence of the redox potential of cytochrome *c*. Bioelect Bioenerg 5 : 196 - 203

Krishtalik LI (1985) The negative cooperativity in cytochrome *c* oxidase redox reactions : The electrostatic effect. Arch Biochem Biophys 243 : 701 - 702

Kuki A, Wolynes PG (1987) Electron tunneling paths in proteins. Science 236 : 1647 - 1652

Kurowski B, Ludwig B (1987) The genes of the *Paracoccus denitrificans* bc_1 complex. J Biol Chem 262 : 13805 - 13811

La Mar GN, Frye JS, Satterlee JD (1976) Proton NMR study of coordinated imidazoles in low-spin ferric heme complexes. Biochim Biophys Acta 428 : 78 - 90

La Mar GN, Jackson JT, Bartsch RG (1981) Analysis of field-dependent relaxation data and hyperfine shifts of cytochrome *c'* from *R. rubrum* in terms of the high-spin iron ligation state. J Am Chem Soc 103 : 4405 - 4410

Labhardt A, Yuen C (1979) X-ray absorption edge fine structure spectroscopy of the active site haem of cytochrome *c*. Nature (London) 277 : 150 - 151

Lake JA (1988) Origins of the eukaryotic nucleus determined by rate invariant analysis of rRNA sequences. Nature (London) 331 : 184 - 186

Lambeth DO, Campbell KL, Zand R, Palmer G (1973) The appearance of transient species of cytochrome *c* upon rapid oxidation or reduction at alkaline pH. J Biol Chem 248 : 8130 - 8136

Landrum JT, Hatano K, Scheidt WR, Reed CA (1980) Imidazolate complexes of iron and manganese tetraphenylporphyrins. J Am Chem Soc 102 : 6729 - 6735

Lang G, Spartalian K, Yonetani T (1976) Mössbauer spectroscopic study of compound ES of cytochrome *c* peroxidase. Biochim Biophys Acta 451 : 250 - 258

Lanir A, Aviram I (1975) Proton magnetic relaxation and anion effect in solutions of acid ferricytochrome *c*. Arch Biochem Biophys 166 : 439 - 445

Lanir A, Yu NT, Felton RH (1979) Conformational transitions and vibronic couplings in acid ferricytochrome *c*: a Resonance Raman study. Biochemistry 18 : 1656 - 1660

Laycock MV (1972) The amino acid sequence of cytochrome *c*-553 from the Chrysophycaean alga, *Monochrysis lutheri*. Can J Biochem 50 : 1311 - 1325

Laycock MV (1975) The amino acid sequence of cytochrome *f* from the brown alga, *Alaria esculenta* (L.) Grev. Biochem J 149 : 271 - 279

Le Gall J, Mazza G, Bruschi-Heriaud M, Der Vartanian DV (1971) EPR and light absorption studies on *c*-type cytochromes of the anaerobic sulfate reducer *Desulfovibrio*. Biochim Biophys Acta 234 : 499 - 512

Lederer F (1972) *Candida krusei* cytochrome *c* a correction to the sequence. Eur J Biochem 31 : 144 - 147

Lederer F, Simon AM (1974) *Neurospora crassa* and *Humicola lanuginosa* cytochromes *c*: more homology in the heme region. Biochem Biophys Res Commun 56 : 317 - 323

Lederer F, Simon AM, Verdiére J (1972) *Saccharomyces cerevisiae* iso-cytochromes *c*: revision of the amino acid sequence between the cysteine residues. Biochem Biophys Res Commun 47 : 55 - 58

Leitch FA, Moore GR, Pettigrew GW (1984) Structural basis for the variation of pH dependent redox potentials of *Pseudomonas* cytochromes *c*-551. Biochemistry 23 : 1831 - 1838

Lemberg R, Barrett J (1973) The cytochromes. Academic Press, London New York

Leonard JJ, Yonetani T (1974) Interaction of cytochrome *c* peroxidase with cytochrome *c*. Biochemistry 13 : 1465 - 1468

Leszczynski JF, Rose GD (1986) Ω loops in globular proteins: a novel category of secondary structure. Science 234 : 849 - 855

Letellier L, Shechter E (1973) Correlations between structure and spectroscopic properties in membrane model system. Fluorescence and circular dichroism of the cytochrome *c* cardiolipin system. Eur J Biochem 40 : 507 - 512

Lewin R (1981) *Prochloron* and the theory of symbiogenesis. Annals NY Acad Sci 361 : 325 - 329

Li WH, Luo CC, Wu CI (1985) Evolution of DNA sequences. In: MacIntyre RJ (ed) Molecular evolutionary genetics. Plenum, New York

Li PM, Gelles J, Chan SI, Sullivan RJ, Scott RA (1987) EXAFS of copper in Cu_A depleted p-hydroxymercuribenzoate modified and native cytochrome *c* oxidase. Biochemistry 26 : 2091 - 2095

Liang N, Pielak GJ, Mauk AG, Smith M, Hoffman BM (1987) Yeast cytochrome *c* with phenylalanine or tyrosine at position 87 transfers electrons to (zinc cytochrome *c* peroxidase)$^+$ at a rate ten thousand times that of the serine-87 or glycine-87 variants. Proc Natl Acad Sci USA 84 : 1249 - 1252

Liang N, Mauk AG, Pielak GJ, Johnson JA, Smith M, Hoffman BM (1988) Regulation of interprotein electron transfer by residue 82 of yeast cytochrome *c*. Proc Natl Acad Sci USA 240 : 311 - 313

Libby WF (1952) Theory of electron exchange reactions in aqueous solutions. J Phys Chem 56 : 863 - 868

Limbach KJ, Wu R (1985) Characterization of two *Drosophila melanogaster* cytochrome *c* genes and their transcripts. Nucl Acids Res 13 : 631 - 644

Lin DK, Niece RL, Fitch WM (1973) The properties and amino acid sequence of cytochrome *c* from *Euglena gracilis*. Nature (London) 241 : 533 - 535

Linderstrøm-Lang KU, Schellman JA (1959) Protein structure and enzyme activity. In: Boyer PD, Lardy H, Myrbäck K (eds) Academic Press, London New York 2nd edn. pp 443 - 510

Lode ET, Murray CL, Rabinowitz JC (1974a) Semisynthetic synthesis and biological activity of a Clostridial-type ferredoxin free of aromatic acid residues. Biochem Biophys Res Commun 61 : 163 - 169

Lode ET, Murray CL, Sweeney WV, Rabinowitz JC (1974b) Synthesis and properties of *Clostridium acidi-urici* [Leu-2]-ferredoxin : A function of the peptide chain and evidence against the direct role of the aromatic residues in electron transfer. Proc Natl Acad Sci USA 71 : 1361 - 1365

Loehr S, Freedman TB, Loehr TM (1974) Oxygen binding to hemocyanin : a resonance Raman spectroscopic study. Biochem Biophys Res Commun 56 : 510 - 515

Looze Y, Polastro E, Gielens C, Léonis J (1976) Iso-cytochrome *c* species from bakers yeast: Analysis of their circular-dichroism spectra. Biochem J 157 : 773 - 775

Looze Y, Polastro E, Deconinck M, Léonis J (1978) Alkaline isomerization of horse and yeast cytochromes *c*. Int J Peptide Prot Res 12 : 233 - 236

Louie GV, Hutcheon WLB, Brayer GD (1988a) Yeast iso-1 cytochrome *c*: a 2.8Å resolution three-dimensional structure determination. J Mol Biol 199 : 295 - 314

Louie GV, Pielak GJ, Smith M, Brayer GD (1988b) Role of Phe-82 in yeast iso-1 cytochrome *c* and remote conformational changes induced by a serine residue at this position. Biochemistry 27 : 7870 - 7876

Louie GV, Brayer GD (1989) A polypeptide chain-refolding event occurs in the Gly 82 variant of yeast iso-1 cytochrome *c*. J Mol Biol 209 : 313 - 322

Lubitz W, Babcock GT (1987) ENDOR spectroscopy. Trends Biochem Sci 12 : 96 - 100

Ludwig ML, Pattridge KA, Powers TB, Dickerson RE, Takano T (1982) Structure analysis of a ferricytochrome *c* from the cyanobacterium, *Anacystis nidulans*. In: Chien Ho (ed) Electron Transport and Oxygen Utilization Elsevier - North Holland Inc. vol. 1 pp27 - 32

Luntz TL, Schejter A, Garber EAE, Margoliash E (1989) Structural significance of an internal water molecule studied by site-directed mutagenesis of Tyr-67 in rat cytochrome *c*. Proc Natl Acad Sci USA 86 : 3524 - 3528

Lyddiatt A, Boulter D (1976a) The amino acid sequence of cytochrome *c* from common starfish. FEBS Lett 67 : 331 - 334

Lyddiatt A, Boulter D (1976b) A comparison of cytochrome *c* from *Macrobrachium malcomsonii* with other invertebrate cytochromes *c*. Comp Biochem Physiol 55B : 337 - 342

Lyddiatt A, Boulter D (1976c) The amino acid sequence of cytochrome *c* from *Eisenia foetida* (Savigny) (common brandling worm). FEBS letters 62 : 85 - 88

Lyddiatt A, Boulter D (1977) The amino acid sequence of cytochrome *c* from the locust, *Schistocerca gregaria* Forskal. Biochem J 163 : 333 - 338

Lyddiatt A, Peacock D, Boulter D (1978) Evolutionary change in invertebrate cytochrome *c*. J Molec Evol 11 : 35 - 45

Mailer C, Taylor CPS (1972) EPR study of single crystals of horse heart ferricytochrome *c* at 4.2K. Can J Biochem 50 : 1048 - 1055

Makinen MW, and Churg AK (1983) Structural and analytical aspects of the electronic spectra of hemeproteins. In: Lever ABP, Gray HB (eds) Iron Porphyrins part one Addison-Wesley, London Amsterdam pp 141 - 235

Makinen MW, Schichman SA, Hill SG, Gray HB (1983) Heme - heme orientation and electron transfer kinetic behaviour of multisite oxidation reduction enzymes. Science 222 : 929 - 931

Malmström BG (1973) Cytochrome *c* oxidase: some current problems. Quart Rev Biophys 6 : 389 - 431

Malmström BG (1989) The mechanism of proton translocation in respiration and photosynthesis. FEBS Lett 250 : 9 - 21

Maltempo MM (1974) Magnetic state of an unusual bacterial heme protein. J Chem Phys 61 : 2540 - 2547

Maltempo MM (1976) Visible absorption spectra of quantum mixed-spin ferric hemeproteins. Biochim Biophys Acta 434 : 513 - 518

Maltempo MM, Moss TH (1976) The spin $^{3}/_{2}$, $^{5}/_{2}$ state and quantum spin mixtures in haem proteins. Quart Rev Biophys 9 : 181 - 215

Maltempo MM, Moss TH, Cusanovich MA (1974) Magnetic studies on the changes in the iron environment in *Chromatium* ferricytochrome *c*. Biochim Biophys Acta 342 : 290 -305

Mandel N, Mandel G, Trus BL, Rosenberg J, Carlson G, Dickerson RE (1977) Tuna cytochrome *c* at 2.0 Å resolution: Coordinate optimization and comparison of structures. J Biol Chem 252 : 4619 - 4636

Marchon JC, Mashiko T, Reed CA (1982) How does nature control cytochrome redox potentials? In: Chien Ho (ed) Electron transport and oxygen utilisation. Elsevier, North Holland, pp 67 - 72

Marcus RA (1964) Chemical and electrochemical electron transfer theory. Ann Rev Phys Chem 15 : 155 - 196

Marcus RA (1979) Electron and nuclear tunneling in chemical and biological systems. In: Chance B, DeVault DC, Fraunfelder H, Marcus RA, Schrieffer JR, Sutin N (eds) Tunneling in biological systems. Academic Press, London New York

Marcus RA, Sutin N (1975) Electron transfer reactions with unusual activation parameters. A treatment of reactions accompanied by large entropy decreases. Inorg Chem 14 : 213 - 216

Marcus RA, Sutin N (1985) Electron transfers in chemistry and biology. Biochim Biophys Acta 811 : 265 - 322

Marcus RA, Sutin N (1986) The relation between the barriers for thermal and optical electron transfer reactions in solution. Comments Inorg Chem 5 : 119 - 133

Margalit R, Schejter A (1973) Cytochrome *c* - a thermodynamic study of the relationships among oxidation state ion binding and structural parameters. Eur J Biochem 32 : 492 - 499

Margoliash E, Smith EL, Kreil G, Tuppy H (1961) Amino acid sequence of horse-heart cytochrome *c*. Nature (London) 192: 1121 - 1127

Margoliash E, Schejter A (1966) Cytochrome *c*. Adv Prot Chem 21 : 113 - 287

Margoliash E, Reichlin M, Nisonoff A (1968) The Immunological Properties of Cytochrome *c*. In: Okunuki K, Kamen MD, Sekuzu I (eds) University Park Press, Baltimore pp 269 - 280

Margoliash E, Fitch WM, Markowitz E, Dickerson RE (1972) In: Ehrenberg A (ed) Oxidation Reduction Enzymes. Winksel, Stockholm pp 5 - 17

Margoliash E, Ferguson-Miller S, Kang CH, Brautigan DL (1976) Do evolutionary changes in cytochrome *c* structure reflect structural adaptations? Fed Proc 35 : 2124 - 2130

Margulis L (1970) Origin of Eukaryotic cells. Yale Univ Press, New Haven Conn.

Marini MA, Martin CJ, Berger RL, Forlani L (1981) Analysis of the ionization constants and heats of ionization of reduced and oxidized horse heart cytochrome *c*. Biopolymers 20 : 2253 - 2261

Marques HM, Baldwin DA, Pratt JM (1987) Hemes and hemoproteins. 3 The reaction of microperoxidase-8 with cyanide: Comparison with aquocobalamin and hemoproteins. J Inorg Biochem 29 : 77 - 91

Martin JP, Fridovich I (1981) Evidence for a natural gene transfer from the ponyfish to its bioluminescent bacterial symbiont, *Photobacter leiognathi*. J Biol Chem 256 : 6080 - 6089

Martin RB (1974) Pyrrole hydrogen ionization of imidazole derivatives in metal ion complexes and carbonic anhydrase. Proc Natl Acad Sci USA 71 : 4346 - 4347

Martinez G, Rochat H, Ducet G (1974) The amino acid sequence of cytochrome *c* from potato. FEBS Lett 47 : 212 - 217

Mashiko T, Reed CA, Haller KJ, Kastner ME, Scheidt WR (1981) Thioether ligation in iron-porphyrin complexes : models for cytochrome *c*. J Am Chem Soc 103 : 5758 - 5767

Mathews FS (1985) The structure, function and evolution of cytochromes. Prog Biophys molec Biol 45 : 1 - 56

Matsubara H, Smith EL (1963) Human heart cytochrome *c*. J Biol Chem 238 : 2732 - 2753

Matsuura Y, Hata Y, Yamaguchi T, Tanaka N, Kakudo M (1979) Structure of bonito heart ferricytochrome *c* and some remarks on molecular interaction in its crystalline state. J Biochem 85 : 729 - 737

Matsuura Y, Takano T, Dickerson RE (1982) Structure of cytochrome *c*-551 from *Ps. aeruginosa* refined at 1.6Å resolution and comparison of the two redox forms. J Mol Biol 156 : 389 - 409

Mauk MR, Reid LS, Mauk AG (1982) Spectrophotometric analysis of the interaction between cytochrome b_5 and cytochrome *c*. Biochemistry 21 : 1843 - 1846

Mayo SL, Ellis Jr WR, Crutchley RJ, Gray HB (1986) Long-range electron transfer in heme proteins. Science 233 : 948 - 952

McDonald CC, Phillips WD (1973) Proton magnetic resonance studies of horse cytochrome *c*. Biochemistry 12 : 3170 - 3186

McDonald CC, Phillips WD, Vinogradov SN (1969) Proton magnetic resonance evidence for methionine-iron coordination in mammalian-type ferrocytochrome *c*. Biochem Biophys Res Commun 36 : 442 _ 449

McDonald CC, Phillips WD, Le Gall J (1974) Proton magnetic resonance studies of *Desulfovibrio* cytochrome c_3. Biochemistry 13 : 1952 - 1959

McDowall MA, Smith EL (1965) Amino acid sequence of dog heart cytochrome *c*. J Biol Chem 240 : 4635 - 4647

McIntire W, Singer TP, Smith AJ, Mathews FS (1986) Amino acid and sequence analysis of the cytochrome and flavoprotein subunits of *p*-cresol methylhydroxylase. Biochemistry 25 : 5975 - 5981

McKenna MC (1969) The origin and early differentiation of therian mammals. Annals NY Acad Sci 167 : 217 - 240

McLaughlin PJ, Dayhoff MO (1973) Eukaryotic evolution: a view based on cytochrome *c* sequence data. J Molec Evol 2 : 99 - 116

McLendon G (1988) Long-distance electron transfer in proteins and model systems. Acc Chem Res 21 : 160 - 167

McLendon G, Smith M (1978) Equilibrium and kinetic studies of unfolding of homologous cytochromes *c*. J Biol Chem 253 : 4004 - 4008

McLendon G, Miller JR (1985) The dependence of biological electron transfer rates on exothermicity: the cytochrome *c*/cytochrome b_5 couple. J Am Chem Soc 107 : 7811 - 7816

McLendon G, Winkler JR, Nocera DG, Mauk MR, Mauk AG, Gray HB (1985) Quenching of zinc-substituted cytochrome *c* excited states by cytochrome b_5. J Am Chem Soc 107 : 739 - 740

McLendon G, Pardue K, Bak P (1987) Electron transfer in the cytochrome *c*/cytochrome b_2 complex : Evidence for conformational gating. J Am Chem Soc 109 : 7540 - 7541

Meade TJ, Gray HB, Winkler JR (1989) Driving-force effects on the rate of long-range electron transfer in ruthenium-modified cytochrome *c*. J Am Chem Soc 111 : 4353 - 4356

Meatyard BT, Boulter D (1974) The amino acid sequence of cytochrome *c* from *Enteromorpha intestinalis*. Phytochem 13 : 2777 - 2782

Merchant S, Bogorad L (1987) The Cu(II) repressible plastidic cytochrome *c*. J Biol Chem 262 : 9062 - 9067

Meyer TE (1980) Cytochromes and ferredoxins in bacterial electron transport and classification. In: Kaplan NO, Robinson AB (eds) From Cyclotrons to Cytochromes. Academic Press, New York pp 157 - 180

Meyer TE, Cusanovich MA (1985) Soluble cytochrome composition of the purple phototrophic bacterium *Rps. sphaeroides* (ATCC 17023). Biochim Biophys Acta 807 : 308 - 319

Meyer TE, Kamen MD (1982) New perspectives on *c*-type cytochromes. Adv Prot Chem 35 : 105 - 212

Meyer TE, Ambler RP, Bartsch RG, Kamen MD (1975) Amino acid sequence of cytochrome *c'* from the purple photosynthetic bacterium *Rhodospirillum rubrum* S1. J Biol Chem 250 : 8416 - 8421

Meyer TE, Przysiecki CT, Watkins JA, Bhattacharyya A, Simondson RP, Cusanovich MA, Tollin G (1983) Correlation between rate constant for reduction and redox potential as a basis for systematic investigation of reaction mechanisms of electron transfer proteins. Proc Natl Acad Sci USA 80 : 6740 - 6744

Meyer TE, Watkins JA, Przysiecki CT, Tollin G, Cusanovich MA (1984) Electron-transfer reactions of photoreduced flavin analogues with *c*-type cytochromes: Quantitation of steric and electrostatic factors. Biochemistry 23 : 4761 - 4767

Meyer TE, Cusanovich MA, Kamen MD (1986) Evidence against the use of bacterial amino acid sequence data for construction of all-inclusive phylogenetic trees. Proc Natl Acad Sci USA 83 : 217 - 220

Michel B, Mauk AG, Bosshard HR (1989a) Binding and oxidation of mutant cytochromes *c* by cytochrome *c* oxidase. FEBS Lett 243 : 149 - 152

Michel B, Proudfoot AEI, Wallace CJA, Bosshard HR (1989b) The cytochrome *c* oxidase-cytochrome *c* complex: spectroscopic analysis of the conformational changes in the protein-protein interaction domain. Biochemistry 28 : 456 - 462

Michel H, Deisenhofer J (1988) Relevance of the photosynthetic reaction centre from purple bacteria to the structure of photosystem II. Biochemistry 27 : 1 - 7

Milkman R (1973) Electrophoretic variation in *E coli* from natural sources. Science 182 : 1024 - 1026

Miller DJ, Nicholas DJ (1986) N-terminal amino acid sequence of cytochrome *c*-552 from *Nitrosomonas europaea*. Biochem Int 12 : 167 - 172

Miller Jr, Calcaterra LT, Closs GL (1984) Intramolecular long-distance electron transfer in radical anions. The effects of free energy and solvent on the reaction rates. J Am Chem Soc 106 : 3047 - 3049

Minday RM, Schmidt LD, Davis HT (1972) Mobility of excess electrons in liquid hydrocarbon mixtures. J Phys Chem 76 : 442 - 446

Mitchell KAR (1969) The use of outer d-orbitals in bonding. Chem Rev 69 : 157 - 178

Mitchell P (1961) Coupling of phosporylation to electron and proton transfer by a chemiosmotic type of mechanism. Nature (London) 191 : 144 - 148

Mitchell P (1980) Proton motive cytochrome system of mitochondria. Annals NY Acad Sci 341 : 564 - 584

Mitchell P, Mitchell R, Moody AJ, West IC, Baum H, Wrigglesworth JM (1985) Chemiosmotic coupling in cytochrome oxidase. FEBS Lett 188 : 1 - 7

Montgomery DL, Leung DW, Smith M, Shalit P, Faye G, Hall BD (1980) Isolation and sequence of the gene for iso-2 cytochrome *c* in *Saccharomyces cerevisiae*. Proc Natl Acad Sci 77 : 541 - 545

Moore GR (1983) Control of redox properties of cytochrome *c* by special electrostatic interactions. FEBS Lett 161 : 171 - 175

Moore GR (1985) ^{1}H NMR studies of the haem and coordinated methionine of class I and class II cytochromes *c*. Biochim Biophys Acta 829 : 425 - 429

Moore GR, Rogers GR (1985) The influence of electrostatic interactions between buried charges on the properties of membrane proteins. J Inorg Biochem 23 : 219 - 226 & 25 : 295

Moore GR, Williams G (1984) Assignment of ^{1}H-NMR resonances of the heme and axial histidine ligand of mitochondrial cytochrome *c*. Biochim Biophys Acta 788 : 147 - 150

Moore GR, Williams RJP (1976) Electron transfer proteins. Coord Chem Rev 18 : 125 - 197

Moore GR, Williams RJP (1977) Structural basis for the variation in redox potential of cytochromes. FEBS Lett 79 : 229 - 23

Moore GR, Williams RJP (1980a) NMR studies of ferrocytochrome *c*: pH and temperature dependence. Eur J Biochem 103 : 513 - 521

Moore GR, Williams RJP (1980b) The stability of ferricytochrome *c*: Temperature dependence of its NMR spectrum. Eur J Biochem 103 : 523 - 532

Moore GR, Williams RJP (1980c) The solution structures of tuna and horse cytochromes *c*. Eur J Biochem 103 : 533 - 541

Moore GR, Williams RJP (1980d) Comparison of the structures of various eukaryotic cytochromes *c* and their antigenic differences. Eur J Biochem 103 : 543 - 550

Moore GR, Pitt RC, Williams RJP (1977) NMR studies of *Ps. aeruginosa* cytochrome *c*-551. Eur J Biochem 77 : 53 - 60

Moore GR, Williams RJP, Chien J, Dickinson LC (1980a) Nuclear Magnetic Resonance studies of metal substituted horse cytochrome *c*. J Inorg Biochem 12 : 1 - 15

Moore GR, De Aguiar ABVP, Pluck ND, Williams RJP (1980b) The properties and function of tryptophan in proteins. In: Hayaishi O, Ishimura Y, Kido R (eds) Biochemical and Medical Aspects of Tryptophan Metabolism. Elsevier, Amsterdam pp 83 - 94

Moore GR, Pettigrew GW, Pitt RC, Williams RJP (1980c) pH dependence of the redox potential of *Ps. aeruginosa* cytochrome *c*-551. Biochim Biophys Acta 590 : 261 - 271
Moore GR, Huang Z-H, Eley CGS, Barker HA, Williams G, Robinson MN, Williams RJP (1982a) Electron transfer in biology: The function of cytochrome *c*. Faraday Discuss Chem Soc 74 : 311 - 329
Moore GR, McClune GJ, Clayden NJ, Williams RJP, Alsaadi BM, Ångström J, Ambler RP, Van Beeuman J, Tempst P, Bartsch RG, Meyer TE, Kamen MD (1982b) Metal coordination centres of class II cytochromes *c*. Eur J Biochem 123 : 73 - 80
Moore GR, Harris DE, Leitch FA, Pettigrew GW (1984a) Characterisation of ionisations that influence the redox potential of mitochondrial cytochrome *c* and photosynthetic bacterial cytochrome c_2. Biochim Biophys Acta 764 : 331 - 342
Moore GR, Eley CGS, Williams G (1984b) Electron transfer reactions of class 1 cytochromes *c*. Adv Inorg Bioinorg Mech 3 : 1 - 96
Moore GR, Robinson MN, Williams G, Williams RJP (1985a) Solution structure of mitochondrial cytochrome *c*: ^{1}H NMR of ferrocytochrome *c*. J Mol Biol 183 : 429 - 446
Moore GR, Williams RJP, Peterson J, Thomson AJ, Mathews FS (1985b) A spectroscopic investigation of the structure and redox properties of *E. coli* cytochrome b_{562}. Biochim Biophys Acta 829 : 83 - 96
Moore GR, Pettigrew GW, Rogers NK (1986a) Factors influencing redox potentials of electron transfer proteins. Proc Natl Acad Sci USA 83 : 4998 - 4999
Moore GR, Leitch FA, Pettigrew GW, Rogers NK, Williams G (1986b) The importance of electrostatic interactions involving buried groups in determining the structure and properties of metalloproteins In: Xavier AV (ed) Frontiers in Bioinorganic Chemistry VCH Publishers, FRG pp 494 - 506
Moore GW, Goodman M, Callahan C, Holmquist R, Moise H (1976) Stochastic versus augmented maximum parsimony method for estimating superimposed mutations in the divergent evolution of protein sequences. J Mol Biol 105 : 15 - 37
Moore WJ (1972) Physical Chemistry. Longman, Harlow Essex New Jersey
Morgan WT, Hensley Jr CP, Riehm JP (1972) Proteins of the thermophilic fungus *Humicola lanuginosa*: isolation and amino acid sequence of a cytochrome *c*. J Biol Chem 247 : 6555 - 6565
Mori E, Morita Y (1980) Amino acid sequence of cytochrome *c* from rice. J Biochem 87 : 249 - 266
Morishima I, Ogawa S, Yonezama T, Iizuka T (1977) NMR studies of hemoproteins: pH dependent features of horse heart cytochrome *c*. Biochim Biophys Acta 495 : 287 - 298
Morton RA (1973) The EPR spectrum of ferricytochrome *c* and lysine-modified derivatives at alkaline pH. Can J Biochem 51 : 465 - 471
Moss TH (1978) Magnetic susceptibility applied to metalloproteins. Meth Enzymol LIV : 380 - 396
Moss T, Ehrenberg A, Bearden AJ (1969) Mössbauer spectroscopic evidence for the electronic configuration of iron in horseradish peroxidase and its peroxide derivatives. Biochemistry 8 : 4159 - 4162
Moss D, Nabedryk E, Breton J, Mäntele W (1990) Redox-linked conformational changes in proteins detected by a combination of infrared spectroscopy and protein electrochemistry. Eur J Biochem 187 : 565 - 572
Motokawa Y, Hayashi N, Kikuchi G (1964) On the reactivity of Hemoglobin M-iwate. Arch Biochem Biophys 105 : 612 - 619

Motonaga K, Misaka E, Nakajima E, Ueda S, Nakanishi K (1965) Structure of yeast cytochrome *c*. J Biochem 57 : 22 - 28

Moura JJG, Xavier AV, Cookson DJ, Moore GR, Williams RJP, Bruschi M, Le Gall J (1977) Redox states of cytochrome c_3 in the absence and presence of ferredoxin. FEBS Lett 81 : 275 - 180

Moura JJG, Xavier AV, Cookson DJ, Moore GR, Williams RJP, Fauque G, Bruschi M, Le Gall J (1978) Electron-transfer mechanisms in multihaem cytochromes. Biochem Soc Trans 6 : 1285 - 1287

Moura JJG, Santos H, Moura I, Le Gall J, Moore GR, Williams RJP, Xavier AV (1982) NMR redox studies of *D. vulgaris* cytochrome c_3. Eur J Biochem 127 : 151 - 155

Moura JJG, Moore GR, Williams RJP, Probst I, LeGall J, Xavier AV (1984) NMR studies of *Desulfuromonas acetoxidans* cytochrome *c*-551.5. Eur J Biochem 144 : 433 - 440

Moura JJG, Liu MY, Costa C, Liu MC, Pai G, Xavier AV, LeGall J, Payne WJ, Moura I (1988) Spectroscopic characterization of a high-potential monohaem cytochrome from *Wolinella succinogenes,* a nitrate-respiring organism. Eur J Biochem 177 : 673 - 682

Mukai K, Yoshida M, Yao Y, Wakabayashi S, Matsubara H (1988) Evidence for a single thioether bond in heme binding of cytochrome c_1 from *Euglena gracilis*. Proc Japan Acad 64, Ser B : 41 - 44

Mukai K, Wakabayashi S, Matsubara H (1989) Molecular cloning and nucleotide sequence of a c-DNA encoding *Euglena gracilis* cytochrome c_1. J Biochem 106 : 479 - 482

Münck E, (1979) Mössbauer spectra of hemoproteins. In : The Porphyrins Dolphin D (ed) vol 4 Academic Press, London New York pp 379 - 423

Murray RGE (1984) The higher taxa, or, a place for everything? In: Krieg NR, Holt JG (eds) Bergeys Manual of Systematic Bacteriology Williams and Wilkins, Baltimore pp 31 - 34

Myer YP (1968) Conformation of cytochromes. Effect of urea, temperature, extrinsic ligands, and pH variation on the conformation of horse ferricytochrome *c*. Biochemistry 7 : 765 - 776

Myer YP (1970) Conformational homology of functionally homologous cytochromes *c*. Biochim Biophys Acta 214 : 94 - 106

Myer YP, Bullock PA (1978) Cytochrome b_{562} from *E. coli*: Conformational, configurational, and spin-state characterization. Biochemistry 17 : 3723 - 3729

Myer YP, Pande AJ (1978) Circular dichroism studies of hemoproteins and heme models. In : Dolphin D (ed) The Porphyrins vol. 3A. Academic Press, London New York pp 271 - 322

Myer YP, Murphy AJ, Harbury HA (1966) Effect of ionic strength on the conformation of horse heart ferricytochrome *c* and its fully esterified derivative. J Biol Chem 241 : 5370 - 5374

Myer YP, Saturno AF, Verma BC, Pande A (1979) Horse heart cytochrome *c*: The oxidation-reduction potential and protein structures. J Biol Chem 254 : 11202 - 11207

Myer YP, MacDonald LH, Verma BC, Pande A (1980a) Urea denaturation of horse heart ferricytochrome *c*. Equilibrium studies and characterization of intermediate forms. Biochemistry 19 : 199 - 207

Myer YP, Thallum KK, Pande A (1980b) Kinetics of the reduction of horse heart ferricytochrome *c*. Ascorbate reduction in the presence and absence of urea. J Biol Chem 255 : 9666 - 9673

Myer YP, Thallum KK, Pande J, Verma BC (1980c) Selectivity of oxidase and reductase activity of horse heart cytochrome *c*. Biochem Biophys Res Commun 94 : 1106 - 1112

Myer YP, Pande A, Saturno AF (1981) Kinetics of folding and unfolding of horse heart ferricytochrome *c*. J Biol Chem 256 : 1576 - 1581

Mylvaganam SE, Paterson Y, Tainer JA, Getzoff ED (1988) 2nd Symp. of the Protein Society. abstract S107

Nabedrylk-Viala E, Thiéry C, Calvet P, Thiéry JM (1976) Hydrogen-isotope exchange of oxidized and reduced cytochrome *c*. Eur J Biochem 61 : 253 - 258

Nakano K, Kikumoto Y, Yagi T (1983) Amino acid sequence of cytochrome *c*-553 from *Desulfovibrio vulgaris*, Miyazaki. J Biol Chem 258:12409-12412

Nakashima T, Higa H, Matsubara H, Benson A, Yasunobu KT (1966) The amino acid sequence of bovine heart cytochrome *c*. J Biol Chem 241 : 1166 - 1177

Nakayama T, Titani K, Narita K (1971) The amino acid sequence of cytochrome *c* from bonito (*Katsuwonus pelamis, Linnaeus*). J Biochem 70 : 311 - 326

Nall BT, Landers TA (1981) Guanidine hydrochloride induced unfolding of yeast iso-2 cytochrome *c*. Biochemistry 20 : 5403 - 5411

Narita K, Titani K (1968) The amino acid sequence of cytochrome *c* from *Candida krusei*. J Biochem 63 : 226 - 241

Narita K, Titani K (1969) The complete amino acid sequence in baker's yeast cytochrome *c*. J Biochem 65 : 259 - 267

Needleman SB, Margoliash E (1966) Rabbit heart cytochrome *c*. J Biol Chem 241 : 853 - 863

Needleman SB, Wunsch CD (1970) A general method, applicable to the search for similarities in the amino acid sequence of two proteins. J Mol Biol 48 : 443 - 453

Nicholls P (1974) Cytochrome *c* binding to enzymes and membranes. Biochim Biophys Acta 346 : 271 - 310

Nicholls P, Peterson LC (1974) Heme-heme interactions in cytochrome aa_3 during the anaerobic-aerobic transition. Biochim Biophys Acta 357 : 462 - 467

Niece RL, Margoliash E, Fitch WM (1977) Complete amino acid sequence of guanaco cytochrome *c*. Biochemistry 16 : 68 - 72

Niki K, Kawasaki Y, Nishimura N, Higuchi Y, Yasuoka N, Kakudo M (1984) Electrochemical and structural studies of tetraheme proteins from *Desulfovibrio* - standard potentials of the redox sites and heme-heme interactions. J Electroanal Chem 168 : 275 - 286.

Nitschke W, Rutherford AW (1989) Tetraheme cytochrome *c* subunit of *Rhodopseudomonas viridis* characterized by EPR. Biochemistry 28 : 3161 - 3168

Nix PT, Warme PK (1979) Semisynthetic analogs of cytochrome *c* reconstructed from natural and synthetic peptides. Biochim Biophys Acta 578 : 413 - 427

Noda S, Keran L (1974) Conduction state energy of electrons in liquid methane and ethane. J Chem Phys 61 : 2467 - 2468

Nolan C, Margoliash E (1966) Primary structure of the cytochrome *c* from the great grey kangaroo, *Macropus canguru*. J Biol Chem 241 : 1049 - 1059

Nolan C, Fitch WM, Uzzell T, Weiss LJ, Margoliash E (1973) Amino acid sequence of a cytochrome *c* from the common Pacific lamprey, *Entosphenus tridentatus*. Biochemistry 12 : 4052 - 4060

Nording M, Young S, Karlsson BG, Lundberg LG (1990) The structural gene for cytochrome *c*-551 from *Pseudomonas aeruginosa*. FEBS Lett 259 : 230 - 232

Norris GE, Anderson BF, Baker EN, Rumball SV (1979) Purification and preliminary crystallographic studies on azurin and cytochrome *c'* from *Alcaligenes denitrificans* and *Alcaligenes* sp. NCIB 11015. J Mol Biol 135 : 309 - 312

Northrup SH, Pear MR, McCammon JA, Karplus M, Takano T (1980) Internal mobility of ferrocytochrome *c*. Nature (London) 287 : 659 - 660

Northrup SH, Pear MR, Morgan JD, McCammon JA, Karplus M (1981) Molecular dynamics of ferrocytochrome *c*. J Mol Biol 153 : 1087 - 1109

Nozaki M, Mizushima H, Horio T, Okunuki K (1958) Further studies on proteinase digestion of bakers yeast cytochrome *c*. J Biochem 45 : 815 - 823

Nunn DN, Anthony C (1988) The nucleotide sequence and deduced amino acid sequence of the cytochrome c_L gene of *Methylobacter extorquens* AM1 : a novel class of cytochrome. Biochem J 256 : 673 - 676

O'Hern DJ, Pal P-K, Myer YP (1975) Conformational and functional studies of chemically modified cytochromes: N-bromosuccinimide and formyl- cytochrome *c*. Biochemistry 14 : 382 - 391

Obar R, Green J (1985) Molecular archaeology of the mitochondrial genome. J Molec Evol 22 : 243 - 251

Ochi H, Hata Y, Tanaka N, Kakudo M, Sakurai T, Aihara S, Morita Y (1983) Structure of rice ferricytochrome *c* at 2.0Å resolution. J Mol Biol 166 : 407 - 418

Ochman H, Wilson AC (1987) Evolution in bacteria - evidence for a universal substitution rate in cellular genomes. J Molec Evol 26 : 74 - 86

Oertle M, Immergluck K, Paterson Y, Bosshard HR (1989) Mapping of four discontiguous antigenic determinants on horse cytochrome *c*. Eur J Biochem 182 : 699 - 704

Ohnishi T, Schägger H, Meinhardt SW, LoBrutto R, Link TA, von Jagow G (1989) Spatial organization of the redox active centers in the bovine heart ubiquinol-cytochrome *c* oxidoreductase. J Biol Chem 264 : 735 - 744

Ohno N, Cusanovich MA (1981) Reaction of *c*-type cytochromes with the iron hexa-cyanides. Biophys J 36 : 589 - 605

Ohta T, Kimura M (1971) On the constancy of the evolutionary rate of cistrons. J Molec Evol 1 : 18 - 25

Ohyama K, Fukuzawa H, Kohchi T, Shirai H, Sano T, Sano S, Umesono K, Shiki Y, Takeuchi M, Chang Z, Aota S, Inokuchi H, Ozeki H (1986) Chloroplast gene organisation deduced from complete sequence of liverwort *Marchantia polymorpha* chloroplast DNA. Nature (London) 322 : 572 - 574

Okamura K, Miyata T, Iwanaga S, Takamiya K, Nishimura M (1987) Complete amino acid sequence of cytochrome *c*-551 from *Erythrobacter* species strain OCh 114. J Biochem 101 : 957 - 966

Olson JM (1981) Evolution of the photosynthetic and respiratory prokaryotes and organelles. Annals NY Acad Sci 361 : 8 - 17

Ono K, Kimura K,Yagi T, Inokuchi H (1975) Mössbauer study of cytochrome c_3. J Chem Phys 63 : 1640 - 1643

Osheroff N, Feinberg BA, Margoliash E, Morrison M (1977) Lactoperoxidase-catalyzed iodination of horse cytochrome *c*: monoiodotyrosyl 74 cytochrome *c*. J Biol Chem 252 : 7743 - 7751

Osheroff N, Borden D, Koppenol WH, Margoliash E (1980) Electrostatic interactions in cytochrome *c*. J Biol Chem 255 : 1689 - 169

Osheroff N, Speck SH, Margoliash E, Veerman EC, Wilms J, Konig BW, Muisjers AO (1983) The reaction of primate cytochrome *c* with cytochrome oxidase. J Biol Chem 258 : 5731 - 5738

Osvath P, Salmon GA, Sykes AG (1988) Preparation, characterization, and intramolecular rate constant for Ru(II) to Fe(III) electron transfer in the pentaammineruthenium histidine modified cytochrome *c*-551 from *Pseudomonas stutzeri*. J Am Chem Soc 110 : 7114 - 7118

Pace CN (1975) The stability of globular proteins. Crit Rev Biochem 3 : 1 - 43

Paleus S (1954) A comparative study of cytochrome *c* from beef, chicken and salmon. Acta Chem Scand 8 : 971 - 984

Palmer G (1979) Electron paramagnetic resonance of hemoproteins. In : Dolphin D (ed) The Porphyrins vol 4 Academic Press, London New York pp 313 - 353

Palmer G (1983) Electron paramagnetic resonance of hemoproteins. In : Lever ABP, Gray HB (eds) Iron Porphyrins, part two. Addison-Wesley, London Amsterdam pp 43 - 88

Palmer G (1985) The electron paramagnetic resonance of metalloproteins. Biochem Soc Trans 13 : 548 - 560

Pan LP, Durham B, Wolinska J, Millett F (1988) Preparation and characterization of singly labeled ruthenium polypyridine cytochrome *c* derivatives. Biochemistry 27 : 7180 - 7184

Pande J, Myer YP (1980) The arginines of cytochrome *c*. J Biol Chem 255 : 11094 - 11097

Papa S (1976) Proton translocation reactions in respiratory chains. Biochim Biophys Acta 456 : 39 - 84

Parker FS (1971) Applications of infrared spectroscopy in biochemistry, biology and medicine. Adam Hilger, London

Parr GR, Hantgen RR, Taniuchi H (1978) Formation of two alternative complementary structures from a cytochrome *c* heme fragment and the apoprotein. J Biol Chem 253 : 5381 - 5388

Patel DJ, Canuel LL (1976) NMR studies of slowly exchanging peptide protons in cytochrome *c*. Proc Natl Acad Sci USA 73 : 1398 - 1402

Paterson Y (1985) Delineation and conformational analysis of two synthetic peptide models of antigenic sites on rodent cytochrome *c*. Biochemistry 24 : 1048 - 1055

Paul K-G (1951) The iron-protein bonds in cytochrome *c*. Acta Chem Scand 5 : 379 - 388

Peacock D, Boulter D (1975) Use of amino acid sequence data in phylogeny and evaluation of methods using computer simulation. J Mol Biol 95 : 513 - 527

Pearce LL, Gärtner AL, Smith M, Mauk AG (1989) Mutation-induced perturbations of the cytochrome *c* alkaline transition. Biochemistry 28 : 3152 - 3156

Penny D (1974) Evolutionary clock - the rate of evolution of rattlesnake cytochrome *c*. J Molec Evol 3 : 179 - 188

Perrin DD (1959) The oxidation reduction potentials of metal complex ions. Reviews Pure Applied Chem 9 : 257 - 285

Perutz MF (1970) Stereochemistry of cooperative effects in haemoglobin. Nature (London) 228 : 726 - 739

Perutz M, Pulsinelli PD, Ranney HM (1972) Structure and subunit interaction of Haemoglobin M Milwaukee. Nature New Biol 237 : 259 - 263

Petersen RL, Gupta RK (1979) Location of Cr(III) in the Cr(III)-cytochrome *c* complex as observed by NMR spectroscopy. FEBS Lett 107 : 427 - 430

Pettigrew GW (1972) The amino acid sequence of a cytochrome *c* from a protozoan *Crithidia oncopelti*. FEBS Lett 22 : 64 - 66

Pettigrew GW (1973)The amino acid sequence of cytochrome *c* from *Euglena gracilis*. Nature 241 : 531 - 533

Pettigrew GW (1974) The purification and amino acid sequence of cytochrome *c*-552 from *Euglena gracilis*. Biochem J 139 : 449 - 459

Pettigrew GW (1979) Structural features of protozoan cytochromes. In: Levandowsky M, Hutner SH (eds) Academic Press: pp 59 - 90

Pettigrew GW, Brown KR (1988) Free and membrane-bound forms of bacterial cytochrome c_4. Biochem J 252 : 427 - 435

Pettigrew GW, Moore GR (1987) Cytochromes *c*: Biological Aspects. Springer-Verlag, New York Heidelberg

Pettigrew GW, Seilman S (1982) Properties of a cross-linked complex between cytochrome *c* and cytochrome *c* peroxidase. Biochem J 201 : 9 - 18

Pettigrew GW, Aviram I, Schejter A (1975a) Physicochemical properties of two atypical cytochromes *c*, *Crithidia* cytochrome *c*-557 and *Euglena* cytochrome *c*-558. Biochemistry 149 : 155 - 167

Pettigrew GW, Meyer TE, Bartsch RG, Kamen MD (1975b) pH dependence of the oxidation-reduction potential of cytochrome c_2. Biochim Biophys Acta 430 : 197 - 208

Pettigrew GW, Aviram I, Schejter A (1976) The role of the lysines in the alkaline heme-linked ionization of ferric cytochrome *c*. Biochem Biophys Res Commun 68 : 807 - 813

Pettigrew GW, Bartsch RG, Meyer TE, Kamen MD (1978) Redox potentials of the photosynthetic bacterial cytochrome c_2 and the structural bases for variability. Biochim Biophys Acta 503 : 509 - 523

Pfennig N, Biebl H (1976) *Desulfuromonas acetoxidans* gen nov and sp nov, a new anaerobic sulfur reducing acetate oxidising bacterium. Arch Microbiol 110 : 3 - 12

Phillips JN (1963) Physico-chemical properties of porphyrins. Comp Biochem 9 : 34 - 72

Phillips CSG, Williams RJP (1966) Inorganic Chemistry Clarendon Press, Oxford

Pielak GJ, Mauk AG, Smith M (1985) Site-directed mutagenesis of cytochrome *c* shows that an invariant Phe is not essential for function. Nature (London) 313 : 152 - 154

Pielak GJ, Boyd J, Moore GR, Williams RJP (1988a) Proton-NMR studies show that the Thr-102 mutant of yeast iso-1 cytochrome *c* is a typical member of the eukaryotic cytochrome *c* family. Eur J Biochem 177 : 167 - 177

Pielak GJ, Atkinson RA, Boyd J, Williams RJP (1988b) Two-dimensional NMR as a probe of structural similarity applied to mutants of cytochrome *c*. Eur J Biochem 177 : 179 - 185

Pierrot M, Haser R, Frey M, Payan F, Astier J-P (1982) Crystal structure and electron-transfer properties of cytochrome c_3. J Biol Chem 257 : 14341 - 14348

Polerio E, Parr GR, Taniuchi H (1986) A study of roles of evolutionarily invariant Proline 30 and Glycine 34 of cytochrome *c*. J Biol Chem 261 : 10976 - 10989

Potasek MJ (1978) Investigation of electron tunneling between cytochrome *c* peroxidase and cytochrome *c*. Science 201 : 151 - 153

Potasek MJ, Hopfield JJ (1977) Fundamental aspects of electron transfer : Experimental verification of vibronically coupled electron tunneling. Proc Natl Acad Sci USA 74 : 3817 - 3820

Poulos TL, Finzel BC (1984) Heme enzyme structure and function. Pept Prot Rev 4 : 115 - 171

Poulos TL, Kraut J (1980) A hypothetical model of the cytochrome *c* peroxidase: cytochrome *c* electron transfer complex. J Biol Chem 255 : 10322 - 10330
Poulos TL, Finzel BC, Howard AJ (1986) Crystal structure of substrate-free *Pseudomonas putida* cytochrome P-450. Biochemistry 25 : 5314 - 5322
Prince RC, Leigh JS, Dutton PL (1974) An electron spin resonance characterization of *Rhodopseudomonas capsulata*. Biochem Soc Trans 2 : 950 - 953
Prince RC, Leigh JS, Dutton PL (1976) Thermodynamic properties of the reaction center of *Rhodopseudomonas viridis*. Biochim Biophys Acta 440 : 622 - 636
Prochaska LJ, Bisson R, Capaldi RA, Steffens GCM, Buse G (1981) Inhibition of cytochrome *c* oxidase function by dicyclohexylcarbodiimide. Biochim Biophys Acta 637 : 360 - 373
Proudfoot AEI, Wallace CJA, Harris DE, Offord RE (1986) A new non-covalent complex of semi-synthetically modified tryptic fragments of cytochrome *c*. Biochemistry 239 : 333 - 337
Provencher SW, Glöckner J (1981) Estimation of globular protein secondary structure from circular dichroism. Biochemistry 20 : 33 - 37
Raff RA, Mahler HR (1972) The non-symbiotic origin of mitochondria. Science 177 : 575 - 582
Ramshaw JAM, Thompson EW, Boulter D (1970) The amino acid sequence of *Helianthus annuus* L. (Sunflower) cytochrome *c* deduced from chymotryptic peptides. Biochem J 119 : 535 - 539
Ramshaw JAM, Richardson M, Boulter D (1971) The amino acid sequence of the cytochrome *c* of *Ginkgo biloba* L. Eur J Biochem 23 : 475 - 483
Raven PH (1970) A multiple origin for plastids and mitochondria. Science 169 : 641 - 646
Rawlings J, Stephens PJ, Nafie LA, Kamen MD (1977) Near-infrared magnetic circular dichroism of cytochrome *c'*. Biochemistry 16 : 1725 - 1729
Redfield AG, Gupta RK (1971) Pulsed NMR study of the structure of cytochrome *c*. Cold Spring Harb Symp Quant Biol 36 : 405 - 411
Reed CA, Mashiko T, Bentley SP, Kastner ME, Scheidt WR, Spartalian K, Lang G (1979) The missing heme spin state and a model for cytochrome *c'*. The mixed $S={}^3/_2,{}^5/_2$ intermediate spin ferric porphyrin:perchlorato (*meso*-tetraphenylporphyrin) iron(III). J Am Chem Soc 101 : 2948 - 2958
Rees DC (1980) Experimental evaluation of the effective dielectric constant of proteins. J Mol Biol 141 : 323 - 326
Rees DC (1985) Electrostatic influence on energetics of electron transfer reactions. Proc Natl Acad Sci USA 82 : 3082 - 085
Reichlin M, Fogel S, Nisonoff A, Margoliash E (1966) Antibodies against cytochromes *c* from vertebrates. J Biol Chem 241 : 251 - 253
Reid GA, Ingledew WJ (1979) Characterisation and phenotypic control of the cytochrome content of *Escherichia coli*. Biochem J 182 : 465 - 472
Rich PR (1982) The kinetics and thermodynamics of the reduction of cytochromes *c* by substituted p-benzoquinols in solution. Faraday Discuss Chem Soc 74 : 349 - 364
Richardson M, Ramshaw JAM, Boulter D (1971) The amino acid sequence of rape (*Brassica napus* L) cytochrome *c*. Biochim Biophys Acta 251 : 331 - 333
Ridge JA, Baldwin RL, Labhardt AM (1981) Nature of the fast and slow refolding reactions of iron(III) cytochrome *c*. Biochemistry 20 1622 - 1630

Rigby SEJ, Moore GR, Gray JC, Gadsby PMA, George SJ, Thomson AJ (1988) NMR, EPR and MCD studies of cytochrome *f*: identity of the heme axial ligands. Biochem J 256 : 571 - 577

Rigby SEJ, Alleyne TA, Wilson MT, Moore GR (1989) A proton NMR study of bovine cytochrome oxidase: paramagnetically shifted resonances of haem *a*. FEBS Lett 257 : 155 - 158

Robinson Jr JB, Strottman JM, Stellwagen E (1983b) A globular, high-spin form of ferricytochrome *c*. J Biol Chem 258 : 6772 - 6776

Robinson MN, Boswell AP, Eley CSG, Huang Z-X, Moore GR (1983a) The conformation of eukaryotic cytochrome *c* around residues 39, 57, 59 and 74. Biochem J 213 : 687 - 700

Roder H, Elöve GA, Englander SW (1988) Structural characterization of folding intermediates in cytochrome *c* by H-exchange labelling and proton NMR. Nature (London) 335 : 700 - 704

Rogers NK (1986) The modelling of electrostatic interactions in the function of globular proteins. Prog Biophys molec Biol 48 : 37 - 66

Rogers NK, Moore GR (1988) On the energetics of conformational changes and pH dependent redox behaviour of electron transfer proteins. FEBS Lett 228 : 69 - 173

Rogers NK, Sternberg MJE (1984) Electrostatic interaction in globular proteins. J Mol Biol 174 : 527 - 542

Rogers NK, Moore GR, Sternberg MJE (1985) Electrostatic interactions in globular proteins - calculations of the pH dependence of the redox potential of cytochrome *c*-551. J Mol Biol 182 : 613 - 616

Romer AS (1966) Vertebrate palaeontology. University of Chicago Press, Chicago

Romisch J, Tropschug M, Sebald W, Weiss H (1987) The primary structure of cytochrome c_1 from *Neurospora crassa*. Eur J Biochem 164 : 111 - 115

Rönnberg M (1987) Specific cleavage of *Pseudomonas* cytochrome *c* peroxidase by elastase from *Pseudomonas aeruginosa*. Biochim Biophys Acta 916 : 112 - 118

Rönnberg M, Kalkkinen N, Ellfolk N (1989) The primary structure of *Pseudomonas* cytochrome *c* peroxidase. FEBS Lett 250 : 175 - 178

Rothfus JA, Smith EL (1965) Amino acid sequence of rhesus monkey heart cytochrome *c*. J Biol Chem 240 : 4277 - 4283

Rowan NS, Storm CB, Rowan III R (1981) Properties of metal-ion coordinated imidazoles : NMR and C-2H exchange in Co(III) complexes. J Inorg Biochem 14 : 59 - 65

Rubinow SC, Kassner RJ (1984) Cytochromes *c'* in their reaction with ethyl isocyanide. Biochemistry 23 : 2590 - 2595

Russell AJ, Fersht AR (1987) Rational modification of enzyme catalysis by engineering surface charge. Nature (London) 328 : 496 - 500

Russell AJ, Warshel A (1985) Calculations of electrostatic energies in proteins. J Mol Biol 185 : 389 - 404

Saari H, Penttila T, Wikstrom M (1980) Interaction of Ca^{2+} and H^+ with heme *a* of cytochrome oxidase. J Bioenerg Biomemb 12 : 325 - 338

Sadler I, Suda K, Schatz G, Kaudewitz F, Haid A (1984) Sequencing of the nuclear gene for the yeast cytochrome c_1 precursor reveals an unusually complex amino-terminal presequence. EMBO J 3 : 2137 - 2143

Saigo S (1981a) A transient spin-state change during alkaline isomerization of ferricytochrome *c*. J Biochem 89 : 1977 - 1980

Saigo S (1981b) Kinetic and equilibrium studies of alkaline isomerization of vertebrate cytochromes *c*. Biochim Biophys Acta 669 : 13 - 20

Salemme FR (1974) Preliminary crystallographic data for cytochrome *c'* of *Rhodopseudomonas palustris*. Arch Biochem Biophys 163 : 423 - 425

Salemme FR (1976) An hypothetical structure for an intermolecular electron transfer complex of cytochrome *c* and cytochrome b_5. J Mol Biol 102 : 563 - 568

Salemme FR (1977) Structure and function of cytochromes *c*. Ann Rev Biochem 46 : 299 - 329

Salemme FR, Freer ST, Xuong NH, Alden RA, Kraut J (1973a) The structure of oxidized cytochrome c_2 of *Rhodospirillum rubrum*. J Biol Chem 248 : 3910 - 3921

Salemme FR, Kraut J, Kamen MD (1973b) Structural bases for function in cytochromes *c*. J Biol Chem 248 : 7701 - 7716

Salerno JC (1984) Cytochrome electron spin resonance line shapes, ligand fields, and components stoichiometry in ubiquinol-cytochrome *c* oxidoreductase. J Biol Chem 259 : 2331 - 2336

Santos H, Turner DL (1987) Proton NMR studies of horse ferricytochrome *c*: completion of the assignment of the well resolved hyperfine shifted resonances. FEBS Lett 226 : 179 - 185

Santos H, Turner DL (1988) Characterization and NMR studies of a novel cytochrome *c* isolated from *Methylophilus methylotrophus* which shows a redox-linked change of spin-state. Biochim Biophys Acta 954 : 277 - 286

Santos H, Moura JJG, Moura I, Le Gall J, Xavier AV (1984) NMR studies of electron transfer mechanisms in a protein with interacting redox centres: *D. gigas* cytochrome c_3. Eur J Biochem 141 : 283 - 296

Satterlee JD, Moench SJ, Erman JE (1987) A proton NMR study of the non-covalent complex of horse cytochrome *c* and yeast cytochrome *c* peroxidase and its comparison with other interacting protein complexes. Biochim Biophys Acta 912 : 87 - 97

Sawyer L, Jones CL, Damas AM, Harding MM, Gould RO, Ambler RP (1981) Cytochrome c_4 from *Pseudomonas aeruginosa*. J Mol Biol 153 : 831 - 835

Schechter E, Saludjian P (1967) Conformation of cytochrome *c*. IV. Relationship between optical absorption and protein conformation. Biopolymers 5 : 788 - 790

Scheidt WR (1977) Trends in metalloporphyrin stereochemistry. Acc Chem Res 10 : 339 - 345

Scheidt WR, Gouterman M (1983) Ligands, spin state, and geometry in hemes and related metalloporphyrins. In : Lever ABP, Gray HB (eds) Iron Porphyrins part one Addison-Wesley, London Amsterdam pp 89 - 139

Scheidt WR, Reed CA (1981) Spin-state/stereochemical relationships in iron porphyrins: Implications for the hemoproteins. Chem Rev 81 : 543 - 555

Scheidt WR, Osvath SR, Lee YJ (1987) Crystal and molecular structure of bis(imidazole)(*meso*-tetraphenylporphinato) iron(III) chloride. A classic molecule revisited. J Am Chem Soc 109 : 1958 - 1963

Schejter A, Aviram I (1970) The effects of alkylation of methionyl residues on the properties of horse cytochrome *c*. J Biol Chem 245 : 1552 - 1557

Schejter A, Eaton WA (1984) Charge transfer optical spectra, electron paramagnetic resonance and redox potentials of cytochromes. Biochemistry 23 : 1081 - 1084

Schejter A, George P (1964) The 695mμ band of ferricytochrome *c* and its relationship to protein conformation. Biochemistry 3 1045 - 1049

Schejter A, Aviram I, Goldkorn T (1982) Contribution of electrostatic factors to the oxidation reduction potentials of *c*-type cytochromes. In: Chien Ho (ed) Electron transport and oxygen utilisation. Elsevier, North Holland pp 95 - 99

Schlauder G, Kassner RJ (1979) Comparative solvent perturbation of horse heart cytochrome *c* and *Rhodospirillum rubrum* cytochrome c_2. J Biol Chem 254 : 4110 - 4113

Schopf JW (1970) Precambrian microorganisms and evolutionary events prior to the origin of vascular plants. Biol Rev 45 : 319 - 352

Schulz CE, Rutter R, Sage JT, Debrunner PG, Hager LP (1984) Mössbauer and EPR studies of horseradish peroxidase and its catalytic intermediates. Biochemistry 23 : 4743 - 4754

Schwartz RM, Dayhoff MO (1978) Origins of prokaryotes, eukaryotes, mitochondria and chloroplasts. Science 199 : 395 - 403

Schwartz RM, Dayhoff MO (1981) Chloroplast origins: inferences from protein and nucleic acid sequences. Annals NY Acad Sci 361 : 260 - 269

Schweingruber ME, Sherman F, Stewart JW (1977) Altered absorption spectra of iso-1 cytochromes *c* from mutants of yeast. J Biol Chem 252 : 6577 - 6580

Schweingruber ME, Sherman F, Stewart JW (1978) Amino acid replacements of the evolutionarily invariant tryptophan at position 64 in mutant forms of iso-1 cytochrome *c* from *S. cerevisiae*. J Mol Biol 118 : 481 - 496

Schweingruber ME, Sherman F, Stewart JW (1979) Primary site and second site revertants of missense mutants of the evolutionarily invariant tryptophan 64 in iso-1 cytochrome *c* from yeast. J Biol Chem 254 ; 4132 - 4143

Scogin R, Richardson M, Boulter D (1972) The amino acid sequence of cytochrome *c* from tomato. Arch Biochem Biophys 150 : 489 - 492

Senn H, Wüthrich K (1983) Individual ^{1}H-NMR assignments for the heme groups and the axially-bound amino acids and determination of the coordination geometry at the heme iron in a mixture of two isocytochromes *c*-551 from *Rhodopseudomonas gelatinosa*. Biochim Biophys Acta 743 : 69 - 81

Senn H, Wüthrich K (1985) Amino acid sequence, haem-iron coordination geometry and functional properties of mitochondrial and bacterial *c*-type cytochromes. Quart Rev Biophys 18 : 111 - 134

Senn H, Keller RM, Wüthrich K (1980) Different chirality of the axial methionine in homologous cytochromes *c* determined by ^{1}H NMR and CD spectroscopy. Biochem Biophys Res Commun 92 : 1362 - 1369

Senn H, Guerlesquin F, Bruschi M, Wüthrich K (1983) Coordination of the heme iron in the low-potential cytochrome *c*-553 from *Desulfovibrio vulgaris* and *Desulfovibrio desulfuricans*. Biochim Biophys Acta 748 : 194 - 204

Shack J, Clark WM (1947) Metalloporphyrins: cycles of changes in systems containing heme. J Biol Chem 171 : 143 - 187

Sharrock M, Münck E, Debrunner PG, Marshall V, Lipscomb JD, Gunsalus IC (1973) Mössbauer studies of cytochrome P-450_{cam}. Biochemistry 12 : 258 - 265

Sharrock M, Debrunner PG, Schulz C, Lipscomb JD, Marshall V, Gunsalus IC (1976) Cytochrome P-450_{cam} and its complexes. Mössbauer parameters of the heme iron. Biochim Biophys Acta 420 : 8 - 26

Shaw DC, Williams KL, Smith E, Birt LM (1978) The amino acid sequence of cytochrome *c* from the blowfly *Lucilia cuprina*. Biochim Biophys Acta 532 : 179 - 184

Shaw RW, Hartzell CR (1976) Hydrogen ion titration of horse heart ferricytochrome *c*. Biochemistry 15 : 1909 - 1914

Shelnutt JA, Rousseau DL, Dethmers JK, Margoliash E (1981) Protein influences on porphyrin structure in cytochrome *c*. Evidence from Raman difference spectroscopy. Biochemistry 20 : 6485 - 6497

Sheridan RP, Levy RM, Salemme FR (1982) α-helix dipole model and electrostatic stabilization of 4-α-helical proteins. Proc Natl Acad Sci USA 79 : 4545 - 4549

Sherwood C, Brayer GD (1985) Crystallization and preliminary diffraction data for iso-1-cytochrome *c* from yeast. J Mol Biol 185 : 209 - 210

Shiao DDF, Sturtevant JM (1976) Heats of binding protons to globular proteins. Biopolymers 15 : 1201 - 1211

Shinkai W, Hase T, Yagi T, Matsubara H (1980) Amino acid sequence of cytochrome c_3 from *Desulfovibrio vulgaris,* Miyazaki. J Biochem 87 : 1747 - 1756

Shinkarev VP, Drachev AL, Dracheva SM (1990) The thermodynamic characteristics of four-heme cytochrome *c* in *Rhodopseudomonas viridis* reaction centers, as derived from a quantitative analysis of the differential absorption spectra in α-domain. FEBS Lett 261 : 11 - 13

Shinozaki K, Ohme M, Tanaka M, Wakasugi T, Hayashida N, Matsubayashi T, Zaita N, Chunwongse J, Obokata J, Yamaguchi-Shinozaki K, Ohto C, Torazawa K, Kusuda J, Takaiwa F, Kato A, Tohdoh N, Shimada H, Sugiura M (1986) The complete nucleotide sequence of the tobacco chloroplast genome : its gene organisation and expression. EMBO J 5 : 2043 - 2049

Siedow JN, Vickery LE, Palmer G (1980) The nature of the axial ligands of spinach cytochrome *f*. Arch Biochem Biophys 203 : 101 - 107

Sievers G, Gadsby PMA, Peterson J, Thomson AJ (1983) Magnetic circular dichroism spectra of soybean leghaemoglobin *a* at room temperature and 4.2K. Biochim Biophys Acta 742 : 637 - 647

Sigel E, Carafoli E (1978) The proton pump of cytochrome *c* oxidase and its stoichiometry. Eur J Biochem 89 : 119 - 123

Sigel E, Carafoli E (1980) Quantitative analysis of the proton and charge stoichiometry of cytochrome *c* oxidase of beef heart reconstituted into phospholipid vesicles. Eur J Biochem 111 : 299 - 306

Silvestri I, Taniuchi H (1988) A study of fine specificity of monoclonal antibodies to yeast iso-1 cytochrome *c*. J Biol Chem 263 : 18702 - 18713

Silvestrini MC, Brunori M, Wilson MT, Darley-Usmar VM (1981) The electron transfer system of *Pseudomonas aeruginosa* - a study of the pH dependent transitions between redox forms of azurin and cytochrome *c*-551. J Inorg Biochem 14 : 327 - 338

Silvestrini MC, Galeotti CL, Gervais M, Schinina E, Barra D, Bossa F, Brunori M (1989) Nitrite reductase from *Pseudomonas aeruginosa* : sequence of the gene and the protein. FEBS Lett 254 : 33 - 38

Simon-Becam AM, Claisse M, Lederer F (1978) Cytochrome *c* from *Schizosaccharomyces pombe*. Eur J Biochem 86 : 407 - 416

Simpkin D, Palmer G, Devlin FJ, McKenna MC, Jensen GM, Stephens PJ (1989) The axial ligands of heme in cytochromes : A near-infrared magnetic circular dichroism study of yeast cytochromes *c*, c_1 and *b* and spinach cytochrome *f*. Biochemistry 28 : 8033 - 8039

Singleton Jr R, Campbell LL, Hawkridge FM (1979) Cytochrome c_3 from the sulfate reducing anaerobe *D. africanus* (Benghazi). J Bact 140 : 893 - 901

Slater EC (1987) The mechanism of the conservation of energy of biological oxidation. Eur J Biochem 166 : 489 - 504

Smith DW, Williams RJP (1970) The spectra of ferric haems and haemoproteins. Structure and Bonding 7 : 1 - 45

Smith EL, Margoliash E (1964) Evolution of cytochrome *c* . Fed Proc 23 : 1243 - 1247

Smith GM (1979) Proton nuclear magnetic resonance studies of *Rhodospirillum rubrum* cytochrome c_2. Biochemistry 18 : 1628 - 1634

Smith HT, Millett F (1980) Involvement of lysines 72 and 79 in the alkaline isomerization of horse heart ferricytochrome *c*. Biochemistry 19 : 1117 - 1120

Smith HT, Staudenmeyer N, Millet F (1977) Use of lysine specific modifications to locate the reaction site of cytochrome *c* with cytochrome *c* oxidase. Biochemistry 16 : 4971 - 4978

Smith L, Nava ME, Margoliash E (1973) In: King TE, Mason HS, Morrison M (eds) Oxidases and Related Redox Systems, Vol 2. Univ Park Press, Baltimore, pp 629 - 638

Smith LF (1972) Amino acid sequences of insulins. Diabetes 21 : 457 - 460

Smith M, McLendon G (1981) Comparative NMR studies of cytochrome *c* and its active site octapeptide. J Amer Chem Soc 103 : 4912 - 4921

Sneath PHA, Sokal RR (1973) Numerical Taxonomy. Freeman, San Francisco

Sokol WF, Evans DH, Niki K, Yagi T (1980) Reversible voltammetric response for a molecule containing four non-equivalent redox sites with application to cytochrome c_3 of *Desulfovibrio vulgaris* strain Miyazaki. J Electroanal Chem 108 : 107 - 115

Sokolovsky M, Moldovan M (1972) Primary structure of cytochrome *c* from the camel, *Camelus dromedarius*. Biochemistry 11 : 145 - 149

Sorrell TN, Martin PK, Bowden EF (1989) A novel, functional variant of cytochrome *c*: replacement of the histidine ligand with arginine via site-directed mutagenesis. J Am Chem Soc 111 : 766 - 767

Spiro TG (1983) The resonance Raman spectroscopy of metalloporphyrins and heme proteins. In: Lever ABP, Gray HB (eds) Iron Porphyrins, part two. Addison-Wesley, London Amsterdam pp 89 - 159

Spiro TG, Stong JD, Stein P (1979) Porphyrin core expansion and doming in heme proteins. New evidence from resonance Raman spectra of six coordinate high spin Fe(III) hemes. J Am Chem Soc 101 : 2648 - 2655

Sreenathan BR, Taylor CPS (1971) The insensitivity of the 695 nm band of horse heart ferricytochrome *c* to protein conformation. Biochem Biophys Res Commun 42 : 1122 - 1126

Staley JT, Krieg NR (1984) Classification of prokaryotic organisms - an overview. In: Krieg NR, Holt JG (eds) Bergeys Manual of Systematic Bacteriology. Williams and Wilkins, Baltimore pp 1 - 4

Stanier RY, Douderoff M, Adelberg EA (1970) General Microbiology. McMillan, London

Staudenmeyer N, Ng S, Smith MB, Millet F (1977) Effect of specific trifluoroacetylation of individual cytochrome *c* lysines on the reaction with cytochrome *c* oxidase. Biochemistry 16 : 600 - 604

Steigemann W, Weber E (1979) Structure of erythrocruorin in different ligand states refined at 1.4Å resolution. J Mol Biol 127 : 309 - 338

Stellwagen E (1964) The spectrophotometric titration of the phenolic groups of horse heart cytochrome *c*. Biochemistry 3 : 919 - 923

Stellwagen E (1968) Carboxymethylation of horse heart ferricytochrome *c* and cyanoferricytochrome *c*. Biochemistry 7 : 2496 - 2501

Stellwagen E (1978) Haem exposure as the determinant of oxidation-reduction potential of haem proteins. Nature (London) 275 : 73 - 74

Stellwagen E, Babul J (1975) Stabilization of the globular structure of ferricytochrome *c* by chloride in acidic solvents. Biochemistry 14 : 5135 - 5140

Stellwagen E, Babul J, Wilgus H (1975) The alkaline isomerisation of lysine-modified ferricytochrome *c*. Biochim Biophys Acta 405 : 115 - 121

Sternberg MJ, Thornton JM (1976) On the conformation of proteins: the handedness of the β-strand, α-helix, β-strand unit. J Mol Biol 105 : 367 - 382

Stevens FC, Glazer AN, Smith El (1967) The amino acid sequence of wheat germ cytochrome *c*. J Biol Chem 242 : 2764 - 2779

Stewart JW, Margoliash E (1965) The primary structure of the cytochrome *c* from various organs of the hog. Can J Biochem 43 : 1187 - 1206

Stewart KD, Mattox KR (1984) The case for a polyphyletic origin of mitochondria: morphological and molecular comparisons. J Molec Evol 21 : 54 - 57

Stonehuerner J, O'Brien P, Geren L, Millet F, Steidl J, Yu L, Yu C (1985) Identification of the binding site on cytochrome c_1 for cytochrome *c*. J Biol Chem 260 : 5392 - 5398

Strickland EH, Horwitz J, Kay E, Shannon LM, Wilchek M, Billups C (1971) Near-ultraviolet absorption bands of tryptophan. Studies using horseradish peroxidase isoenzymes, bovine and horse heart cytochrome *c*, and *N*-stearyl-L-tryptophan *n*-hexyl ester. Biochemistry 10 : 2631 - 2638

Strydom DJ, van der Walt SJ, Botes DP (1972) The amino acid sequence of bat cytochrome *c*. Comp Biochem Physiol 43B : 21 - 24

Sugeno K, Narita K, Titani K (1971) The amino acid sequence of cytochrome *c* from *Debaromyces kloeckeri*. J Biochem 70 : 659 - 682

Sugimura Y, Hase T, Matsubara H, Shimokoriyama M (1981) Studies on algal cytochromes. III Amino acid sequence of cytochrome *c* from a brown alga *Petalonia fascia*. J Biochem 90 : 1213 - 1219

Summerville DA, Cohen IA, Hatano K, Scheidt WR (1978) Preparation and physical and stereochemical characterization of the tricyanomethanide salt of 5, 10, 15, 20 - tetraphenylporphinatoiron(III). Inorg Chem 17 : 2906 - 2910

Sutherland JC (1978) The magnetic optical activity of porphyrins. In : Dolphin D (ed) The Porphyrins vol 3A Academic Press, London New York pp 225 - 248

Sutherland JC, Klein MP (1972) Magnetic circular dichroism of cytochrome *c*. J Chem Phys 57 : 76 - 86

Sutin N (1977) Electron transfer reactions of cytochrome *c*. Adv Chem Ser 162 : 156 - 172

Sutin N (1982) Nuclear, electronic, and frequency factors in electron-transfer reactions. Acc Chem Res 15 : 275 - 282

Sutin N, Yandell JK (1972) Mechanisms of the reactions of cytochrome *c*: Rate and equilibrium constants for ligand binding to horse heart ferricytochrome *c*. J Biol Chem 247 : 6932 - 6936

Swanson R, Trus BL, Mandel N, Mandel G, Kallai OB, Dickerson RE (1977) Tuna cytochrome *c* at 2.0Å resolution: Ferricytochrome *c* structure analysis. J Biol Chem 252 : 759 - 775

Swanson MS, Zieminn SM, Miller DD, Garber EAE, Margoliash E (1985) Developmental expression of nuclear genes that encode mitochondrial proteins : Insect cytochrome *c*. Proc Natl Acad Sci USA 82 : 1964 - 1968

Sykes AG (1988) Electron transfer in biological systems. Chemistry in Britain 24 : 551 - 554

Sylvanen M (1987) Molecular clocks and evolutionary relationships - possible distortions due to horizontal gene flow. J Mol Evol 26 : 16 - 23

Taborsky G, McCollum K (1979) Phosphate binding by cytochrome *c* : specific binding site involved in the formation and reactivity of a complex of ferricytochrome *c*, ferrous ion and phosphate. J Biol Chem 254 : 7069 - 7075

Takano T, Dickerson RE (1981a) Conformation change of cytochrome *c*: Ferrocytochrome *c* structure refined at 1.5Å resolution. J Mol Biol 153 : 79 - 94

Takano T, Dickerson RE (1981b) Conformation change of cytochrome *c*: Ferricytochrome *c* refinement at 1.8Å and comparison with the ferrocytochrome structure. J Mol Biol 153 : 95 - 115

Takano T, Dickerson RE (1982) Conformational differences between ferri- and ferrocytochrome *c*. In: Chien Ho (ed) Electron Transport and Oxygen Utilisation Elsevier-North Holland Inc vol. 1 pp 17 - 26

Takano T, Kallai OB, Swanson R, Dickerson RE (1973) The structure of ferrocytochrome *c* at 2.45Å resolution. J Biol Chem 248 : 5234 - 5255

Takano T, Swanson R, Kallai OB, Dickerson RE (1971) Conformational changes upon reduction of cytochrome *c*. Cold Spring Harbour Symp Quant Biol 36 : 397 - 404

Takano T, Trus BL, Mandel N, Mandel G, Kallai OB, Swanson R, Dickerson RE (1977) Tuna cytochrome *c* at 2.0Å resolution: ferrocytochrome structure analysis. J Biol Chem 252 : 776 - 785

Tanaka N, Yamane T, Tsukihara T, Ashida T, Kakudo M (1975) The crystal structure of bonito ferrocytochrome *c* at 2.3Å resolution. J Biochem 77 : 147 - 162

Tanaka Y, Fukumori Y, Yamanaka T (1982) The complete amino acid sequence of *Nitrobacter agilis* cytochrome *c*-550. Biochim Biophys Acta 707 : 14 - 20

Tanford C (1961) Physical chemistry of macromolecules. Wiley, NY

Taniguchi S, Kamen MD (1963) On the anomalous interactions of ligands with *Rhodospirillum* haem protein. Biochim Biophys Acta 74 : 438 - 455

Taniguchi VT, Sailasuta-Scott N, Anson FC, Gray HB (1980) Thermodynamics of metalloprotein electron transfer reactions. Pure and Applied Chem 52 : 2275 - 2281

Tarr GE, Fitch WM (1976) Amino acid sequence from *Tetrahymena pyriformis* phenoset A. Biochem J 159 : 193 - 199

Tasaki A, Otsuka J, Kotani M (1967) Magnetic susceptibility measurements on hemoproteins down to 4.2K. Biochim Biophys Acta 140 : 284 - 290

Taube H (1984) Electron transfer between metal complexes: Retrospective. Science 226 : 1028 - 1036

Taylor CPS (1977) The EPR of low spin heme complexes. Biochim Biophys Acta 491 : 137 - 149

Taylor DS (1939) The magnetic properties of myoglobin and ferrimyoglobin, and their bearing on the problem of the existence of magnetic interactions in hemoglobin. J Am Chem Soc 61 : 2150 - 2154

Teissie J (1981) Interaction of cytochrome *c* with phospholipid monolayers. Orientation and penetration of protein as functions of the packing density of film, nature of the phospholipids, and ionic content of the aqueous phase. Biochemistry 20 : 1554 - 1560

Tempst P, van Beeumen J (1983) The amino acid sequence of cytochrome *c*-556 from *Agrobacterium tumefaciens* strain Apple 185. Eur J Biochem 135 : 321 - 330

Ten Kortenaar PBW (1983) Investigation into the synthesis of cytochrome *c* analogues. Doctoral Thesis: Catholic University of Nijmegen.

Ten Kortenaar PBW, Adams PJHM, Tesser GI (1985) Semisynthesis of horse heart cytochrome *c* analogues from two or three fragments. Proc Natl Acad Sci USA 82 : 8279 - 8283

Teraoka J, Kitagawa T (1980) Raman characterization of axial ligands for penta and hexa-coordinate ferric high and intermediate-spin (Octaethylporphyrinato) iron(III) complexes. Elucidation of unusual resonance Raman spectra of cytochrome *c'*. J Phys Chem 84 : 1928 - 1935

Terner J, Sitter AJ, Reczek CM (1985) Resonance Raman spectroscopic characterizations of horseradish peroxidase. Observation of the Fe(IV)=O stretching vibration of Compound II. Biochim Biophys Acta 828 : 73 - 80

Tervoort MJ, van Gelder BF (1983) The interaction between heme and protein in cytochrome c_1. Biochim Biophys Acta 722 : 137 - 143

Theorell H (1938) Uber die chemische konstitution des cytochroms *c*. Biochem Z 298 : 242 - 267

Theorell H, Åkesson Å (1941) Studies on cytochrome *c*. J Amer Chem Soc 63 : 1804 - 1820

Thomas MA, Gervais M, Favaudon V, Valat P (1983) Study of the *Hansenula anomala* yeast flavocytochrome b_2 - cytochrome *c* complex. Eur J Biochem 135 : 577 - 581

Thomas MA, Delsuc MA, Beloeil JC, Lallemand JY (1987) ^{1}H-NMR investigation of yeast cytochrome *c*. Interaction with the corresponding specific reductase (L-lactate cytochrome). Biochem Biophys Res Commun 145 : 1098 - 1104

Thompson EW, Laycock MV, Ramshaw JAM, Boulter D (1970) The amino acid sequence of *Phaseolus aureus* L. (mung-bean) cytochrome *c*. Biochem J 117 : 183 - 192

Thompson EW, Notton BA, Richardson M, Boulter D (1971a) The amino acid sequence of cytochrome *c* from *Abutilon theophrasti* Medic, and *Gossypium barbadense* L. (cotton). Biochem J 124 : 787 - 791

Thompson EW, Richardson M, Boulter D (1971b) The amino acid sequence of sesame (*Sesamum indicum* L.) and castor (*Ricinus communis* L.) cytochrome *c*. Biochem J 121 : 439 - 446

Thompson EW, Richardson M, Boulter D (1971c) The amino acid sequence of cytochrome *c* from *Cucurbita maxima* L. (pumpkin). Biochem J 124 : 779 - 781

Thompson EW, Richardson M, Boulter D (1971d) The amino acid sequence of cytochrome *c* of *Fagopyrum esculentum* Moench (buckwheat) and *Brassica oleracea* L. (cauliflower). Biochem J 124 : 783 - 785

Thompson RB, Borden D, Tarr GE, Margoliash E (1978) Heterogeneity of amino acid sequence in hippopotamus cytochrome *c*. J Biol Chem 253 : 8957 - 8961

Thomson AJ (1990) Natural and magnetic circular dichroism spectroscopies. In : Andrews DL (ed) Perspectives in Modern Chemical Spectroscopy. Springer - Verlag New York London *in press*.

Thomson AJ, Gadsby PMA (1990) A theoretical model of the intensity of the near infrared porphyrin-to-ferric charge-transfer transitions in low-spin Fe(III) haemoprotein: a correlation between the intensity of the MCD bands and the rhombic distortion parameter of iron. J Chem Soc Dalton Trans *in press*

Thomson AJ, Johnson MK (1980) Magnetization curves of haemoproteins measured by low-temperature magnetic circular dichroism spectroscopy. Biochem J 191 : 411 - 420

Thornber JP, Olson JM, Williams DM, Clayton ML (1969) Isolation of the reaction center of *Rhodopseudomonas viridis*. Biochim Biophys Acta 172 : 351 - 354

Thornber JP, Cogdell RJ, Seftor REB, Webster GD (1980) Further studies on the composition and spectral properties of the photochemical reaction centers of bacteriochlorophyll *b*-containing bacteria. Biochim Biophys Acta 593 : 60 - 75

Thornton JM, Sibanda BL (1983) Amino and carboxy-terminal regions in globular proteins. J Mol Biol 167 : 443 - 460

Tiede DM, Leigh JS, Dutton PL (1978) Structural organization of the redox groups in a bacterial photosynthetic reaction center complex. In : Dutton PL, Leigh JS, Scarpa A (eds) Frontiers of biological energetics. Academic Press, London New York 1 : 45 - 53

Timasheff SN, Susi H, Rupley JA (1973) Difference infrared spectrophotometric titration of protein carboxyls. Methods Enzymol 27 : 548 - 557

Timkovich R (1979) The architecture of a protein-porphyrin complex. In: Dolphin D (ed) The Porphyrins vol 7 Academic Press, London New York pp 241 - 294

Timkovich R (1980) Chemical modification of the haem propionate of cytochrome *c*. Biochem J 185 : 47 - 57

Timkovich R, Cork MS (1984) Proton NMR spectroscopy of cytochrome c_{554} from *Alcaligenes faecalis*. Biochemistry 23 : 851 - 860

Timkovich R, Dickerson RE (1976) The structure of *Paracoccus denitrificans* cytochrome c_{550}. J Biol Chem 251 : 4033 - 4046

Timkovich R, Dickerson RE, Margoliash E (1976) Amino acid sequence of *Paracoccus denitrificans* cytochrome *c*-550. J Biol Chem 251 : 2197 - 2206

Timkovich R, Cork MS, Taylor PV (1985) ^{1}H-NMR spectroscopy of cytochrome cd_1 derivatives. Arch Biochem Biophys 240 : 689 - 697

Titani K, Ericsson LH, Hon-nami K, Miyazawa T (1985) Amino acid sequence of cytochrome *c*-552 from *Thermus thermophilus* HB8. Biochem Biophys Res Commun 128 : 781 - 787

Tobe ML (1972) Inorganic reaction mechanisms. Nelson, London

Tollin G, Brown K, Francesco RD, Edmondson DE (1987) Flavodoxin-cytochrome *c* interactions: CD and NMR studies. Biochemistry 26 : 5042 - 5048

Tonge P, Wharton CW, Moore GR (1989) Fourier-transform infra-red studies of the alkaline isomerization of mitochondrial cytochrome *c* and the ionization of carboxylic acids. Biochem J 258 : 599 - 605

Torii K, Iizuka T, Ogura Y (1970) Magnetic susceptibility and EPR measurements of catalase and its derivatives. J Biochem 68 : 837 - 841

Trousil EB, Campbell LL (1974) Amino acid sequence of cytochrome c_3 from *Desulfovibrio vulgaris*. J Biol Chem 249 : 386 - 393

Tsong TY (1973) Detection of three kinetic phases in the thermal unfolding of ferricytochrome *c*. Biochemistry 12 : 2209 - 2214

Tsong TY (1974) The Trp 59 fluorescence of ferricytochrome *c* as a sensitive measure of the over-all protein conformation. J Biol Chem 249 : 1988 - 1990

Tsong TY (1975) An acid induced conformational transition of denatured cytochrome *c* in urea and guanidine hydrochloride solutions. Biochemistry 14 : 1542 - 1547

Tsuboi M, Nakanishi M (1979) Overall and localized fluctuation in the structure of a protein molecule. Adv Biophysics 12 : 101 - 130

Tsugita A, Gregor I, Kubota I, Broek RvD (1979) In: King TE, Orii Y, Chance B, Okunuki K (eds) Cytochrome Oxidase. Elsevier-North Holland New York pp 67 - 77

Tsukihara T, Yamane T, Tanaka N, Ashida T, Kakudo M (1973) Oxidation of a ferrocytochrome *c* in the crystalline state. J Biochem 73 : 1163 - 1167

Ulmer DD (1965) Optical rotatory dispersion of oxidized and reduced cytochrome *c*. Biochemistry 4 : 902 - 907

Ulmer DD, Kägi JHR (1968) Hydrogen-deuterium exchange of cytochrome *c*: Effect of oxidation state. Biochemistry 7 : 2710 - 2717

Ulrich EL, Krogmann DW, Markley JL (1982) Structure and heme environment of ferrocytochrome c_{553} from ^{1}H NMR studies. J Biol Chem 257 : 9356 - 9364

Uno T, Nishimura Y, Tsuboi M (1984) Time-resolved resonance Raman study of alkaline isomerization of ferricytochrome *c*. Biochemistry 23 : 6802 - 6808

Urbanski GJ, Margoliash E (1977) Topographic determinants on cytochrome *c*. J Immunol 118 : 1170 - 1180

Urry DW (1968) Circular dichroism of cytochromes and hemoproteins in the aromatic region. In : Okunuki K, Kamen MD, Sekuzu I (eds) Structure and function of cytochromes. Univ Park Press, Baltimore pp 311 - 317

Utuno M, Ono K, Kimura K, Inokuchi H, Yagi T (1980) Kinetics of cytochrome c_3 reduction with hydrogenase. A Mössbauer effect study. J Chem Phys 72 : 2264 - 2266

Uzzell T, Spolsky C (1981) Two data sets: alternative explanations and interpretations. Annals NY Acad Sci 361 : 481 - 499

Valentine JS, Sheridan RP, Allen LC, Kahn PC (1979) Coupling between oxidation state and hydrogen bond formation in heme proteins. Proc Natl Acad Sci USA 76 : 1009 - 1013

Vallee BL, Holmquist B (1980) Circular dichroism and magnetic circular dichroism. In: Darnell DW, Wilkins RG (eds) Methods for Determining Metal Ion Environments in Proteins. Adv Inorg Biochem 2 : 27 - 74 Elsevier-North Holland, New York

Vallee BL, Williams RJP (1968) Metalloenzymes: the entatic nature of their active sites. Proc Natl Acad Sci USA 59 : 498 - 505

van Beeumen J (1980) On the evolution of bacterial cytochromes *c*. In: Peeters H (ed) Protides of the Biological Fluids. Vol 28: pp 61 - 68

van Beeumen J, Ambler RP, Meyer TE, Kamen MD, Olson JM, Shaw EK (1976) The amino acid sequences of the cytochromes *c*-555 from two green sulfur bacteria of the genus *Chlorobium*. Biochem J 159 : 757 - 774

van Beeumen J, van den Branden C, Tempst P, de Ley J (1980) Cytochromes c_{556} from three genetic races of *Agrobacterium*. Eur J Biochem 107 : 475 - 483

van den Berg A, Beintema JJ (1975) Nonconstant evolution rates in amino acid sequences of guinea pig, chinchilla and coypu pancreatic ribonucleases. Nature (London) 253 : 207 - 210

van den Branden C, van Beeumen J, de Ley J (1975) Purification and properties of cytochrome *c*-556 from *Agrobacterium tumefaciens* B2a. Hoppe-Seyler's Z Physiol Chem 356 : 1251 - 1258

Vanderkooi J, Erecińska M (1974) Cytochrome *c* interaction with membranes: interaction of cytochrome *c* with isolated membrane fragments and purified enzymes. Arch Biochem Biophys 162 : 385 - 391

Vanderkooi JM, Erecińska M (1975) Cytochrome *c* interaction with membranes. Absorption and emission spectra and binding characteristics of iron-free cytochrome *c*. Eur J Biochem 60 : 199 - 207

Vanderkooi J, Erecińska M, Chance B (1973) Comparative study of the interaction of cytochrome *c* with the mitochondrial membrane. Arch Biochem Biophys 157 : 531 - 540

Vanderkooi J, Adar F, Erecinska M (1976) Metallocytochromes *c* : Characterization of electronic absorption and emission spectra of Sn^{4+} and Zn^{2+} cytochromes *c*. Eur J Biochem 64 : 381 - 387
Vanderkooi J, Landesberg R, Hayden GW, Owen CS (1977) Metal free and metal substituted cytochromes *c*. Use in characterization of the cytochrome *c* binding site. Eur J Biochem 81 : 339 - 347
Vanderkooi J, Glatz P, Casadel J, Woodrow GV (1980) Cytochrome *c* interaction with yeast cytochrome b_2. Eur J Biochem 110 : 189 - 196
van Rooijen GJH, Bruschi M, Voordouw G (1989) Cloning and sequencing the gene encoding cytochrome *c*-553 from *Desulfovibrio vulgaris* Hildenborough. J Bact 171 : 3575 - 3578
Veerman ECI, Wilms J, Casteleijn G, Van Gelder BF (1980) The pre-steady state reaction of ferrocytochrome *c* with the cytochrome *c* - cytochrome aa_3 complex. Biochim Biophys Acta 590 : 117 - 127
Veerman ECI, Leeuwen JW van, Buuren KJH van, Gelder BF van (1982) Reaction of cytochrome aa_3 with porphyrin-cytochrome *c* as studied by pulse radiolysis. Biochim Biophys Acta 680 : 134 - 141
Venkatachalan CM (1968) Stereochemical criteria for polypeptides and proteins. V Conformation of a system of three linked peptide units. Biopolymers 6 : 1425 - 1436
Verma AL, Kimura K, Nakamura A, Yagi T, Inokuchi H, Kitigawa T (1988) Resonance Raman studies of hydrogenase-catalyzed reduction of cytochrome c_3 by hydrogen. Evidence for heme-heme interactions. J Am Chem Soc 110 : 6617 - 6623
Vermeglio A, Richaud P, Breton J (1989) Orientation and assignment of the four cytochrome hemes in *Rhodopseudomonas viridis* reaction centers. FEBS Lett 243 : 259 - 263
Vernon LP, Kamen MD (1954) Hematin compounds in photosynthetic bacteria. J Biol Chem 211 : 643 - 663
Vickery L, Nozawa T, Sauer K (1976) MCD studies of low-spin cytochromes. Temperature dependence and effects of axial coordination on the spectra of cytochrome *c* and cytochrome b_5. J Am Chem Soc 98 : 351 - 357
Villalain J, Moura I, Liu MC, Payne WJ, LeGall J, Xavier AV, Moura JJG (1984) NMR and EPR studies of a dihaem cytochrome from *Pseudomonas stutzeri* (ATCC 11607)(cytochrome *c* peroxidase). Eur J Biochem 141 : 305 - 312
Vinogradov SN (1970) Some properties of the 695 mμ band of *Pseudomonas* cytochrome *c*. Biopolymers 9 : 507 - 509
Voordouw G, Brenner S (1986) Cloning and sequencing of the gene encoding cytochrome c_3 from *Desulfovibrio vulgaris* (Hildenborough). Eur J Biochem 159 : 347 - 351
Wakabayashi S, Matsubara H, Kim CH, King TE (1982) Structural studies of bovine heart cytochrome c_1. J Biol Chem 257 : 9335 - 9344
Walker FA, Huynh BH, Scheidt WR, Osvath SR (1986) Models of the cytochromes *b*. Effect of axial ligand plane orientation on the EPR and Mössbauer spectra of low-spin ferrihemes. J Am Chem Soc 108 : 5288 - 5297
Walker JE, Saraste M, Gay NJ (1984) The UNC operon - nucleotide sequence regulation and structure of ATP synthase. Biochim Biophys Acta 768 : 164 - 200
Wallace CJA (1984) Modulation of the alkaline transition in cytochrome *c* and cytochrome *c*-T by full or specific partial acetimidylation. Biochem J 217 : 601 - 604
Wallace CJA (1987) Functional consequences of the excision of an Ω loop, residues 40 - 55, from mitochondrial cytochrome *c*. J Biol Chem 262 : 16767 - 16770

Wallace CJA, Corthésy BE (1986) Protein engineering of cytochrome *c* by semisynthesis. Substitutions at glutamic acid 66. Protein Eng 1 : 23 - 27

Wallace CJA, Offord RE (1979) The semisythesis of fragments corresponding to residues 66-104 of horse heart cytochrome *c*. Biochem J 179 : 169 - 182

Wallace CJA, Proudfoot AEI (1987) On the relationship between oxidation-reduction potential and biological activity in cytochrome *c* analogues. Biochem J 245 : 773 - 779

Wallace CJA, Rose K (1983) The semisynthesis of analogues of cytochrome *c*. Modifications of arginine residues 38 and 91. Biochem J 215 : 651 - 658

Wallace CJA, Mascagni P, Chait BT, Collawn JF, Paterson Y, Proudfoot AEI, Kent SBH (1989) Substitutions engineered by chemical synthesis at three conserved sites in mitochondrial cytochrome *c*. J Biol Chem 264 : 15199 - 15209

Wallace DG, Brown RH, Boulter D (1973) The amino acid sequence of *Cannabis sativa* cytochrome *c*. Phytochem 12 : 2617 - 2622

Wand AJ, Roder H, Englander SW (1986) Two dimensional ^{1}H NMR studies of cytochrome *c* hydrogen exchange in the N-terminal helix. Biochemistry 25 : 1107 - 1114

Wand AJ, Stefano DLD, Feng Y, Roder H, Englander SW (1989) Proton resonance assignments of horse ferricytochrome *c*. Biochemistry 28 : 186 - 194

Warme PK (1975) The influence of amino acid substitutions on the conformational energy of cytochrome *c*. Biochemistry 14 : 3518 - 3526

Warshel A, Russell ST (1984) Calculations of electrostatic interactions in biological systems and in solutions. Quart Rev Biophysics 17 : 283 - 422

Warshel A, Russell ST, Churg AK (1984) Macroscopic models for studies of electrostatic interactions in proteins-limitations and applicability. Proc Natl Acad Sci USA 81 : 4785 - 4789

Warwicker J, Watson HC (1982) Calculation of the electric potential in the active site cleft due to α-helix dipoles. J Mol Biol 157 : 671 - 679

Watt GD, Sturtevant JM (1969) The enthalpy change accompanying the oxidation of ferrocytochrome *c* in the pH range 6 - 11 at 25°C. Biochemistry 8 : 4567 - 4571

Wayland BB, Fitzgerald RJ, Drago RS (1966) Investigations of the cobalt(II) complexes of N, N - dimethylacetamide. J Am Chem Soc 88 : 4600 - 4604

Weber PC (1982) Correlations between structural and spectroscopic properties of the high-spin heme protein cytochrome *c′*. Biochemistry 21 : 5116 - 5119

Weber PC, Salemme FR (1980) Structural and functional diversity in 4-α-helical proteins. Nature (London) 287 : 82 - 84

Weber PC, Bartsch RG, Cusanovich MA, Hamlin RC, Howard A, Jordan SR, Kamen MD, Meyer TE, Weatherford DW, Xuong NG, Salemme FR (1980) Structure of cytochrome *c′*: a dimeric, high-spin haem protein . Nature (London) 286 : 302 - 304

Weber PC, Howard A, Xuong NH, Salemme FR (1981a) Crystallographic structure of *R. molischianum* ferricytochrome *c*/ at 2.5Å resolution. J Mol Biol 153 : 399 - 424

Weber PC, Salemme FR, Mathews FS, Bethge PH (1981b) On the evolutionary relationship of the 4-α-helical heme proteins. The comparison of cytochrome *b*-562 and cytochrome *c′*. J Biol Chem 256 : 7702 - 7704

Westerhuis LW, Tesser GI, Nivard RJF (1979) Formation of a biologically active complex from two complementary fragments of horse heart cytochrome *c*. Recl Trav Chim Pays Bas 98 : 109 - 112

Westerhuis LW, Tesser GI, Nivard RJF (1980) Enzymatic synthesis of a peptide bond between a tryptic fragment of horse heart cytochrome *c* and a synthetic model peptide. Recl Trav Chim Pays Bas 99 : 400 - 403.

Westerhuis LW, Tesser GI, Nivard RJF (1982) Functioning complex of two cytochrome *c* fragments with deletion of the (39-58) eicosapeptide. Int J Pept Prot Res 19 : 290 - 299

Weyer KA, Schäfer W, Lottspeich F, Michel H (1987a) The cytochrome subunit of the photosynthetic reaction center from *Rhodopseudomonas viridis* is a lipoprotein. Biochemistry 26 : 2909 - 2914

Weyer KA, Lottspeich F, Gruenberg H, Lang F, Oesterhelt D, Michel H (1987b) Amino acid sequence of the cytochrome subunit of the photosynthetic reaction centre from the purple bacterium *Rhodopseudomonas viridis*. The EMBO J 6 : 2197 - 2202

Wharton CW (1986) Infra-red and Raman spectroscopic studies of enzyme structure and function. Biochem J 233 : 25 - 36

Whatley FR (1981) The establishment of mitochondria, *Paracoccus* and *Rhodopseudomonas*. Annals NY Acad Sci 361 : 330 - 340

Wherland S, Gray HB (1977) Electron transfer mechanisms employed by metalloproteins. In : Addison AW, Cullen WR, Dolphin D, James BR (eds) Biological aspects of inorganic chemistry. Wiley-Interscience, New York London pp 289 - 368

Wherland S, Pecht I (1978) Protein - protein electron transfer. A Marcus theory analysis of reactions between *c* type cytochromes and blue copper proteins. Biochemistry 17 : 2585 - 2591

Wikström M (1977) Proton pump coupled to cytochrome *c* oxidase in mitochondria. Nature (London) 266 : 271 - 273

Wikström M (1982a) Proton translocation by cytochrome oxidase. Curr Topics Memb Transport 16 : 303 - 321

Wikström M (1982b) Properties of cytochrome oxidase relevant to its proton translocating function. In: Chien Ho (ed) Electron transport and oxygen utilisation. Elsevier, North Holland

Wikström M (1988) Protonic sidedness of the binuclear iron-copper centre in cytochrome oxidase. FEBS Lett 231 : 247 - 252

Wikström M (1989) Identification of the electron transfers in cytochrome oxidase that are coupled to proton-pumping. Nature (London) 338 : 776 - 778

Wikström M, Casey RP (1985) What is the essential proton translocating molecular machinery in cytochrome *c* oxidase? J Inorg Biochem 23 : 327 - 334

Wikström M, Krab K (1979) Proton pumping cytochrome *c* oxidase. Biochim Biophys Acta 549 : 177 - 222

Wikström M, Saari H (1975) A spectral shift in cytochrome *c* induced by Ca^{2+} ions. Biochim Biophys Acta 408 : 170-1 79

Wikström M, Hanson HJ, Ingledew WJ, Chance B (1976) A re-evaluation of the spectral, potentiometric and energy linked properties of cytochrome *c* oxidase in mitochondria. FEBS Lett 65 : 259 - 277

Wikström M, Krab K, Saraste M (1981) Cytochrome oxidase: a synthesis. Academic Press, London New York

Wilgus H, Ranweiler JS, Wilson GS, Stellwagen E (1978) Spectral and electrochemical studies of cytochrome *c* peptide complexes. J Biol Chem 253 : 3265 - 3272

Willey DL, Howe CJ, Auffret AD, Bowman CM, Dyer TA, Gray JC (1984a) Location and nucleotide sequence of the gene for cytochrome *f* in wheat chloroplast DNA. Mol Gen Genet 194 : 416 - 422

Willey DL, Auffret AD, Gray JC (1984b) Structure and topology of cytochrome *f* in pea chloroplast membranes. Cell 36 : 555 - 562

Williams G, Eley CGS, Moore GR, Robinson MN, Williams RJP (1982) The reduction of cytochrome *c* with $[Fe(edta)]^{2-}$. FEBS Lett 150 : 293 - 299

Williams G, Moore GR, Porteous E, Robinson MN, Soffe N, Williams RJP (1985a) Solution structure of mitochondrial cytochrome *c*: ^{1}H NMR of ferricytochrome *c*. J Mol Biol 183 : 409 - 428

Williams G, Clayden NJ, Moore GR, Williams RJP (1985b) Comparison of the solution and solid-state structures of mitochondrial cytochrome *c*: Analysis of paramagnetic shifts in the NMR spectrum of ferricytochrome *c*. J Mol Biol 183 : 447 - 460

Williams G, Moore GR, Williams RJP (1985c) Biological electron transfer: the structure, dynamics and reactivity of cytochrome *c*. Comments Inorg Chem IV : 55 - 98

Williams RJP (1959) Coordination, chelation, and catalysis. In : Boyer PD, Lardy H, Myrbäck K (eds) The Enzymes 2nd Ed Academic Press, London New York pp 391 - 441

Williams RJP (1961a) Nature and properties of metal ions of biological interest and their coordination compounds. Fed Proc 20 (suppl) : 5 - 14

Williams RJP (1961b) Possible functions of chains of catalysts. J Theoret Biol 1 : 1 - 17

Williams RJP (1969) Electron transfer and energy conservation. Curr Topics in Bioenergetics 3 : 79 - 156

Williams RJP (1971) Catalysis by metalloenzymes : The entatic state. Inorg Chim Acta 5 : 137 - 155

Williams RJP (1987) The mechanism of cytochrome oxidase and other reaction centres for electron/proton pumping. FEBS Lett 226 : 1 - 7

Wilson AC, Carlson SS, White TJ (1977) Biochemical evolution. Ann Rev Biochem 46 : 573 - 639

Wilson DF, Lindsay JG, Brocklehurst ES (1972) Heme-heme interactions in cytochrome *c* oxidase. Biochim Biophys Acta 256 : 277 - 286

Wilson DF, Nelson D (1982) Coulometric and potentiometric evaluation of the redox components of cytochrome *c* oxidase *in situ*. Biochim Biophys Acta 680 : 233 - 241

Wilson MT, Greenwood C (1971) A correlation between reducibility and the possession of the 695 nm absorption band of ferricytochrome *c*. Eur J Biochem 22 : 11 - 18

Wilson MT, Brunori M, Rotilio GC, Antonini E (1973) Properties of modified cytochromes. II. Ligand binding to reduced carboxymethyl cytochrome *c*. J Biol Chem 248 : 8162 - 8169

Wilson MT, Silvestrini MC, Morpurgo L, Brunori M (1979) Electron transfer kinetics between *Rhus vernicifera* stellacyanin and cytochrome *c* (horse heart cytochrome *c* and *Pseudomonas* cytochrome *c*-551). J Inorg Biochem 11 : 95 - 100

Winkler JR, Nocera DG, Yocom KM, Bordignon E, Gray HB (1982) Electron-transfer kinetics of pentaammineruthenium(III)(His-33) ferricytochrome *c*. J Am Chem Soc 104 : 5798 - 5800

Winfield MW (1965) Electron transfer within and between haemoprotein molecules. J Mol Biol 12 : 600 - 611

Woese CR (1981) Archaebacteria. Sci Amer 244 : 94 - 106

Woese CR, Gibson J, Fox GE (1980) Do geneological patterns in purple sulfur bacteria reflect interspecific gene transfer? Nature (London) 283 : 212 - 214

Woese CR, Stackebrandt E, Macke TJ, Fox GE (1985) A phylogenetic definition of the major eubacterial taxa. System Appl Microbiol 6 : 143 - 151

Wood LC, Muthukrishnan K, White TB, Ramdas L, Nall BT (1988) Construction and characterization of mutant iso-2 cytochromes *c* with replacement of conserved prolines. Biochemistry 27 : 8554 - 8561

Wood PM (1974) Rate of electron transfer between plastocyanin. cytochrome *f*, related proteins and artificial redox reagents in solution. Biochim Biophys Acta 357 : 370 - 379

Wood PM (1984) Bacterial proteins with CO-binding *b*- or *c*-type haem functions and absorption spectroscopy. Biochim Biophys Acta 768 : 293 - 317

Woodward C, Hilton BD (1979) Hydrogen exchange kinetics and internal motions in proteins and nucleic acids. Ann Rev Biophys Bioeng 8 : 99 - 127

Woolley KJ (1987) The soluble *c*-type cytochromes from the bacterium, *Aquaspirillum itersonii*. Eur J Biochem 166 : 131 - 137

Wooten JB, Cohen JS, Vig I, Schejter A (1981) pH induced conformational transitions of ferricytochrome *c*: A carbon-13 and deuterium NMR study. Biochemistry 20 : 5394 - 5402

Wright PE (1989) What can two-dimensional NMR tell us about proteins? Trends Biochem Sci 14 : 255 - 260

Wright S (1932) The roles of mutation, inbreeding, crossbreeding and selection in evolution. Proc VI Int Cong Genet 1 : 356 - 366

Wu CI, Li WH (1985) Evidence for higher rates of nucleotide substitution in rodents than in man. Proc Natl Acad Sci USA 82 : 1741 - 1745

Wu N, Cote JC, Wu R (1986a) Nucleotide sequence of the rice cytochrome *f* gene and the presence of sequence variation near this gene. Gene 50 : 271 - 278

Wu CI, Li WH, Shen JJ, Scarpulla RC, Limbach KJ, Wu R (1986b) Evolution of cytochrome *c* genes and pseudogenes. J Molec Evol 23 : 61 - 75

Wu TT, Kabat EA (1970) An analysis of the sequences of the variable regions of Bence-Jones proteins and myeloma light chains and their implications for antibody complementarity. J Exp Med 132 : 211 - 250

Wüthrich K (1969) High-resolution proton nuclear magnetic resonance spectroscopy of cytochrome *c*. Proc Natl Acad Sci USA 63 : 1071 - 1078

Wüthrich K (1976) NMR in Biological Research : Peptides and Proteins. North-Holland, Amsterdam Oxford

Wüthrich K (1986) NMR of proteins and nucleic acids. Wiley-Interscience, New York Chichester

Xavier AV, Czerwinski EW, Bethge PH, Mathews FS (1978) Identification of the haem ligands of cytochrome b_{562} by x-ray and NMR methods. Nature (London) 275 : 245 - 247

Xavier AV, Moura JJG, Le Gall J, DerVartanian DV (1979) Oxidation-reduction potentials of the hemes in cytochrome c_3 from *D. gigas* in the absence and presence of ferredoxin by EPR spectroscopy. Biochimie 61 : 689 - 695

Xia Z, Shamala N, Bethge PH, Lim LW, Bellamy HD, Lederer F, Mathews FS (1987) 3-dimensional structure of flavocytochrome b_2 from baker's yeast at 3.0Å resolution. Proc Natl Acad Sci USA 84 : 2629 - 2633

Yagil G (1967) The proton dissociation constant of pyrrole, indole and related compounds. Tetrahedron 23 : 2855 - 2861

Yamanaka T, Mizushima H, Nozaki M, Horio T, Okunuki K (1959) Studies on cytochrome *c*. J Biochem 46 : 121 - 132

Yamanaka T, Inoue S, Hiroyoshi T (1980) Structural difference between larval and adult cytochromes *c* of the housefly, *Musca domestica*. J Biochem 88 : 601 - 604

Yandell JK, Fay DP, Sutin N (1973) Mechanisms of the reaction of cytochrome *c*. II. The rate of reduction of horse-heart ferricytochrome *c* by Chromium(II). J Am Chem Soc 95 : 1131 - 1137

Yang D, Oyaizu Y, Oyaizu H, Olsen GJ, Woese CR (1985) Mitochondrial origins. Proc Natl Acad Sci USA 82 : 4443 - 4447

Yaoi Y (1967) Comparison of the primary structures of cytochromes *c* from wild and respiration-deficient mutant yeasts. J Biochem 61 : 54 - 58

Yasui M, Harada S, Kai Y, Kasai N (1985) Structure of ferricytochrome *c'* from *R. rubrum* at 6Å resolution. J Biochem 98 : 77 - 80

Yonetani T (1976) Cytochrome *c* peroxidase. In: Boyer PD (ed) The Enzymes. vol 13 Academic Press, London New York pp 345 - 361

Yong FC, King TE (1970) Optical rotatory dispersion of some *c'* and *c''* cytochromes. J Biol Chem 245 : 2457 - 2464

Yoshimura T, Ozaki T (1984) Imidazole, imidazolate, and hydroxide complexes of (protoporphyrin IX) and its dimethyl ester as model systems for ferric hemoproteins : EPR and electronic spectral study. Arch Biochem Biophys 230 : 466 - 482

Yoshimura T, Matsushima A, Aki K (1980) Relation of the structure and function of ferricytochrome *c* bound to the phosphoprotein phosvitin. Biochim Biophys Acta 625 : 100 - 108

Young JZ (1962) The life of the vertebrates. University Park Press, Chicago

Yu CA, Yu L, King TE (1972) Preparation and properties of cardiac cytochrome c_1. J Biol Chem 247 : 1012 - 1019

Yu L, Dong JH, Yu CA (1986) Characterization of purified cytochrome c_1 from *Rhodobacter sphaeroides* R-26. Biochim Biophys Acta 852 : 203 - 211

Yu LP, Smith GM (1988) ^{15}N and ^{1}H NMR studies of *Rhodospirillum rubrum* cytochrome c_2. Biochemistry 27 : 1949 - 1956

Zand R, Vinogradov S (1968) The far ultraviolet circular dichroism of cytochrome *c*. Arch Biochem Biophys 125 : 94 - 97

Zuckerkandl E (1976) Evolutionary processes and evolutionary noise at the molecular level. J Mol Evol 7 : 269 - 312

Zuckerkandl E, Pauling L (1962) In: Kasha M, Pullamn B (eds) Horizons in Biochemistry. Academic Press, New York, pp 189 - 225

Zuckerkandl E, Pauling L (1965) Evolutionary divergence and convergence in proteins. In: Bryson V, Vogel HJ (eds) Evolving Genes and Proteins. Academic Press, New York, pp 97 - 166

Zuniga EH, Nall BT (1983) Folding of yeast iso-1-AM cytochrome *c*. Biochemistry 22 : 1430 - 1437

Subject Index

MIX
Papier aus verantwortungsvollen Quellen
Paper from responsible sources
FSC® C105338

If you have any concerns about our products,
you can contact us on
ProductSafety@springernature.com

In case Publisher is established outside the EU,
the EU authorized representative is:
Springer Nature Customer Service Center GmbH
Europaplatz 3, 69115 Heidelberg, Germany

Printed by Libri Plureos GmbH
in Hamburg, Germany